Phosphodiesterase Inhibitors

The Handbook of Immunopharmacology

Series Editor: Clive Page
King's College London, UK

Titles in this series

Cells and Mediators

Immunopharmacology of Eosinophils
(edited by H. Smith and R. Cook)

The Immunopharmacology of Mast Cells and Basophils
(edited by J.C. Foreman)

Lipid Mediators
(edited by F. Cunningham)

Immunopharmacology of Neutrophils
(edited by P.G. Hellewell and T.J. Williams)

Immunopharmacology of Macrophages and other Antigen-Presenting Cells
(edited by C.A.F.M. Bruijnzeel-Koomen and E.C.M. Hoefsmit)

Adhesion Molecules
(edited by C.D. Wegner)

Immunopharmacology of Lymphocytes
(edited by M. Rola-Pleszczynski)

Immunopharmacology of Platelets
(edited by M. Joseph)

Immunopharmacology of Free Radical Species
(edited by D. Blake and P.G. Winyard)

Cytokines
(edited by A. Mire-Sluis, forthcoming)

Systems

Immunopharmacology of the Gastrointestinal System
(edited by J.L. Wallace)

Immunopharmacology of Joints and Connective Tissue
(edited by M.E. Davies and J. Dingle)

Immunopharmacology of the Heart
(edited by M.J. Curtis)

Immunopharmacology of Epithelial Barriers
(edited by R. Goldie)

Immunopharmacology of the Renal System
(edited by C. Tetta)

Immunopharmacology of the Microcirculation
(edited by S. Brain)

Immunopharmacology of the Respiratory System
(edited by S.T. Holgate)

The Kinin System
(edited by S. Farmer, forthcoming)

Drugs

Immunotherapy for Immune-related Diseases
(edited by W.J. Metzger, forthcoming)

Phosphodiesterase Inhibitors
(edited by C. Schudt, G. Dent and K. Rabe)

Immunopharmacology of AIDS
(forthcoming)

Immunosuppressive Drugs
(forthcoming)

Glucocorticosteroids
(forthcoming)

Angiogenesis
(forthcoming)

Phosphodiesterase Inhibitors

edited by

Christian Schudt
Department of Biochemistry
Byk Gulden Pharmaceuticals
Konstanz, Germany

Gordon Dent
and
Klaus F. Rabe
Krankenhaus Großhansdorf
Zentrum für Pneumologie und Thoraxchirurgie
Großhansdorf, Germany

ACADEMIC PRESS
Harcourt Brace and Company, Publishers
San Diego London New York
Boston Sydney Tokyo Toronto

This book is printed on acid-free paper

A catalogue record for this book
is available from the British Library

ISBN 0-12-210720-9

ACADEMIC PRESS LIMITED
24/28 Oval Road
London NW1 7DX

United States Edition published by
ACADEMIC PRESS INC.
San Diego, CA 92101

Typeset by Mathematical Composition Setters Ltd, Salisbury, Wiltshire
Printed and bound in Great Britain by The Bath Press

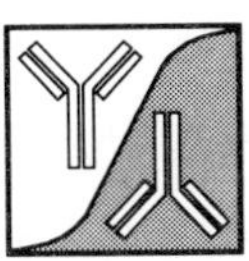

Contents

Contributors ix

Series Preface xiii

Preface xv

1. *Identification and Quantification of PDE Isoenzymes and Subtypes by Molecular Biological Methods* 1

Kate Loughney *and* Ken Ferguson

1. Introduction 1
2. The PDE Gene Family 1
3. Molecular Cloning and Localization of the Mammalian PDEs 3
   3.1 PDE1 3
   3.2 PDE2 5
   3.3 PDE3 6
   3.4 PDE4 7
   3.5 PDE5 12
   3.6 PDE6 12
   3.7 PDE7 13
   3.8 Additional PDEs 14
4. Summary 14
5. References 14

2. *Analysis of PDE Isoenzyme Profiles in Cells and Tissues by Pharmacological Methods* 21

Hermann Tenor *and* Christian Schudt

1. Introduction 21
2. Analysis of PDE Isoenzyme Activities in Cells and Tissues 22
   2.1 Procedure for Establishment of PDE Isoenzyme Activity Profiles 22
   2.2 PDE Isoenzyme Activity Profiles 23
3. Regulation of PDE Isoenzyme Activities 30
   3.1 Regulation of PDE Activity by Cyclic Nucleotide Concentrations and Ca^{2+}/Calmodulin 30
   3.2 Short-term Regulation of PDE Activity by Phosphorylation 32
   3.3 Long-term PDE4 Induction by cAMP-elevating Agents 32
   3.4 Elevated PDE Activity in Atopic Mononuclear Cells 33
4. Conclusions 34
5. References 34

3. *Effects of Theophylline and Non-selective Xanthine Derivatives on PDE Isoenzymes and Cellular Function* 41

Gordon Dent *and* Klaus F. Rabe

1. Introduction 41
2. Inhibition of PDE 42
   2.1 Inhibition of Tissue PDE Activity 42
   2.2 Inhibition of Isoenzymes 42
3. Effects on Cell Function 44
   3.1 Immune Cells 45
   3.2 Other Cells 54
4. Summary and Directions for Future Research 55
5. References 56

4. Ca^{2+}/Calmodulin-dependent Cyclic Nucleotide Phosphodiesterase (PDE1) 65

Rajendra K. Sharma *and* Robert A. Hickie

1. Introduction 65
2. Purification and Characterization 66
3. Isoenzymes of CaM-PDE 66
   3.1 Demonstration of Isoenzymes 66
   3.2 Kinetic Properties 68
   3.3 Differential Activation by Calmodulin and Ca^{2+} 68
   3.4 Regulation by Phosphorylation 69
   3.5 Inhibitors 73
4. Activity in Cancer Cells 74
5. Conclusions 74
6. Acknowledgements 74
7. References 75

5. EHNA as an Inhibitor of PDE2: a Pharmacological and Biochemical Study in Cardiac Myocytes 81

Pierre-François Méry, Catherine Pavoine, Françoise Pecker *and* Rodolphe Fischmeister

1. Introduction 81
   1.1 Cardiac Ca^{2+} Current is Inhibited by cGMP via Activation of PDE2 81
   1.2 Use of the Cardiac I_{Ca} in Determining the Effects of EHNA on PDE2 82
   1.3 Use of Purified Cardiac PDE Isoforms to Determine the Selectivity of Action 82
2. Methods 82
   2.1 Electrophysiology 82
   2.2 PDE Assays for cAMP 83
3. Results 83
   3.1 EHNA has no effect on I_{ca} in the Absence of cGMP 83
   3.2 EHNA Antagonizes the Inhibitory Effect of cGMP on I_{Ca} 84
   3.3 EHNA Antagonizes the Inhibitory Effect of Nitric Oxide Donors on I_{Ca} 84
   3.4 EHNA Inhibits a cGMP-stimulated PDE in the Crude Particulate Fraction 85
   3.5 EHNA Selectively Inhibits the Purified Soluble PDE2 86
   3.6 Participation of Adenosine Deaminase in the Effects of EHNA? 86
4. Discussion 86
   4.1 EHNA Acts as a Selective Inhibitor of PDE2 in Cardiac Myocytes 86
   4.2 EHNA Should be Useful in Evaluating the Role of PDE2 in Various Tissues 87
5. Acknowledgements 87
6. References 87

6. cGMP-Inhibited Phosphodiesterases (PDE3) 89

Narcisse Komas, Matthew Movsesian, Sasko Kedev, Eva Degerman, Per Belfrage *and* Vincent C. Manganiello

1. Introduction 89
2. Purification and Characterization 89
3. Molecular Cloning and Domain Organization 92
4. Structure/Function Relationships 93
   4.1 Catalytic Domain 93
   4.2 Membrane-Association Domain 95
   4.3 Regulatory Domain Phosphorylation/Activation 95
5. Pharmacology and Potential Therapeutic Usage of PDE3 Inhibitors 96
   5.1 Inotropic Agents 97
   5.2 Vasodilators 98
   5.3 Relaxation of Airway Smooth Muscle 99
   5.4 Antithrombotic Agents 100
   5.5 Anti-inflammatory Agents 100
6. Therapeutic Usage of PDE3 Inhibitors 101
7. Acknowledgements 101
8. References 101

7. Interaction of PDE4 Inhibitors with Enzymes and Cell Functions 111

Gordon Dent *and* Mark A. Giembycz

1. The PDE4 Isoenzyme Family 111
   1.1 Enzyme Characteristics 111
   1.2 Enzyme Distribution 112
   1.3 Selective Inhibitors 112

2. Pharmacology of PDE4 Inhibitors 115
 2.1 *In Vitro* 115
 2.2 *In Vivo* 117
3. Adverse Effects of PDE4 Inhibitors 119
4. PDE4 Alterations in Allergic Diseases 119
5. Summary and Future Directions 120
6. References 121

8. *Inhibition of Phosphodiesterase Isoenzymes and Cell Function by Selective PDE5 Inhibitors* 127

Paul J. Silver

1. Introduction 127
2. Scientific Rationale for PDE5 Inhibitors 127
3. PDE5 Inhibition and Vasorelaxation 128
4. Potential Therapeutic Applications of PDE5 Inhibitors 130
5. Additional Indications for PDE5 Inhibitors 131
6. Newer PDE5 Inhibitors 131
7. Combination Inhibitors 132
8. Summary 132
9. Acknowledgement 132
10. References 132

9. *Design and Synthesis of Xanthines and Cyclic GMP Analogues as Potent Inhibitors of PDE5* 135

Konjeti R. Sekhar, Pascal Grondin, Sharron H. Francis *and* Jackie D. Corbin

1. Introduction 135
2. Strategy for the Design of PDE5 Inhibitors 136
3. Synthesis of IBMX and cGMP Analogues as PDE Inhibitors 136
4. Selectivity of the IBMX Analogues as PDE Inhibitors 138
 4.1 IBMX Analogues as PDE5 Inhibitors 138
 4.2 Effects of Hydrophobic Substitutions on IBMX 141
 4.3 IBMX Analogues as Inhibitors of Other PDEs 141
5. cGMP Analogues as PDE Inhibitors 142
6. Smooth Muscle Relaxation by IBMX and cGMP Analogues 143
7. Conclusions 144
8. Acknowledgements 145
9. References 145

10. *Enzymatic and Functional Aspects of Dual-selective PDE3/4 Inhibitors* 147

Armin Hatzelmann, Renate Engelstätter, John Morley *and* Lazzarro Mazzoni

1. Introduction 147
2. Dual Inhibitors of PDE3/4 for Asthma Therapy 148
 2.1 Airway Smooth Muscle Relaxation 148
 2.2 Modulation of Inflammatory Cell Functions 148
 2.3 Side-effects 148
3. Preclinical Pharmacology of PDE3/4 Inhibitors 149
 3.1 AH 21–132 (Benafentrine) 149
 3.2 Zardaverine 151
 3.3 Others (Tolafentrine, Org 20241, Org 30029, EMD 54622) 151
4. Clinical Experience with PDE3/4 Inhibitors 152
 4.1 AH 21–132 152
 4.2 Zardaverine 153
 4.3 Other Selective PDE Inhibitors (Tibenelast, Enoximone/Isomazole/SDZ-MKS 492, Zaprinast) 157
5. Conclusions 157
6. References 158

11. *An Isoform-selective Inhibitor of Cyclic AMP-specific Phosphodiesterase (PDE4) with Anti-inflammatory Properties* 161

Robert Alvarez, Donald V. Daniels, Earl R. Shelton, Preston A. Baecker, T. Annie T. Fong, Bruce Devens, Robert Wilhelm, Richard M. Eglen *and* Marco Conti

1. Introduction 161
2. Materials and Methods 162
 2.1 Cyclic Nucleotide PDE Assays 162
 2.2 Accumulation of Cyclic AMP in Intact 43D Cells 163
 2.3 Preparation of Recombinant Human PDE4 Isoforms 163
 2.4 Phosphorylation of Human PDE4D3 164
 2.5 Recombinant Human PDE7 164
 2.6 Inflammatory Assays 164

2.7 Bronchoconstriction and Cell Infiltration in Guinea Pigs 165
2.8 Statistical Analysis 165
2.9 Chemicals 165
3. Results 165
3.1 Biochemical Studies 165
3.2 Recombinant PDE4 Isoforms 166
3.3 Biological Responses 168
4. Discussion 169
5. Acknowledgements 170
6. References 170

12. *Characterization of Different States of PDE4 by Rolipram and RP 73401* 173

John E. Souness

1. Introduction 173
2. Pharmacology of RP 73401 173
3. Interactions of RP 73401 and Rolipram with Eosinophil PDE4 175
3.1 Poor Correlation Between PDE4 Inhibition and Suppression of Eosinophil Functions 175
3.2 Effects of Solubilization and Vanadate/Glutathione Complex 176
3.3 Possible Role of the High-Affinity Rolipram-binding Site 178
4. Inhibition of PDE4 from Other Cells and Tissues 178
4.1 Enzyme Data 178
4.2 Whole Cell Data 180
5. Possible Therapeutic Implications 181
6. Conclusions 182
7. References 182

13. *Molecular Aspects of Inhibitor Interaction with PDE4* 185

Siegfried B. Christensen, Walter E. DeWolf, Jr, M.D. Ryan *and* Theodore J. Torphy

1. Introduction 185
1.1 Chemistry and Characteristics of the Phosphodiesterases 185
1.2 Historical Perspective on Isoenzyme-selective PDE Inhibitors 186
2. Molecular Biology of PDE4 190
2.1 Subtypes and mRNA Splice Variants 190
2.2 Functional Domains of PDE4 191
2.3 Role of Conserved Histidines 192
3. Rolipram-Binding Site 193
3.1 Historical Perspective 193
3.2 Nature and Function of the High-Affinity Rolipram-binding Site 194
3.3 Proposals on the Nature and Function of the High-Affinity Rolipram-binding Site 194
3.4 Biological Significance of High-Affinity Rolipram Binding 195
4. Mechanistic Enzymology 195
4.1 Kinetic Behaviour of PDE4s 195
4.2 Inhibition by *R*-Rolipram 197
4.3 Binding of *R*-Rolipram to $Met^{265-886}$ 199
5. Structure–Activity Relationships 199
5.1 Introduction 199
5.2 Rolipram and Lead PDE4 Inhibitors 200
5.3 Rolipram and Derivatives 201
5.4 Overlay Model of PDE4 Inhibition 201
6. Summary and Conclusions 202
7. References 203

A colour plate section appears between pp. 208–209

Glossary 209

Key to Illustrations 219

Index 225

Contributors

Robert Alvarez
Department of Gynecology and Obstetrics,
Stanford University School of Medicine,
Stanford,
CA 94305,
USA

Preston A. Baecker
Roche Bioscience,
Palo Alto,
CA 94304,
USA

Per Belfrage
Department of Medical and Physiological Chemistry,
University of Lund,
S–22100 Lund,
Sweden

Siegfried B. Christensen
Department of Medicinal Chemistry,
SmithKline Beecham Pharmaceuticals,
King of Prussia,
PA 19406,
USA

Marco Conti
Department of Gynecology and Obstetrics,
Stanford University School of Medicine,
Stanford,
CA 94305,
USA

Jackie D. Corbin
Department of Molecular Physiology and Biophysics,
Vanderbilt University School of Medicine,
Nashville,
TN 37232,
USA

Donald V. Daniels
Roche Bioscience,
Palo Alto,
CA 94304,
USA

Eva Degerman
Department of Medical and Physiological Chemistry,
University of Lund,
S–22100 Lund,
Sweden

Gordon Dent
Krankenhaus Großhansdorf,
Zentrum für Pneumologie und Thoraxchirurgie,
D–22927 Großhansdorf,
Germany

Bruce Devens
Targetted Genetics,
Seattle WA,
USA

Walter E. De Wolf Jr
Department of Medicinal Chemistry,
SmithKline Beecham Pharmaceuticals,
King of Prussia,
PA 19406,
USA

Richard M. Eglen
Roche Bioscience,
Palo Alto,
CA 94304,
USA

Renate Engelstätter
Department of Clinical Research,
Byk Gulden Pharmaceuticals,
D–78403 Konstanz,
Germany

Ken Ferguson
ICOS Corporation,
Bothell,
WA 98021,
USA

Rodolphe Fischmeister
Laboratoire de Cardiologie Cellulaire et Moléculaire,
INSERM CJP 92–11,
Université de Paris-Sud,
F–92296 Châtenay-Malabry,
France

T. Annie T. Fong
Sugen Corporation,
Redwood City, CA
USA

Sharron H. Francis
Department of Molecular Physiology and Biophysics,
Vanderbilt University School of Medicine,
Nashville,
TN 37232,
USA

Mark A. Giembycz
Department of Thoracic Medicine,
National Heart and Lung Institute,
Imperial College of Science, Technology and Medicine,
London SW3 6LY,
UK

Pascal Grondin
Glaxo France Centre de Récherche,
Z.A. de Courtaboeuf,
F–91951 Les Ulis Cedex,
France

Armin Hatzelmann
Department of Biochemistry,
Byk Gulden Pharmaceuticals,
D–78403 Konstanz,
Germany

Robert A. Hickie
Saskatoon Cancer Center and
Department of Pharmacology,
University of Saskatchewan College of Medicine,
Saskatoon,
Canada S7N 5E5

Sasko Kedev
Pulmonary and Critical Care Medicine Branch,
National Heart, Lung and Blood Institute,
Bethesda,
MD 20892,
USA

Narcisse Komas
Pulmonary and Critical Care Medicine Branch,
National Heart, Lung and Blood Institute,
Bethesda,
MD 20892,
USA

Kate Loughney
ICOS Corporation,
Bothell,
WA 98021,
USA

Vincent C. Manganiello
Pulmonary and Critical Care Medicine Branch,
National Heart, Lung and Blood Institute,
Bethesda,
MD 20892,
USA

Lazzarro Mazzoni
Preclinical Research,
Sandoz AG,
CH–4002 Basel,
Switzerland

Pierre-François Méry
Laboratoire de Cardiologie Cellulaire et Moléculaire,
INSERM CJF 92–11,
Université de Paris-Sud,
F–92296 Châtenay-Malabry,
France

John Morley
Department of Applied Pharmacology,
National Heart and Lung Institute,
Imperial College of Science, Technology and Medicine,
London SW3 6LY,
UK

Matthew Movsesian
Division of Cardiology,
University of Utah School of Medicine,
Salt Lake City,
UT 84132,
USA

Catherine Pavoine
INSERM U–99,
Hôpital Henri-Mondor,
F–94010 Créteil,
France

Françoise Pecker
INSERM U–99,
Hôpital Henri-Mondor,
F–94010 Créteil,
France

Klaus F. Rabe
Krankenhaus Großhansdorf,
Zentrum für Pneumologie und Thoraxchirurgie,
D–22927 Großhansdorf,
Germany

M.Dominic Ryan
Department of Medicinal Chemistry,
SmithKline Beecham Pharmaceuticals,
King of Prussia,
PA 19406,
USA

Christian Schudt
Department of Biochemistry,
Byk Gulden Pharmaceuticals,
D–78403 Konstanz,
Germany

Konjeti R. Sekhar
Department of Molecular Physiology and Biophysics,
Vanderbilt University School of Medicine,
Nashville,
TN 37232,
USA

Rajendra K. Sharma
Saskatoon Cancer Center and
Departments of Pathology and Pharmacology,
University of Saskatchewan College of Medicine,
Saskatoon,
Canada S7N 5E5

Earl R. Shelton
Roche Bioscience,
Palo Alto,
CA 94304,
USA

Paul J. Silver
154 Barton Drive,
Spring City,
PA 19475,
USA

John E. Souness
Rhône-Poulenc Rorer Ltd.,
Dagenham Research Centre,
Dagenham,
Essex RM10 7XS,
UK

Hermann Tenor
Department of Biochemistry,
Byk Gulden Pharmaceuticals,
D–78403 Konstanz,
Germany

Theodore J. Torphy
Department of Inflammation and Respiratory
Pharmacology,
SmithKline Beecham Pharmaceuticals,
King Of Prussia,
PA 19406,
USA

Robert Wilhelm
Roche Bioscience,
Palo Alto,
CA 94304,
USA

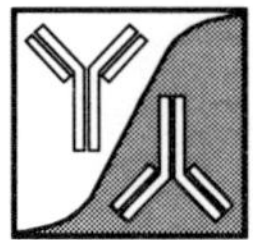

Series Preface

The consequences of diseases involving the immune system such as AIDS, and chronic inflammatory diseases such as bronchial asthhma, rheumatoid arthritis and atherosclerosis, now account for a considerable economic burden to governments worldwide. In response to this, there has been a massive research effort investigating the basic mechanisms underlying such diseases, and a tremendous drive to identify novel therapeutic applications for the prevention and treatment of such diseases. Despite this effort, however, much of it within the pharmaceutical industries, this area of medical research has not gained the prominence of cardiovascular pharmacology or neuropharmacology. Over the last decade there has been a plethora of research papers and publications on immunology, but comparatively little written about the implications of such research for drug development. There is also no focal information source for pharmacologists with an interest in diseases affecting the immune system or the inflammatory response to consult, whether as a teaching aid or as a research reference. The main impetus behind the creation of this series was to provide such a source by commissioning a comprehensive collection of volumes on all aspects of immunopharmacology. It has been a deliberate policy to seek editors for each volume who are not only active in their respective areas of expertise, but who also have a distinctly *pharmacological* bias to their research. My hope is that *The Handbook of Immunopharmacology* will become indispensable to researchers and teachers for many years to come, with volumes being regularly updated.

The series follows three main themes, each theme represented by volumes on individual component topics. The first covers each of the major cell types and classes of inflammatory mediators. The second covers each of the major organ systems and the diseases involving the immune and inflammatory responses that can affect them. The series will thus include clinical aspects along with basic science. The third covers different classes of drugs that are currently being used to treat inflammatory disease or diseases involving the immune system, as well as novel classes of drugs under development for the treatment of such diseases.

To enhance the usefulness of the series as a reference and teaching aid, a standardized artwork policy has been adopted. A particular cell type, for instance, is represented identically throughout the series. An appendix of these standard drawings is published in each volume. Likewise, a standardized system of abbreviations of terms has been implemented and will be developed by the editors involved in individual volumes as the series grows. A glossary of abbreviated terms is also published in each volume. This should facilitate cross-referencing between volumes. In time, it is hoped that the glossary will be regarded as a source of standard terms.

While the series has been developed to be an integrated whole, each volume is complete in itself and may be used as an authoritative review of its designated topic.

I am extremely grateful to the officers of Academic Press, and in particular to Dr Carey Chapman, for their vision in agreeing to collaborate on such a venture, and greatly hope that the series does indeed prove to be invaluable to the medical and scientific community.

C.P. Page

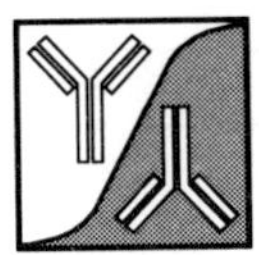

Preface

Non-selective inhibitors of cyclic nucleotide phosphodiesterase (PDE), such as theophylline, have been used extensively in biochemical and pharmacological research since cyclic AMP and its degrading enzyme were discovered in 1958. Theophylline and papaverine were the original plant-derived drugs which served as crucial tools for the investigation of the signalling function of cyclic AMP and the involvement of PDE in cellular regulation. Whereas papaverine was an effective vasodilator, theophylline acted as a cardiotonic or bronchodilating drug. Several xanthine derivatives, such as 3-isobutyl-1-methylxanthine (IBMX), were synthesized chemically but – although their potency at the target enzyme increased – their clinical efficacy did not exceed that of theophylline. These early non-selective substances are classed as 'first generation' PDE inhibitors.

In the period from 1970 to 1980 various PDE isoenzymes were defined by differences in substrate specificities and/or regulation characteristics; these were designated PDE families 1 to 5. Subsequently, an extensive research and development effort led to the synthesis, characterization and clinical study of a series of new compounds that were selective for one or two isoenzymes. These mono- and dual-selective 'second generation' PDE inhibitors were isoenzyme-targeted, especially to PDE3 and 4, and they offered both new clinical uses and the possibility for analysis of the tissue distribution and functional role of individual isoenzymes. It emerged that these compounds exhibited a limited organ specificity and – importantly, from a clinical viewpoint – a therapeutic potential against heart failure and depression. Currently, a variety of new substances are under test as potential anti-inflammatory drugs and, if the theoretical concept holds true and agents with low potential for side-effects are discovered, a bright future can be predicted for this family of drugs.

As a result of progress in molecular biology in the past five years, a superfamily of PDE isoenzymes – including seven families, 16 genes and (to date) 33 individual enzyme proteins – has been defined. This complexity raises questions regarding the regulation and function as well as the distribution of the multiple enzyme subtypes in different cells and tissues. Indications exist that some subtypes are involved in signal transduction under normal metabolic conditions whereas others participate selectively in pathological situations such as inflammation, hyperreactivity, hyperplasia or hypertension. In view of the different functions and the tissue localization of isoenzyme subtypes, synthesis of 'third generation' – or subtype-selective – PDE inhibitors has already started and provides a challenge to chemists, biochemists and pharmacologists. Again, such compounds may bear an enormous potential for research in physiology and pathology on the one hand, and on the other for the therapy of a broad range of diseases.

We consider that a summary of the present state of knowledge, as well as a comprehensive description of the available compounds, should provide a useful synopsis for everybody who wants to choose the most suitable PDE inhibitor for their research or who is dealing with such drugs in a clinical setting and wishes to be informed in the background and scientific developments.

C. Schudt
G. Dent
K.F. Rabe

A Note on Nomenclature

The rapidly expanding list of new PDE isoenzymes has led to some divergence in systems of nomenclature. Although occasional references will be found in the text to earlier nomenclature systems, we have tried, on the whole, to adhere to the system presented by the PDE nomenclature group at the 1994 meeting of the American Society for Pharmacology and Experimental Therapeutics (ASPET). This system, published by Beavo *et al.* (1994) and accessible at the University of Washington via the Internet (gopher://www.hs.washington.edu), is described in Chapter 1 and summarized here.

Short name	PDE isoenzyme gene family	Alternative names
PDE1	Calmodulin-dependent PDEs	CaM-PDE, PDE I
PDE2	Cyclic GMP-stimulated PDEs	cGS-PDE, PDE II
PDE3	Cyclic GMP-inhibited PDEs	cGI-PDE, PDE III
PDE4	Cyclic AMP-specific PDEs	RD, DPD, PDE IV
PDE5	Cyclic GMP-specific PDEs	cGB-PDE, PDE V
PDE6	Photoreceptor PDEs	ROS/COS-PDE, PDE VI
PDE7	High affinity, cyclic AMP-specific PDE	HCP1, PDE VII

Reference

Beavo, J.A., Conti, M. and Heaslip, R.J. (1994). ASPET meeting report: multiple cyclic nucleotide phosphodiesterases. Mol. Pharmacol. 46, 399–405.

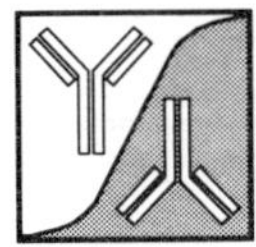

1. Identification and Quantification of PDE Isoenzymes and Subtypes by Molecular Biological Methods

Kate Loughney *and* Ken Ferguson

1.	Introduction	1
2.	The PDE Gene Family	1
3.	Molecular Cloning and Localization of the Mammalian PDEs	3
	3.1 PDE1	3
	3.2 PDE2	5
	3.3 PDE3	6
	3.4 PDE4	7
	3.5 PDE5	12
	3.6 PDE6	12
	3.7 PDE7	13
	3.8 Additional PDEs	14
4.	Summary	14
5.	References	14

1. Introduction

Cyclic nucleotide phosphodiesterases (PDEs) are essential regulators of cyclic nucleotide-dependent signal transduction processes. They terminate the action of the second messengers adenosine 3′:5′-cyclic monophosphate (cAMP) and guanosine 3′:5′-cyclic monophosphate (cGMP) by hydrolysing them to their respective 5′-nucleoside monophosphates. The PDEs form a biochemically and structurally diverse family of proteins, which has driven a search for therapeutic agents designed to inhibit specific mammalian PDEs and thereby affect specific cellular functions.

The PDEs fall into two major classes depending on which of two amino acid sequence motifs they possess. The first class contains one of the two genes in *Saccharomyces cerevisiae* (Sass *et al.*, 1986), the "dunce" gene product from *Drosophila melanogaster* (Chen *et al.*, 1986) and all of the identified mammalian PDEs. The second class includes PDEs found in *Vibrio fischeri* (Dunlap and Callahan, 1993), *S. cerevisiae* (Nikawa *et al.*, 1987), *Candida albicans* (Hoyer *et al.*, 1994), *Dictyostelium discoideum* (Lacombe *et al.*, 1986) and *Schizosaccharomyces pombe* (Matviw *et al.*, 1993). No mammalian counterparts of this second class have been discovered and this class will not be considered further here.

The biochemical and molecular biological investigation of the phosphodiesterases has led to a greater appreciation of the complexity and diversity of this enzyme family. This chapter will briefly review the properties of these enzymes and then review the expression patterns of individual isoenzymes in tissues and cells.

2. The PDE Gene Family

Mammalian PDEs have historically been classified by their preference or affinity for cAMP or cGMP, their kinetic parameters of cyclic nucleotide hydrolysis, their relative sensitivity to inhibition by various compounds, their allosteric regulation by other molecules, and their chromatographic behaviour on anion exchange columns (Fig. 1.1). This classification system has been augmented with amino acid sequence data, obtained either

Phosphodiesterase Inhibitors
ISBN 0-12-210720-9

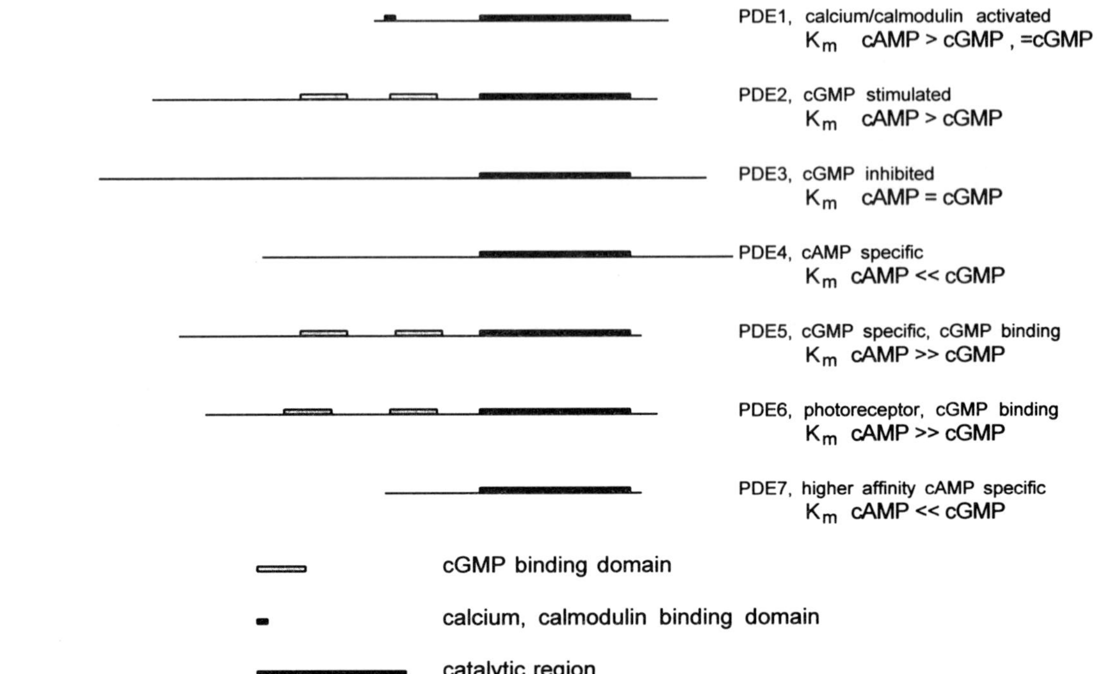

Figure 1.1 Regulatory and catalytic domains of the seven PDE families. The line drawings represent the amino acid sequences of the PDEs. They are drawn approximately to scale with the initiator methionine represented by the left end of the lines. The two K_m values for PDE1 represent the diversity within the family.

by sequencing purified proteins or, more commonly, by inference from complementary DNA (cDNA) clones. Seven families or types of PDE are recognized. Some of the families contain more than one gene and some genes are alternatively spliced.

The families are designated PDE1 to PDE7, whereas genes within a particular family are designated by a letter (A, B, C, etc.). Different splice variants from the same gene are numbered in order of discovery (Beavo *et al.*, 1994). For example, PDE1A1 and PDE1A2 are two different splice variants of gene A in the Ca^{2+}/calmodulin (CaM)-regulated family (PDE1; see Chapter 4). Multiple reports of the same splice variants are lettered consecutively, for example PDE1A1a and PDE1A1b. A current list of all PDE genes and splice variants is available on the Internet (uniform resource locator: gopher://www.hs.washington.edu; Beavo *et al.*, 1994).

The PDEs of all seven families share in common an arrangement of structural domains (Fig. 1.1). The catalytic region is localized in the carboxy-terminal portion of the protein. It is approximately 270 amino acids in length and contains the conserved amino acids that make up the motif found in all members of this class of PDEs (Charbonneau, 1990). The amino-terminal portions of some PDEs contain regulatory domains. These vary among the seven different families and include Ca^{2+}/CaM binding domains, cGMP binding sites, membrane localization domains and sites for phosphorylation by a number of different protein kinases.

It is apparent from studies of PDE expression patterns that cells can express more than one PDE and even more than one PDE from the same family. A major question under current investigation is what role each of these PDEs plays and whether these roles are unique or redundant.

The presence of multiple family members and multiple splice variants of individual family members raises several issues that complicate the analysis of the PDEs. First, identifying the enzymes present is, on occasion, difficult. Biochemical, pharmacological and immunological methods are required to distinguish among these enzymes. Years of work on the biochemistry of PDE has made it reasonably straightforward to separate and identify the enzyme families present in a given sample; the existence of specific, selective inhibitors for some of the families aids this categorization (see Chapter 2). However, it can be difficult to distinguish proteins that are in the same family and to identify proteins that arise from alternative splicing of transcripts from the same gene. Another confounding factor arises from proteolysis, which can generate smaller proteins that remain enzymatically active. This further complicates their identification.

A second issue, the identification and validation of different splice variants, can also prove problematic. Isolation of a cDNA from a library is not always sufficient to define a new splice variant, since many libraries contain cDNAs that represent partially spliced precursors to the mature messenger RNA (mRNA). Many of the known splice variants diverge from each other at the 5′ or 3′ ends of the cDNA and yield proteins that differ completely from each other at the amino or carboxy terminus. Thus, it is difficult to use homology to the family as a whole to determine if a new splice variant is valid. These considerations imply that data other than the original cDNA clone must be used to verify that the predicted sequence encodes an enzyme expressed by a cell. Methods for characterizing nucleic acids include Northern analysis, ribonuclease (RNase) protection analysis, reverse transcription of mRNA followed by polymerase chain reaction amplification (RT-PCR), *in situ* hybridization, and studies of the gene's structure. In addition, recombinant proteins can be expressed and the protein size, immunological reactivity and kinetic properties compared to those of the native cellular PDEs.

A third issue arises when studying this diverse yet homologous family of PDEs. The presence of multiple enzymes makes it difficult to assign a specific role in the cell's signal transduction network to a given enzyme. Selective inhibitors, the best tools for assessing this, are lacking for some of the enzymes. An understanding of the specific roles that individual PDE isoenzymes fill in cells should facilitate the exploration of potential therapeutic uses of PDE inhibitors.

Table 1.1 PDE cDNAs isolated

Gene	*Human*	*Bovine*	*Rat*	*Mouse*	*Chicken*	*Dog*
PDE1A	✓	✓	✓			
PDE1B	✓	✓	✓	✓		
PDE1C	✓		✓	✓		
PDE2A	✓	✓	✓			
PDE3A	✓		✓			
PDE3B	✓		✓			
PDE4A	✓		✓	✓		
PDE4B	✓		✓	✓		
PDE4C	✓		✓			
PDE4D	✓		✓	✓		
PDE5	✓	✓				
PDE6-α	✓	✓		✓		
PDE6-α'	✓	✓			✓	
PDE6-β	✓	✓		✓		✓
PDE6-γ	✓	✓		✓		
PDE6-δ		✓				
PDE6-13 kD		✓				
PDE7	✓					

✓ indicates that a cDNA has been isolated; references are in the text.

3. *Molecular Cloning and Localization of the Mammalian PDEs*

3.1 PDE1

Three different PDE1 genes have been identified (Table 1.1). The products of two, PDE1A and PDE1B, have greater affinity for cGMP than for cAMP whereas the product of the third, PDE1C, has high affinity for both cAMP and cGMP. The activity of each of these PDEs is stimulated by the binding of Ca^{2+}/CaM.

3.1.1 Molecular Cloning

3.1.1.1 PDE1A

Three slightly different PDE1A proteins have been identified through a combination of protein and cDNA sequencing (Fig. 1.2). Each contains an amino-terminal Ca^{2+}/CaM binding domain and a PDE catalytic region towards the carboxy terminus. The PDE1A1 protein, also known as the bovine 59 kD CaM-PDE, has been sequenced (Novack *et al.*, 1991) and a bovine cDNA has been isolated (Sonnenburg *et al.*, 1995). Bovine PDE1A2 cDNAs (61 kD CaM-PDE) have been isolated (Sonnenburg *et al.*, 1993) and the protein has been sequenced (Charbonneau *et al.*, 1991). PDE1A2 differs from PDE1A1 at the amino terminus as a result of alternative splicing of the mRNAs. The two proteins diverge from each other in the Ca^{2+}/CaM binding region. The affinities of these two enzymes for Ca^{2+}/CaM differ, presumably owing to the different amino termini (Sharma and Kalra, 1994; Sonnenburg *et al.*, 1995). PDE1A3 is a human cDNA (Loughney *et al.*, 1994, 1996) that differs from bovine PDE1A2 by the presence of a 42-nucleotide insertion and a divergent 3′ end (Fig. 1.2). When examined by RNase protection, both the 42-nucleotide insertion and the divergent 3′ end found in the PDE1A3 cDNA were present in mRNA isolated from a variety of human tissues (Loughney *et al.*, 1994, 1996). The insertion is also present in rat PDE1A cDNA clones (K. Loughney and K. Ferguson, unpublished data).

3.1.1.2 PDE1B

PDE1B cDNAs have been cloned from bovine, human, mouse and rat sources (Bentley *et al.*, 1992; Polli and Kincaid, 1992; Repaske *et al.*, 1992; Usui *et al.*, 1994; Yu *et al.*, 1994). They encode a protein referred to as the 63 kD CaM-PDE, which has a linear domain structure similar to PDE1A. Evidence for a possible amino-terminal splice variant has been reported (Bentley *et al.*, 1992). The PDE1B genes from cow, human, mouse and rat encode proteins that are 93% identical.

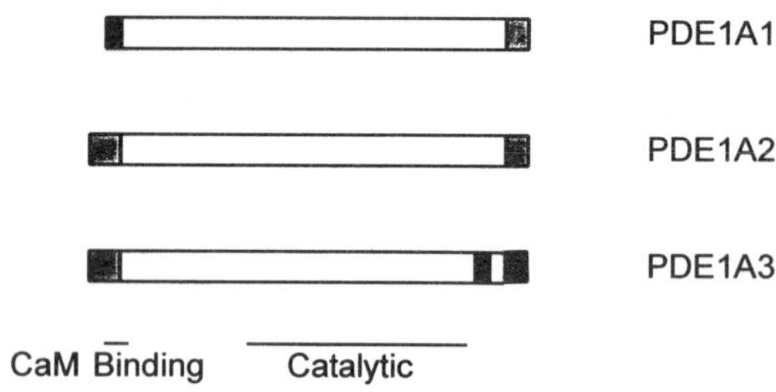

Figure 1.2 PDE1A protein structure. The PDE1A proteins are represented by the rectangles with the hatched ends representing the two divergent amino termini and two divergent carboxy termini. PDE1A1 (bovine, Novack *et al.*, 1991), PDE1A2 (bovine, Sonnenburg *et al.*, 1993) and PDE1A3 (human, Loughney *et al.*, 1996) contain different combinations of the divergent termini. The human sequence (PDE1A3) also contains an insertion (black box) not present in the bovine sequences.

3.1.1.3 PDE1C

Human, mouse and rat PDE1C cDNAs have been isolated (Loughney *et al.*, 1994, 1996; Yan and Beavo, 1994; Yan *et al.*, 1995). They encode a family of proteins that differ from each other at both the amino and carboxy termini owing to alternative splicing of the PDE1C mRNAs. The proteins predicted from the cloned transcripts range in size from 72 to 87 kD. RNase protection analysis and Northern blots have been used to validate the different splice variants. The PDE1C linear domain structure is similar to that of PDE1A and PDE1B in that the PDE1C proteins have a Ca^{2+}/CaM binding region and catalytic region. However, the PDE1C proteins have longer carboxy termini than do the PDE1A and PDE1B proteins, which accounts for their larger sizes.

3.1.2 Homology

PDE1 genes are sufficiently conserved among mammals that one can identify the particular PDE1 protein encoded by a specific cDNA by sequence comparison. For example, excluding the divergent termini, the predicted PDE1A protein of human shares greater than 90% amino acid identity with its bovine counterpart but has only 60–70% identity with human PDE1B or PDE1C. The same is true when one compares PDE1B or PDE1C to their counterparts in other species.

It is not known whether there are any additional genes in the PDE1 family. Many of the biochemically characterized PDE activities that belong to this family can be accounted for by the properties of the proteins encoded by the cDNAs corresponding to PDE1A, 1B, and 1C (Wang *et al.*, 1990). However, one partially characterized brain PDE activity (75 kD, affinity for cGMP greater than for cAMP) appears to be distinct and may represent the product of an additional gene (Shenolikar *et al.*, 1985).

3.1.3 Cell and Tissue Expression

The distribution of PDE1 proteins has been examined in rat brain by immunocytochemistry. The specificity of the antibody used was not fully characterized and thus the immunological staining may represent the localization of one or more type I PDEs (Kincaid *et al.*, 1987b). PDE1 is most abundant in neuronal dendrites and soma in the hippocampus (CA1–2 pyramidal cells), cerebellum (Purkinje cells), and cerebral cortex (pyramidal cells) (Kincaid *et al.*, 1987a,b). This distribution suggests a role for PDE1 in neurones that integrate different inputs. PDE1 expression in Purkinje cells was reduced following treatment of rats with 3-acetylpyridine, a neurotoxin that destroys the climbing fibre input to Purkinje cells. Thus, synaptic input appears to affect PDE1 levels (Balaban *et al.*, 1989; Kincaid *et al.*, 1987a). Subcellular localization by electron microscopy revealed PDE1 immunoreactivity in specific cytoplasmic regions, in dendrites, and in some postsynaptic densities (Ludvig *et al.*, 1991; Ariano and Appleman, 1979). The protein was not found in the axons or the nucleus.

The pattern of PDE1 expression in the rat brain changes during development with some regions showing increased PDE1 expression as neural differentiation proceeded and other regions showing high levels of PDE1 expression only early in development. Throughout development most of the rat brain PDE1 activity is cytosolic, the remainder being associated with brain synaptosomes (Billingsley *et al.*, 1990). This analysis did not distinguish products of the three PDE1 genes from each other. Thus, it is not clear if the activity associated with synaptosomes represents a specific gene product or a mixture of gene products.

The expression of PDE1A, PDE1B and PDE1C transcripts has been examined by RNase protections and Northern blot analyses in a variety of tissues and species (Table 1.2). PDE1A appears to be most abundant in bovine brain and heart (Sonnenburg *et al.*, 1993) and in human brain, heart and kidney (Loughney *et al.*, 1994, 1996). Low but detectable levels are seen in bovine liver and lung and in human lung and pancreas. The bovine heart, aorta, kidney, trachea, liver and lung appear to express a different 5′ splice variant (possibly the 59 kD or PDE1A1) to the bovine brain and spinal cord, which express PDE1A2 (61 kD). PDE1A expression has been observed throughout the bovine brain (Sonnenburg *et al.*, 1993).

The distribution of PDE1A mRNA in mouse brain has been examined by *in situ* hybridization (Yan *et al.*, 1994). Cerebral cortex and hippocampal pyramidal cells displayed the highest levels of expression and throughout the brain the PDE1A appears to be confined to neurones. These authors commented that the distribution differed somewhat from that seen in the bovine brain by Northern and RNase protection analyses

Table 1.2 Expression of PDE1 mRNAs in bovine, human, mouse and rat tissues

	PDE1A		*PDE1B*				*PDE1C*	
	Bovine[a]	*Human*[b]	*Bovine*[c]	*Human*[d]	*Mouse*[e]	*Rat*[f,g]	*Human*[b,h]	*Rat*[e]
Adrenal	−		+					
Aorta	+	+	−				−	
Brain	+	+	+	+	+	+	+	+
Heart	+	+	−		−	−	+	+
Kidney	+	+	+		−	−	+	
Liver	+	+	−		−	−	+	
Lung	+	+	−		+		+	+
Muscle, skeletal		+			−		+	
Pancreas		+					−	
Placenta		−					−	
Spleen	−	−			+		−	
Testis		+			+	−	+	+
Uterus		+					+	

+, RNA present; −, RNA, absent.
[a]Sonnenburg *et al.* (1993); [b]Loughney *et al.* (1996); [c]Bentley *et al.* (1992); [d]Yu *et al.* (1994); [e]Polli and Kincaid (1992); [f]Repaske *et al.* (1992); [g]Usui *et al.* (1994); [h]Loughney *et al.* (1994); Yan *et al.* (1995).

(Sonnenburg *et al.*, 1993) and noted specifically the absence of the PDE1A mRNA in the mouse cerebellum. PDE1A transcripts were also detected in human cerebellum by RNase protection analyses (Loughney *et al.*, 1996). Thus, there may be species differences in the expression pattern of this gene.

The distribution of PDE1B expression differs from that of PDE1A and PDE1C (Table 1.2). The highest levels of PDE1B transcripts are detected in the brain. Levels in mouse lung, spleen and testis are very low. The varied pattern of PDE1B expression in these tissues in the different species may reflect expression near the level of detection rather than a species difference (Table 1.2). Similarly, the expression in the bovine kidney is low and localized to the papillae; it may not be sufficiently abundant to allow detection in the mouse and rat kidney samples. When different regions of the brain were examined PDE1B mRNA levels were highest in bovine basal ganglia (Bentley *et al.*, 1992) and, more specifically, in rat and mouse striatum (Usui *et al.*, 1994; Polli and Kincaid, 1992). PDE1B transcripts have also been detected in the mouse T-lymphoma cell lines S49.1 and K30a-3.3 (Repaske *et al.*, 1992), although the size of the mRNA detected differs from that seen in mouse brain (Polli and Kincaid, 1992) and thus may represent a splice variant.

PDE1B expression in mouse brain has also been localized by *in situ* hybridization (Polli and Kincaid, 1994; Yan *et al.*, 1994). The highest levels of expression were found in caudate putamen, nucleus accumbens and olfactory tubercle. Expression appears to be primarily neuronal; the authors noted that much of the pattern of PDE1B expression is similar to that of dopamine receptors (Polli and Kincaid, 1994; Yan *et al.*, 1994) and suggested that PDE1B may play a role in modulating dopamine signalling. A similar distribution was observed using an antibody that recognizes PDE1B but not PDE1A (Polli and Kincaid, 1994). Cross-reactivity of this antibody to PDE1C was not examined. A similar distribution of PDE1B expression is seen in rat brain (Furuyama *et al.*, 1994).

PDE1C is expressed in many tissues (Table 1.2). A 5′ splice variant of this gene (PDE1C2) is found in rat olfactory epithelium but not in six other tissues (Yan and Beavo, 1994; Yan *et al.*, 1995) and is proposed as playing a role in odour sensing. Two of the different 3′ splice variants of PDE1C (PDE1C1 and PDE1C2) are readily detected in RNA from human heart (Loughney *et al.*, 1994, 1996). This result is in contrast to that obtained using rat heart RNA, where little PDE1C expression was detected by RNase protection (Yan *et al.*, 1995). This difference may reflect the differential localization of this PDE in the two organisms.

3.2 PDE2

The type II or cGMP-stimulated PDEs hydrolyse both cAMP and cGMP. They contain a carboxy-terminal catalytic region with homology to the other PDEs and a pair of cGMP binding domains located towards the amino terminus (Charbonneau *et al.*, 1990) (see Fig. 1.1). The binding of cGMP to these sites increases the hydrolysis of cAMP at low substrate concentration, which suggests that this enzyme plays a role integrating the cellular response to cAMP and cGMP levels (see Chapter 5).

3.2.1 Molecular Cloning

In contrast to the type I PDEs, only one gene encoding PDE2 has been identified (Table 1.1). The bovine

PDE2 protein has been sequenced (LeTrong *et al.*, 1990). A bovine cDNA encoding PDE2 has been isolated which encodes a 103 kD protein (PDE2A1) (Sonnenburg *et al.*, 1991). RNase protections imply the existence of a second bovine 5′ splice variant (Sonnenburg *et al.*, 1991) and cDNAs possessing a divergent 5′ end have been isolated from rat brain (PDE2A2) (Yang *et al.*, 1994). Genomic DNA encoding the 5′ splice variants has also been isolated. Additional rat and human cDNAs have been identified (Repaske *et al.*, 1992; G.J. Rosman, K. Loughney and K. Ferguson, unpublished observations).

3.2.2 Homology

The bovine and rat PDE2A proteins encoded by the cDNAs are more than 90% identical except at their amino termini, where the terminal 37 amino acids of the rat PDE2A2 differ completely from the terminal 25 amino acids of the bovine PDE2A1. The amino terminus of rat PDE2A2 is hydrophobic and may localize the protein in the membrane; this would be consistent with the biochemical evidence for a membrane-associated PDE2 (Murashima *et al.*, 1990; Sonnenburg *et al.*, 1991; Yang *et al.*, 1994). There is no evidence for additional family members.

3.2.3 Cell and Tissue Expression

Expression of PDE2A in bovine tissues was examined by Northern blot and RNase protection analyses by Sonnenburg *et al.* (1991). The highest levels of PDE2A expression were demonstrated in the adrenal cortex and medulla, heart, hippocampus, and renal cortex, medulla and papillae. Moderate levels are present in cerebral cortex, basal ganglia, trachea, lung, liver, spleen and T lymphocytes; low levels were detected in bovine cerebellum, medulla, spinal cord and aorta. RNase protection analyses revealed that bovine liver contains only the PDE2A1 splice variant. Bovine cerebral cortex, basal ganglia and hippocampus contain little or no PDE2A1 transcript but do contain a different amino-terminal splice variant, presumably PDE2A2. Bovine heart was shown to contain some PDE2A1 but predominantly another variant (presumably PDE2A2) whereas the remainder of the samples examined (adrenal cortex and medulla, kidney cortex, medulla and papillae, trachea, lung, spleen, lymphocytes) contained predominantly PDE2A1 (Sonnenburg *et al.*, 1991).

Expression of PDE2A has been detected in rat brain, heart, liver and kidney by Northern blot analysis but was not detected in testis (Repaske *et al.*, 1993). The level of expression in the brain is greater than in the other four tissues examined. When different regions of the brain were examined, expression was highest in the hippocampus and cerebral cortex but could also be detected in cerebellum, midbrain and forebrain, though not in the hindbrain. *In situ* hybridization was used to look in further detail at the pattern of PDE2 expression in rat brain and many of the regions in which PDE2 expression was detected were noted as being part of the limbic system. This raised the possibility that the limbic system shares some common pathway in which PDE2 plays a role (Repaske *et al.*, 1993).

Using immunocytochemical methods PDE2 expression has also been detected in a subset of olfactory neurones (Juilfs *et al.*, 1995).

3.3 PDE3

The type III PDE isoenzymes are also known as the cGMP-inhibited PDEs. They hydrolyse both cAMP and cGMP but the hydrolysis of cAMP is inhibited by very low concentrations of cGMP, probably by competitive inhibition. PDE3 has similar affinity for cAMP and cGMP but a higher V_{max} value for cAMP; thus it preferentially hydrolyses cAMP. Type III PDEs are the targets of cardiotonic agents such as amrinone, milrinone and vesnarinone. Complementary DNAs corresponding to two different PDE3 genes have been cloned (Table 1.1).

3.3.1 Molecular Cloning

3.3.1.1 PDE3A

Complementary DNAs encoding PDE3A (also known as cGIP2) have been isolated from human (Meacci *et al.*, 1992) and rat (Taira *et al.*, 1993). The human PDE3A encodes a 125 kD protein containing an amino-terminal membrane association domain and a catalytic region located towards the carboxy terminus. PDE3A yields more than one size of mRNA on a Northern blot (Taira *et al.*, 1993). Two of these have been characterized: one encodes the 125 kD protein described above, the other an 80 kD protein that results from internal initiation (Kasuya *et al.*, 1995).

3.3.1.2 PDE3B

PDE3B cDNAs have been isolated from rat and human cells (Taira *et al.*, 1993; Manganiello *et al.*, 1995). Rat PDE3B (RcGIP1) encodes a 123 kD protein with a structure similar to PDE3A. Deletion of the N-terminal hydrophobic domain results in a cytosolic localization when recombinant proteins are expressed (Murata *et al.*, 1995). No splice variants have been reported for the PDE3B cDNAs. Rat PDE3B hybridizes with several differently sized mRNAs on a Northern blot, however; thus there may be different splice variants in the mRNA population (Taira *et al.*, 1993).

3.3.2 Homology

When the available cDNA sequences are compared (Meacci *et al.*, 1992; Taira *et al.*, 1993) rat and human PDE3A share much more homology that do rat PDE3A and rat PDE3B. Thus this family appears similar to the type I PDEs, where homology between species for the same gene is greater than the homology between different family members in the same organism. All the PDE3 cDNAs analysed thus far contain a 44 amino acid insertion in the catalytic region when compared to any other mammalian PDE. Although many of the biochemical activities could be accounted for by PDE3A and PDE3B it is not yet clear whether there are any additional family members (Manganiello *et al.*, 1995).

3.3.3 Gene Mapping

Human PDE3A maps to chromosome 12 and human PDE3B to chromosome 11 (Table 1.3; Manganiello *et al.*, 1995).

3.3.4 Cell and Tissue Expression

Northern blot analysis shows that rat PDE3A is expressed in heart and brain, as well as adipose tissue, but not adipocytes or in undifferentiated (fibroblast) or differentiated (adipocyte) 3T3-L1 cells (Taira *et al.*, 1993). Rat PDE3B is expressed in rat adipose tissue but not in heart, brain or liver (Taira *et al.*, 1993); it is also expressed in adipocytes and in differentiated 3T3-L1 cells, although not in fibroblast-like undifferentiated 3T3-L1 cells.

The distribution of PDE3A and PDE3B in developing and mature rats has been examined by *in situ* hybridization (Reinhardt *et al.*, 1995). Throughout development PDE3A was found to be abundant in the cardiovascular system, including the myocardium and smooth muscle. It was found in developing gastrointestinal smooth muscle and epithelium, in large cells presumed to be megakaryocytes in the foetal liver, in venous and arterial endothelium and smooth muscle, in the ganglion cells of the developing eye, and in oocytes. PDE3B has a very different pattern of distribution. Throughout development it is strongly expressed in adipose tissue but is also found in hepatocytes and oocytes. It localizes to the developing neuroepithelium but is much less abundant in the adult nervous system. In the adult, PDE3B is also found in primary spermatocytes and renal collecting ducts.

Table 1.3 Chromosome localization of PDE genes; references are in the text

Gene	Human	Mouse	Rat
PDE3A	12		
PDE3B	11		
PDE4A	19p13.1–q12	9	
PDE4B	1p31	4	5
PDE4C	19	8B3-C1	
PDE4D	5q12	13	2
PDE6-α	5q31.2-q34	18	
PDE6-α'	10q24		
PDE6-β	4p16.3	5	
PDE6-γ	17q25	11	
PDE7	8q13-q22		

These localization studies are consistent with previous biochemical analyses suggesting that PDE3A is important in regulating smooth muscle and myocardial contractility and platelet aggregation whereas PDE3B plays a role in insulin inhibition of lipolysis and glycogenolysis (Manganiello *et al.*, 1995; Reinhardt *et al.*, 1995).

3.4 PDE4

Type IV PDE isoenzymes are often referred to as rolipram-sensitive, low K_m, cAMP-specific PDEs. Because PDE4 seems to be a predominant PDE in cells of the immune system, PDE4 inhibitors are currently being developed as anti-inflammatory agents (Banner and Page, 1995). Rolipram, a selective PDE4 inhibitor, has also been tested clinically for depression (Hebenstreit *et al.*, 1989; see also Chapter 13). PDE4s are the mammalian PDEs most homologous to the "dunce" gene of *Drosophila*, which was the first PDE gene isolated (Chen *et al.*, 1986). Four PDE4 genes have been identified in humans and rats (Table 1.1). They are now called PDE4A, B, C and D. In some earlier nomenclature systems PDE4A was called rat PDE2 or human DPDE2, PDE4B was called rat PDE4 or human DPDE4, PDE4C was called rat PDE1 or human DPDE1 and PDE4D was called rat PDE3 or human DPDE3.

3.4.1 Molecular Cloning

3.4.1.1 PDE4A

PDE4A cDNAs have been isolated from rat (Davis *et al.*, 1989; Swinnen *et al.*, 1989a; Repaske *et al.*, 1992; Bolger *et al.*, 1994), human (Livi *et al.*, 1990; Colicelli *et al.*, 1991; Repaske *et al.*, 1992; Bolger *et al.*, 1993; Sullivan *et al.*, 1994; Horton *et al.*, 1995) and mouse (Cherry and Davis, 1995). A total of six different putative splice variants have been identified, five in the rat and two in human (Fig. 1.3). Two rat variants have been isolated only once (PDE4A2, PDE4A3) (Davis *et al.*, 1989) and may not represent bona fide splice variants. This leaves four splice variants that result in proteins with different amino termini (PDE4A1, PDE4A5, PDE4A5?, PDE4A6; Fig. 1.3). These range in size from 68 to 98 kD. No full-length cDNAs have been isolated for PDE4A5? but its 5′ end sequence differs from that of PDE4A5 and PDE4A6. Recently, two additional cDNAs have been reported (PDE4A7 and PDE4A8). PDE4A7 appears to be a truncated cDNA that differs from PDE4A5 and PDE4A5? only by scattered nucleotide changes (Sullivan *et al.*, 1994; Horton *et al.*, 1995). PDE4A8 (Horton *et al.*, 1995) diverges from sequences common to all other PDE4A splice variants at its 5′ end and contains a 34 base pair (bp) insertion in the region of

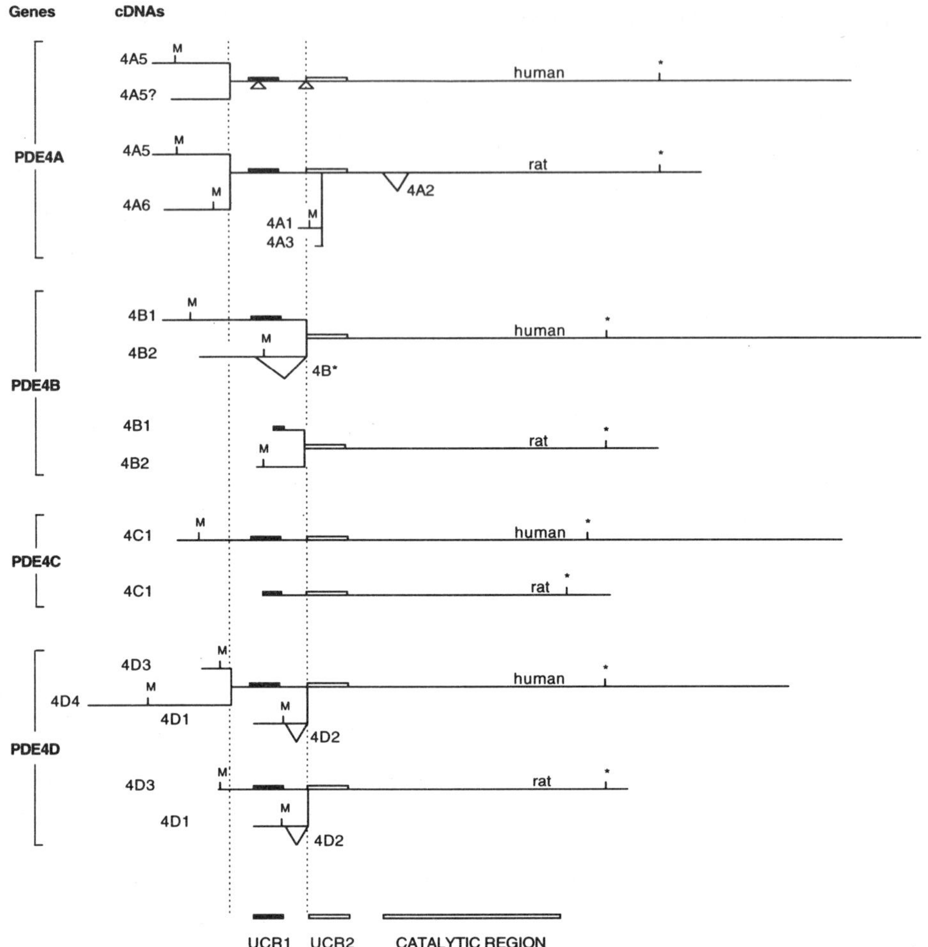

Figure 1.3 PDE4 cDNA structures. Human and rat PDE4 cDNAs are shown as lines with the initiator methionine indicated by an M and the stop codon by an asterisk (*). The cDNAs lacking a M are incomplete; they do not extend through the initiator methionine. For each gene (PDE1A, PDE1B, PDE1C and PDE1D), the known splice variants are diagrammed and labelled. Sequences deleted in splice variants 4A2, 4B* and 4D2 are represented by triangles. Black and white boxes indicate UCR1 and UCR2, respectively (UCR = upstream conserved region). PDE4A5? contains two insertions, indicated with small triangles, that are believed to be introns (Bolger *et al.*, 1993).

the cDNA that encodes the catalytic portion of the PDE. Comparison to genomic DNA sequence revealed that the 34 bp insertion was generated by alternative splicing. The insertion changes the reading frame such that PDE4A8 produces an inactive, truncated protein, the function of which is unknown.

For most of the proteins derived from the PDE4A locus the role of the unique amino terminus is unknown. However, the unique amino terminus encoded by rat splice variant PDE4A1 (Davis *et al.*, 1989; Bolger *et al.*, 1994) directs the PDE to the plasma membrane (Shakur *et al.*, 1993; Scotland and Houslay, 1995).

3.4.1.2 PDE4B

PDE4B cDNAs have been isolated from rat (Colicelli *et al.*, 1989; Swinnen *et al.*, 1989a, 1991; Bolger *et al.*, 1994), human (Bolger *et al.*, 1993; M.M. McLaughlin *et al.*, 1993; Obernolte *et al.*, 1993) and mouse (Cherry and Davis, 1995). A total of three different splice variants have been identified (Fig. 1.3). Two variants (PDE4B1 and PDE4B2) are seen in both rat and human cDNAs and encode proteins with different amino termini. A third variant has been seen in humans (Obernolte *et al.*, 1993). We refer to it here as PDE4B* because it has not yet been given a number in the PDE nomenclature table described in section 2. Splice variant PDE4B* differs from PDE4B2 by the deletion of 248 nucleotides. This changes the reading frame and truncates the amino terminus of the protein. Both PDE4B2 and PDE4B* have been found in human leukocyte RNA using a RT-PCR based approach (Obernolte *et al.*, 1993). The three splice variants encode proteins that range in size from 59 to 83 kD.

A 35 kb portion of the rat gene for PDE4B has been isolated and sequence analysis has identified 11 exons that encode PDE4B2 and PDE4B* (Monaco *et al.*, 1994). The exon or exons that encode the 5′ end of PDE4B1 have not yet been identified.

3.4.1.3 PDE4C

PDE4C cDNAs have been isolated from rat (Swinnen *et al.*, 1989a; Bolger *et al.*, 1994) and human cells (Bolger *et al.*, 1993; Obernolte *et al.*, 1994; Engels *et al.*, 1995b). An 80 kD protein is encoded by the human cDNA. In contrast to the other PDE4 genes no 5′ splice variants have been isolated (Fig. 1.3).

3.4.1.4 PDE4D

PDE4D cDNAs have been isolated from rat (Swinnen *et al.*, 1989a,b, 1991; Bolger *et al.*, 1994), human (Bolger *et al.*, 1993; Baecker *et al.*, 1994; Némoz *et al.*, 1995) and mouse (Repaske *et al.*, 1992; Cherry and Davis, 1995). A total of four different splice variants have been identified, three in the rat and four in the human (Fig. 1.3). The rat splice variant PDE4D2 differs from rat variant PDE4D1 by the removal of an additional intron (Fig. 1.3). These sequences serve as exons for variant PDE4D1 but as introns for variant PDE4D2. The consequence of their removal in PDE4D2 is to alter the reading frame and produce an amino-truncated protein (Swinnen *et al.*, 1989b; Sette *et al.*, 1994b). Both of these splice variants were detected in rat Sertoli cells using RT-PCR (Monaco *et al.*, 1994), although PDE4D1 appeared more prevalent in rat heart as most of the mRNA appeared to retain the intron. The splice variants encode proteins that range in size from 58 to 91 kD.

A portion of the rat PDE4D gene has been isolated. It contains 11 exons spanning more than 25 kb of genomic DNA (Monaco *et al.*, 1994). These exons encode PDE4D1 and PDE4D2. The exons that encode the 5′ ends of PDE4D3 and PDE4D4 have not been isolated.

The functions of the different PDE4D amino-terminal splice variants are being explored. PDE4D1 and PDE4D2 mRNA levels rise in response to increased cAMP and this transcription is driven by a cAMP responsive promoter (Swinnen *et al.*, 1991; Monaco *et al.*, 1994; Vicini *et al.*, 1994; Conti *et al.*, 1995b). Whether the amino termini of PDE4D1 and PDE4D2 have additional specific functions or whether they serve only to associate PDE4 proteins with a regulated promoter is unknown. Expression of PDE4D3 is mediated by a distinct promoter located further upstream. Phosphorylation of the unique amino terminus of PDE4D3 by cAMP-dependent protein kinase (PKA) activates the PDE (Sette *et al.*, 1994a; Conti *et al.*, 1995b) and this activation may underlie the rapid rise in PDE4 activity that occurs following elevation of intracellular cAMP levels (see Chapter 11).

3.4.2 Homology

Despite the amino-terminal sequence variation associated with alternative splicing, the PDEs encoded by the four PDE4 genes are very similar in structure (Fig. 1.3). The catalytic region is found in the carboxy-terminal portion of the proteins (Jin *et al.*, 1992). This is flanked by amino-terminal regions containing two conserved motifs (Bolger *et al.*, 1993; Bolger, 1994) and by carboxy-terminal regions that differ for each gene. The two conserved motifs found near the amino termini of these proteins are called upstream conserved regions 1 and 2 and their function is unknown. Some of the alternative splicing serves to generate proteins that lack upstream conserved region 1 and contain only region 2 (Fig. 1.3).

Compared pairwise, the PDE4 proteins show 60–75% amino acid identity. Within the catalytic domain the amino acid identity is more than 90% when PDE4s are compared pairwise and greater than 80% when all four PDE4s are compared.

Additional PDE4 genes have not been detected. A PCR-based approach using degenerate primers and genomic DNA did not reveal any additional type IV genes (Bolger *et al.*, 1993).

3.4.3 Gene Mapping

PDE4 genes have been mapped in human, rat and mouse (Table 1.3). PDE4A maps to chromosome 19p13.1–q12 in human and chromosome 9 in mouse (Milatovich *et al.*, 1994; Horton *et al.*, 1995). PDE4B maps to 1p31 in humans (Milatovich *et al.*, 1994; Szpirer *et al.*, 1995), to chromosome 4 (near *Pmv-23*) in mouse (Milatovich *et al.*, 1994) and to rat chromosome 5 (Szpirer *et al.*, 1995). PDE4C maps to chromosome 19 in human and to chromosome 8B3-C1 (near *Lpl* and *Aprt*-pseudogene) in mouse (Milatovich *et al.*, 1994). PDE4D maps to 5q12 in human (Milatovich *et al.*, 1994), to chromosome 13 in mouse (Milatovich *et al.*, 1994) and to chromosome 2 in the rat (Szpirer *et al.*, 1995). No association with a particular disease locus has been established for any of the PDE4 genes.

3.4.4 Cell and Tissue Expression

Expression of the four type IV genes has been examined in a number of different tissues. The four genes show widespread expression and in many tissues more than one of them is expressed (Table 1.4). There are several examples where expression of a gene has been detected by RT-PCR but not by Northern blotting (Table 1.4); this may reflect the difference in sensitivity of the two techniques. There are also examples of differences in expression patterns in rat and human tissues. These differences need confirming as much of the data is based on analysis of a single tissue sample from each organism. Expression patterns for the four PDE4 genes have also been examined by *in situ* hybridization in the rat brain where distinct, but overlapping, patterns of expression have been observed (Engels *et al.*, 1995a). High levels of PDE4 expression are detectable in the rat cerebellum

Table 1.4 Expression of PDE4 mRNAs in human and rat tissues

		PDE4A		PDE4B		PDE4C		PDE4D	
Tissue	*Method*	*Rat*	*Human*	*Rat*	*Human*	*Rat*	*Human*	*Rat*	*Human*
Blood	RT-PCR	–[a]	+[a,b]	+[a]	+[a,c]	–[a]	–[a,b]	+[a]	+[a]
	Northern		+[c,d]		+[c]				
Brain	RT-PCR	+[a,e]	+[a,b,f]	+[a,e]	+[a,f]	±[a,e]	+[a,b,f]	+[a,e]	+[a,f]
	Northern	+[g]	+[f]	+[g]	+[f,h]	–[g]	+[b], –[f]	+[j]	–[f], ±[k]
	Protection	+[l]		+[l]		–[l]		+[l]	
Heart	RT-PCR	+[a]	+[a,b]	+[a]	+[a]	+[a]	+[a,b]	+[a]	+[a]
	Northern	+[g]		+[g]	+[h]	–[g]	+[b]	+[j]	+[k]
Kidney	RT-PCR	+[a]	+[a]	+[a]	+[a]	+[a]	+[a]	+[a]	+[a]
	Northern	+[g]		+[g]	–[h]	+[g]	+[b]	+[j]	+[k]
Liver	RT-PCR	+[a]	+[a,b]	+[a]	+[a]	+[a]	+[a,b]	+[a]	+[a]
	Northern	–[b,g]		+[g]	–[h]	–[g]	+[b]	+[j]	–[k]
Lung	RT-PCR	+[a]	+[a,b]	+[a]	+[a]	±[a]	+[a,b]	+[a]	+[a]
	Northern				+[h]		+[b]		±[k]
Muscle, skeletal	Northern				+[h]		+[b]		+[k]
Pancreas	Northern				–[h]		+[b]		–[k]
Placenta	RT-PCR		+[a,b]		+[a]		+[a,b]		+[a]
	Northern		+[d]		–[h]		+[b]		+[k]
Testis	Northern	+[g,m]		+[g]		+[g,m]		+[g,j]	
	Protection		–[f]		–[f]	+[f]	–[f]		–[f]

[a]Engels *et al.* (1994); [b]Engels *et al.* (1995b); [c]Obernolte *et al.* (1993); [d]Livi *et al.* (1990); [e]Engels *et al.* (1995a); [f]Bolger *et al.* (1993); [g]Swinnen *et al.* (1989a); [h]M.M. McLaughlin *et al.* (1993); [j]Swinnen *et al.* (1989b); [k]Baecker *et al.* (1994); [l]Bolger *et al.* (1994); [m]Welch *et al.* (1992).

(PDE4A, B and D), hippocampus (PDE4A, B and D), cortex (PDE4A) and olfactory system (PDE4A, D). Little PDE4C expression is detected in the brain. In mouse brain the expression of PDE4A was examined by Northern blots and Western analysis as well as by immunocytochemistry (Cherry and Davis, 1995). High levels of PDE4A were detected in the olfactory neuroepithelium, specifically in the dendrites and axons of the olfactory receptor neurones.

In Northern blots a wide variety of mRNA sizes are observed, even from the same gene. Some of this can probably be accounted for by the different splice variants that have been identified but in other cases the number of mRNAs exceeds the number of splice variants currently known. Thus there are probably more splice variants. Some of these may result in different proteins being produced and some may reflect differences in promoter or polyadenylation site usage.

When the patterns of PDE4A, B, C and D expression are compared, one of the most striking differences is the absence of PDE4C expression in blood-derived cells (see also Chapter 13). PDE4C expression cannot be detected by RT-PCR in human neutrophils, eosinophils, Jurkat (human T lymphocyte line), U937 (human monocytic cell line), Namwala (human B lymphocyte line), or HL-60 (human promyelocyte line) (Engels *et al.*, 1994, 1995b). Expression of at least one of the other PDE4 genes can be detected in each of these cell types.

All four PDE4 genes have been detected in the rat testis (Swinnen *et al.*, 1989a,b; Welch *et al.*, 1992; Morena *et al.*, 1995), where their expression has been very thoroughly examined. The cellular distribution and developmental profile of expression differs for each of the four genes (Swinnen *et al.*, 1989a,b; Welch *et al.*, 1992; Morena *et al.*, 1995). PDE4A is present in germ cells and is detected by Northern analysis and by *in situ* hybridization primarily in rat maturing round spermatids. Five different PDE4A transcripts are induced between days 26 and 29; these are present until at least day 72. The PDE4B found in rat testis is localized to the Sertoli cells. PDE4C expression is located in the germ cells, predominantly in middle-late pachytene spermatocytes. PDE4C transcript levels vary throughout development. PDE4D transcripts are found in Sertoli cells and the transcript levels vary during the seminiferous cycle.

Thus, although all four PDE4 genes are co-expressed in the testis, when examined at a cellular level each of the four genes appears to be differentially regulated and may play a unique role in testicular function.

There are few examples of whole tissues in which only a single PDE4 gene is expressed (Table 1.4), although when specific regions of the rat brain are examined by *in situ* hybridization some regions appear to express only one of the PDE4 genes (Engels *et al.*, 1995a). At the cellular level some cell lines express all four PDE4 genes and some express only a subset (Bolger *et al.*,

1993; Engels *et al.*, 1994). One cell line expressing a single PDE4 gene (Jurkat JKE6-1) has been described (Engels *et al.*, 1994). However, although unstimulated Jurkat cells express only PDE4A, following treatment with dibutyryl (diBu)-cAMP both PDE4A and PDE4D are detected.

For some of the PDE4 genes distinct patterns of splice variant expression have also been seen (Bolger *et al.*, 1994, 1995). In the rat brain PDE4B2 is uniformly expressed in frontal, temporal, parietal and occipital cortex, cerebellum and brain stem. In contrast, PDE4B1 is preferentially expressed in the temporal cortex and cerebellum. In the same set of tissue samples PDE4D3 and PDE4D1 were most abundant in temporal cortex and parietal cortex, respectively. Although PDE4A1 and PDE4A5 are co-expressed in many brain tissues, only PDE4A1 is found in cerebellum and brain stem.

In some cells PDE4 mRNA levels increase following treatment of the cells with agents that elevate intracellular cAMP levels (Table 1.5). This increase in PDE mRNA levels may underlie a portion of the phenomenon of desensitization (reviewed in Conti *et al.*, 1995b). Desensitization involves a reduced cellular response to repeated hormonal signals. Elevation of PDE in response to initial increases in cAMP would reduce the magnitude of the cAMP increase following subsequent hormonal stimulation and thus dampen the cellular response. Although PDE4A, 4B and 4D have all been shown to respond to elevations in cAMP with increased mRNA levels, the details of this regulation appear to differ with the gene and probably also with the cell type (Table 1.5).

Following cAMP elevation the time course of the PDE4 mRNA increase can differ for the individual PDE4 genes. During treatment of rat Sertoli cells with follicle stimulating hormone (FSH) PDE4B and PDE4D mRNA levels reach maximal levels at different times. Nuclear run-off assays have been used to establish that some of the PDE4D mRNA increase is due to an increase in transcription. In contrast, the increase in PDE4B message is not transcriptional and may be due to increased mRNA stability (Swinnen *et al.*, 1991). Likewise, an increase in transcription accounts for at least some of the PDE4D mRNA increase observed when L6 myoblasts were treated with dibutyryl-cAMP and the PDE inhibitor, 3-isobutyl-1-methylxanthine (IBMX); this increase in transcription required PKA (Kovala *et al.*, 1994). In MonoMac6 cells treated with dibutyryl-cAMP the time course of PDE4A, 4B and 4D mRNA and protein accumulation also differs for the three genes (Verghese *et al.*, 1995). Transcripts corresponding to only some of the PDE4 splice variants are elevated in response to increases in cAMP levels (Swinnen *et al.*, 1989b; Obernolte *et al.*, 1993; Verghese *et al.*, 1995). This could occur if particular splice variants are transcribed from promoters whose activities are positively regulated by cAMP. Rat PDE4D is an example of this. A cAMP-responsive promoter has been identified for splice variants PDE4D1 and PDE4D2 and these are the two variants whose mRNA levels rise after treatment of the cells with agents that raise intracellular cAMP levels (Swinnen *et al.*, 1991; Monaco *et al.*, 1994; Vicini *et al.*, 1994; Conti *et al.*, 1995a,b; Verghese *et al.*, 1995).

Antibodies to the PDE4 gene products have recently been generated and used in Western blots of tissues and cells from human, rat and mouse (Swinnen *et al.*, 1991;

Table 1.5 Elevation of PDE4 mRNA levels following treatments that elevate cellular cAMP levels

Cell line	*Treatment*	*Time*	*PDE4A*	*PDE4B*	*PDE4C*	*PDE4D*	*Reference*
U-937 (human)	diBu-cAMP	18 h	↑(>2 x)[a]	↑(>2 x)	same	same	Engels *et al.* (1994)
Jurkat JKE6-1 (human)	diBu-cAMP	18 h	↑(2 x)	same	same	↑(>2 x)	Engels *et al.* (1994)
SH-SY5Y (human)	diBu-cAMP	18 h	same	same	same	same	Engels *et al.* (1994)
43D-Cl_2 (human)	diBu-cAMP	5 h	↑[b]	↑(4 x)			Obernolte *et al.* (1993)
C6 glioma (rat)	diBu-cAMP	5 h				↑(10 x)[c]	Swinnen *et al.* (1989b)
L6 myoblasts (rat)	diBu-cAMP/IBMX	2 h				↑	Kovala *et al.* (1994)
	forskolin/IBMX	2 h				↑	Kovala *et al.* (1994)
Mono Mac6 (human)	diBu-cAMP	2 h	↑[d]	↑[e]		↑[f]	Verghese *et al.* (1995)
Sertoli (rat)	diBu-cAMP	24 h		↑(5 x)		↑(100 x)	Swinnen *et al.* (1989b and 1991)
	FSH	20 h		↑		↑(100 x)	Swinnen *et al.* (1989b, 1991)

[a] Increase in mRNA level (magnitude in parentheses); same, no change in mRNA level.
[b] 4.6 kb mRNA elevated, 3.0 kb not.
[c] 6.8 kb mRNA elevated, 4.6 kb not.
[d] PDE4A5 elevated (see Fig. 1.3).
[e] PDE4B2 elevated (see Fig. 1.3).
[f] PDE4D1 elevated (see Fig. 1.3).
Abbreviations: diBu-cAMP, dibutyryl cAMP; FSH, follicle stimulating hormone.

Lobban *et al.*, 1994; Sette *et al.*, 1994a,b; Shakur *et al.*, 1995; Cherry and Davis, 1995; Conti *et al.*, 1995a; Verghese *et al.*, 1995). Expression of PDE4 proteins has been detected in human and rat cell lines, in rat brain, and in mouse brain, testis and olfactory system. In many cases the antibodies recognize epitopes that are common to more than one of the known splice variants. This makes it difficult to determine which particular splice variant is being expressed, especially as all of the splice variants have probably not yet been identified. In addition, the protein sizes are not known for splice variants for which only a partial cDNA has been isolated. These proteins may also migrate differently depending on their phosphorylation state (Sette *et al.*, 1994a). As more specific antibodies continue to be generated and characterized it will be of interest to determine the cellular and subcellular localization of the different type IV proteins.

3.5 PDE5

The type V PDE is also called the cGMP-binding, cGMP-specific PDE. As seen in other PDE isoenzymes, PDE5 contains a catalytic region located towards the carboxy terminus (Fig. 1.1). PDE5 also contains a pair of cGMP binding sites located in the amino-terminal portion of the protein (Thomas *et al.*, 1990; McAllister-Lucas *et al.*, 1993) which share some sequence homology with the cGMP binding sites of PDE2 and PDE6.

3.5.1 Molecular Cloning

Bovine PDE5 cDNAs encoding a 100 kD protein have been isolated (McAllister-Lucas *et al.*, 1993). Human (K. Loughney, G.J. Rosman, E.A.S. Harris and K. Ferguson, unpublished observations) and mouse (Burns *et al.*, 1995) cDNAs have also been isolated (Table 1.1). No splice variants have been reported and only a single size of mRNA was seen in a Northern blot using bovine lung mRNA (McAllister-Lucas *et al.*, 1993). It is not known if there are additional family members.

3.6 PDE6

The type VI PDEs are found in the photoreceptors and are involved in visual signalling pathways. There are two types of photoreceptors, rods and cones, and each contains a specific PDE6. Membrane-associated rod PDE6 is composed of two large catalytic subunits (α, 88 kD and β, 84 kD) and two copies of a smaller, inhibitory subunit (γ, 11 kD). Cone PDE6 is a homodimer of two catalytic subunits (α', 94 kD) and is associated with three smaller proteins (15, 13 and 11 kD). Two of these, the 13 kD and 11 kD, are believed to be inhibitory subunits. The third, the 15 kD subunit or δ, is not unique to cone PDE6 but is also found in association with soluble rod PDE6 (reviewed in Gillespie, 1990).

In the rods, rhodopsin absorbs light and, following a conformational change, binds to the G-protein, transducin. GTP replaces the GDP bound to transducin, whose α-subunit then activates rod PDE6 by binding to its γ-subunit. The resultant drop in cGMP levels leads to a closing of the cGMP-gated cation channels in the plasma membrane followed by a transient hyperpolarization of the cell.

3.6.1 Molecular Cloning

3.6.1.1 Rod PDE6

cDNAs encoding the rod PDE6 α-subunit have been isolated from bovine (Ovchinnikov *et al.*, 1987; Pittler *et al.*, 1990), human (Pittler *et al.*, 1990) and murine (Baehr *et al.*, 1991) sources (Table 1.1). cDNAs encoding the rod PDE6 β-subunit have been isolated from bovine (Lipkin *et al.*, 1990), human (Collins *et al.*, 1992; Khramtsov *et al.*, 1993), canine (Suber *et al.*, 1993) and murine sources (Baehr *et al.*, 1991; Pittler and Baehr, 1991; Bowes *et al.*, 1990). Rod PDE6 γ-subunit cDNAs have been isolated from bovine (Ovchinnikov *et al.*, 1986), human (Cotran *et al.*, 1991; Tuteja *et al.*, 1990; Piriev *et al.*, 1994) and murine sources (Tuteja and Farber, 1988).

The gene for the human rod PDE6 β-subunit has been isolated (Weber *et al.*, 1991; Khramtsov *et al.*, 1993). It comprises 22 exons and extends over 43 kb of DNA. The gene for the human rod PDE6 γ-subunit has also been isolated (Piriev *et al.*, 1994). It consists of four exons distributed over 6 kb.

An alternative splice variant has been reported for the PDE6 rod β-subunit of the mouse (β'; Baehr *et al.*, 1991). A different splice acceptor site is used to produce a mRNA that would encode a cDNA with a different carboxy terminus. This variant lacks the carboxy-terminal Cys-Ala-Ala-X (CAAX) motif. The cysteine in this motif is believed to undergo post-translational addition of an isoprenoid lipid followed by proteolytic cleavage – to remove the AAX – and carboxymethylation of the Cys-COOH. These modifications result in a hydrophobic tail that can promote PDE6 anchoring to the rod outer segment membrane (Qin *et al.*, 1992). No evidence was found for the same variant in human retinal mRNA and the sequence of human genomic DNA in that region made the production of human β' seem unlikely (Khramtsov *et al.*, 1993).

Several different sizes of rod PDE6 α and β-subunit mRNAs are seen in Northern blots using human or bovine mRNA (Li *et al.*, 1990; Pittler *et al.*, 1990; Weber *et al.*, 1991; Collins *et al.*, 1992). It is not clear what accounts for the differences.

3.6.1.2 Cone PDE6

cDNAs encoding the cone PDE α'-subunit have been isolated from bovine (Charbonneau *et al.*, 1990; Li *et al.*, 1990), chicken (Semple-Rowland and Green,

1994) and human cells (Piriev *et al.*, 1995) (Table 1.1). cDNAs for the bovine 13 kD subunit (Hamilton and Hurley, 1990) and the 15 kD or δ-subunit (Florio *et al.*, 1994) have been isolated. There are two different α' cone PDE6 mRNAs found in the chicken which represent the use of different polyadenylation sites (Semple-Rowland and Green, 1994).

The gene for the human α'-subunit has been isolated (Piriev *et al.*, 1995). It consists of 22 exons extending over 48 kb.

3.6.2 Homology

The PDE6 rod α and β-subunits and the cone α'-subunit share a common organization (Fig. 1.1). They contain two non-catalytic cGMP binding regions in the amino-terminal portion of the protein and a catalytic domain in the carboxy-terminal portion (Charbonneau *et al.*, 1990). When the bovine, human and mouse cDNAs for PDE6 rod α-subunits are compared the amino acid identity is greater than 90%. This is also true when rod β-subunits or cone α'-subunits are compared. When the sequences of the different subunits are compared to each other the amino acid identity is 60–70%. The bovine cone 13 kD subunit is more than 80% identical to the bovine rod PDE6 γ-subunit. It is not known if there are additional family members.

3.6.3 Gene Mapping

The rod PDE6 α-subunit has been mapped to human chromosome 5q31.2-q34 (Pittler *et al.*, 1990) and to mouse chromosome 18 (Danciger *et al.*, 1990). The rod PDE6 β-subunit has been mapped to human chromosome 4p16.3 (Weber *et al.*, 1991; Altherr *et al.*, 1992; Bateman *et al.*, 1992) and to mouse chromosome 5, 6.1 centimorgan (cM) distal of *Mgsa* (Altherr *et al.*, 1992). The rod PDE6 γ-subunit has been mapped to human chromosome 17q25 (Tuteja *et al.*, 1990; Cotran *et al.*, 1991; Dollfus *et al.*, 1993) and in the mouse to the telomeric region of chromosome 11, 1.2 cM distal to *Es3* (Danciger *et al.*, 1989; Kozak *et al.*, 1995). The human cone PDE6 α'-subunit has been mapped to chromosome 10q24 (Piriev *et al.*, 1995).

3.6.4 Association with Genetic Diseases

Mice with the genetic marker *rd* display hereditary retinal degeneration. By 4 weeks of age no photoreceptors remain. This degeneration is preceded by an accumulation of cGMP and a lack of PDE6 activity. Mutations in the rod PDE6 β-subunit have been identified as causing retinal degeneration in the *rd* mouse (Bowes *et al.*, 1990, 1993; Pittler and Baehr, 1991). Canine rod–cone dysplasia-1 in the Irish setter is also due to mutations in the rod PDE6 β-subunit (Aguirre *et al.*, 1978; Farber *et al.*, 1992; Suber *et al.*, 1993).

Retinitis pigmentosa is a group of hereditary human diseases that result in retinal degeneration. They can be autosomal dominant, recessive or X-linked. Mutation in the rod PDE6 β-subunit is the underlying defect in some cases of this disease (M.E. McLaughlin *et al.*, 1993, 1995; Gal *et al.*, 1994a,b; Bayes *et al.*, 1995). No genetic linkage of retinitis pigmentosa with human rod PDE γ-subunit (Cotran *et al.*, 1991) or chick cone PDE α'-subunit (Semple-Rowland and Green, 1994) has been discovered.

3.6.5 Cell and Tissue Expression

Expression of the bovine and human PDE6 rod α-subunit has been detected in retina (Li *et al.*, 1990; Pittler *et al.*, 1990). Rod PDE6 β-subunit expression has been detected in human, bovine and mouse retina and, at much lower levels, in human frontal cortex, basal ganglia and caudate nucleus (Bowes *et al.*, 1990; Li *et al.*, 1990; Weber *et al.*, 1991; Collins *et al.*, 1992). It is not known what role PDE6 plays in these tissues. No expression is seen in adrenal gland, oesophagus, skeletal muscle, kidney, liver or lung. Expression of rod PDE γ-subunit is also seen in bovine, human and mouse retina but not in a number of other tissues (Tuteja and Farber, 1988; Hamilton and Hurley, 1990; Tuteja *et al.*, 1990). Rod PDE6 γ-subunit expression in the retina is observed in mouse, rat, dog, cow and human but not in ground squirrel (Tuteja *et al.*, 1990). Cone PDE6 α'-subunit can be detected in chicken and bovine retina but not in bovine brain (Li *et al.*, 1990; Semple-Rowland and Green, 1994). Expression of the cone PDE6 13 kD subunit is also detected in bovine retina but not in brain, kidney, adrenal gland, spleen or liver (Hamilton and Hurley, 1990). Immunocytochemical analysis has revealed that the cone PDE6 13 kD subunit maps specifically to blue cones. Cone PDE6 15 kD or δ-subunit is detected in bovine retina, brain and adrenal gland (Florio *et al.*, 1994).

3.7 PDE7

The type VII PDE was first isolated in a functional screen for human proteins that complemented a deficiency in endogenous PDE in *Saccharomyces cerevisiae* (Michaeli *et al.*, 1993). PDE7 hydrolyses only cAMP and its affinity for cAMP is higher than that of the PDE4 isoenzymes, which are also specific for cAMP. Unlike the PDE4s, PDE7 is not sensitive to inhibition by rolipram (see Chapter 2).

3.7.1 Molecular Cloning

The PDE7 cDNAs isolated by Michaeli *et al.* (1993) from the human glioblastoma cell line U118-MG were not full length. They encoded a 57 kD protein. The structures of the cDNAs resemble those of other mammalian PDEs, with a catalytic region in the carboxy-terminal portion of the protein (Fig. 1.1). Two differently sized mRNAs are observable by Northern blotting, suggesting the possibility of alternate splice forms (Michaeli *et al.*, 1993). It is not known whether there are additional genes in this family.

3.7.2 Gene Mapping

The PDE7 gene maps to human chromosome 8q13-q22 and is not associated with any known disease markers (Milatovich *et al.*, 1994).

3.7.3 Cell and Tissue Expression

Expression of PDE7 has been described in human skeletal muscle, kidney, heart, brain and pancreas (Michaeli *et al.*, 1993). No signal was detected in liver or lung. Activity believed to be PDE7 is also observed in the human T lymphocyte cell line, Hut78 (Bloom and Beavo, 1994), and in the human keratinocyte cell line HaCaT (V. Gekeler, cited in Tenor *et al.*, 1995a; see Chapter 2).

3.8 ADDITIONAL PDEs

There are reports of biochemical activities that may correspond to novel PDE activities (Ichimura and Kase, 1993; Burns *et al.*, 1994; Mukai *et al.*, 1994; Medvedeva and Bobruskin, 1994; Tenor *et al.*, 1995b). However, there is no molecular characterization of these activities. It seems likely that some of these novel activities correspond to PDE7. In addition, a novel cDNA has been proposed as a potential PDE (Vambutas and Wolgemuth, 1994). It was isolated by low stringency hybridization using a rat PDE4A probe and it shares 20% amino acid identity with the rat PDE4A protein. It lacks most of the conserved residues that all known mammalian PDEs contain and its biochemical activity as a PDE has not been reported.

4. *Summary*

Combining the data from human, bovine, rat, murine, chicken and canine sources, 18 different PDEs or PDE subunits have been identified (Table 1.1). Although they have not yet all been isolated from each of these sources there are no examples where any given PDE gene could not be found in a particular species. The PDE amino acid sequences are highly conserved between species with the individual PDEs having more amino acid homology to their counterparts in other species than to the other members of their own PDE family. Many of the genes are alternatively spliced and, although some of the splice variants are found in multiple species, it is not clear whether they are all conserved between species. Many of the PDE activities that have been biochemically identified can be accounted for by the known PDE genes but it is still possible that there are additional PDEs to be found.

Chromosomal map positions have been determined for more than half of the PDEs (Table 1.3). Mutations in PDE6 β-subunits have been identified as underlying certain retinal degenerative diseases in mice, dogs and humans. None of the other PDEs has yet been associated with a disease.

It is clear from the PDE localization work that most tissues contain more than one of the seven PDE families and many contain more than one member of an individual PDE family. However, when PDE localization is examined more closely there are numerous examples where individual PDEs are found in very specific locations relative to other PDEs. One example is the PDE4 distribution in the rat testis where each of the four PDE4s has a specific localization and presumably plays a specific role (Swinnen *et al.*, 1989a,b; Welch *et al.*, 1992; Morena *et al.*, 1995). In the olfactory system PDE1C (Yan *et al.*, 1995) and PDE4A (Cherry and Davis, 1995) are highly expressed but with different patterns of subcellular localization (Juilfs *et al.*, 1995). It will be interesting to determine the role that each plays in olfactory sensing. The distribution of PDE1B is far more restricted than that of PDE1A and PDE1C; PDE1B localizes to dopaminergic regions of the brain (Polli and Kincaid, 1994; Yan *et al.*, 1994). PDE2 is abundant in the glomerulosa cells of the adrenal cortex, where it is believed to play a role in the regulation of aldosterone biosynthesis (MacFarland *et al.*, 1991). PDE3A and PDE3B show different patterns of expression and this is consistent with the different role each is believed to play (Manganiello *et al.*, 1995; Reinhardt *et al.*, 1995).

The roles of different splice variants are also currently being explored. Roles in transcriptional regulation, intracellular localization and enzymatic activation have been assigned to particular variants but there are many variants whose roles have not been characterized.

As PDEs are localized it is becoming apparent that their distributions, although similar, are not identical in different species. This implies that care must be taken in generalizing from data obtained using PDE inhibitors in one species. Although the localization work provides clues as to what role individual PDEs play in cellular physiology, the development of inhibitors that are specific for each PDE will allow their roles in animal and cellular model systems to be determined. The cloning and expression studies have provided the materials to assess the specificity of particular inhibitors. This should aid in drawing conclusions from inhibitor studies about the cellular roles of individual PDEs.

5. *References*

Aguirre, G., Farber, D., Lolley, R., Fletcher, R.T. and Chader, G.J. (1978). Rod–cone dysplasia in Irish setters: a defect in cyclic GMP metabolism in visual cells. Science 201, 1133–1134.

Altherr, M.R, Wasmuth, J.J., Seldin, M.F., Nadeau, J.H.,

Baehr, W. and Pittler, S.J. (1992). Chromosome mapping of the rod photoreceptor cGMP phosphodiesterase β-subunit gene in mouse and human: tight linkage to the Huntington disease region (4p16.3). Genomics 12, 750–754.

Ariano, M.A. and Appleman, M.M. (1979). Biochemical characterization of postsynaptically localized cyclic nucleotide phosphodiesterase. Brain Res. 177, 301–309.

Baecker, P.A., Obernolte, R., Bach, C., Yee, C. and Shelton, E.R. (1994). Isolation of a cDNA encoding a human rolipram-sensitive cyclic AMP phosphodiesterase (PDE IVD). Gene 138, 253–256.

Baehr, W., Champagne, M.S., Lee, A.K. and Pittler, S.J. (1991). Complete cDNA sequences of mouse rod photoreceptor cGMP phosphodiesterase α- and β-subunits, and identification of β'-, a putative β-subunit isozyme produced by alternative splicing of the β-subunit gene. FEBS Lett. 278, 107–114.

Balaban, C.D., Billingsley, M.L. and Kincaid, R.L. (1989). Evidence for transsynaptic regulation of calmodulin-dependent cyclic nucleotide phosphodiesterase in cerebellar Purkinje cells. J. Neurosci. 9, 2374–2381.

Banner, K.H. and Page, C.P. (1995). Theophylline and selective phosphodiesterase inhibitors as anti-inflammatory drugs in the treatment of bronchial asthma. Eur. Respir. J. 8, 996–1000.

Bateman, J.B., Klisak, I., Kojis, T., Mohandas, T., Sparkes, R.S., Li, T., Applebury, M.L., Bowes, C. and Farber, D.B. (1992). Assignment of the β-subunit of rod photoreceptor cGMP phosphodiesterase gene PDEB (homolog of the mouse *rd* gene) to human chromosome 4p16. Genomics 12, 601–603.

Bayes, M., Giordano, M., Balcells, S., Grinberg, D., Vilageliu, L., Martinez, I., Ayuso, C., Benitez, J., Ramos-Arroyo, M.A., Chivelet, P., Solans, T., Valverde, D., Amselem, S., Goossens, M., Baiget, M., Gonzalez-Duarte, R. and Besmond, C. (1995). Homozygous tandem duplication within the gene encoding the β-subunit of rod phosphodiesterase as a cause for autosomal recessive retinitis pigmentosa. Hum. Mutat. 5, 228–234.

Beavo, J.A., Conti, M. and Heaslip, R.J. (1994). Multiple cyclic nucleotide phosphodiesterases. Mol. Pharmacol. 46, 399–405.

Bentley, J.K., Kadlecek, A., Sherbert, C.H., Seger, D., Sonnenburg, W.K., Charbonneau, H., Novack, J.P. and Beavo, J.A. (1992). Molecular cloning of cDNA encoding a 63-kDa calmodulin-stimulated phosphodiesterase from bovine brain. J. Biol. Chem. 267, 18676–18682.

Billingsley, M.L., Polli, J.W., Balaban, C.D. and Kincaid, R.L. (1990). Developmental expression of calmodulin-dependent cyclic nucleotide phosphodiesterase in rat brain. Dev. Brain Res. 53, 253–263.

Bloom, T.J. and Beavo, J.A. (1994). Identification of type VII PDE in HUT 78 T-lymphocyte cells. FASEB J. 8, A372. [Abstract]

Bolger, G., Michaeli, T., Martins, T., St John, T., Steiner, B., Rodgers, L., Riggs, M., Wigler, M. and Ferguson, K. (1993). A family of human phosphodiesterases homologous to the *dunce* learning and memory gene product of *Drosophila melanogaster* are potential targets for antidepressant drugs. Mol. Cell. Biol. 13, 6558–6571.

Bolger, G.B. (1994). Molecular biology of the cyclic AMP-specific cyclic nucleotide phosphodiesterases: a diverse family of regulatory enzymes. Cell Signal. 6, 851–859.

Bolger, G.B., Rodgers, L. and Riggs, M. (1994). Differential CNS expression of alternative mRNA isoforms of the mammalian genes encoding cAMP-specific phosphodiesterases. Gene 149, 237–244.

Bolger, G.B., McPhee, I., Pooley, L., Jones, R.E. and Houslay, M.D. (1995). Regulation of biochemical properties and cellular localization of cAMP-specific PDEs by alternative mRNA splicing. FASEB J. 9, A1262. [Abstract]

Bowes, C., Li, T., Danciger, M., Baxter, L.C., Applebury, M.L. and Farber, D.B. (1990). Retinal degeneration in the *rd* mouse is caused by a defect in the β subunit of rod cGMP-phosphodiesterase. Nature 347, 677–680.

Bowes, C., Li, T., Frankel, W.N., Danciger, M., Coffin, J.M., Applebury, M.L. and Farber, D.B. (1993). Localization of a retroviral element within the *rd* gene coding for the β subunit of cGMP phosphodiesterase. Proc. Natl Acad. Sci. USA 90, 2955–2959.

Burns, F., Stevens, P.A. and Pyne, N.J. (1994). The identification of apparently novel cyclic AMP and cyclic GMP phosphodiesterase activities in guinea-pig tracheal smooth muscle. Br. J. Pharmacol. 113, 3–4.

Burns, F., Sonnanburg, W.K., Francis, S.H., Corbin, J.D. and Beavo, J.A. (1995). Isolation of murine cyclic GMP-binding phosphodiesterase cDNAS. FASEB J. 9, A1261. [Abstract]

Charbonneau, H. (1990). Structure–function relationships among cyclic nucleotide phosphodiesterases. In "Cyclic Nucleotide Phosphodiesterases: Structure, Function, Regulation and Drug Action" (eds. J. Beavo and M.D. Houslay), pp. 267–296. Wiley, Chichester.

Charbonneau, H., Prusti, R.K., LeTrong, H., Sonnenburg, W.K., Mullaney, P.J., Walsh, K.A. and Beavo, J.A. (1990). Identification of a noncatalytic cGMP-binding domain conserved in both the cGMP-stimulated and photoreceptor cyclic nucleotide phosphodiesterases. Proc. Natl Acad. Sci. USA 87, 288–292.

Charbonneau, H., Kumar, S., Novack, J.P., Blumenthal, D.K., Griffin, P.R., Shabanowitz, J., Hunt, D.F. and Beavo, J.A. (1991). Evidence for domain organization within the 61-kDa calmodulin-dependent cyclic nucleotide phosphodiesterase from bovine brain. Biochemistry 30, 7931–7940.

Chen, C.-N., Denome, S. and Davis, R.L. (1986). Molecular analysis of cDNA clones and the corresponding genomic coding sequences of the Drosophila *dunce*$^+$ gene, the structural gene for cAMP phosphodiesterase. Proc. Natl Acad. Sci. USA 83, 9313–9317.

Cherry, J.A. and Davis, R.L. (1995). A mouse homolog of *dunce*, a gene important for learning and memory in *Drosophila*, is preferentially expressed in olfactory receptor neurons. J. Neurobiol. 28, 102–113.

Colicelli, J., Birchmeier, C., Michaeli, T., O'Neill, K., Riggs, M. and Wigler, M. (1989). Isolation and characterization of a mammalian gene encoding a high-affinity cAMP phosphodiesterase. Proc. Natl Acad. Sci. USA 86, 3599–3603.

Colicelli, J., Nicolette, C., Birchmeier, C., Rodgers, L., Riggs, M. and Wigler, M. (1991). Expression of three mammalian cDNAs that interfere with RAS function in *Saccharomyces cerevisiae*. Proc. Natl Acad. Sci USA 88, 2913–2917.

Collins, C., Hutchinson, G., Kowbel, D., Riess, O., Weber, B. and Hayden, M.R. (1992). The human β-subunit of rod photoreceptor cGMP phosphodiesterase: complete retinal cDNA sequence and evidence for expression in brain. Genomics 13, 698–704.

Conti, M., Iona, S., Cuomo, M., Swinnen, J.V., Odeh, J. and

Svoboda, M.E. (1995a). Characterization of a hormone-inducible, high affinity adenosine 3′-5′-cyclic monophosphate phosphodiesterase from the rat Sertoli cell. Biochemistry 34, 7979–7987.

Conti, M., Nemoz, G., Sette, C. and Vicini, E. (1995b). Recent progress in understanding the hormonal regulation of phosphodiesterases. Endocr. Rev. 16, 370–389.

Cotran, P.R., Bruns, G.A.P., Berson, E.L. and Dryja, T.P. (1991). Genetic analysis of patients with retinitis pigmentosa using a cloned cDNA probe for the human gamma subunit of cyclic GMP phosphodiesterase. Exp. Eye Res. 53, 557–564.

Danciger, M., Tuteja, N., Kozak, C.A. and Farber, D.B. (1989). The gene for the γ-subunit of retinal cGMP-phosphodiesterase is on mouse chromosome 11. Exp. Eye Res. 48, 303–308.

Danciger, M., Kozak, C.A., Li, T., Applebury, M.L. and Farber, D.B. (1990). Genetic mapping demonstrates that the α-subunit of retinal cGMP-phosphodiesterase is not the site of the *rd* mutation. Exp. Eye Res. 51, 185–189.

Davis, R.L., Takayasu, H., Eberwine, M. and Myres, J. (1989). Cloning and characterization of mammalian homologs of the *Drosophila* dunce$^+$ gene. Proc. Natl Acad. Sci. USA 86, 3604–3608.

Dollfus, H., Mattei, M.-G., Rozet, J.-M., Delrieu, O., Munnich, A. and Kaplan, J. (1993). Physical and genetic localization of the γ subunit of the cyclic GMP phosphodiesterase on the long arm of chromosome 17 (17q25). Genomics 17, 526–528.

Dunlap, P.V. and Callahan, S.M. (1993). Characterization of a periplasmic 3′:5′-cyclic nucleotide phosphodiesterase gene, cpdP, from the marine symbiotic bacterium *Vibrio fischeri*. J. Bacteriol. 175, 4615–4624.

Engels, P., Fichtel, K. and Lubbert, H. (1994). Expression and regulation of human and rat phosphodiesterase type IV isogenes. FEBS Lett. 350, 291–295.

Engels, P., Abdel'Al, S., Hulley, P. and Lubbert, H. (1995a). Brain distribution of four rat homologues of the *Drosophila dunce* cAMP phosphodiesterase. J. Neurosci. Res. 41, 169–178.

Engels, P., Sullivan, M., Muller, T. and Lubbert, H. (1995b). Molecular cloning and functional expression in yeast of a human cAMP-specific phosphodiesterase subtype (PDE IV-C). FEBS Lett. 358, 305–310.

Farber, D.B., Danciger, J.S. and Aguirre, G. (1992). The β subunit of cyclic GMP phosphodiesterase mRNA is deficient in canine rod-cone dysplasia 1. Neuron 9, 349–356.

Florio, S.K., Prusti, R.K. and Beavo, J.A. (1994). Sequencing, cloning, expression, and localization of the 15 kDa (δ) subunit of retinal phosphodiesterases. FASEB J. 8, A372. [Abstract]

Furuyama, T., Iwahashi, Y., Tano, Y., Takagi, H. and Inagaki, S. (1994). Localization of 63-kDa calmodulin-stimulated phosphodiesterase mRNA in the rat brain by *in situ* hybridization histochemistry. Mol. Brain Res. 26, 331–336.

Gal, A., Orth, U., Baehr, W., Schwinger, E. and Rosenberg, T. (1994a). Heterozygous missense mutation in the rod cGMP phosphodiesterase β-subunit gene in autosomal dominant stationary night blindness. Nat. Genet. 7, 64–68.

Gal, A., Orth, U., Baehr, W., Schwinger, E. and Rosenberg, T. (1994b). Heterozygous missense mutation in the rod cGMP phosphodiesterase β-subunit gene in autosomal dominant stationary night blindness. Nat. Genet. 7, 551. [Correction]

Gillespie, P.G. (1990). Phosphodiesterases in visual transduction by rods and cones. In "Cyclic Nucleotide Phosphodiesterases: Structure, Function, Regulation and Drug Action" (eds. J. Beavo and M.D. Houslay), pp. 163–184. Wiley, Chichester.

Hamilton, S.E. and Hurley, J.B. (1990). A phosphodiesterase inhibitor specific to a subset of bovine retinal cones. J. Biol. Chem. 265, 11259–11264.

Hebenstreit, G.F., Fellerer, K., Fichte, K., Fischer, G., Geyer, N., Meya, U., Sastre-y-Hernandez, M., Schony, W., Schratzer, M., Soukop, W., Trampitsch, E., Varosanec, S., Zawada, E. and Zochling, R. (1989). Rolipram in major depressive disorder: results of a double-blind comparative study with imipramine. Pharmacopsychiatry 22, 156–160.

Horton, Y.M., Sullivan, M. and Houslay, M.D. (1995). Molecular cloning of a novel splice variant of human type IVA (PDE-IVA) cyclic AMP phosphodiesterase and localization of the gene to the p13.1-q12 region of human chromosome 19. Biochem. J. 308, 683–691.

Hoyer, L.L., Cieslinski, L.B., McLaughlin, M.M., Torphy, T.J., Shatzman, A.R. and Livi, G.P. (1994). A *Candida albicans* cyclic nucleotide phosphodiesterase: cloning and expression in *Saccharomyces cerevisiae* and biochemical characterization of the recombinant enzyme. Microbiology 140, 1533–1542.

Ichimura, M. and Kase, H. (1993). A new cyclic nucleotide phosphodiesterase isozyme expressed in the T-lymphocyte cell lines. Biochem. Biophys. Res. Commun. 193, 985–990.

Jin, S.-L.C., Swinnen, J.V. and Conti, M. (1992). Characterization of the structure of a low K_m rolipram-sensitive cAMP phosphodiesterase: mapping of the catalytic domain. J. Biol. Chem. 267, 18929–18939.

Juilfs, D., Yan, C., Houslay, M.D. and Beavo, J.A. (1995). Differential localization of PDE1C, PDE2 and PDE4A in olfactory neurons. FASEB J. 9, A1262. [Abstract]

Kasuya, J., Goko, H. and Fujita-Yamaguchi, Y. (1995). Multiple transcripts for the human cardiac form of the cGMP-inhibited cAMP phosphodiesterase. J. Biol. Chem. 270, 14305–14312.

Khramtsov, N.V., Feshchenko, E.A., Suslova, V.A., Shmukler, B.E., Terpugov, B.E., Rakitina, T.V., Atabekova, N.V. and Lipkin, V.M. (1993). The human rod photoreceptor cGMP phosphodiesterase β-subunit: structural studies of its cDNA and gene. FEBS Lett. 327, 275–278.

Kincaid, R.L., Balaban, C.D. and Billingsley, M.L. (1987a). Differential localization of calmodulin-dependent enzymes in rat brain: evidence for selective expression of cyclic nucleotide phosphodiesterase in specific neurons. Proc. Natl Acad. Sci. USA 84, 1118–1122.

Kincaid, R.L., Balaban, C.D. and Billingsley, M.L. (1987b). Regulated expression of calmodulin-dependent cyclic nucleotide phosphodiesterase in the central nervous system. J. Cyclic Nucleotide Protein Phosphor. Res. 11, 473–486.

Kovala, T., Lorimer, I.A.J., Brickenden, A.M., Ball, E.H. and Sanwal, B.D. (1994). Protein kinase A regulation of cAMP phosphodiesterase expression in rat skeletal myoblasts. J. Biol. Chem. 269, 8680–8685.

Kozak, C.A., Danciger, M., Bowes, C., Adamson, M.C., Palczewski, K., Polans, A.S. and Farber, D.B. (1995). Localization of three genes expressed in retina on mouse chromosome 11. Mamm. Genome 6, 142–144.

Lacombe, M.-L., Podgorski, G.J., Franke, J. and Kessin, R.H. (1986). Molecular cloning and developmental expression of the cyclic nucleotide phosphodiesterase gene of *Dictyostelium discoideum*. J. Biol. Chem. 261, 16811–16817.

Le Trong, H., Beier, N., Sonnenburg, W.K., Stroop, S.D., Walsh, K.A., Beavo, J.A. and Charbonneau, H. (1990). Amino acid sequence of the cyclic GMP stimulated cyclic nucleotide phosphodiesterase from bovine heart. Biochemistry 29, 10280–10288.

Li, T., Volpp, K. and Applebury, M.L. (1990). Bovine cone photoreceptor cGMP phosphodiesterase structure deduced from a cDNA clone. Proc. Natl Acad. Sci. USA 87, 293–297.

Lipkin, V.M., Khramtsov, N.V., Vasilevskaya, I.A., Atabekova, N.V., Muradov, K.G., Gubanov, V.V., Li, T., Johnson, J.P., Volpp, K.J. and Applebury, M.L. (1990). β-Subunit of bovine rod photoreceptor cGMP phosphodiesterase. J. Biol. Chem. 265, 12955–12959.

Livi, G.P., Kmetz, P., McHale, M.M., Cieslinski, L.B., Sathe, G.M., Taylor, D.P., Davis, R.L., Torphy, T.J. and Balcarek, J.M. (1990). Cloning and expression of cDNA for a human low-K_m, rolipram-sensitive cyclic AMP phosphodiesterase. Mol. Cell. Biol. 10, 2678–2686.

Lobban, M., Shakur, Y., Beattie, J. and Houslay, M.D. (1994). Identification of two splice variant forms of type-IV(B) cyclic AMP phosphodiesterase, DPD (rPDE-IVB1) and PDE-4 (rPDE-IVB2) in brain: selective localization in membrane and cytosolic compartments and differential expression in various brain regions. Biochem. J. 304, 399–406.

Loughney, K., Martins, T., Sonnenburg, B., Beavo, J. and Ferguson, K. (1994). Isolation and characterization of cDNAs corresponding to two human calcium calmodulin stimulated 3′,5′ cyclic nucleotide phosphodiesterase genes. FASEB J. 8, A81. [Abstract]

Loughney, K., Martins, T.J., Harris, E.A.S., Sadhu, K., Hicks, J.B., Sonnenburg, W.K., Beavo, J.A. and Ferguson, K. (1996). Isolation and characterization of cDNAs corresponding to two human calcium, calmodulin-regulated, 3′,5′ cyclic nucleotide phosphodiesterases. J. Biol. Chem. 271, 796–806.

Ludvig, N., Burmeister, V., Jobe, P.C. and Kincaid, R.L. (1991). Electron microscopic immunocytochemical evidence that the calmodulin-dependent cyclic nucleotide phosphodiesterase is localized predominantly at postsynaptic sites in the rat brain. Neuroscience 44, 491–500.

MacFarland, R.T., Zelus, B.D. and Beavo, J.A. (1991). High concentrations of a cGMP-stimulated phosphodiesterase mediate ANP-induced decreases in cAMP and steroidogenesis in adrenal glomerulosa cells. J. Biol. Chem. 266, 136–142.

Manganiello, V.C., Taira, M., Degerman, E. and Belfrage, P. (1995). Type III cGMP-inhibited cyclic nucleotide phosphodiesterases (PDE3 gene family). Cell Signal. 7, 445–455.

Matviw, H., Li, J. and Young, D. (1993). The *Schizosaccharomyces pombe* pdel/cgs2 gene encodes a cyclic AMP phosphodiesterase. Biochem. Biophys. Res. Commun. 194, 79–82.

McAllister-Lucas, L.M., Sonnenburg, W.K., Kadlecek, A., Seger, D., LeTrong, H., Colbran, J.L., Thomas, M.K., Walsh, K.A., Francis, S.H., Corbin, J.D. and Beavo, J.A. (1993). The structure of a bovine lung cGMP-binding, cGMP-specific phosphodiesterase deduced from a cDNA clone. J. Biol. Chem. 268, 22863–22873.

McLaughlin, M.E., Sandberg, M.A., Berson, E.L. and Dryja, T.P. (1993). Recessive mutations in the gene encoding the β-subunit of rod phosphodiesterase in patients with retinitis pigmentosa. Nat. Genet. 4, 130–133.

McLaughlin, M.E., Ehrhart, T.L., Berson, E.L. and Dryja, T.P. (1995). Mutation spectrum of the gene encoding the β subunit of rod phosphodiesterase among patients with autosomal recessive retinitis pigmentosa. Proc. Natl Acad. Sci. USA 92, 3249–3253.

McLaughlin, M.M., Cieslinski, L.B., Burman, M., Torphy, T.J. and Livi, G.P. (1993). A low-K_m, rolipram-sensitive, cAMP-specific phosphodiesterase from human brain: cloning and expression of cDNA, biochemical characterization of recombinant protein, and tissue distribution of mRNA. J. Biol. Chem. 268, 6470–6476.

Meacci, E., Taira, M., Moos, M., Jr, Smith, C.J., Movsesian, M.A., Degerman, E., Belfrage, P. and Manganiello, V.C. (1992). Molecular cloning and expression of human myocardial cGMP-inhibited cAMP phosphodiesterase. Proc. Natl Acad. Sci. USA 89, 3721–3725.

Medvedeva, M.V. and Bobruskin, I.D. (1994). Ca^{2+}-dependent regulation of cGMP-stimulated phosphodiesterase from the soluble fraction of the human brain. Biokhimiia (Moscow) 59, 866–872.

Michaeli, T., Bloom, T.J., Martins, T., Loughney, K., Ferguson, K., Riggs, M., Rodgers, L., Beavo, J.A. and Wigler, M. (1993). Isolation and characterization of a previously undetected human cAMP phosphodiesterase by complementation of cAMP phosphodiesterase-deficient *Saccharomyces cerevisiae*. J. Biol. Chem. 268, 12925–12932.

Milatovich, A., Bolger, G., Michaeli, T. and Francke, U. (1994). Chromosome localizations of genes for five cAMP-specific phosphodiesterases in man and mouse. Somat. Cell Mol. Genet. 20, 75–86.

Monaco, L., Vicini, E. and Conti, M. (1994). Structure of two rat genes coding for closely related rolipram-sensitive cAMP phosphodiesterases. J. Biol. Chem. 269, 347–357.

Morena, A.R., Boitani, C., De Grossi, S., Stefanini, M. and Conti, M. (1995). Stage and cell-specific expression of the adenosine 3′,5′-monophosphate-phosphodiesterase genes in the rat seminiferous epithelium. Endocrinology 136, 687–695.

Mukai, J., Asai, T., Naka, M. and Tanaka, T. (1994). Separation and characterization of a novel isoenzyme of cyclic nucleotide phosphodiesterase from rat cerebrum. Br. J. Pharmacol. 111, 389–390.

Murashima, S., Tanaka, T., Hockman, S. and Manganiello, V. (1990). Characterization of particulate cyclic nucleotide phosphodiesterases from bovine brain: purification of a distinct cGMP-stimulated isoenzyme. Biochemistry 29, 5285–5292.

Murata, T., Taira, M. and Manganiello, V.C. (1995). Structure/function analysis of recombinant type III cGMP-inhibited cyclic nucleotide phosphodiesterase (PDE3): use of N-terminal region deletion recombinants to study PDE3 membrane-association. FASEB J. 9, A1262. [Abstract]

Némoz, G., Zhang, R.B., Sette, C. and Conti, M. (1995). Identification and characterization of two variants of the PDE4D phosphodiesterase gene expressed in human peripheral mononuclear leukocytes. FASEB J. 9, A1262. [Abstract]

Nikawa, J.-I., Sass, P. and Wigler, M. (1987). Cloning and characterization of the low-affinity cyclic AMP phosphodiesterase gene of *Saccharomyces cerevisiae*. Mol. Cell. Biol. 7, 3629–3636.

Novack, J.P., Charbonneau, H., Bentley, J.K., Walsh, K.A. and Beavo, J.A. (1991). Sequence comparison of the 63-, 61-, and 59-kDa calmodulin-dependent cyclic nucleotide phosphodiesterases. Biochemistry 30, 7940–7947.

Obernolte, R., Bhakta, S., Alvarez, R., Bach, C., Zuppan, P., Mulkins, M., Jarnagin, K. and Shelton, E.R. (1993). The cDNA of a human lymphocyte cyclic-AMP phosphodiesterase (PDE IV) reveals a multigene family. Gene 129, 239–247.

Obernolte, R., Baecker, P.A., Ratzliff, J.R., Kasten, G., Daniels, D. and Shelton, E.R. (1994). cDNA cloning and expression of a rare human cAMP-specific, rolipram-sensitive phosphodiesterase, PDE IV C. FASEB J. 8, A368. [Abstract]

Ovchinnikov, Y.A., Lipkin, V.M., Kumarev, V.P., Gubanov, V.V., Khramtsov, N.V., Akhmedov, N.B., Zagranichny, V.E. and Muradov, K.G. (1986). Cyclic GMP phosphodiesterase from cattle retina: amino acid sequence of the γ-subunit and nucleotide sequence of the corresponding cDNA. FEBS Lett. 204, 288–292.

Ovchinnikov, Y.A., Gubanov, V.V., Khramtsov, N.V., Ischenko, K.A., Zagranichny, V.E., Muradov, K.G., Shuvaeva, T.M. and Lipkin, V.M. (1987). Cyclic GMP phosphodiesterase from bovine retina: amino acid sequence of the α-subunit and nucleotide sequence of the corresponding cDNA. FEBS Lett. 223, 169–173.

Piriev, N.I., Khramtsov, N.V. and Lipkin, V.M. (1994). Cloning and characterization of the gene encoding the cGMP-phosphodiesterase γ-subunit of human rod photoreceptor cells. Gene 151, 297–301.

Piriev, N.I., Viczian, A.S., Ye, J., Kerner, B., Korenberg, J.R. and Farber, D.B. (1995). Gene structure and amino acid sequence of the human cone photoreceptor and cGMP-phosphodiesterase α' subunit (PDEA2) and its chromosomal localization to 10q24. Genomics 28, 429–435.

Pittler, S.J. and Baehr, W. (1991). Identification of a nonsense mutation in the rod photoreceptor cGMP phosphodiesterase β-subunit gene of the *rd* mouse. Proc. Natl Acad. Sci. USA 88, 8322–8326.

Pittler, S.J., Baehr, W., Wasmuth, J.J., McConnel, D.G., Champagne, M.S., van Tuinen, P., Ledbetter, D. and Davis, R.L. (1990). Molecular characterization of human and bovine rod photoreceptor cGMP phosphodiesterase α-subunit and chromosomal localization of the human gene. Genomics 6, 272–283.

Polli, J.W. and Kincaid, R.L. (1992). Molecular cloning of DNA encoding a calmodulin-dependent phosphodiesterase enriched in striatum. Proc. Natl Acad. Sci. USA 89, 11079–11083.

Polli, J.W. and Kincaid, R.L. (1994). Expression of a calmodulin-dependent phosphodiesterase isoform (PDE1B1) correlates with brain regions having extensive dopaminergic innervation. J. Neurosci. 14, 1251–1261.

Qin, N., Pittler, S.J. and Baehr, W. (1992). *In vitro* isoprenylation and membrane association of mouse rod photoreceptor cGMP phosphodiesterase α and β subunits expressed in bacteria. J. Biol. Chem. 267, 8458–8463.

Reinhardt, R.R., Chin, E., Zhou, J., Taira, M., Murata, T., Manganiello, V.C. and Bondy, C.A. (1995). Distinctive anatomical patterns of gene expression for cGMP-inhibited cyclic nucleotide phosphodiesterases. J. Clin. Invest. 95, 1528–1538.

Repaske, D.R., Swinnen, J.V., Jin, S.-L. Catherine, Van Wyk, J.J. and Conti, M. (1992). A polymerase chain reaction strategy to identify and clone cyclic nucleotide phosphodiesterase cDNAs. J. Biol. Chem. 267, 18683–18688.

Repaske, D.R., Corbin, J.G., Conti, M. and Goy, M.F. (1993). A cyclic GMP-stimulated cyclic nucleotide phosphodiesterase gene is highly expressed in the limbic system of the rat brain. Neuroscience 56, 673–686.

Sass, P., Field, J., Nikawa, J., Toda, T. and Wigler, M. (1986). Cloning and characterization of the high-affinity cAMP phosphodiesterase of *Saccharomyces cerevisiae*. Proc. Natl Acad. Sci. USA 83, 9303–9307.

Scotland, G. and Houslay, M.D. (1995). Chimeric constructs show that the unique N-terminal domain of the cyclic AMP phosphodiesterase RD1 (RNPDE4A1A; rPDE-IVA1) can confer membrane association upon the normally cytosolic protein chloramphenicol acetyltransferase. Biochem. J. 308, 673–681.

Semple-Rowland, S.L. and Green, D.A. (1994). Molecular characterization of the α'-subunit of cone photoreceptor cGMP phosphodiesterase in normal and *rd* chicken. Exp. Eye Res. 59, 365–372.

Sette, C., Iona, S. and Conti, M. (1994a). The short-term activation of a rolipram-sensitive, cAMP-specific phosphodiesterase by thyroid-stimulating hormone in thyroid FRTL-5 cells is mediated by a cAMP-dependent phosphorylation. J. Biol. Chem. 269, 9245–9252.

Sette, C., Vicini, E. and Conti, M. (1994b). The rat PDE3/IVd phosphodiesterase gene codes for multiple proteins differentially activated by cAMP-dependent protein kinase. J. Biol. Chem. 269, 18271–18274.

Shakur, Y., Pryde, J.G. and Houslay, M.D. (1993). Engineered deletion of the unique N-terminal domain of the cyclic AMP-specific phosphodiesterase RD1 prevents plasma membrane association and the attainment of enhanced thermostability without altering its sensitivity to inhibition by rolipram. Biochem. J. 292, 677–686.

Shakur, Y., Wilson, M., Pooley, L., Lobban, M., Griffiths, S.L., Campbell, A.M., Beattie, J., Daly, C. and Houslay, M.D. (1995). Identification and characterization of the type IVA cyclic AMP-specific phosphodiesterase RD1 as a membrane-bound protein expressed in cerebellum. Biochem. J. 306, 801–809.

Sharma, R.K. and Kalra, J. (1994). Characterization of calmodulin-dependent cyclic nucleotide phosphodiesterase isoenzymes. Biochem. J. 299, 97–100.

Shenolikar, S., Thompson, W.J. and Strada, S.J. (1985). Characterization of a Ca^{2+}-calmodulin-stimulated cyclic GMP phosphodiesterase from bovine brain. Biochemistry 24, 672–678.

Sonnenburg, W.K., Mullaney, P.J. and Beavo, J.A. (1991). Molecular cloning of a cyclic GMP-stimulated cyclic nucleotide phosphodiesterase cDNA: identification and distribution of isozyme variants. J. Biol. Chem. 266, 17655–17661.

Sonnenburg, W.K., Seger, D. and Beavo, J.A. (1993). Molecular cloning of a cDNA encoding the "61 kDa" calmodulin-stimulated cyclic nucleotide phosphodiesterase:

tissue-specific expression of structurally related isoforms. J. Biol. Chem. 268, 645–652.

Sonnenburg, W.K., Seger, D., Kwak, K.S., Huang, J., Charbonneau, H. and Beavo, J.A. (1995). Identification of inhibitory and calmodulin-binding domains of the PDE1A1 and PDE1A2 calmodulin-stimulated cyclic nucleotide phosphodiesterases. J. Biol. Chem. 270, 30989–31000.

Suber, M.L., Pittler, S.J., Qin, N., Wright, G.C., Holcombe, V., Lee, R.H., Craft, C.M., Lolley, R.N., Baehr, W. and Hurwitz, R.L. (1993). Irish setter dogs affected with rod/cone dysplasia contain a nonsense mutation in the rod cGMP phosphodiesterase β-subunit gene. Proc. Natl Acad. Sci. USA 90, 3968–3972.

Sullivan, M., Egerton, M., Shakur, Y., Marquardsen, A. and Houslay, M.D. (1994). Molecular cloning and expression, in both COS-1 cells and *S. cerevisiae*, of a human cytosolic type-IVA, cyclic AMP specific phosphodiesterase (hPDE-IVA-h6.1). Cell Signal. 6, 793–812.

Swinnen, J.V., Joseph, D.R. and Conti, M. (1989a). Molecular cloning of rat homologues of the *Drosophila melanogaster* dunce cAMP phosphodiesterase: evidence for a family of genes. Proc. Natl Acad. Sci. USA 86, 5325–5329.

Swinnen, J.V., Joseph, D.R. and Conti, M. (1989b). The mRNA encoding a high-affinity cAMP phosphodiesterase is regulated by hormones and cAMP. Proc. Natl Acad. Sci. USA 86, 8197–8201.

Swinnen, J.V., Tsikalas, K.E. and Conti, M. (1991). Properties and hormonal regulation of two structurally related cAMP phosphodiesterases from the rat Sertoli cell. J. Biol. Chem. 266, 18370–18377.

Szpirer, C., Szpirer, J., Riviere, M., Swinnen, J., Vicini, E. and Conti, M. (1995). Chromosomal localization of the human and rat genes (PDE4D and PDE4B) encoding the cAMP-specific phosphodiesterases 3 and 4. Cytogenet. Cell Genet. 69, 11 -14.

Taira, M., Hockman, S.C., Calvo, J.C., Taira, M., Belfrage, P. and Manganiello, V.C. (1993). Molecular cloning of the rat adipocyte hormone-sensitive cyclic GMP-inhibited cyclic nucleotide phosphodiesterase. J. Biol. Chem. 268, 18573–18579.

Tenor, H., Hatzelmann, A., Wendel, A. and Schudt, C. (1995a). Identification of phosphodiesterase IV activity and its cyclic adenosine monophosphate-dependent up-regulation in a human keratinocyte cell line (HaCaT). J. Invest. Dermatol. 105, 70–74.

Tenor, H., Staniciu, L., Schudt, C., Hatzelmann, A., Wendel, A., Djukanović, R., Church, M.K. and Shute, J.K. (1995b). Cyclic nucleotide phosphodiesterase from purified human $CD4^+$ and $CD8^+$ T lymphocytes. Clin. Exp. Allergy 25, 616–624.

Thomas, M.K., Francis, S.H. and Corbin, J.D. (1990). Characterization of a purified bovine lung cGMP-binding cGMP phosphodiesterase. J. Biol. Chem. 265, 14964–14970.

Tuteja, N. and Farber, D.B. (1988). γ-Subunit of mouse retinal cyclic-GMP phosphodiesterase: cDNA and corresponding amino acid sequence. FEBS Lett. 232, 182–186.

Tuteja, N., Danciger, M., Klisak, I., Tuteja, R., Inana, G., Mohandas, T., Sparkes, R.S. and Farber, D.B. (1990). Isolation and characterization of cDNA encoding the gamma-subunit of cGMP phosphodiesterase in human retina. Gene 88, 227–232.

Usui, H., Falk, J.D., Dopazo, A., de Lecea, L., Erlander, M.G. and Sutcliffe, J.G. (1994). Isolation of clones of rat striatum-specific mRNAs by directional tag PCR subtraction. J. Neurosci. 14, 4915–4926.

Vambutas, V. and Wolgemuth, D.J. (1994). Identification and characterization of the developmentally regulated pattern of expression in the testis of a mouse gene exhibiting similarity to the family of phosphodiesterases. Biochim. Biophys. Acta 1217, 203–206.

Verghese, M.W., McConnell, R.T., Lenhard, J.M., Hamacher, L. and Jin, S.-L.C. (1995). Regulation of distinct cyclic AMP-specific phosphodiesterase (phosphodiesterase type 4) isozymes in human monocytic cells. Mol. Pharmacol. 47, 1164–1171.

Vicini, E., Sette, C. and Conti, M. (1994). Differential expression of three variants of the cAMP-specific, rolipram-sensitive rat PDE3/IVD phosphodiesterase. FASEB J. 8, A369. [Abstract]

Wang, J.H., Sharma, R.K. and Mooibroek, M.J. (1990). Calmodulin-stimulated cyclic nucleotide phosphodiesterases. In "Cyclic Nucleotide Phosphodiesterases: Structure, Function, Regulation and Drug Action" (eds. J. Beavo and M.D. Houslay), pp. 19–60. Wiley, Chichester.

Weber, B., Riess, O., Hutchinson, G., Collins, C., Lin, B., Kowbel, D., Andrew, S., Schappert, K. and Hayden, M.R. (1991). Genomic organization and complete sequence of the human gene encoding the β-subunit of the cGMP phosphodiesterase and its localisation to 4p16.3. Nucleic Acids Res. 19, 6263–6268.

Welch, J.E., Swinnen, J.V., O'Brien, D.A., Eddy, E.M. and Conti, M. (1992). Unique adenosine 3′,5′ cyclic monophosphate phosphodiesterase messenger ribonucleic acids in rat spermatogenic cells: evidence for differential gene expression during spermatogenesis. Biol. Reprod. 46, 1027–1033.

Yan, C. and Beavo, J.A. (1994) Molecular cloning and cellular distribution of the 75 kDa calmodulin-dependent phosphodiesterase. FASEB J. 8, A81. [Abstract]

Yan, C., Bentley, J.K., Sonnenburg, W.K. and Beavo, J.A. (1994). Differential expression of the 61 kDa and 63 kDa calmodulin-dependent phosphodiesterases in the mouse brain. J. Neurosci. 14, 973–984.

Yan, C., Zhao, A.Z., Bentley, J.K., Loughney, K., Ferguson, K. and Beavo, J.A. (1995). Molecular cloning and characterization of a calmodulin-dependent phosphodiesterase enriched in olfactory sensory neurons. Proc. Natl Acad. Sci. USA 92, 9677–9681.

Yang, Q., Paskind, M., Bolger, G., Thompson, W.J., Repaske, D.R., Cutler, L.S. and Epstein, P.M. (1994). A novel cyclic GMP stimulated phosphodiesterase from rat brain. Biochem. Biophys. Res. Commun. 205, 1850–1858.

Yu, J., Frazier, A., Florio, V., Martins, T., Ferguson, K., Bentley, K., Beavo, J. and Gelinas, R. (1994). Cloning and expression of a calmodulin regulated human PDE HCAM2. FASEB J. 8, A81. [Abstract]

2. *Analysis of PDE Isoenzyme Profiles in Cells and Tissues by Pharmacological Methods*

Hermann Tenor *and* Christian Schudt

1. Introduction 21
2. Analysis of PDE Isoenzyme Activities in Cells and Tissues 22
 2.1 Procedure for Establishment of PDE Isoenzyme Activity Profiles 22
 2.2 PDE Isoenzyme Activity Profiles 23
3. Regulation of PDE Isoenzyme Activities 30
 3.1 Regulation of PDE Activity by Cyclic Nucleotide Concentrations and Ca^{2+}/Calmodulin 30
 3.2 Short-term Regulation of PDE Activity by Phosphorylation 32
 3.3 Long-Term PDE4 Induction by cAMP-elevating Agents 32
 3.4 Elevated PDE Activity in Atopic Mononuclear Cells 33
4. Conclusions 34
5. References 34

1. *Introduction*

Theophylline and papaverine are non-selective phosphodiesterase (PDE) inhibitors which have been used therapeutically in a variety of diseases. Papaverine proved to ameliorate pathological conditions related to vasoconstriction, decreased blood perfusion or reduced tissue oxygenation. Theophylline, however, has been used as a diuretic, a cardiotonic and – for the last three decades – predominantly as an anti-asthmatic drug (see Chapter 3). As regards the treatment of asthma it has been recognized during the last 10 years that acute exacerbations of a chronic airway inflammation underlie the periodic bronchoconstrictory attacks. Theophylline was long assumed to be a weak bronchodilator and thus served as a rescue medication in asthma which was partly replaced by more efficient bronchodilating β_2-receptor agonists. However, total replacement or withdrawal of theophylline often results in a deterioration in the condition of asthmatic patients, suggesting this drug may have activities other than simple bronchodilation. Recent studies have demonstrated a considerable attenuation of the inflamed state of the airway mucosa after theophylline treatment, supporting the view that theophylline also interferes with activation of inflammatory cells. The clinical use of theophylline, however, has been limited by the central nervous and gastrointestinal side-effects, such as nausea, dizziness and headache, occurring at serum levels above 80 μM. Carefully performed earlier studies as well as recent evaluations support the view that most effects of theophylline are mechanistically best explained by PDE inhibition, strongly suggesting that PDE isoenzymes are a relevant target for a variety of therapies.

From the experience with theophylline it can be concluded that PDE inhibition in general may interfere with many physiological and pathological functions, particularly with the regulation of (i) immune, autoimmune, allergic and inflammatory responses, (ii) smooth muscle contractile tone, (iii) cell proliferation and (iv) neurotransmitter release or other tightly controlled syntheses and secretions. Furthermore, pathological dysregulations may be reversed. Translated to the cellular level this means that cyclic nucleotide PDEs are involved in the network of signal transduction in nearly every cell and the question arises of whether interference with a selected target cell or process without simultaneously affecting others is, in fact, feasible.

Phosphodiesterase Inhibitors
ISBN 0-12-210720-9

At present 27 PDE isoenzymes, including splice variants, have been identified in human tissue (see Chapter 1) and information is accumulating that shows these enzymes have distinct distribution patterns in different cell types. Reverse transcriptase/polymerase chain reaction (RT-PCR) and Western blot analysis of PDE isoenzymes have been performed in several cells, including human monocytes (Verghese *et al.*, 1995a) and B lymphocytes (see Chapter 11). However, since the tools for this type of analysis have only recently become available, another, simpler procedure had been chosen to study the isoenzyme activity profiles of different cells. Using different substrates, allosteric activators and selective pharmacological inhibitors for PDE isoenzyme families 1–5 it was possible differentially to inhibit or increase the basal cyclic nucleotide-hydrolysing activity in a cellular homogenate (Fig. 2.1). From these data the activities of the individual PDE isoenzymes could be calculated even in an isoenzyme mixture. The resulting bar diagrams (see, for example, Fig. 2.2) reflect the basal activities of PDE3, 4 and 5 and the activated activities of PDE1 and 2. PDE7 could be roughly estimated by non-inhibitable activity.

Such PDE isoenzyme profile measurements have the following advantages:

1. Determination of each individual isoenzyme in a mixture is possible;
2. Information about subcellular distribution of isoenzymes is obtained;
3. Analysis is fast, thereby reducing the risk of proteolytic degradation;
4. Activity of different isoenzymes is determined simultaneously, giving quantitative indication of their participation in cyclic nucleotide synthesis;
5. A comparison of PDE isoenzyme profiles between different cells and tissues – with regard to specific activities and isoenzyme ratios – is possible.

The corresponding disadvantages are:

1. Activities are determined in mixtures and not in separated and purified fractions;
2. The substrate concentrations are fixed and do not correspond to those in the local intracellular environment;
3. Concentrations of regulating substances in the extracts will be diluted but are not necessarily zero and may affect PDE activities;
4. Enzyme phosphorylation (and parallel activation) will be greatly reduced during cell isolation and homogenization procedures: these effects have not yet been systematically investigated.

In any case, this method has yielded a variety of highly reproducible specific cell PDE isoenzyme profiles. From these profiles it can be deduced qualitatively – and in part quantitatively – which isoenzymes participate in cyclic nucleotide hydrolysis. From the comparison of these findings with data on the inhibition of cell functions by selective PDE inhibitors and their combinations it can be deduced whether the ratios between PDE activities *in situ* are similar or are greatly modified. Quantitative knowledge of the PDE content of cells stimulates questions about the different functions of the individual members of the PDE family and may suggest strategies for the more rational design of new PDE inhibitors in addition to furthering our understanding of intracellular signalling.

2. *Analysis of PDE Isoenzyme Activities in Cells and Tissues*

2.1 Procedure for Establishment of PDE Isoenzyme Activity Profiles

The PDE isoenzyme activity profiles presented in this chapter are based on the evaluation of the effects of PDE

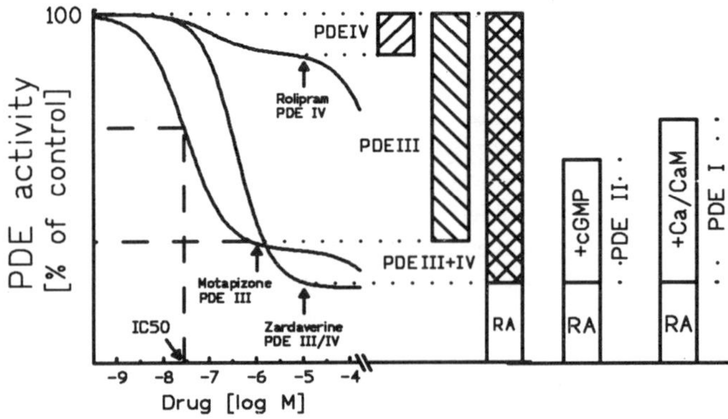

Figure 2.1 Procedure for calculation of PDE isoenzyme activities. From concentration–inhibition curves, concentrations of rolipram and motapizone are derived which completely and selectively inhibit their corresponding isoenzyme; these concentrations are used to evaluate PDE3 and PDE4 activities. In parallel, PDE5 activity is defined as the cGMP-catalysing activity inhibited by 10 μM zaprinast (not shown). PDE1 and PDE2 activities are defined from their characteristic activation by Ca^{2+}/calmodulin and cGMP, respectively (for details see text).

isoenzyme-selective inhibitors and activators on total cyclic nucleotide hydrolysis. Studies were performed at 0.5 μM cyclic nucleotide substrate concentration in cytosolic and particulate fractions of tissue and cell homogenates. It was decided that the PDE inhibitors and activators should be used at concentrations which selectively and completely inhibited their corresponding isoenzyme, this ensured that a limited panel of PDE inhibitors and activators (Table 2.1) could optimally discriminate between different PDE isoenzyme activities. For this purpose concentration–inhibition curves for each PDE inhibitor used were constructed and, from these curves, appropriate concentrations were derived as illustrated in Fig. 2.1. These concentrations of PDE inhibitors were then applied to calculate activities of PDE3, PDE4 and PDE5 as shown in Fig. 2.1 and Table 2.1(b). To define PDE1 and PDE2 activities the PDE activity increments induced by Ca^{2+}/calmodulin (CaM) and cGMP were used. PDE1 activity was defined as that part of Ca^{2+} (1 mM)/CaM (150 nM)-stimulated cGMP hydrolysis that was inhibitable by 1 mM EGTA. PDE2 activity was defined as the increase by 5 μM cGMP of a residual cAMP PDE activity remaining after complete inhibition of PDE3 and PDE4 by the selective inhibitors, motapizone and rolipram. These concentrations of Ca^{2+}/CaM and cGMP have been demonstrated previously to maximally activate PDE1 and PDE2 activities (H. Tenor and C. Schudt, unpublished observations). Since no selective inhibitor of PDE7 is known, the evaluation of PDE7 by this method is difficult. However, a residual cAMP PDE activity, insensitive to zardaverine, cGMP and Ca^{2+}/CaM, would suggest the presence of PDE7.

2.2 PDE ISOENZYME ACTIVITY PROFILES

2.2.1 Eosinophils and Neutrophils

Eosinophils are involved in inflammatory diseases of lung, skin and gut, and in parasitic diseases. They represent terminal effector and immunomodulatory cells because they elaborate cytotoxic proteins (eosinophil cationic protein (ECP), eosinophil-derived neurotoxin (EDN), eosinophil peroxidase (EPO)), reactive oxygen species, sulphidopeptide leukotrienes (LTC_4) and cytokines (tumour necrosis factor-α (TNF-α), granulocyte-macrophage colony-stimulating factor (GM-CSF), interleukins (IL-) 3, 4, 5, 6 and 8). It was, therefore of interest to find out the PDE isoenzyme

Table 2.1 Procedure for calculation of PDE isoenzyme activities. (a) Concentrations of selective inhibitors that completely inhibit the corresponding isoenzymes at 0.5 µM cyclic nucleotide substrate concentration are derived from concentration–inhibition curves (Fig. 2.1). Concentrations of Ca^{2+}/CaM or cGMP are selected for maximal activation of PDE1 and PDE2, respectively. (b) Activity of individual isoenzymes is calculated under conditions of full inhibition of other isoenzymes and, where appropriate, activation of the isoenzyme under investigation

(a)

Drug addition	*Isoenzyme specificity*	*Inhibitor (I)/ Activator (A)*
None	Control	
1 µM motapizone(M)	PDE3	I
10 µM rolipram (R)	PDE4	I
10 µM zaprinast (Zp)	PDE5	I
10 µM zardaverine	PDE3/PDE4	I
1 mM $CaCl_2$ + 150 nM calmodulin (Ca^{2+}/CaM)	PDE1	A
1 mM EGTA	PDE1	I
5 µM cGMP	PDE2	A

(b)

Isoenzyme	*Substrate*	*Calculation of activity*
PDE1	cGMP	Ca^{2+}/CaM – EGTA
PDE2	cAMP	(R + M + cGMP) – (R + M)
PDE3	cAMP	Control – M
PDE4	cAMP	Control – R
PDE5	cGMP	M – (M + Zp)

profile of eosinophils and the effects of isoenzyme-selective PDE inhibitors on eosinophil functions.

Initial studies with guinea-pig peritoneal eosinophils demonstrated that these cells exclusively contained a membrane-associated PDE4 activity, since the total cAMP PDE activity was almost completely blocked by the PDE4-selective inhibitor rolipram (Souness *et al.*, 1991; Dent *et al.*, 1991). Using RT-PCR, PDE4 in guinea-pig peritoneal eosinophils could be further characterized as PDE4D (Souness *et al.*, 1995). Although only one PDE4 subtype was identified in guinea-pig eosinophils it was found that cAMP hydrolysis showed non-linear kinetics. In addition, concentration–response curves for inhibition of cAMP hydrolysis by rolipram were very shallow (Dent *et al.*, 1991; Souness *et al.*, 1991). These observations might indicate the existence of distinct substrate or inhibitor binding sites on the enzyme, possibly due to post-translational modification. This hypothesis is supported by recent findings that human recombinant monocyte PDE4 expressed in yeast exhibits a high affinity rolipram binding site which is different from the catalytic site (Torphy *et al.*, 1992b). Furthermore, different splice variants of PDE4D have now been described and one of these can be phosphorylated, resulting in its activation (Sette *et al.*, 1994b); this phosphorylation has been reported to affect the inhibition of the hydrolytic activity by selective PDE4 inhibitors (Conti, 1995; Alvarez *et al.*, 1995; see also Chapter 11).

More recent investigations focused on human peripheral blood eosinophils. These cells are now readily accessible using a new immunomagnetic purification procedure which produces eosinophil preparations with a purity of almost 100% (Hansel *et al.*, 1991). As in guinea-pig eosinophils, PDE4 was identified as the predominant PDE isoenzyme activity of human blood eosinophils (Fig. 2.2) (Dent *et al.*, 1994; Hatzelmann *et al.*, 1995). RT-PCR revealed the presence of the subtypes PDE4A, 4B and 4D in human eosinophils (Engels *et al.*, 1994). The subcellular localization of the eosinophil PDE4 activity is controversial at present. Using different homogenization procedures, both membrane-associated (Dent *et al.*, 1994) and cytosolic (Fig. 2.2; Hatzelmann *et al.*, 1995) PDE4 activities have been found.

Corresponding to the finding that PDE4 predominates in eosinophils, selective (rolipram, RP 73401) and non-selective (zardaverine, theophylline) PDE4 inhibitors were shown to inhibit degranulation and superoxide anion (O_2^-) generation of human peripheral blood and guinea-pig peritoneal eosinophils whereas, in contrast, selective inhibitors of PDE3 and PDE5 were ineffective (Dent *et al.*, 1991, 1994; Kita *et al.*, 1991; Souness *et al.*, 1991, 1995; Hatzelmann *et al.*, 1995; Hadjokas *et al.*, 1995).

Neutrophils are involved in many inflammatory diseases due to their ability to release reactive oxygen species, the chemoattractant LTB_4 and lysosomal enzymes and other proteins (neutral proteases, acid hydrolases, lactoferrin). Similarly to eosinophils, neutrophils contain exclusively PDE4 activity (Nielson *et al.*, 1990; Wright *et al.*, 1990; Schudt *et al.*, 1991). At 0.5 μM cAMP substrate concentration, the specific activity of PDE4 in neutrophils and eosinophils is almost identical (Fig. 2.2).

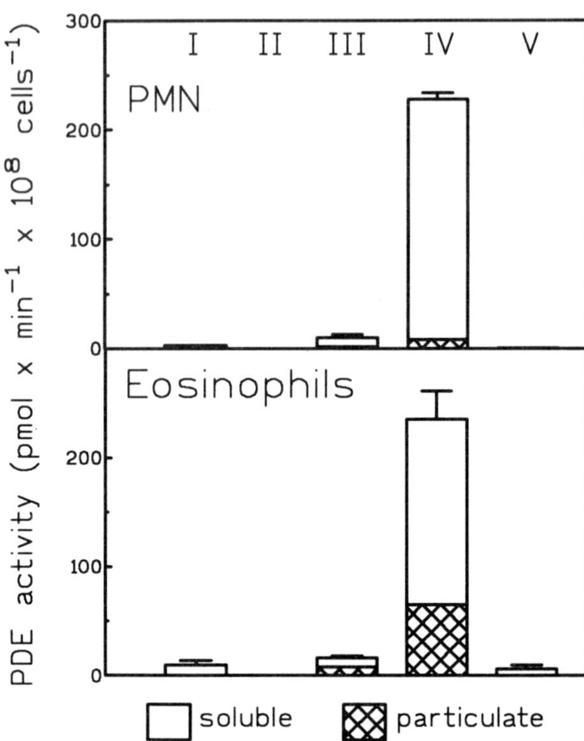

Figure 2.2 PDE isoenzyme profiles of human peripheral blood neutrophils (PMN) and eosinophils (>99.5% purity).

Functional studies demonstrating inhibition of degranulation and mediator release (PAF, LTB_4, O_2^-) of neutrophils by selective PDE4 inhibitors, but not by inhibitors of PDE3 or PDE5, confirm the biochemical finding of the exclusivity of PDE4 activity in neutrophils (Nourshargh and Hoult, 1986; Nielson *et al.*, 1990; Wright *et al.*, 1990; Schudt *et al.*, 1991; Fonteh *et al.*, 1993).

2.2.2 Monocytes and Macrophages

Peripheral blood monocytes represent the precursor cells of tissue macrophages. Monocytes and macrophages are antigen-presenting cells and generate several pro-inflammatory mediators (e.g. IL-1; TNF-α; IL-8; reactive oxygen species; 5-lipoxygenase metabolites). In human peripheral blood monocytes PDE4 has been identified as the predominant isoenzyme activity but in addition a minor PDE3 activity which accounts for about 15–25% of total cAMP hydrolysis was described.

Whereas the PDE4 activity was located in the cytosol, PDE3 activity was mainly membrane-associated (Fig. 2.3) (Elliott and Leonard, 1989; Torphy *et al.*, 1992b; Tenor *et al.*, 1995a; Verghese *et al.*, 1995a,b). A similar PDE isoenzyme profile was found in the monocytic cell lines U937 (Torphy *et al.*, 1992c) and MonoMac 6 (Verghese *et al.*, 1995a). RT-PCR revealed the presence of PDE4A and PDE4B subtypes in human monocytes and U937 cells (Barnette *et al.*, 1993). More recently, Verghese *et al.* (1995a) detected mRNA and protein of PDE4A, 4B and 4D subtypes in monocytic MonoMac 6 cells.

In parallel to the maturation of monocytes to macrophages *in vivo*, culture of peripheral blood monocytes for several days results in the generation of monocyte-derived macrophages. This *in vitro* maturation of monocytes is accompanied by a distinct change of the PDE isoenzyme profile (Fig. 2.3). In monocyte-derived macrophages an additional PDE1 activity was detected. Moreover, the ratio of PDE3/PDE4 activity was clearly increased (Tenor *et al.*, 1995a). This PDE isoenzyme profile of monocyte-derived macrophages is comparable to the isoenzyme profile of human alveolar macrophages (Fig. 2.4) (Dent *et al.*, 1993; Tenor *et al.*, 1995a).

The PDE isoenzyme profile of macrophages was found to be species- and tissue-specific. In contrast to human alveolar macrophages, guinea-pig peritoneal

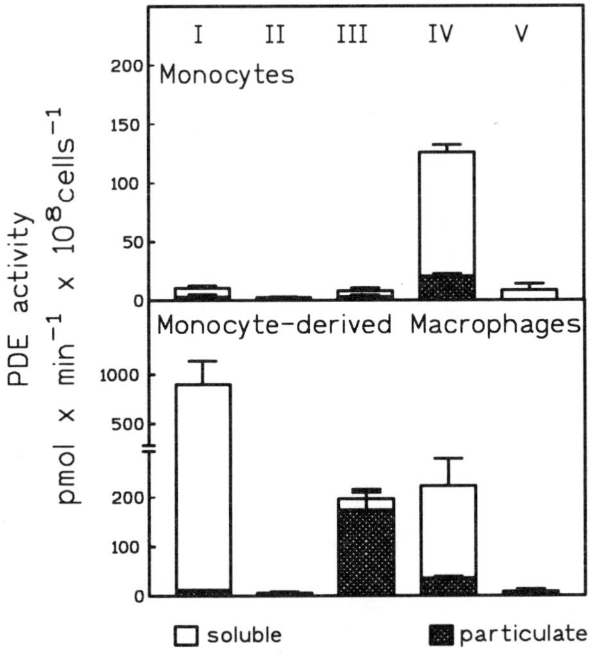

Figure 2.3 PDE isoenzyme profiles of freshly prepared human peripheral blood monocytes (purified by elutriation; purity > 88%) and monocytes cultured for 1 week (monocyte-derived macrophages). Note that the PDE isoenzyme profile has changed dramatically during the culture period.

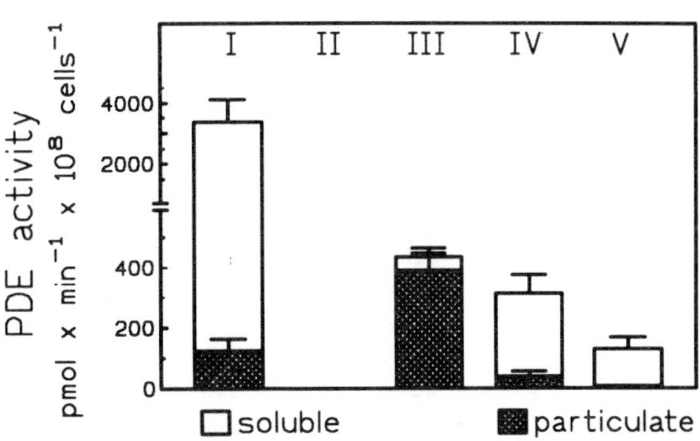

Figure 2.4 PDE isoenzyme profile of human alveolar macrophages obtained by bronchoalveolar lavage and purified by adherence steps (purity > 95%). Note that the PDE isoenzyme profile of alveolar macrophages is qualitatively similar to that of monocyte-derived macrophages.

macrophages were described as containing a membrane-bound PDE4 activity and a cytosolic PDE1 activity. PDE2, PDE3 and PDE5 activities were absent (Turner *et al.*, 1993). Whereas PDE2 was not found in human alveolar and guinea-pig peritoneal macrophages, a substantial PDE2 activity has been detected in murine peritoneal macrophages (Fig. 2.5) (Okonogi *et al.*, 1991; Prpic *et al.*, 1993). In parallel, although the PDE isoenzyme profile of human peritoneal macrophages is unknown at present, the occurrence of PDE2 in these cells might be hypothesized in view of the fact that cGMP triggers a fall in their intracellular cAMP concentration (Houdjik *et al.*, 1990).

2.2.3 T Lymphocytes

T lymphocytes may orchestrate the course of many inflammatory diseases through their ability to elaborate a variety of cytokines. It has been shown that different subsets of human T lymphocytes are committed to the production of restricted cytokine profiles. T_H1-cells are characterized by their IL-2 and interferon-γ (IFNγ) production and are involved in autoimmune reactions, contact dermatitis and graft versus host reactions. T_H2-cells characteristically synthesize IL-4, IL-5 and IL-10 and promote allergic inflammation. Furthermore, T lymphocytes have been divided according to their surface antigens into $CD4^+$, $CD8^+$ and $CD25^+$ (activated) cells.

For the investigation of PDE isoenzyme profiles in human T lymphocytes it is crucial to use highly purified lymphocyte preparations, avoiding monocyte and platelet contamination. Early studies were mostly performed with mixed lymphocyte preparations (containing T, B and natural killer cells). In cytosolic fractions of human lymphocytes, Epstein and Hachisu (1984) described a cAMP PDE activity sensitive to Ro 20–1724 which could therefore be assigned to PDE4. Several other authors described cAMP PDE activities in

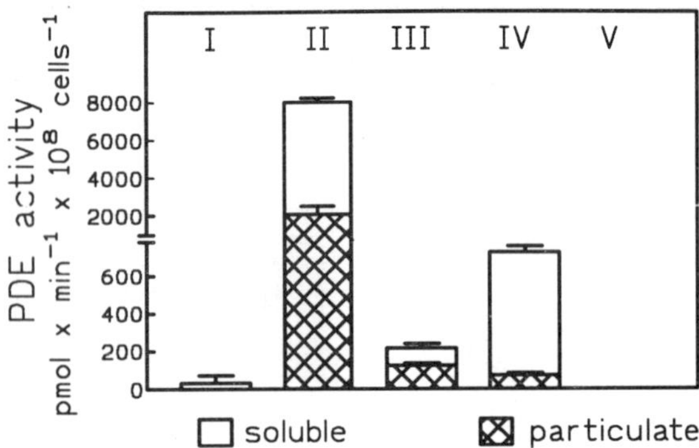

Figure 2.5 PDE isoenzyme profile of mouse peritoneal macrophages. Note the presence of PDE2, which is absent from human alveolar macrophages.

human lymphocyte preparations with K_m values of about 0.4–1.5 μM, whereas only minor cGMP PDE activities were found (Thompson *et al.*, 1976; Wedner *et al.*, 1979). PDE activities in a T-cell enriched lymphocyte population were first investigated by Takemoto *et al.* (1978). DEAE-cellulose chromatography of their T lymphocyte preparations revealed the presence of a single cAMP-hydrolysing peak (K_m = 1 μM). Moreover, cAMP PDE activity in whole cell homogenates was strongly inhibited by cGMP, indicating the presence of PDE3 in these cells. In addition, from their data the presence of PDE2, 1 and 5 can be excluded. Robiscek *et al.* (1989, 1991) confirmed the presence of PDE4 and PDE3 in human T lymphocyte preparations which were differentially located in the soluble and particulate subcellular fractions, respectively. Using an immunomagnetic separation technique, the PDE isoenzyme profiles of highly purified peripheral blood CD4$^+$ and CD8$^+$ T lymphocytes were analysed (Tenor *et al.*, 1995c). As shown in Fig. 2.6, CD4$^+$ and CD8$^+$ T cells exhibit comparable PDE isoenzyme activity profiles with predominantly membrane-associated PDE3 and soluble PDE4 activities. Moreover, in the soluble fractions of both T lymphocyte subsets about 20% of cAMP hydrolysis could not be assigned to PDE1–5 owing to their insensitivity to specific activators and inhibitors of these isoenzymes. Since, in addition, this residual activity hydrolysed cAMP with a K_m = 0.05–0.08 μM it was speculated that this activator/inhibitor-insensitive cAMP PDE activity might represent the recently described PDE7 (Michaeli *et al.*, 1993). This hypothesis is supported by findings from other groups reporting the identification of an activator/inhibitor-insensitive cAMP PDE activity in human T lymphocytes (Robiscek *et al.*, 1989) as well as in human T cell lines (Ichimura and Kase, 1993). Finally, using RT-PCR, PDE7 was recently identified in the human T cell line HUT-78 (Bloom and Beavo, 1994).

With respect to PDE4 subtypes, the presence of PDE4A was demonstrated in the Jurkat T-cell line

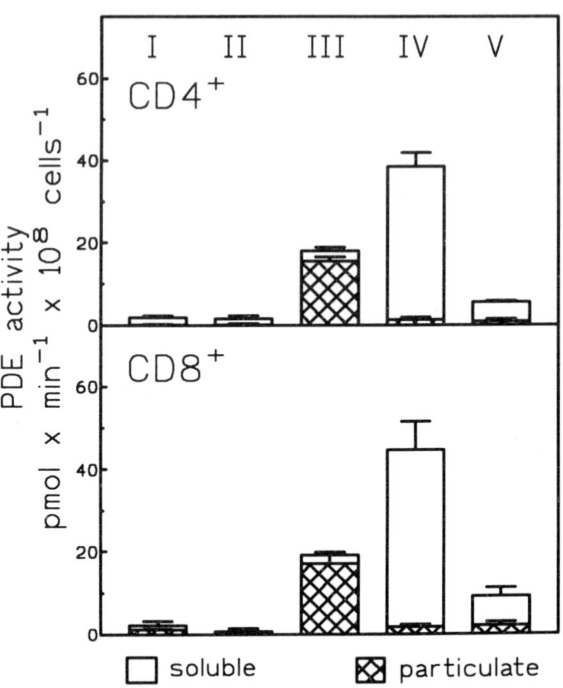

Figure 2.6 PDE isoenzyme profiles of human peripheral blood CD4$^+$ and CD8$^+$ T lymphocytes purified by immunomagnetic selection (purity > 98%). The isoenzyme profiles of CD4$^+$ and CD8$^+$ T cells are not different.

(Engels *et al.*, 1994). Correspondingly, a highly purified PDE4 from human mononuclear cells exhibited an N-terminal amino acid sequence similar to that deduced from cDNA sequences of PDE4A and PDE4D (Truong and Müller, 1994).

At present the PDE isoenzyme profile of T_H1 and T_H2 cells is not known. However, functional studies investigating the inhibition of proliferative responses of T_H1 cells (tetanus toxoid-stimulated) and T_H2 cells (ragweed-stimulated) by selective PDE3 and PDE4 inhibitors might indicate a difference in (functional) PDE isoenzyme profile between these subtypes. It was found that inhibition of PDE4 – but not PDE3 – attenuated T_H2 cell proliferation whereas T_H1 cell proliferation was most effectively diminished by combined PDE3/4 inhibitors (Essayan *et al.*, 1994).

2.2.4 Mast Cells and Basophils

Mast cells elaborate cytokines (e.g. IL-3, IL-4, IL-5, TNF-α, GM-CSF), histamine, heparin, LTC_4 and prostaglandin D_2 (PGD_2). The release of these mediators in response to antigen stimulation enables mast cells to play an important role in allergic inflammation. Therefore, mast cells represent an interesting target for PDE inhibitor action. Regrettably little is known about PDE isoenzymes in these cells, however. Recently, soluble PDE3 and PDE4 activities have been identified as the main isoenzymes from highly purified human lung mast cells (Fig. 2.7) (H. Tenor and Y. Okayama, unpublished observations). Correspondingly, functional studies by Anderson and Peachell (1994) demonstrated that PDE3 and PDE4 inhibitors, but not PDE5 inhibitors, increased cAMP concentrations and inhibited histamine release from human lung mast cells. However, very high concentrations (300 μM) of PDE inhibitors were used in this study. Similarly, PDE4 inhibitors attenuated histamine release from human nasal polyp mast cells (Lau *et al.*, 1993) suggesting the presence of PDE4 in these cells.

Due to the methodological difficulties involved in obtaining highly purified mast cells from human tissues, investigations have been performed with mast cells from other sources. In the murine mast cell lines FB1/PT18, Torphy *et al.* (1992a) found PDE1 and PDE4 activities using DEAE-Sepharose and CaM-affinity chromatography. In rat peritoneal mast cells PDE5 activity was identified (Bergstrand *et al.*, 1978). However, the difference between these PDE isoenzyme profiles and that of human lung mast cells emphasizes the importance of studying cells from human sources.

Basophils produce a profile of mediators similar to mast cells (e.g. LTC_4, histamine, IL-4) and may therefore be involved in allergic inflammation. In human peripheral blood basophils about 72% and 16% of total cAMP PDE activity was classified as PDE4 and PDE3, respectively. Furthermore, PDE5 activity was identified but PDE1 and PDE2 activities were not found (Peachell *et al.*, 1992). Functional studies have shown that PDE4 and PDE3/4 inhibitors attenuate anti-IgE-stimulated histamine and LTC_4 release whereas PDE5 inhibitors are ineffective. This might indicate that in human basophils PDE4 and PDE3 are involved in cyclic nucleotide metabolism (Peachell *et al.*, 1992; Kleine-Tebbe *et al.*, 1992).

2.2.5 Platelets

Platelets are involved in thrombosis and haemostasis and anti-platelet drugs have been suggested to be useful in myocardial infarction and cerebral and peripheral vascular disease. Furthermore, platelets may show abnormalities in inflammatory diseases. For example, an increase in circulating platelet aggregates is found in asthmatic patients. Analysis of human platelet PDE isoenzymes in homogenates and separation of PDE isoenzymes from platelet extracts using DEAE-Sepharose chromatography revealed the presence of PDE3, PDE5 and minor PDE1 and PDE2 activities. PDE4 was absent. All PDE isoenzymes were found almost exclusively in the cytosol (Fig. 2.8) (Simpson *et al.*, 1988; Murray *et al.*, 1990).

2.2.6 Vascular Endothelial Cells

Vascular endothelial cells provide a monolayer barrier of the blood vessel wall. Following stimulation, endothelial cell layers exhibit an enhanced permeability for plasma proteins and the expression of adhesion molecules is increased. By these mechanisms endothelial cells affect oedema formation and cellular infiltration which are characteristic features of inflammation. Therefore endothelial cells represent an important pharmacological target.

Selective PDE isoenzyme inhibitors have been shown to inhibit hydrogen peroxide-induced permeability of porcine pulmonary artery endothelial cells (Suttorp *et al.*, 1993). These results stimulated analysis of PDE isoenzyme activities in endothelial cells from different sources. In human umbilical vein endothelial cells (HUVEC) PDE2, PDE3 and PDE4 were identified and about 70% of the isoenzyme activities were found in the cytosol (Fig. 2.9). Using DEAE-Sepharose chromatography, Tani *et al.* (1992) confirmed the presence of PDE3 and PDE4 in cytosolic fractions of HUVEC cells. However, they found an additional PDE5 activity whereas PDE2 activity was absent from their preparations. It remains an open question whether differences in the culture passage number between these HUVEC cell preparations dramatically affect PDE isoenzyme profiles, as reported for bovine aortic endothelial cells (Ashikaga *et al.*, 1993).

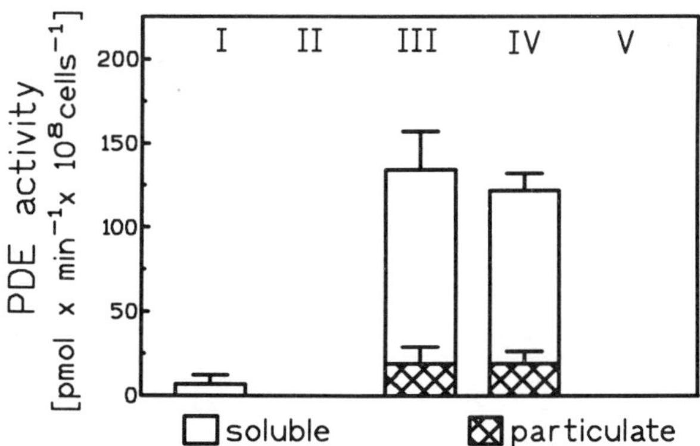

Figure 2.7 PDE isoenzyme profiles of human lung mast cells purified using c-*kit* antibodies with immunomagnetic selection (purity > 90%). (For details of the purification procedure see Okayama *et al.*, 1994.)

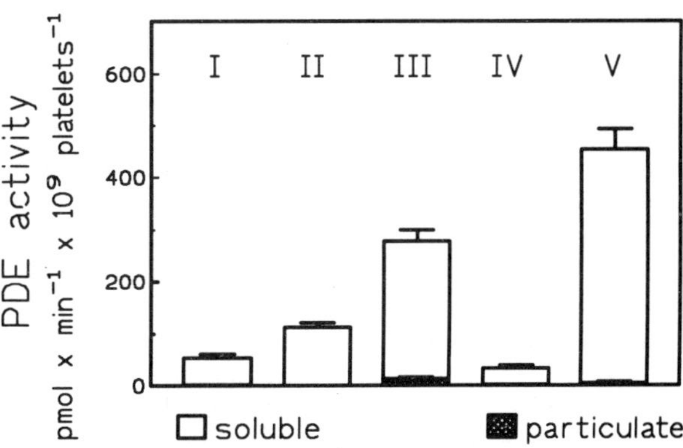

Figure 2.8 PDE isoenzyme profile of human platelets obtained from peripheral blood.

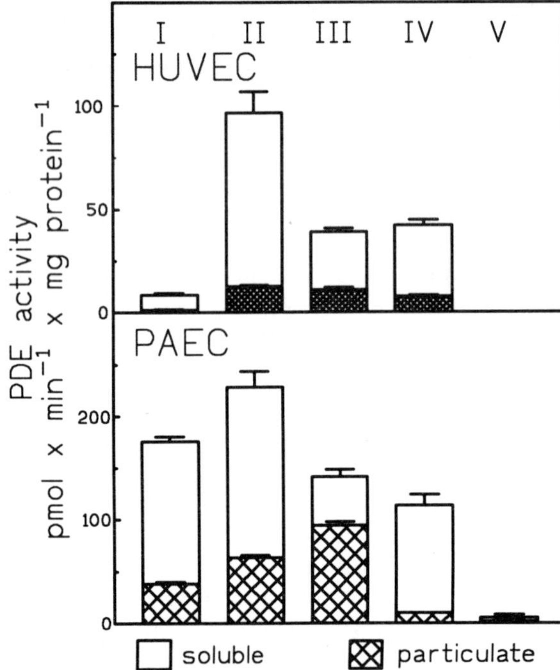

Figure 2.9 PDE isoenzyme profiles of human umbilical vein endothelial cells (HUVEC) and porcine pulmonary artery endothelial cells (PAEC).

Several authors have investigated PDE isoenzyme profiles from endothelial cells of non-human origin. Porcine pulmonary endothelial cells contain PDE1–4 (Fig. 2.9) (Suttorp *et al.*, 1993) and in pig aortic and bovine aortic endothelial cells PDE2 and PDE4 activities were detected (Lugnier and Schini, 1990; Souness *et al.*, 1990; Kishi *et al.*, 1992; Ashikaga *et al.*, 1993). Interestingly, Ashikaga *et al.* (1993) reported that, in bovine aortic endothelial cells, the PDE isoenzyme profile changed with increasing culture passage number. Whereas PDE2 and PDE4 were detected in early culture passages, PDE1, PDE3 and PDE4 were found in later passages. Taken together, PDE4 and PDE2 activities have been identified in endothelial cells from most sources and some cells also contain PDE3.

2.2.7 Epithelial Cells

Epithelial cells form a barrier that protects against exogenous agents but, in addition, depending on their localization, they have more specialized functions. For example, bronchial epithelial cells regulate mucociliary clearance by their ciliary beat frequency and may produce a variety of mediators, thus enabling them to be involved in the inflammatory process. Ocular ciliary epithelium is the site of aqueous humour production, thus regulating the intraocular pressure which may be critical in glaucoma. Skin keratinocytes have been demonstrated to produce a variety of cytokines and are involved in psoriasis and contact dermatitis.

Cyclic nucleotide-modulating agents such as PDE inhibitors were shown to increase ciliary beat frequency (Yang *et al.*, 1989; Di Benedetto *et al.*, 1991) and to inhibit prostanoid generation (Rabe *et al.*, 1994b) in bronchial epithelial cells. Moreover, aqueous humour production from ocular ciliary epithelium (Wax, 1992) and keratinocyte proliferation (Green, 1978) were demonstrated to be affected by cyclic nucleotides.

PDE isoenzyme profiles have been identified in bovine tracheal (Rousseau *et al.*, 1994) and human bronchial (Rabe *et al.*, 1994b) epithelial cells, in ocular ciliary epithelium (Bode *et al.*, 1993) and in the human keratinocyte cell line, HaCaT (Tenor *et al.*, 1995b). In airway epithelium from bovine trachea, HPLC separation of cellular supernatants revealed soluble PDE1, 2, 4 and 5 activities. Analysis of microsomal fractions indicated the presence of a major PDE4 activity but PDE3 and PDE2 were also detected (Rousseau *et al.*, 1994). PDE4 and PDE1 were the predominant activities in primary cultured human bronchial epithelial cells but minor PDE2, 3 and 5 activities were also found. More than 80% of the PDE activities were located in the cytosol (Fig. 2.10a) (Rabe *et al.*, 1994b). A similar PDE isoenzyme profile was exhibited by the human airway epithelial cell line A549, although approximately 10-fold higher activities were found (Fig. 2.10b)

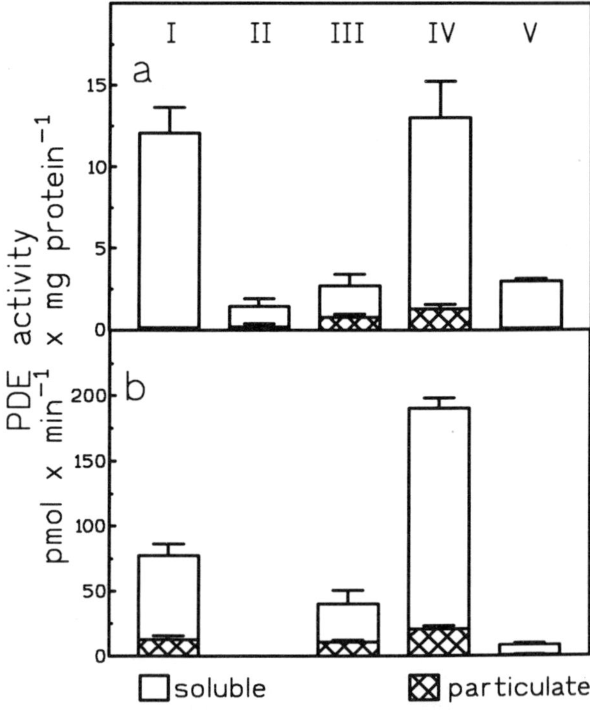

Figure 2.10 PDE isoenzyme profiles of (a) primary cultured human bronchial epithelial cells and (b) human pulmonary epithelial A549 cells. Note that A549 cells exhibit about 10-fold higher PDE activities than primary culture cells.

(Tenor *et al.*, 1994). In addition, Drumm *et al.* (1992) also identified PDE4 activity in the human airway epithelial cystic fibrosis cell line T43. This finding is of particular interest in view of the fact that the defective chloride conductance in cystic fibrosis can be ameliorated by cAMP-elevating agents (Drumm *et al.*, 1991).

Ocular epithelial cells can be differentiated into pigmented and non-pigmented cells, which differ in their PDE isoenzyme profiles. Using Mono Q chromatography, PDE1 was the only isoenzyme isolated from non-pigmented cells whereas pigmented cells contained PDE4 and PDE5 (Bode *et al.*, 1993). HaCaT cells express PDE4 and PDE5 activities that are mostly located in the cytosolic fractions (Fig. 2.11) (Tenor *et al.*, 1995b).

In summary, epithelial cells from most sources contain PDE4 whereas the expression of other isoenzymes differs between cells from different sources.

2.2.8 Neuronal Cells

PDE inhibitors are known to have various effects on the central nervous system (CNS) that are mostly revealed as side-effects (e.g. nausea, vomiting). However, they may also act as anti-depressants (e.g. rolipram). Only recently have PDE isoenzymes in neuronal cells been investigated. Kincaid *et al.* (1992), using immunocytochemistry, demonstrated the presence of PDE1 in certain classes of neurone from rat brain. Giorgi *et al.* (1993) and Tohda *et al.* (1994) isolated PDE5 and PDE4 activities from the murine neuroblastoma N18TG2 cell line. A predominant PDE4 and minor PDE3 and PDE5 activities have been found in human neuroblastoma SH-SY-5Y cells (Fig. 2.12) (H. Tenor and C. Schudt, unpublished observation). In these cells, PDE4 subtypes A–D were identified by RT-PCR (Engels *et al.*, 1994). It is noteworthy that, in contrast to neuronal cells, glial C6Bu-1 cells only contained minor PDE4 activity (Tohda *et al.*, 1994).

2.2.9 Smooth Muscle Tissues

PDE inhibitors trigger smooth muscle relaxation. They may, therefore, be useful in bronchial obstruction and

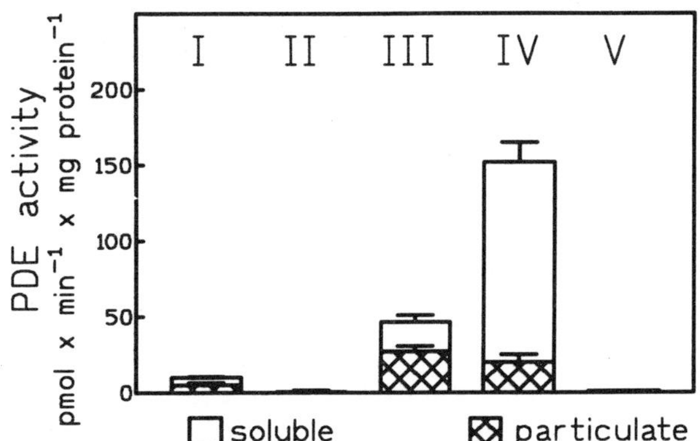

Figure 2.11 PDE isoenzyme profile of the human keratinocyte cell line, HaCaT.

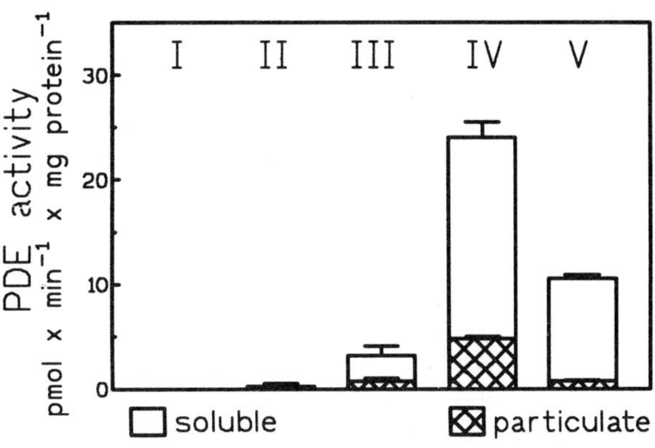

Figure 2.12 PDE isoenzyme profile of SH-SY-5Y human neuroblastoma cells.

hypertension. PDE isoenzyme activities in bronchial and vascular smooth muscle tissues from human and animal origin have been intensively investigated using anion exchange chromatography or by assessing effects of isoenzyme-selective inhibitors and activators on PDE activities in crude subcellular fractions. In human bronchus, most authors have found PDE1–5 activities (Bergstrand and Lundquist, 1978; de Boer *et al.*, 1992; Rabe *et al.*, 1993; Torphy *et al.*, 1993). However, Cortijo *et al.* (1993) could not detect PDE3 activity. Because human bronchial homogenates may also contain non-smooth muscle cellular constituents, Man *et al.* (1994) analysed PDE isoenzymes in cytosolic fractions of primary cultured human bronchial smooth muscle cells. Using Mono Q chromatography they identified PDE4, PDE3 and PDE5 activities. It might, therefore, be speculated that PDE1 and PDE2 activities obtained from whole bronchus preparations originate from other sources (e.g. endothelial cells). With regard to airway smooth muscle from other species, PDEs 1–5 were isolated from bovine (Giembycz and Barnes, 1991; Shahid *et al.*, 1991) and canine (Torphy and Cieslinksi, 1990) tracheal smooth muscle. In guinea-pig trachea PDEs 2–5 were identified (Harris *et al.*, 1989; H. Tenor and C. Schudt, unpublished observation).

PDE1, 3, 4 and 5 – but not PDE2 – activities have been found in most vascular smooth muscle tissue. Human pulmonary artery (Rabe *et al.*, 1994a), rat mesenteric artery (Komas *et al.*, 1991a), rat aorta (Lugnier *et al.*, 1986; Komas *et al.*, 1991b), bovine aorta (Lugnier and Komas, 1993) and bovine coronary artery (Weishaar *et al.*, 1986) contained PDE1, 3, 4 and 5. In pig aorta PDE2 activity was also detected (Saeki and Saito, 1993) but in cultured pig aortic smooth muscle cells PDE1, 3, 4 and 5 (Souness *et al.*, 1992) or PDE1, 3 and 4 (Xiong *et al.*, 1995) were described. Again non-smooth muscle cellular constituents may account for these differences in the PDE isoenzyme profiles between whole aortic smooth muscle preparations and cultured smooth muscle cells.

2.2.10 Cardiac Muscle

PDE inhibitors, and in particular selective inhibitors of PDE3, have positive inotropic and chronotropic effects on heart muscle. Therefore, PDE3 inhibitors were originally developed for treatment of chronic heart failure in order to improve cardiac performance. However, this concept has now been severely compromised by results from the PROMISE study which demonstrated that long-term therapy with milrinone significantly increased the mortality rate in chronic heart failure (Packer *et al.*, 1991). From this and other studies the authors suggested that an increased mortality from chronic heart failure might be a general effect of chronically administered drugs which enhance cAMP concentrations in failing hearts. For this reason the results of the PROMISE study should be considered when contemplating the development of new PDE inhibitors for non-cardiac diseases.

PDE isoenzyme profiles for heart muscle of human or other origins have been described by several authors. In cytosolic fractions of human heart muscle PDEs 1–4 were chromatographically separated and PDE3 represented the main cAMP-hydrolysing isoenzyme (Reeves *et al.*, 1987; Leyen *et al.*, 1991; Sugioka *et al.*, 1994). The PDE isoenzyme profiles obtained from cytosolic fractions of normal and failing human hearts were not different (Leyen *et al.*, 1991). Particulate fractions contained mainly PDE3 (Sugioka *et al.*, 1994) which was associated with the sarcoplasmic reticulum (SaR) (Movsesian *et al.*, 1991; Lugnier *et al.*, 1993). There was no difference in SaRPDE3 activities or inhibitor sensitivities between normal and failing human heart preparations (Movsesian *et al.*, 1991). In addition to the SaR PDE3 activity, Lugnier *et al.*(1993), using sucrose density centrifugation of human heart microsomal fractions, detected a PDE4 activity that was associated with surface membranes.

PDE isoenzyme profiles from the heart muscle of other species was similar. PDE1–4 activities were found in the cytosolic fractions of cardiac muscle from rabbits (Kithas *et al.*, 1988), dogs (Komas *et al.*, 1989), guinea-pigs (Reeves *et al.*, 1987) and rats (Tenor *et al.*, 1987; Bode *et al.*, 1991). SaR-associated PDE3 was found in rabbit (Kithas *et al.*, 1988) and canine (Kauffman *et al.*, 1986; Lugnier *et al.*, 1993) heart muscle.

3. *Regulation of PDE Isoenzyme Activities*

Measuring PDE activities in cellular homogenates can only be a snapshot and may reflect merely the effects of the cellular microenvironment on the PDE activity at the time of homogenization. In fact, PDE activities have been found to be regulated by a variety of endogenous or exogenous factors, including β_2-agonists and prostanoids (Torphy *et al.*, 1992c), cytokines (Li *et al.*, 1992), FSH (Conti *et al.*, 1981) and atrial natriuretic peptides (ANP) (MacFarland *et al.*, 1991). Therefore, regulation of PDE isoenzyme activities might also be involved in the generation of certain pathophysiological conditions. For example, in mononuclear cells of atopic individuals an increased PDE activity has been described which may possibly be related to the enhanced occurrence of T_H2-cell-derived cytokines in atopy (Chan *et al.*, 1993a). This section is dedicated to a brief review of the mechanisms of PDE activity regulation. Fully comprehensive reviews of this topic have recently been published (Conti *et al.*, 1991, 1995; Beltman *et al.*, 1993; Sette *et al.*, 1994c).

3.1 Regulation of PDE Activity by Cyclic Nucleotide Concentrations and Ca^{2+}/Calmodulin

3.1.1 Substrate Regulation

PDE isoenzymes exhibit different substrate-dependent activity regulation due to their different K_m values. This is illustrated schematically in Fig. 2.13 for endothelial PDE isoenzyme content (for comparison see Fig. 2.9; PDE isoenzymes in porcine pulmonary endothelial cells) composed of PDE1–5. For example, comparison of PDE3 and PDE4 activities at 5 μM and 0.5 μM cAMP substrate concentration reveals that PDE3 activity is almost unchanged whereas PDE4 activity is increased and, as a result, the PDE3/PDE4 activity ratio has decreased. These theoretical considerations have been experimentally confirmed, as shown in Fig. 2.14 for the substrate-dependent regulation of PDE3 and PDE4 activities of the human alveolar macrophage. At the higher substrate concentration there was an approximately three-fold increase in PDE4 activity whereas PDE3 activity was unchanged. These findings might give rise to speculation that the synergy between inhibitors of PDE4 and PDE3, which has been described

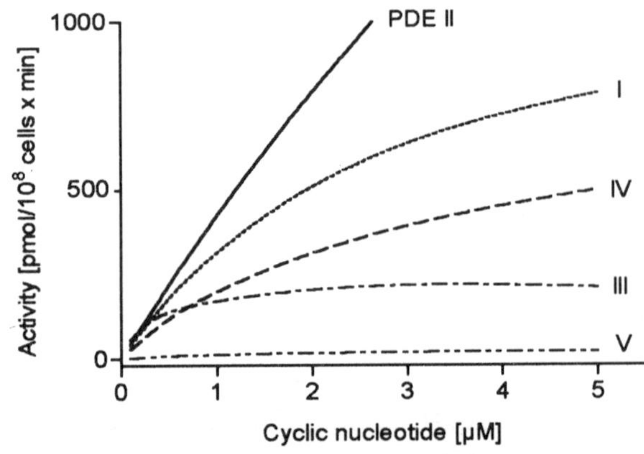

Figure 2.13 Substrate-dependent activity regulation of PDE1–5, schematically illustrated for porcine pulmonary artery endothelial cells (for comparison see Fig. 2.9).

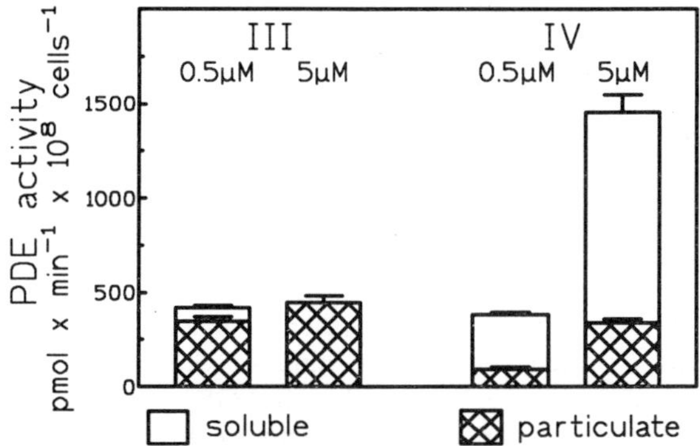

Figure 2.14 PDE3 and PDE4 activities from human alveolar macrophages measured at 0.5 μM and 5 μM cAMP substrate concentration. Note that, at the higher substrate concentration, the PDE4/PDE3 ratio has increased as would be expected from the K_m values of PDE3 and PDE4.

(Shahid and Nicholson, 1990; Giembycz *et al.*, 1994), could partly be explained by these effects.

3.1.2 Regulation of PDE1 by Ca^{2+}/calmodulin

PDE1 activity can be stimulated by Ca^{2+}/CaM by about 6–20-fold. PDE1 isolated from different sources may be differentially activated by CaM (see Chapter 4). For example, half-maximal activation of PDE1 from bovine brain and bovine heart were attained with 0.9 nM and 0.15 nM CaM, respectively (Sharma, 1991). With regard to the role of PDE1 in cyclic nucleotide metabolism in intact cellular systems, it has been suggested by Erneux *et al.* (1985) that Ca^{2+}/CaM-dependent activation of PDE1 is involved in the muscarinic inhibition of β-receptor-mediated cAMP accumulation in a human astrocytoma cell line. Furthermore, it could be speculated that the Ca^{2+}-dependent activation of PDE1 in human alveolar macrophages may account for the observed platelet activating factor (PAF)/Ca^{2+} ionophore-induced decrease of cAMP levels in these cells (Bachelet *et al.*, 1993). Finally, Saitoh *et al.* (1985) have demonstrated activation of PDE1 in porcine coronary artery strips under various conditions of increasing Ca^{2+} *ex vivo*.

3.1.3 Allosteric Regulation of PDE2

Cyclic AMP hydrolysis by PDE2 is allosterically activated by cGMP and hydrolysis of cAMP and cGMP by this isoenzyme displays positive co-operative kinetics (see Chapter 5). For example, cAMP hydrolysis by PDE2 from calf liver exhibits a maximum 30-fold activation by cGMP (K_{act} = 0.5 μM cGMP) and a Hill coefficient for cAMP hydrolysis of 1.6–1.8 (Yamamoto *et al.*, 1983). More recently, it has been shown that the allosteric cGMP binding site on the PDE2 molecule is distinct from the catalytic site and that cGMP binding results in a conformational change of the molecule (Stroop and Beavo, 1991). Regulation of PDE2 by cyclic nucleotides is now considered to have functional implications in a variety of systems. MacFarland *et al.* (1991) demonstrated that in adrenal cortex zona glomerulosa cells, where PDE2 accounts for more than 90% of total PDE activity, PDE2 is involved in the regulation of aldosterone synthesis by adrenocorticotrophin hormone (ACTH) and ANP. In these cells ACTH caused a stimulation of aldosterone synthesis that was mimicked by membrane-permeant cAMP derivatives. In contrast, ANP, by elevating cGMP, antagonizes this aldosterone synthesis and, in parallel, decreases cellular cAMP content. From these results it was hypothesized that the effects of ANP on aldosterone synthesis in adrenal cortex glomerulosa cells were triggered by cGMP-dependent activation of PDE2. Similarly, in bovine aortic endothelial cells (Kishi *et al.*, 1994) and PC12 cells (Whalin *et al.*, 1991), from which predominantly PDE2 activities were isolated, an ANP-triggered cAMP decrease was suggested to be mediated by cGMP-dependent activation of PDE2. In frog cardiac myocytes, which contain a particulate PDE2, cGMP stimulation of cAMP hydrolysis has been suggested to trigger the cGMP-induced reduction of the trans-sarcolemmal Ca^{2+} current previously elevated by cAMP (Simmons and Hartzell, 1988; Fischmeister and Hartzell, 1990). Furthermore, cGMP-mediated activation of PDE2 might be involved in the ANP- and sodium nitroprusside (SNP)-triggered cAMP decrease in human peritoneal macrophages, although the PDE isoenzyme profile of these cells is not known at present (Houdjik *et al.*, 1990). Finally, Barber *et al.* (1992) demonstrated that, in mouse transfected L-WTβ2AR cells, the positive co-operative hydrolysis of cAMP by PDE2 accounts for the decreased cAMP response to adrenaline observed in these cells.

3.1.4 Regulation of PDE3 Activity by cGMP

One of the main characteristics of PDE3 is its potent inhibition by cGMP; this mechanism has recently been suggested to be involved in functional effects of a cGMP increase. Kirstein *et al.* (1995) have demonstrated that the nitric oxide (NO) donor, 3-morpholinosydnonimine (SIN-1), enhances trans-sarcolemmal Ca^{2+} current in human atrial myocytes, an effect usually considered to be mediated by cAMP. To explain their results the authors suggested that a NO-dependent cGMP increase might inhibit PDE3 and hence elevate cAMP levels. In rabbit platelets, Maurice and Haslam (1990) demonstrated that SNP enhances cAMP levels and this effect was synergically enhanced by PGE_1. Again it was hypothesized that these results could be attributed to cGMP inhibition of PDE3. Finally, in rat thymocytes a combination of selective PDE4 and PDE5 inhibitors synergically augmented the cAMP response and synergically inhibited concanavalin A-induced proliferation; this effect could be due to a cGMP-mediated PDE3 inhibition by PDE5 inhibitors (Marcoz *et al.*, 1993). Endogenous PDE3 inhibition must also be considered, in view of high NO levels in the airways under conditions of atopy and asthma (Kharitonov *et al.*, 1994).

3.2 SHORT-TERM REGULATION OF PDE ACTIVITY BY PHOSPHORYLATION

3.2.1 Phosphorylation of PDE1

Phosphorylation of PDE1 by cAMP-dependent (PKA) and Ca^{2+}/CaM-dependent protein kinases has been described (Sharma and Wang, 1985, 1986; Sharma, 1991; see also Chapter 4). Such phosphorylation results in a decreased affinity of the enzyme for Ca^{2+}/CaM. On the other hand, Ca^{2+}/CaM inhibits the PKA-dependent phosphorylation of PDE1 (Sharma and Wang, 1985; Hashimoto *et al.*, 1989; Sharma, 1991). Although Wang *et al.* (1990) have recently proposed a model of time-dependent regulation of PDE1 by phosphorylation and Ca^{2+}/CaM there are no data available at present concerning the physiological implications of PDE1 phosphorylation.

3.2.2 Phosphorylation of PDE3

Cyclic AMP-dependent and insulin-dependent phosphorylation of PDE3 in adipocytes, platelets and liver cells has been demonstrated. Particulate PDE3 activity from rat fat-cells was enhanced by about 50–100% following incubation of these cells with 300 nM isoprenaline for 3 min or 1 nM insulin for 12 min (Degerman *et al.*, 1990; Smith *et al.*, 1991). This increase in PDE3 activity was suggested to be due to phosphorylation since it was parallelled by an augmented ^{32}P incorporation into the enzyme and was mimicked by the catalytic subunit of PKA (Gettys *et al.*, 1988; Degerman *et al.*, 1990; Smith *et al.*, 1991). The phosphorylation site of adipocyte particulate PDE3 by PKA has recently been identified (Rascon *et al.*, 1994). Interestingly, the combined administration of insulin and isoprenaline to rat adipocytes resulted in a synergic increase in PDE3 activity and phosphorylation state. These results indicate that insulin and isoprenaline trigger the phosphorylation of PDE3 at different sites (Smith *et al.*, 1991). In platelets, Grant *et al.* (1988) and MacPhee *et al.* (1988) have reported an increase of PDE3 activity following incubation with forskolin or prostacyclin for 5 min that was due to PKA-dependent phosphorylation. More recently, it was demonstrated that insulin also triggered phosphorylation and activation of PDE3 from platelets. In addition, an insulin-stimulated PDE3 serine kinase, which was shown to phosphorylate and activate PDE3 in a cell-free system, has been partly purified from platelet extracts (Lopez-Aparicio *et al.*, 1992, 1993).

3.2.3 Phosphorylation of PDE4

A PKA-dependent phosphorylation of PDE4 accompanied by an increase in PDE4 activity *in vitro* and *ex vivo* has recently been shown in rat follicular FRTL-5 cells (Sette *et al.*, 1994a). Incubation of FRTL-5 cells with TSH, dibutyryl cAMP (diBu-cAMP) or forskolin for 10–15 min resulted in an approximately two-fold increase in PDE4 activity which was not affected by cycloheximide but was further enhanced by okadaic acid, a protein phosphatase inhibitor. Similarly, incubation of monocytic U937 cells with PGE_2 (Torphy *et al.*, 1992c) or rat aortic vascular smooth muscle cells with forskolin (Haider *et al.*, 1995) for 15 min triggered a significant up-regulation of PDE4 activity which might be due to cAMP-dependent phosphorylation. In view of the different PDE4 subtypes, it is relevant that Sette *et al.* (1994b) and Conti (1995) have recently described differential regulation of PDE4D splice variants. PDE4D3 was subject to cAMP-dependent phosphorylation resulting in enzyme activation whereas PDE4D1 and PDE4D2 did not show any increase in activity under the same experimental conditions. It is interesting to note that phosphorylation of PDE4D3 dramatically enhances the potency of selective PDE4 inhibitors (Alvarez *et al.*, 1995; Conti, 1995; see also Chapter 11).

In addition to PKA-dependent short-term PDE4 activation, protein kinase C activation has been suggested to be involved in short-term PDE4 up-regulation in rat medullary collecting ducts (Tetsuka *et al.*, 1995) and human granulosa cells (Michael and Webley, 1991).

3.2.4 Phosphorylation of PDE5

PDE5 has been demonstrated to be phosphorylated by both cGMP-dependent protein kinase (PKG) and PKA (Robichon, 1991). Burns and colleagues incubated PDE5 purified from guinea-pig lung with the catalytic subunit of PKA and found PDE5 phosphorylation that was parallelled by an increase in PDE activity (Burns and Pyne, 1992; Burns *et al.*, 1992). Furthermore, the sensitivity of the phosphorylated enzyme to the PDE5-selective inhibitor zaprinast was diminished. However, the relevance of these findings under *in vivo* conditions is not known at present.

3.3 LONG-TERM PDE4 INDUCTION BY cAMP-ELEVATING AGENTS

There is evidence from many studies for a cAMP-mediated long-term up-regulation of PDE4. In an early study, Conti *et al.* (1981) demonstrated an approximately 10-fold increase in PDE4 activity following incubation of rat Sertoli cells with diBu-cAMP, FSH or 3-isobutyl-1-methylxanthine (IBMX). In human monocytic U937 cells, an increase in PDE4 activity of up to four-fold was observed after incubation of the cells for 4 hours with a combination of 1 μM salbutamol and 30 μM rolipram (Torphy *et al.*, 1992c). Verghese *et al.* (1995a) found a 2–3-fold increase in PDE4 activity in human peripheral blood monocytes and human monocytic MonoMac6 cells incubated with diBu-cAMP for 18–24 hours. Incubation of rat aortic vascular smooth muscle cells with forskolin resulted in a long-lasting PDE4 up-regulation (Haider *et al.*, 1995). Finally, incubation of cells of the human keratinocyte HaCaT line for 6 hours with rolipram and salbutamol resulted in an up-regulation of about three-fold in their PDE4 activity. This PDE4 up-regulation in HaCaT cells was partly prevented by the PKA inhibitor (Rp)8-bromoadenosine 3′:5′-cyclic monophosphorothioate (Rp-8-Br-cAMPS) and mimicked by the PKA activator (Sp)-5,6-dichloro-1-β-D-ribofuranosylbenzimidazole 3′:5′-cyclic monophosphorothioate (Sp-5,6-diCl-cBIMPS), indicating that PKA activation was involved in PDE4 up-regulation (Fig. 2.15) (Tenor *et al.*, 1995b). More direct evidence for a role of PKA in long-term PDE4 up-regulation came from experiments with L6 myoblasts expressing a mutant regulatory subunit of PKA which acted as an inhibitor of the native enzyme. In myoblasts expressing the PKA mutant, a long-term cAMP-dependent up-regulation that was observed in normal cells was abolished (Kovala *et al.*, 1994).

With regard to the mechanism of cAMP-dependent long-term PDE4 up-regulation it was speculated that PKA-mediated phosphorylation of nuclear proteins might enhance mRNA transcription of PDE4 genes and, hence, PDE protein expression. This hypothesis was based on the observation that actinomycin D and cycloheximide

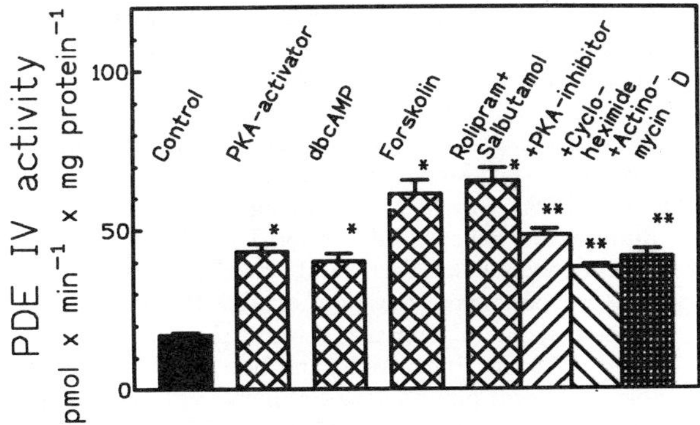

Figure 2.15 HaCaT cells (human keratinocyte cell line) were incubated for 6 hours with cAMP-mimetic agents (500 μM 5,6-diCl-cBIMPS, PKA activator; 500 μM diBu-cAMP; 100 μM forskolin; 10 μM rolipram + 1 μM salbutamol), resulting in an approximately 2–4-fold up-regulation of PDE4 activity. This up-regulation was partially reversed by pre-incubation rolipram/salbutamol-treated cells with the PKA inhibitor Rp-8-Br-cAMPS (500 μM), the mRNA inhibitor actinomycin D (2.5 μg/ml) or the protein synthesis inhibitor cycloheximide (25 μM).

attenuated the PDE4 up-regulation (Conti *et al.*, 1981; Torphy *et al.*, 1992c; Tenor *et al.*, 1995b). In fact, Northern blot analysis of mRNA extracts from rat Sertoli cells incubated with FSH for up to 24 hours revealed a dramatic concentration- and time-dependent increase in PDE4 mRNA. (Swinnen *et al.*, 1989). This enhanced mRNA transcription of PDE4 in FSH-treated Sertoli cells preceded the increased expression of PDE4 protein which was parallelled by an up-regulation of PDE4 activity (Swinnen *et al.*, 1991). Analysis of PDE4 subtypes demonstrated an increase in PDE4D mRNA whereas PDE4B was less affected (Swinnen *et al.*, 1991). To extend these investigations with rat Sertoli cells, studies were performed with other cell types to assess the effect of cAMP-elevating agents on PDE4 subtype mRNA levels in these cells. Engels *et al.* (1994), by incubating Jurkat cells and U937 cells with 500 μM diBu-cAMP for 18 hours, found an increase in PDE4A and 4D mRNA and PDE4A and 4B mRNA, respectively. In contrast, in SH-SY-5Y human neuronal cells PDE4A–4D mRNAs were not affected by an 18 hour exposure to diBu-cAMP. Incubation of rat vascular smooth muscle cells with forskolin resulted in an increase in PDE4B and 4D mRNA (Haider *et al.*, 1995). In human MonoMac6 monocytic cells an up-regulation of PDE4A, 4B and 4D mRNA was observed after incubation of cells with diBu-cAMP for 18 hours that was parallelled by an increased expression of PDE4A, 4B and 4D protein, verified by Western blotting analysis using subtype-specific PDE4 antibodies (Verghese *et al.*, 1995a). These data indicate a cell-specific regulation of PDE4 subtype mRNA levels. In rat Sertoli cells it was shown that refractory steroidogenesis following long-term treatment with FSH, inducing PDE4 up-regulation was prevented by the presence of the PDE inhibitor IBMX (Conti *et al.*, 1986).

3.4 ELEVATED PDE ACTIVITY IN ATOPIC MONONUCLEAR CELLS

Monocytes and T lymphocytes of atopic individuals show an enhanced PDE activity compared to normals (Butler *et al.*, 1983; Cooper *et al.*, 1985; Hanifin and Chan, 1988; Chan *et al.*, 1993c; Townley, 1993; Sawai *et al.*, 1995). The nature of this PDE activity increment has not yet been fully characterized but may involve PDE1 and/or PDE4 up-regulation (Chan *et al.*, 1993c). It has been speculated that enhanced PDE activity in atopic mononuclear cells could be the cause of increased histamine, immunoglobulin E (IgE) and IL-4 release by these cells, since the release of these factors was reversed by the PDE4 inhibitor Ro 20–1724 (Butler *et al.*, 1983; Cooper *et al.*, 1985; Chan *et al.*, 1993b). Interestingly, it was demonstrated that IL-4 production by mononuclear cells from atopics is more sensitive to inhibition by PDE4 inhibitors than IL-4 production by mononuclear cells from normals (Chan *et al.*, 1993b). On the other hand it was shown that incubation of normal human monocytes with IL-4 triggers an elevation of PDE activity (Li *et al.*, 1992). However, IFNγ, which counteracts some effects of IL-4, reversed the enhanced PDE activity in atopic monocytes (Li *et al.*, 1993). Since IL-4 levels have been shown to be elevated in the serum of atopic subjects, whereas atopic mononuclear cells release less IFNγ, (Chan *et al.*, 1993a) it might be hypothesized that the increased IL-4/IFNγ ratio partly accounts for the enhanced PDE activity in atopic mononuclear cells. Alternatively, an elevated PGE_2 production – as has been described from atopic mononuclear cells (Chan *et al.*, 1993a) – might contribute to PDE up-regulation in atopic monocytes (Torphy *et al.*, 1992c).

4. *Conclusions*

PDE inhibitors have been proposed as drugs for a variety of diseases (asthma, vascular disease, diabetes, AIDS, rheumatic disease, multiple sclerosis). To target a specific disease and to limit adverse events caused by a PDE inhibitor, the ubiquity of PDE and the heterogeneity of its isoenzymes both require and facilitate the development of cell-specific PDE inhibitors. To meet these challenges it is essential to establish PDE isoenzyme profiles of target cells and of those cells which could potentially contribute to adverse events. In this chapter PDE isoenzyme activity profiles obtained by pharmacological methods have been reviewed. These pharmacological methods used selective inhibitors and activators of PDE isoenzyme families to establish PDE isoenzyme family activity profiles. It has been shown with these methods that, for example, most human inflammatory cells contain PDE4 activities but, in addition, PDE3 activity is found in T lymphocytes and macrophages. From these results it can be concluded that PDE4/PDE3 inhibitors might be beneficial in inflammatory diseases. However, neuronal cells also contain PDE4 activity and this finding may explain why PDE4 inhibitors frequently induce CNS-related adverse events (vomiting, nausea). Recent evidence for the occurrence of PDE4 subtypes (see Chapters 1, 11 and 13) would suggest that PDE4 subtype-specific inhibitors might provide a new tool to circumvent CNS-related adverse effects of PDE4 inhibitors. However, the search for subtype-specific inhibitors is only just beginning. Therefore, PDE4 subtype activities cannot be analysed using the pharmacological methods described in this chapter.

A principal problem of PDE isoenzyme activity profiling is that it is still not known whether there is a difference between the PDE activities measured in cellular homogenates and the PDE activities in the intact cell. Such a difference can be anticipated in view of the fact that isoenzyme activities are subject to a complex network of regulation. In particular, the extent of short-term regulation in the intact cell by phosphorylation, cyclic nucleotides and Ca^{2+}/CaM cannot be reliably assessed after cell homogenization. For example, the importance of cGMP inhibition of PDE3 and cGMP-mediated activation of PDE2 in the intact cell can only be derived from functional experiments. Furthermore, the recent findings that phosphorylation of PDE4 and PDE5 isoenzymes may change their sensitivity to drug inhibition could call into question whether the potency of inhibitors in the cell-free system truly reflects the potency of these inhibitors in the intact cell.

These problems aside, however, a combined approach of PDE activity measurements, detection of PDE proteins with subtype- and splice variant-specific monoclonal antibodies and detection of PDE mRNA with RT-PCR as described in this volume represents the most powerful approach for characterizing PDE isoenzymes in cells and tissues.

5. *References*

Alvarez, R., Sette, C., Yang, D., Eglen, R.M., Wilhelm, R., Shelton, E.R. and Conti, M. (1995). Activation and selective inhibition of a cyclic AMP-specific phosphodiesterase PDE-4D3. Mol. Pharmacol 48, 616–622.

Anderson, N. and Peachell, P.T. (1994). Effect of isoenzyme selective inhibitors of phosphodiesterase (PDE) on human lung mast cells (HLMC). FASEB J. 8, A239. [Abstract]

Ashikaga, T., Strada, S.J. and Thompson, W.J. (1993). Altered expression of bovine aortic endothelial cell (BAEC) cyclic nucleotide (CN) phosphodiesterase (PDE) isozymes during culture. FASEB J. 7, A759. [Abstract]

Bachelet, M., Chouaid, C., Havet, N., Masliah, J., Barre, A., Housset, B. and Vargaftig, B.B. (1992). Modulation of arachidonic acid metabolism and cyclic AMP content of human alveolar macrophages. Eicosanoids 5, 185–190.

Barber, R., Goka, T.J. and Butcher, R.W. (1992). Positively cooperative cAMP phosphodiesterase attenuates cellular cAMP response. Second Messengers Phosphoproteins 14, 77–97.

Barnette, M.S., Grous, M., Manning, C., Cieslinski, L.B., Burman, M., Bell, D. and Torphy, T.J. (1993). Modulation of phosphodiesterase IV (PDE IV) activity in inflammatory cells by prolonged elevation of cAMP content. World Conference on Inflammation, Vienna, p. 112. [Abstract]

Beltman, J., Sonnenburg, W.K. and Beavo, J.A. (1993). The role of protein phosphorylation in the regulation of cyclic nucleotide phosphodiesterases. Mol. Cell. Biochem. 127/128, 239–253.

Bergstrand, H. and Lundquist, B. (1978). Partial purification and characterization of cyclic nucleotide phosphodiesterases from human bronchial tissue. Mol. Cell. Biochem. 21, 9–15.

Bergstrand, H., Lundquist, B. and Schurmlann, A. (1978). Rat mast cell high affinity cyclic nucleotide phosphodiesterases: separation and inhibitory effects of two antiallergic agents. Mol. Pharmacol. 14, 848–855.

Bloom, T.J. and Beavo, J.A. (1994). Identification of type VII PDE in HUT T-lymphocytes cells. FASEB J. 8, A372. [Abstract]

Bode, D.C., Kanter, J.R. and Brunton, L.L. (1991). Cellular distribution of phosphodiesterase isoforms in rat cardiac tissue. Circ. Res. 68, 1070–1079.

Bode, D.C., Hamel, L.T. and Wax, M.B. (1993). Distinct profiles of phosphodiesterase isozymes in cultured cells derived from nonpigmented and pigmented ocular ciliary epithelium. J. Pharmacol. Exp. Ther. 267, 1286–1291.

Burns, F. and Pyne, N.J. (1992). Interaction of the catalytic subunit of protein kinase A with the lung type V cyclic GMP phosphodiesterase: modulation of non-catalytic binding sites. Biochem. Biophys. Res. Commun. 189, 1389–1396.

Burns, F., Rodgers, I.W. and Pyne, N.J. (1992). The catalytic subunit of protein kinase A triggers activation of the type V cyclic GMP-specific phosphodiesterase from guinea pig lung. Biochem. J. 283, 487–491.

Butler, J.M., Chan, S.C., Stevens, S. and Hanifin J.M. (1983). Increased leucocyte histamine release with elevated cyclic AMP-phosphodiesterase activity in atopic dermatitis. J. Allergy Clin. Immunol. 71, 490–497.

Chan, S.C., Kim, J.-W., Henderson, W.R. and Hanifin, J.M. (1993a). Altered prostaglandin E_2 regulation of cytokine production in atopic dermatitis. J. Immunol. 151, 3345–3352.

Chan, S.C., Li, S.-H. and Hanifin, J.M. (1993b). Increased interleukin-4 production by atopic mononuclear leucocytes correlates with increased cyclic adenosine monophosphate-phosphodiesterase activity and is reversible by phosphodiesterase inhibition. J. Invest. Dermatol. 100, 681–684.

Chan, S.C., Reifsnyder, D., Beavo, J.A. and Hanifin, J.M. (1993c). Immunochemical characterization of the distinct monocyte cyclic AMP-phosphodiesterase from patients with atopic dermatitis. J. Allergy Clin. Immunol. 91, 1179–1188.

Conti, M. (1995). Structure, regulation and drug interaction of type IV cAMP-specific phosphodiesterases expressed in endocrine cells. Harvey Conference on Phosphodiesterases. 19. [Abstract]

Conti, M., Geremia, R., Adamo, S. and Stefanini, M. (1981). Regulation of Sertoli cell cyclic adenosine 3′: 5′-monophosphate phosphodiesterase activity by follicle stimulating hormone and dibutyryl cyclic AMP. Biochem. Biophys. Res. Commun. 98, 1044–1050.

Conti, M., Monaco, I., Geremia, R. and Stefanini, M. (1986). Effect of phosphodiesterase inhibitors on sertoli cell refractoriness: reversal of the impaired androgen aromatization. Endocrinology 118, 901–908.

Conti, M., Jin, S.-L.C., Monaco, L., Repaske, D.R. and Swinnen, J.V. (1991). Hormonal regulation of cyclic nucleotide phosphodiesterases. Endocr. Rev. 12, 218–234.

Conti, M., Némoz, G., Sette, C. and Vicini, E. (1995). Recent progress in understanding the hormonal regulation of phosphodiesterases. Endocr. Rev. 16, 370–389.

Cooper, K.D., Kang, K., Chan, S.C. and Hanifin, J.M. (1985). Phosphodiesterase inhibition by Ro 20–1724 reduces hyper-IgE synthesis by atopic dermatitis cells *in vitro*. J Invest. Dermatol. 84, 477–482.

Cortijo, J., Bou, J., Beleta, J., Cardelús, I., Llenas, J., Morcillo, E. and Gristwood, R.W. (1993). Investigation into the role of phosphodiesterase IV in bronchorelaxation, including studies with human bronchus. Br. J. Pharmacol. 108, 562–568.

de Boer, J., Philpot, A.J., Amsterdam, R.G.M., Shahid, M., Zaagsma, J. and Nicholson, C.D. (1992). Human bronchial cyclic nucleotide phosphodiesterase isoenzymes: biochemical and pharmacological analysis using selective inhibitors. Br. J. Pharmacol. 106, 1028–1034.

Degerman, E., Smith, C.J., Tornqvist, H., Vasta, V., Belfrage, P. and Manganiello, V.C. (1990). Evidence that insulin and isoprenaline activate the cGMP-inhibited low-K_m cAMP phosphodiesterase in rat fat cells by phosphorylation. Proc. Natl Acad. Sci. USA 87, 533–537.

Dent, G., Giembycz, M.A., Rabe, K.F. and Barnes, P.J. (1991). Inhibition of eosinophil cyclic nucleotide PDE activity and opsonized zymosan-stimulated respiratory burst by "type IV"-selective PDE inhibitors. Br. J. Pharmacol. 103, 1339–1346.

Dent, G., Giembycz, M.A., Wolf, B., Rabe, K.F., Barnes, P.J. and Magnussen, H. (1993). Identification of PDE III and PDE IV isoenzymes of cyclic AMP phosphodiesterase in human alveolar macrophages. Am. Rev. Respir. Dis. 147, A184. [Abstract]

Dent, G., Giembycz, M.A., Evans, P.M., Rabe, K.F. and Barnes, P.J. (1994). Suppression of human eosinophil respiratory burst and cyclic AMP hydrolysis by inhibitors of type IV phosphodiesterase: interaction with the *beta* adrenoceptor agonist albuterol. J. Pharmacol. Exp. Ther. 271, 1167–1174.

Di Benedetto, G., Manara-Shediac, F.S. and Mehta, A. (1991). Effect of cyclic AMP on ciliary activity of human respiratory epithelium. Eur. Respir. J. 4, 789–795.

Drumm, M.L., Wilkinson, D.J., Smit, L.S., Worell, R.T., Strong, T.V., Frizzell, R.A., Dawson, D.C. and Collins, F.S. (1991). Chloride conductance expressed by F508 and other mutant CFTRs in *Xenopus* oocytes. Science 254, 1797–1799.

Drumm, M.L., Saltiel, A., Levine, H., Wilkinson, D.J., Dawson, D.C. and Collins, F.S. (1992). Comparison of phosphodiesterase activities and their inhibition in CF epithelial cell lines. Pediatr. Pulmonol. 8, Suppl. D, P96. [Abstract]

Elliott, K.R.F. and Leonard, E.J. (1989). Interactions of formyl-leucyl-phenylalanine, adenosine, and phosphodiesterase inhibitors in human monocytes. FEBS Lett. 254, 94–98.

Engels, P., Fichtel, K. and Lübbert, H. (1994). Expression and regulation of human and rat phosphodiesterase type IV isogenes. FEBS Lett. 350, 291–295.

Epstein, P.M. and Hachisu, R. (1984). Cyclic nucleotide phosphodiesterase in normal and leukemic human lymphocytes and lymphoblasts. Adv. Cyclic Nucleotide Protein Phosphorylation Res. 16, 303–324.

Erneux, C., Van Sande, J., Miot, F., Cochaux, P., Decoster, C. and Dumont, J.E. (1985). A mechanism in the control of intracellular cAMP level: the activation of a calmodulin-sensitive phosphodiesterase by a rise of intracellular calcium. Mol. Cell. Endocrinol. 43, 123–134.

Essayan, D.M., Huang, S.-K., Undem, B.J., Kagey-Sobotka, A. and Lichtenstein, L.M. (1994). Modulation of antigen- and mitogen-induced proliferative responses of peripheral blood mononuclear cells by nonselective and isozyme selective cyclic nucleotide phosphodiesterase inhibitors. J. Immunol. 153, 3408–3416.

Fischmeister, R. and Hartzell, H.C. (1990). Regulation of calcium current by low K_m cyclic AMP phosphodiesterases in cardiac cells. Mol. Pharmacol. 38, 426–433.

Fonteh, A.N., Winkler, J.D., Torphy, T.J., Heravi, J., Undem, B.J. and Chilton, F.H. (1993). Influence of isoproterenol and phosphodiesterase inhibitors on platelet activating factor biosynthesis in the human neutrophil. J. Immunol. 151, 339–350.

Gettys, T.W., Vine, A.J., Simonds, M.F. and Corbin, J.D. (1988). Activation of the particulate low K_m phosphodiesterase of adipocytes by addition of cAMP-dependent protein kinase. J. Biol. Chem. 263, 10359–10363.

Giembycz, M.A. and Barnes, P.J. (1991). Selective inhibition of a high affinity type IV cyclic AMP phosphodiesterase in bovine trachealis by AH 21-132. Biochem. Pharmacol. 42, 663–677.

Giembycz, M.A., Corrigan, C.J., Kay, A.B. and Barnes, P.J. (1994). Inhibition of CD4 and CD8 T-lymphocyte (T-LC) proliferation and cytokine secretion by isoenzyme-selective phosphodiesterase (PDE) inhibitors: correlation with intracellular cyclic AMP (cAMP) concentrations. J. Allergy Clin. Immunol. 93, 167. [Abstract]

Giorgi, M., Caniglia, C., Scarsella, G. and Augusti-Tocco, G. (1993). Characterization of 3′: 5′cyclic nucleotide phosphodiesterase activities of mouse neuroblastoma N18TG2 cells. FEBS Lett. 324, 76–80.

Grant, P.G., Mannario, A.P. and Colman, R.W. (1988). cAMP-mediated phosphorylation of the low-K_m cAMP phosphodiesterase markedly stimulates its catalytic activity. Proc. Natl Acad. Sci. USA 85, 9071–9075.

Green, H. (1978). Cyclic AMP in relation to proliferation of the epidermal cell: a new view. Cell 15, 801–811.

Hadjokas, N.E., Crowley, J.J., Bayer, C.R. and Nielson, C.P. (1995). β-Adrenergic regulation of the eosinophil respiratory burst as detected by lucigenin-dependent chemoluminescence. J. Allergy Clin. Immunol. 95, 735–741.

Haider, S.R., Smith, C.J., Cui, Y., Ding, J., Bentley-Hibbert, S., Kaml, G., Moggio, R.A. and Stemerman, M.B. (1995). Cyclic AMP-mediated induction of low K_M cyclic AMP phosphodiesterase in rat aortic smooth muscle. FASEB J. 9, A678. [Abstract]

Hanifin, J.M. and Chan, S.C. (1988). Characterization of cAMP-phosphodiesterase as a possible laboratory marker of atopic dermatitis. Drug Dev. Res. 13, 123–136.

Hansel, T.T., DeVries, I.J.M., Iff, T., Rihs, S., Wandzilak, M., Betz, S., Blaser, K. and Walker, C. (1991). An improved immunomagnetic procedure for the isolation of highly purified human blood eosinophils. J. Immunol. Methods 145, 105–110.

Harris, A.L., Connell, M.J., Fergusson, E.W., Wallace, A.M., Gordon, R.J., Pagani, E.D. and Silver, P.J. (1989): Role of low K_M cyclic AMP phosphodiesterase inhibition in tracheal relaxation and bronchodilation in the guinea pig. J. Pharmacol. Exp. Ther. 251, 199–206.

Hashimoto, Y., Sharma, R.K. and Soderling, T.R. (1989). Regulation of calcium-calmodulin dependent cyclic nucleotide phosphodiesterase by the autophosphorylated form of calcium-calmodulin dependent protein kinase II. J. Biol. Chem. 264, 10884–10887.

Hatzelmann, A., Tenor, H. and Schudt, C. (1995). Differential effects of non-selective and selective phosphodiesterase inhibitors on human eosinophil functions. Br. J. Pharmacol. 114, 821–831.

Houdjik, A.P., Adolfs, M.J., Bonta, I.L. and DeJonge, H.R. (1990). Atriopeptins and nitroprusside provoke opposite changes in cGMP and cAMP levels in human macrophages. Eur. J. Pharmacol. 179, 413–417.

Ichimura, M. and Kase, H. (1993). A new cyclic nucleotide phosphodiesterase isozyme expressed in the T-lymphocyte cell lines. Biochem. Biophys. Res. Commun. 193, 985–990.

Kauffman, R.F., Crowe, V.G., Utterback, B.G. and Robertson, D.W. (1986). LY195115: a potent, selective inhibitor of cyclic nucleotide phosphodiesterase located in the sarcoplasmic reticulum. Mol. Pharmacol. 30, 609–616.

Kharitonov, S.A., Yates, D., Robbins, R.A., Logan-Sinclair, R., Shinebourne, E.A. and Barnes, P.J. (1994). Increased nitric oxide in exhaled air of asthmatic patients. Lancet 343, 133–135.

Kincaid, R.L., Balaban, C.D. and Billingsley, M.L. (1992). Regional and developmental expression of calmodulin-dependent cyclic nucleotide phosphodiesterase in rat brain. Adv. Second Messenger Phosphoprotein Res. 25, 111–122.

Kirstein, M., Rivet-Bastide, M., Hatem, S., Benardeau, A., Mercadier, J.-J. and Fischmeister, R. (1995). Nitric oxide regulates the calcium current in isolated human atrial myocytes. J. Clin. Invest. 95, 794–802.

Kishi, Y., Ashikaga, T. and Numano, F. (1992). Phosphodiesterases in vascular endothelial cells. Adv. Second Messenger Phosphoprotein Res. 25, 201–213.

Kishi, Y., Ashikaga, T., Watanabe, R. and Numano, F. (1994). Atrial natriuretic peptide reduces cyclic AMP by activating cyclic GMP-stimulated phosphodiesterase in vascular endothelial cells. J. Cardiovasc. Pharmacol. 24, 351–357.

Kita, H., Abu-Ghazaleh, R.I., Gleich, G.J. and Abraham, R.T. (1991). Regulation of Ig-induced eosinophil degranulation by adenosine 3′,5′-cyclic monophosphate. J. Immunol. 146, 2712–2718.

Kithas, P.A., Artman, M., Thompson, W.J. and Strada, S.J. (1988). Subcellular distribution of high-affinity type IV cyclic AMP phosphodiesterase activity in rabbit ventricular myocardium: relations to the effects of cardiotonic drugs. Circ. Res. 62, 782–789.

Kleine-Tebbe, J., Wicht, L., Gagne, H., Friese, A., Schunack, W., Schudt, C. and Kunkel, G. (1992). Inhibition of IgE-mediated histamine release from human peripheral leukocytes by selective PDE inhibitors. Agents Actions 36, 200–206.

Komas, N., Lugnier, C., LeBec, A., Serradeil-LeGal, C., Barthelemy, G. and Stoclet, J.-C. (1989). Differential sensitivity to cardiotonic drugs of cyclic AMP phosphodiesterases isolated from canine ventricular and sinoatrial-enriched tissues. J. Cardiovasc. Pharmacol. 14, 213–220.

Komas, N., Lugnier, C., Andriantsitohaina, R. and Stoclet, J.-C. (1991a). Characterization of cyclic nucleotide phosphodiesterases from rat mesenteric artery. Eur. J. Pharmacol. 208, 85–87.

Komas, N., Lugnier, C. and Stoclet, J.-C. (1991b). Endothelium-dependent and independent relaxation of the rat aorta by cyclic nucleotide phosphodiesterase inhibitors. Br. J. Pharmacol. 104, 495–503.

Kovala, T., Lorimer, I.A.J., Brickenden, A.M., Ball, E.H. and Sanwal, B.D. (1994). Protein kinase A regulation of cAMP phosphodiesterase expression in rat skeletal myoblasts. J. Biol. Chem. 269, 8680–8685.

Lau, L.C.K., Howarth, P.H., Walls, A.F. and Church, M.K. (1993). Effect of phosphodiesterase inhibitors on the degranulation of human nasal mast cells. Clin. Exp. Allergy 23, 30. [Abstract]

Leyen, H., Mende, U., Meyer, W., Neumann, J., Nose, M., Schmitz, W., Scholz, H., Starbatty, J., Stein, B., Wenzlaff, H., Döring, V., Kalmar, P. and Haverich, A. (1991). Mechanism underlying the reduced positive inotropic effects of the phosphodiesterase III inhibitors pimobendan, adibendan and saterinone in failing as compared to nonfailing human cardiac muscle preparations. Naunyn-Schmiedbergs Arch. Pharmacol. 344, 90–100.

Li, S.-H., Chan, S.C., Toshitani, A., Leung, D.Y.M. and Hanifin, J.M. (1992). Synergistic effects of interleukin 4 and interferon-gamma on monocyte phosphodiesterase activity. J. Invest. Dermatol. 99, 65–70.

Li, S.-H., Chan, S.C., Kramer, S.M. and Hanifin, J.M. (1993). Modulation of leukocyte cyclic AMP phosphodiesterase activity by recombinant interferon-γ: evidence for a differential effect on atopic monocytes. J. Interferon Res. 13, 197–202.

Lopez-Aparicio, P., Racson, A., Manganiello, V.C., Andersson, K.-E., Belfrage, P. and Degerman, E. (1992). Insulin induced phosphorylation and activation of the cGMP-inhibited cAMP phosphodiesterase in human platelets. Biochem. Biophys. Res. Commun. 186, 517–523.

Lopez-Aparicio, P., Belfrage, P., Manganiello, V.C., Kono, T. and Degerman, E. (1993). Stimulation by insulin of a serine kinase in human platelets that phosphorylates and activates the cGMP-inhibited cAMP phosphodiesterase. Biochem. Biophys. Res. Commun. 193, 1137–1144.

Lugnier, C. and Komas, N. (1993). Modulation of vascular cyclic nucleotide phosphodiesterases by cyclic GMP: role in vasdodilation. Eur. Heart J. 14, 141–148.

Lugnier, C. and Schini, V.B. (1990). Characterization of cyclic nucleotide phosphodiesterases from cultured bovine aortic endothelial cells. Biochem. Pharmacol. 39, 75–84.

Lugnier, C., Schoeffter, P., LeBec, A., Strouthou, E. and Stoclet, J.-C. (1986). Selective inhibition of cyclic nucleotide phosphodiesterases of human, bovine and rat aorta. Biochem. Pharmacol. 35, 1743–1751.

Lugnier, C., Muller, B., LeBec, A., Beaudry, C. and Rousseau, E. (1993). Characterization of indolidan- and rolipram-sensitive cyclic nucleotide phosphodiesterases in canine and human cardiac microsomal fractions. J. Pharmacol. Exp. Ther. 265, 1142–1151.

MacFarland, R.T., Zelus, B.D. and Beavo, J.A. (1991). High concentrations of a cyclic GMP-stimulated phosphodiesterase mediate ANP-induced decrease in cAMP and steroidogenesis in adrenal glomerulosa cells. J. Biol. Chem. 266, 136–142.

MacPhee, C.H., Reifsnyder, D.H., Moore, T.A., Lerea, K.M. and Beavo, J.A. (1988). Phosphorylation results in activation of a cAMP phosphodiesterase in human platelets. J. Biol. Chem. 263, 10353–10358.

Man, Y., West, M.R., Madden, J.A., Patel, M.T., Harreman, M. and Myles, D.D. (1994). The effect of phosphodiesterase inhibitors on intracellular cyclic AMP levels in cultures of human bronchial smooth muscle cells. Am. J. Respir. Crit. Care Med. 149, A468. [Abstract]

Marcoz, P., Prigent, A.F. and Némoz, G. (1993). Modulation of rat thymocyte proliferative response through the inhibition of different cyclic nucleotide phosphodiesterase isoforms by means of selective inhibitors and cGMP-elevating agents. Mol. Pharmacol. 44, 1027–1035.

Maurice, D.H. and Haslam, R.J. (1990). Molecular basis of the synergistic inhibition of platelet function by nitrovasodilators and activators of adenylate cyclase: inhibition of cyclic AMP breakdown by cyclic GMP. Mol. Pharmacol. 37, 671–681.

Michael, A.E. and Webley, G.E. (1991). Prostaglandin $F_{2\alpha}$ stimulates cAMP phosphodiesterase via protein kinase C in cultured human granulosa cells. Mol. Cell. Endocrinol. 82, 207–214.

Michaeli, T., Bloom, T.J., Martins, T., Loughney, K., Ferguson, K., Riggs, M., Rodgers, L., Beavo, J.A. and Wigler, M. (1993). Isolation and characterization of a previously undetected human cAMP phosphodiesterase by complementation of cAMP phosphodiesterase-deficient *Saccharomyces cerevisiae*. J. Biol. Chem. 268, 12925–12932.

Movsesian, M.A., Smith, C.J., Krall, J., Bristow, M.R. and Manganiello, V.C. (1991). Sarcoplasmic reticulum-associated cyclic adenosine 5′-monophosphate phosphodiesterase activity in normal and failing human hearts. J. Clin. Invest. 88, 15–19.

Murray, K.J., England, P.J., Hallam, T.J., Maguire, J., Moores, K., Reeves, M.L., Simpson, A.W.M. and Rink, T.J. (1990). The effects of siguazodan, a selective phosphodiesterase inhibitor, on human platelet function. Br. J. Pharmacol. 99, 612–616.

Nielson, C.P., Vestal, R.E., Sturm, R.J. and Heaslip, R. (1990). Effects of selective phosphodiesterase inhibitors on the polymorphonuclear leucocyte respiratory burst. J. Allergy Clin. Immunol. 86, 801 –808.

Nourshargh, S. and Hoult, J.R.S. (1986). Inhibition of human neutrophil degranulation by forskolin in the presence of phosphodiesterase inhibitors. Eur. J. Pharmacol. 122, 205–212.

Okayama, Y., Hunt, T.C., Kassel, O., Ashman, L.K. and Church, M.K. (1994). Assessment of the anti-c-*kit* monoclonal antibody YB5.B8 in affinity magnetic enrichment of human lung mast cells. J. Immunol. Methods 169, 153–161.

Okonogi, K., Gettys, T.W., Uhig, R.J., Tarry, W.C., Adams, D.O. and Prpic, V. (1991). Inhibition of prostaglandin E_2-stimulated cAMP accumulation by lipopolysaccharide in murine peritoneal macrophages. J. Biol. Chem. 266, 10305–10312.

Packer, M., Carver, J.R., Rodeheffer, R.J., Ivanhoe, R.J., DiBianco, R., Zeldis, S.M., Hendrix, G.H., Bommer, W.J., Elkayam, U., Kukin, M.L., The PROMISE Study Research Group (1991). Effects of oral milrinone on mortality in severe chronic heart failure. N. Engl. J. Med. 325, 1468–1475.

Peachell, P.T., Undem, B.J., Schleimer, R.P., MacGlashan, D.W., Jr, Lichtenstein, L.M., Cieslinski, L.B. and Torphy, T.J. (1992). Preliminary identification and role of phosphodiesterase isozymes in human basophils. J. Immunol. 148, 2503–2510.

Prpic, V., Uhig, R.J. and Gettys, T.W. (1993). Separation and assay of phosphodiesterase isoforms in murine peritoneal macrophages using membrane matrix DEAE chromatography and [^{32}P]cAMP. Anal. Biochem. 208, 155–160.

Rabe, K.F., Tenor, H., Dent, G., Schudt, C., Liebig, S. and Magnussen, H. (1993). Phosphodiesterase isozymes modulating inherent tone in human airways: identification and characterization. Am. J. Physiol. 264, L458-L464.

Rabe, K.F., Tenor, H., Dent, G., Schudt, C., Nakashima, M. and Magnussen, H. (1994a). Identification of PDE isozymes in human pulmonary artery and effect of selective PDE inhibitors. Am. J. Physiol. 266, L536-L543.

Rabe, K.F., Tenor, H., Spaethe, S.M., Dent, G., Schudt, C., Magnussen, H. and White, S.R. (1994b). Identification of the PDE isoenzymes in human bronchial epithelial cells and the effects of PDE inhibition on PGE_2 secretion. Am. J. Respir. Crit. Care Med. 149, A986. [Abstract]

Rascon, A., Degerman, E., Taira, M., Meacci, E., Smith, C.J., Manganiello, V.C., Belfrage, P. and Tornqvist, H. (1994). Identification of the phosphorylation site *in vitro* for cAMP-dependent protein kinase on the rat adipocyte cGMP-inhibited cAMP phosphodiesterase. J. Biol. Chem. 269, 11962–11966.

Reeves, M.L., Leigh, B.K. and England, P.J. (1987). The identification of a new cyclic nucleotide phosphodiesterase activity in human and guinea-pig cardiac ventricle: implications for the mechanism of action of selective phosphodiesterase inhibitors. Biochem. J. 241, 535–541.

Robichon, A. (1991). A new cGMP phosphodiesterase isolated from bovine platelets is substrate for cAMP- and cGMP-dependent protein kinases: evidence for a key role in the process of platelet activation. J. Cell. Biochem. 47, 147–157.

Robiscek, S.A., Krzanowski, J.J., Szentivanyi, A. and Polson, J.B. (1989). High pressure chromatography of cyclic nucleotide phosphodiesterase from purified human T lymphocytes. Biochem. Biophys. Res. Commun. 163, 554–560.

Robiscek, S.A., Blanchard, D.K., Djeu, J.Y., Krzanowski, J.J., Szentivanyi, A. and Polson, J.B. (1991). Multiple high affinity cAMP-phosphodiesterases in human T-lymphocytes. Biochem. Pharmacol. 42, 869–877.

Rousseau, E., Gagnon, J. and Lugnier, C. (1994). Biochemical and pharmacological characterization of cyclic nucleotide phosphodiesterase in airway epithelium. Mol. Cell. Biochem. 140, 171–175.

Saeki, T. and Saito, I. (1993). Isolation of cyclic nucleotide phosphodiesterase isozymes from pig aorta. Biochem. Pharmacol. 46, 833–839.

Saito, Y., Hardman, J.G. and Wells, J.N. (1985). Differences in the association of calmodulin with cyclic nucleotide phosphodiesterase in relaxed and contracted arterial strips. Biochemistry 24, 1613–1620.

Sawai, T., Ikai, K. and Uehara, M. (1995). Elevated cyclic adenosine monophosphate phosphodiesterase activity in peripheral blood mononuclear leucocytes from children with atopic dermatitis. Br. J. Dermatol. 132, 22–24.

Schudt, C., Winder, S., Forderkunz, S., Hatzelmann, A. and Ulrich, V. (1991). Influence of selective phosphodiesterase inhibitors on human neutrophil functions and levels of cAMP and Ca_i. Naunyn-Schmiedebergs Arch. Pharmacol. 344, 682–690.

Sette, C., Iona, S. and Conti, M. (1994a). The short-term activation of a rolipram-sensitive, cAMP-specific phosphodiesterase by thyroid-stimulating hormone in thyroid FRTL-5 cells is mediated by a cAMP-dependent phosphorylation. J. Biol. Chem. 269, 9245–9252.

Sette, C., Vicini, E. and Conti, M. (1994b). The rat PDE3/IVd phosphodiesterase gene codes for multiple proteins differentially activated by cAMP-dependent protein kinase. J. Biol. Chem. 269, 18271–18274.

Sette, C., Vicini, E. and Conti, M. (1994c). Modulation of cellular responses by hormones: role of cAMP specific, rolipram-sensitive phosphodiesterases. Mol. Cell. Endocrinol. 100, 75–79.

Shahid, M. and Nicholson, C.D. (1990). Comparison of cyclic nucleotide phosphodiesterase isoenzymes in rat and rabbit ventricular myocardium: positive inotropic and phosphodiesterase inhibitory effects of Org 30029, milrinone and rolipram. Naunyn-Schmiedebergs Arch. Pharmacol. 342, 698–705.

Shahid, M., Amsterdam, R.G.M., Boer, J., ten Berge, R.E., Nicholson, C.D. and Zaagsma, J.A. (1991). The presence of five cyclic nucleotide phosphodiesterase isoenzyme activities in bovine tracheal smooth muscle and the functional effects of selective inhibitors. Br. J. Pharmacol. 104, 471–477.

Sharma, R.K. (1991). Phosphorylation and characterization of bovine heart calmodulin-dependent phosphodiesterase. Biochemistry 30, 5963–5968.

Sharma, R.K. and Wang, J.H. (1985). Differential regulation of bovine brain calmodulin-dependent cyclic nucleotide phosphodiesterase isozymes by cyclic AMP-dependent protein kinase and calmodulin-dependent phosphatase. Proc. Natl Acad. Sci. USA 82, 2603–2607.

Sharma, R.K. and Wang, J.H. (1986). Calmodulin and Ca-dependent phosphorylation and dephosphorylation of 63 kDa subunit containing bovine brain calmodulin-stimulated cyclic nucleotide phosphodiesterase isozyme. J. Biol. Chem. 261, 1322–1328.

Simmons, M.A. and Hartzell, H.C. (1988). Role of phosphodiesterases in regulation of calcium current in isolated cardiac myocytes. Mol. Pharmacol. 33, 664–671.

Simpson, A.W.M., Reeves, M.L. and Rink, T.J. (1988). Effects of SK&F 94120, an inhibitor of cyclic nucleotide phosphodiesterase type III, on human platelets. Biochem. Pharmacol. 37, 2315–2320.

Smith, C.J., Vasta, V., Degerman, E., Belfrage, P. and Manganiello, V.C. (1991). Hormone-sensitive cyclic GMP-inhibited cyclic AMP phosphodiesterase in rat adipocytes. J. Biol. Chem. 266, 13385–13390.

Souness, J.E., Diocee, B.K., Martin, W. and Moodie, S.A. (1990). Pig aortic endothelial-cell cyclic nucleotide phosphodiesterases. Biochem. J. 266, 127–132.

Souness, J.E., Carter, C.M., Diocee, B.K., Hassall, G.H., Wood, L.J. and Turner, N.C. (1991). Characterization of guinea-pig eosinophil phosphodiesterase activity: assessment of its involvement in regulating superoxide generation. Biochem. Pharmacol. 42, 937–945.

Souness, J.E., Hassall, G.A. and Parrott, D.P. (1992). Inhibition of pig aortic smooth muscle cell DNA synthesis by selective type III and type IV cyclic AMP phosphodiesterase inhibitors. Biochem. Pharmacol. 44, 857–866.

Souness, J.E., Maslen, C., Webber, S., Foster, M., Raeburn, D., Palfreyman, M.N., Ashton, M.J. and Karlsson, J.A. (1995). Suppression of eosinophil function by RP 73401, a potent and selective inhibitor of cAMP-specific phosphodiesterase: comparison with rolipram. Br. J. Pharmacol. 115, 39–46.

Stroop, S.D. and Beavo, J.A. (1991). Structure and function studies of the cGMP-stimulated phosphodiesterase. J. Biol. Chem. 266, 23802–23809.

Sugioka, M., Ito, M., Masuoka, H., Ichikawa, K., Konishi, T., Tanaka, T. and Nakano, T. (1994). Identification and characterization of isoenzymes of cyclic nucleotide phosphodiesterases in human kidney and heart, and the effect of new cardiotonic agents on these isoenzymes. Naunyn-Schmiedebergs Arch. Pharmacol. 350, 284–293.

Suttorp, N., Weber, U., Welsch, T. and Schudt, C. (1993). Role of phosphodiesterases in the regulation of endothelial permeability *in vitro*. J. Clin. Invest. 91, 1421–1428.

Swinnen, J.V., Joseph, D.R. and Conti, M. (1989). The mRNA encoding a high affinity cAMP phosphodiesterase is regulated by hormones and cAMP. Proc. Natl Acad. Sci. USA 86, 8197–8201.

Swinnen, J.V., Tsikalas, K.E. and Conti, M. (1991). Properties and hormonal regulation of two structurally related cAMP phosphodiesterases from the rat Sertoli cell. J. Biol. Chem. 266, 18370–18377.

Takemoto, D.J., Lee, W.-N., P., Kaplan, S.A. and Appleman, M.M. (1978). Cyclic AMP phosphodiesterase in human lymphocytes and lymphoblasts. J. Cyclic Nucl. Res. 4, 123–132.

Tani, T., Sakurai, K., Kimura, Y., Ishikawa, T. and Hidaka, H. (1992). Pharmacological manipulation of tissue cyclic AMP by inhibitors: effects of phosphodiesterase inhibitors on the functions of platelets and vascular endothelial cells. Adv. Second Messenger Phosphoprotein Res. 25, 215–227.

Tenor, H., Bartel, S. and Krause, E.-G. (1987). Cyclic nucleotide phosphodiesterase activity in the rat myocardium: evidence of four different subtypes. Biomed. Biochim. Acta 46, S749–S753.

Tenor, H., Rabe, K.F., Hatzelmann, A., Wendel, A. and Schudt, C. (1994). Characterization of phosphodiesterase isoenzyme pattern in human pulmonary epithelial A549 cells. Eur. Respir. J. 7, 186s. [Abstract]

Tenor, H., Hatzelmann, A., Kupferschmidt, R., Stanciu, L., Djukanović, R., Schudt, C., Wendel, A., Church, M.K. and Shute, J.K. (1995a). Cyclic nucleotide phosphodiesterase isoenzyme activities in human alveolar macrophages. Clin. Exp. Allergy 25, 625–633.

Tenor, H., Hatzelmann, A., Wendel, A. and Schudt, C. (1995b). Identification of phosphodiesterase IV activity and its cyclic adenosine monophosphate-dependent up-regulation in a human keratinocyte cell line (HaCaT). J. Invest. Dermatol. 105, 70–74.

Tenor, H., Staniciu, L., Schudt, C., Hatzelmann, A., Wendel, A., Djukanović, R., Church, M.K. and Shute, J.K. (1995c). Cyclic nucleotide phosphodiesterases from purified human $CD4^+$ and $CD8^+$ T lymphocytes. Clin. Exp. Allergy 25, 616–624.

Tetsuka, T., Kusano, E., Takeda, S., Homma, S., Yoshida, I., Ando, Y. and Asano, Y. (1995). Activation of protein kinase C stimulates cAMP phosphodiesterase in rat renal collecting tubes. Am. J. Physiol. 268, F808-F814.

Thompson, W.J., Ross, C.P., Pledger, W.J., Strada, S.J., Banner, R.L. and Hersh, E.M. (1976). Cyclic adenosine 3′:5′-monophosphate phosphodiesterase: distinct forms in human lymphocytes and monocytes. J. Biol. Chem. 251, 4922–4929.

Tohda, M., Murayama, T., Hasegawa, H., Nogiri, S. and Nomura, Y. (1994). [^{3}H]Rolipram binding and phosphodiesterase activity in neuroblastoma N18TG-2 and glioma C6Bu-1. Neurosci. Lett. 175, 89–91.

Torphy, T.J. and Cieslinski, L.B. (1990). Characterization and selective inhibition of cyclic nucleotide phosphodiesterase isozymes in canine tracheal smooth muscle. Mol. Pharmacol. 37, 206–214.

Torphy, T.J., Livi, G.P., Balcarek, J.M., White, J.R., Chilton, F.H. and Undem, B.J. (1992a). Therapeutic potential of isozyme-selective phosphodiesterase inhibitors in the treatment of asthma. Adv. Second Messenger Phosphoprotein Res. 25, 289–305.

Torphy, T.J., Stadel, J.M., Burman, M., Cieslinski, L.B., McLaughlin, M.M., White, J.R. and Livi, G.P. (1992b). Coexpression of human cAMP-specific phosphodiesterase activity and high affinity rolipram binding in yeast. J. Biol. Chem. 267, 1789–1804.

Torphy, T.J., Zhou, H.-L. and Cieslinski, L.B. (1992c). Stimulation of *beta* adrenoceptors in a human monocyte cell line (U937) up-regulates cyclic AMP-specific phosphodiesterase activity. J. Pharmacol. Exp. Ther. 263, 1195–1205.

Torphy, T.J., Undem, B.J., Cieslinski, L.B., Luttmann, M.A., Reeves, M.L. and Hay, D.W.P. (1993). Identification, characterization and functional role of phosphodiesterase isozymes in human airway smooth muscle. J. Pharmacol. Exp. Ther. 265, 1213–1223.

Townley, R.G. (1993). Elevated cAMP-phosphodiesterase in atopic disease: cause or effect? J. Lab. Clin. Med. 121, 15–17.

Truong, V.H. and Müller, T. (1994). Isolation, biochemical characterization and N-terminal sequence of rolipram-sensitive cAMP phosphodiesterase from human mononuclear leukocytes. FEBS Lett. 252, 113–118.

Turner, N.C., Wood, L.J., Burns, F., Gueremy, T. and Souness, J.E. (1993). The effect of cyclic AMP and cyclic GMP phosphodiesterase inhibitors on the superoxide burst of guinea pig peritoneal macrophages. Br. J. Pharmacol. 108, 876–883.

Verghese, M.W., McConell, R.T., Lenhard, J.M., Hamacher, L. and Jin, S.-L.C. (1995a). Regulation of distinct cyclic AMP-specific phosphodiesterase (phosphodiesterase type IV) isozymes in human monocytic cells. Mol. Pharmacol. 47, 1164–1171.

Verghese, M.W., McConell, R.T., Strickland, A.B., Gooding, R.C., Stimpson, S.A., Yarnell, D.P., Taylor, J.D. and Furdon, P.J. (1995b). Differential regulation of human monocyte derived TNF and IL-1β by type IV cAMP-phosphodiesterase (cAMP-PDE) inhibitors. J. Pharmacol. Exp. Ther. 272, 1313–1320.

Wang, J.H., Sharma, R.K. and Mooibroek, M.J. (1990). Calmodulin-stimulated cyclic nucleotide phosphodiesterases. In "Cyclic Nucleotide Phosphodiesterases: Structure, Regulation and Drug Action" (eds. J. Beavo and M.D. Houslay), pp. 19–59. Wiley, Chichester.

Wax, M. (1992). Signal transduction in the ciliary epithelium. In "Pharmacology of Glaucoma" (eds. S. Drance, M. Van Buskirk and A. Neufeld), pp. 184–210. Williams & Wilkins, Baltimore.

Wedner, H.J., Chan, B.Y., Parker, C.S. and Parker, C.W. (1979). Cyclic nucleotide phosphodiesterase activity in human peripheral blood lymphocytes and monocytes. J. Immunol. 123, 725–732.

Weishaar, R.E., Burrows, S.D., Kobylarz, D.C., Quade, M.M. and Evans, D.B. (1986). Multiple molecular forms of cyclic nucleotide phosphodiesterases in cardiac and smooth muscle and in platelets. Biochem. Pharmacol. 35, 787–800.

Whalin, M.E., Scammell, J.G., Strada, S.J. and Thompson, W.J. (1991). Phosphodiesterase II, the cGMP-activatable cyclic nucleotide phosphodiesterase, regulates cyclic AMP metabolism in PC12 cells. Mol. Pharmacol. 39, 711–717.

Wright, C.D., Kuipers, P.J., Kobylarz-Singer, D., Devall, L.J., Klinjefus, B.A. and Weishaar, R.E. (1990). Differential inhibition of human neutrophil functions: role of cyclic AMP-specific, cyclic GMP-insensitive phosphodiesterase. Biochem. Pharmacol. 40, 699–707.

Xiong, Y., Westhead, E.W. and Slakey, L.L. (1995). Role of phosphodiesterase isoenzymes in regulating intracellular cyclic AMP in adenosine-stimulated smooth muscle cells. Biochem. J. 305, 627–633.

Yamamoto, T., Manganiello, V.C. and Vaughan, M. (1983). Purification and characterization of cyclic GMP-stimulated cyclic nucleotide phosphodiesterase from calf liver. J. Biol. Chem. 258, 12526–12533.

Yang, C., Luh, K.-T., Lee, Y.-C. and Wu, R. (1989). Regulation of ciliary activity in cultured human bronchial epithelial cells. Eur. Respir. J. 2, 283s. [Abstract]

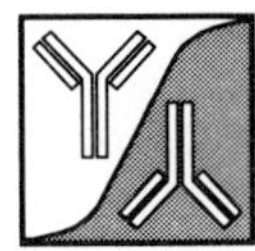

3. Effects of Theophylline and Non-selective Xanthine Derivatives on PDE Isoenzymes and Cellular Function

Gordon Dent *and* Klaus F. Rabe

1. Introduction 41
2. Inhibition of PDE 42
   - 2.1 Inhibition of Tissue PDE Activity 42
   - 2.2 Inhibition of Isoenzymes 42
3. Effects on Cell Function 44
   - 3.1 Immune Cells 45
   - 3.2 Other Cells 54
4. Summary and Directions for Future Research 55
5. References 56

1. Introduction

The archetypical cyclic nucleotide phosphodiesterase (PDE) inhibitors are the methylxanthines – a family of plant alkaloids including caffeine, theobromine and theophylline, derived from the heterocyclic compound, xanthine (2,6-dihydroxypurine; Fig. 3.1) – and papaverine, an isoquinoline derivative isolated from opium. These compounds were identified as smooth muscle relaxants, relieving the bronchospasm of asthma, in the case of the methylxanthines (see section 3.2.1 and Chapter 2), or intestinal colic, in the case of papaverine. Xanthine – a metabolite of the purine nucleosides adenosine and inosine – and its derivatives are based on the purine core of adenosine (Fig. 3.1) and, perhaps as a result of this structural analogy, theophylline and caffeine are weak antagonists at most classes of adenosine receptors. Similarly, papaverine, in common with dipyridamole (an inhibitor of the PDE5 isoenzymes and a relaxant of coronary blood vessels), inhibits cellular uptake of adenosine. These interactions with adenosine binding sites have been suggested to mediate some muscle-relaxing actions of the drugs but a large body of evidence implicates the inhibition of PDE in the mechanism of action of drugs such as theophylline both in smooth muscle and in other cells sensitive to intracellular cyclic nucleotide elevations, including those of the immune system (Sullivan *et al.*, 1994b; Banner and Page, 1995a,b). The possible role of adenosine binding site interactions in the pharmacological profile of methylxanthines is discussed elsewhere (Fredholm, 1980; Persson and Karlsson, 1987).

The purpose of this chapter is to review briefly the available data on PDE inhibition by theophylline and other xanthine derivatives (Fig. 3.2) and to describe the pharmacological actions of these drugs on immune cells *in vitro* and *in vivo*. A brief description of the documented actions of these drugs in other cell types will also be given. This summary of the actions of theophylline and related drugs may represent the biological spectrum and therapeutic potential of PDE inhibitors in general. On the other hand, chemical variants of xanthine have been synthesized with selectivity for individual PDE isoenzyme families, including PDE3 (Miyamoto *et al.*, 1994), PDE4 (Buckle *et al.*, 1994; Miyamoto *et al.*, 1994) and PDE5 (Buckle *et al.*, 1994; see also Chapter 9), which resemble the inhibitors discussed in later chapters. It therefore seems appropriate here to summarize the functional effects of the non-selective alkylxanthines such as theophylline,

Phosphodiesterase Inhibitors
ISBN 0-12-210720-9

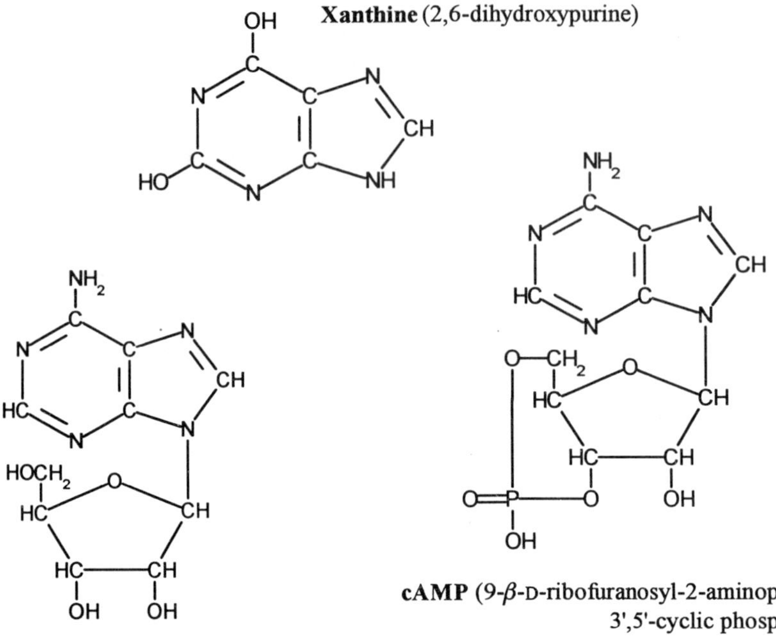

Xanthine (2,6-dihydroxypurine)

cAMP (9-β-D-ribofuranosyl-2-aminopurine 3',5'-cyclic phosphate)

Adenosine (9-β-D-ribofuranosyl-2-aminopurine)

Figure 3.1 Xanthine is a metabolite of cyclic AMP and adenosine and a structural analogue of their purine core.

3-isobutyl-1-methylxanthine (IBMX) and pentoxifylline. Finally, the important questions regarding the potential for PDE inhibitors in the treatment of disease and the directions that research into this class of drugs should follow will be raised. These questions are addressed in subsequent chapters.

2. *Inhibition of PDE*

2.1 INHIBITION OF TISSUE PDE ACTIVITY

The widespread use of theophylline in the treatment of bronchial asthma led to studies of the drug's biochemical actions in lung tissues. It had been established that smooth muscle relaxation could result from an elevation of intracellular cyclic AMP and the demonstration that theophylline inhibits PDE (Butcher and Sutherland, 1962) was assumed to explain theophylline's bronchodilator action. This assumption was tested by several studies comparing the inhibition of lung PDE activity with airway smooth muscle relaxation *in vitro* (Triner *et al.*, 1977; Newman *et al.*, 1978; Polson *et al.*, 1978, 1979). The results of these experiments indicated that similar concentrations of theophylline were required to inhibit cAMP hydrolysis in lung homogenates and to relax tracheal or bronchial smooth muscle to similar extents, although these concentrations were fairly high when compared with the serum theophylline concentrations required to effect bronchodilation *in vivo* (Bergstrand, 1980). Similarly high concentrations of theophylline and other methylxanthine PDE inhibitors are often required to suppress functional responses of isolated inflammatory cells (see following sections) and it appears likely that the presence *in vivo* of mediators that activate adenylate cyclase (AC) or guanylate cyclase (GC) confers a greater functional efficacy on PDE inhibitors, which prolong the stimulated elevation of intracellular cyclic nucleotide concentrations. In many cells studied, the co-administration of a methylxanthine PDE inhibitor with an AC-activating drug, such as a β-adrenoceptor agonist, E-series prostaglandin or forskolin, leads to an additive or supra-additive increase in intracellular cAMP, indicating the co-operation of AC activation and PDE inhibition.

Interestingly, it was observed more than ten years ago that treatment of platelets with IBMX plus an adenylate cylase-activating drug, such as prostacyclin (PGI_2) or forskolin, led to a rapid increase in platelet PDE activities hydrolysing both cAMP and cGMP (Hamet *et al.*, 1983; Tremblay *et al.*, 1985). Prolonged exposure of lymphocyte cultures to IBMX also leads to a marked increase in the activity of a distinct PDE enzyme (Thompson *et al.*, 1980). Similar findings in monocytes are of substantial current interest (see Chapters 1, 2 and 11).

2.2 INHIBITION OF ISOENZYMES

The early studies in lung and airway homogenates revealed the presence of distinct PDE activities that

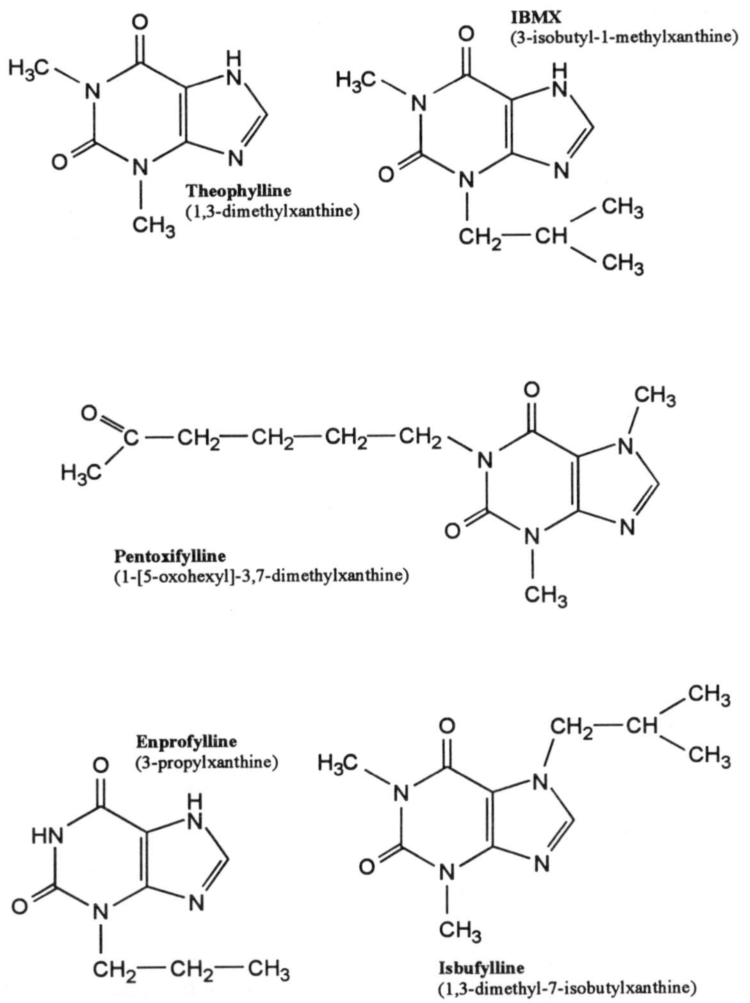

Theophylline
(1,3-dimethylxanthine)

IBMX
(3-isobutyl-1-methylxanthine)

Pentoxifylline
(1-[5-oxohexyl]-3,7-dimethylxanthine)

Enprofylline
(3-propylxanthine)

Isbufylline
(1,3-dimethyl-7-isobutylxanthine)

Figure 3.2 Alkylxanthine PDE inhibitors. Substitution of the xanthine molecule at N-1 and N-3 leads to localization of bonding electrons to the C—O bonds at positions 2 and 6 and confers PDE inhibitory and adenosine antagonistic potency. Substitution at N–9 is also possible: this leads to localization of the 8,9 double bond to 7,8, with consequent loss of biological activity (Persson and Karlsson, 1987). Enprofylline does not bear a methyl substituent and is not, therefore, classified as a methylxanthine.

appeared as four separate peaks eluted from anion-exchange chromatography columns (Bergstrand, 1980). These peaks differed in substrate specificity and Ca^{2+}/calmodulin sensitivity and probably correspond to the PDE1, 2, 3 and 4 isoenzyme families. Although PDE3 and PDE4 could not be fully separated, all of the fractions appeared to be inhibited by theophylline with similar potency with IC_{50} values ranging from 100 to 220 μM (Bergstrand, 1980).

With the development of techniques for the separation of isoenzymes and the pharmacological study of the individual families (see Chapter 2), it has become possible to evaluate the inhibitory potency of non-selective PDE inhibitors against the separate isoenzymes in a wide range of cells. As summarized in Table 3.1, the methylxanthines theophylline and IBMX inhibit PDE isoenzymes of families 1 to 5 with roughly equal potencies in the cells and tissues studied to date. Similarly, pentoxifylline has been shown to inhibit three distinct PDE isoenzymes (corresponding to PDE1, PDE2 and PDE4) purified from human bronchial smooth muscle with approximately equal potencies (IC_{50} values of 65 μM, 98 μM and 45 μM, respectively) (Cortijo *et al.*, 1993). The effects of these drugs on PDE6 and PDE7 have not yet been investigated in depth. It appears unlikely, however, that PDE6 will be found to occur in immune cells. PDE7 may prove to be important in the regulation of lymphocyte function (see Chapter 2) and details of the actions of methylxanthines on this isoenzyme family are awaited.

Table 3.1 Inhibition of PDE isoenzymes from various cell sources, including immune cells, by theophylline and IBMX

Isoenzyme	*Cell/ tissue*	*IC$_{50}$*[a]	*Reference*
Theophylline			
1	Brain, bovine	280 μM[e]	Schudt *et al.* (1991c)
2	Heart, rat	270 μM[b]	Schudt *et al.* (1991c)
	Bronchus, human	55 μM[c]	Cortijo *et al.* (1993)
3	Heart, rat	390 μM[b]	Schudt *et al.* (1991c)
	Platelets, human	98 μM[b]	Schudt *et al.* (1991c)
4	Trachea, dog	155 μM[b]	Schudt *et al.* (1991c)
	Bronchus, human	150 μM[c]	Cortijo *et al.* (1993)
	Eosinophils, human	290 μM[b]	Hatzelmann *et al.* (1995)
5	Platelets, human	630 μM[e]	Schudt *et al.* (1991c)
IBMX			
1	Heart, bovine	2.5 μM	Beavo (1988)
	Brain, bovine	10 μM[e]	Schudt *et al.* (1991a)
		8.9 μM[b]	Ukena *et al.* (1993)
	Trachea, bovine	5 μM	Shahid *et al.* (1991)
2	Heart, bovine	50 μM	Beavo (1988)
	Heart, rat	6 μM[b]	Schudt *et al.* (1991a)
		6.3 μM[b]	Ukena *et al.* (1993)
	Trachea, bovine	4 μM	Shahid *et al.* (1991)
3	Heart, bovine	2 μM	Beavo (1988)
	Heart, rabbit	5 μM	Shahid *et al.* (1991)
	Heart, guinea pig	2 μM[b]	Galvan and Schudt (1990)
	Heart, rat	10 μM[b]	Ukena *et al.* (1993)
	Platelets, human	4 μM[b]	Schudt *et al.* (1991a)
	Platelets, human	10 μM	Bray and Mueller (1994)
4	Heart, bovine	15 μM	Beavo (1988)
	Trachea, bovine	5 μM	Shahid *et al.* (1991)
	Trachea, dog	9 μM[b]	Galvan and Schudt (1990)
	Neutrophils, human	8 μM[b]	Schudt *et al.* (1991a)
	Neutrophils, human	10 μM[b]	Schudt *et al.* (1991b)
	Neutrophils, human	20 μM	Bray and Mueller (1994)
	Eosinophils, human	14 μM[b]	Hatzelmann *et al.* (1995)
5	Trachea, bovine	4.5 μM[d]	Shahid *et al.* (1991)
	Platelets, human	10 μM[e]	Schudt *et al.* (1991a)
	Lung, human	1.8 μM[d]	Bray and Mueller (1994)

[a] IC$_{50}$ is the concentration causing 50% inhibition of enzyme activity.
The substrate is cAMP 1 μM except: [b] cAMP, 0.5 μM; [c] cAMP 0.25 μM; [d] cGMP 1 μM; [e] cGMP 0.5 μM.

3. *Effects on Cell Function*

The immunopharmacology of theophylline has become a subject of some interest in recent years, with renewed consideration of the contribution made to the drug's therapeutic effects in bronchial asthma by its anti-inflammatory actions. The ability of drugs that inhibit certain PDE isoenzymes to suppress both immune cell function *in vitro* and allergic pulmonary inflammation *in vivo* is also believed to present an opportunity for the development of novel anti-asthma and anti-inflammatory drugs with greater potency and selectivity than theophylline.

The actions of PDE-inhibitory alkylxanthines upon immune cell function has been studied in some depth during the past two decades and many of these actions would support the hypothesis that such drugs can suppress allergic responses and leucocyte-dependent inflammation. The actions of the drugs in cells of other systems, on the other hand, might explain some of the undesired effects of non-selective PDE inhibitors in general and of methylxanthines in particular.

3.1 IMMUNE CELLS

Over the last 20 years, several studies have suggested that theophylline and other methylxanthines in use in clinical therapeutics (principally pentoxifylline, see section 3.1.2) might exert suppressive actions on cells of the immune system that may contribute to the beneficial effects of these drugs. In the field of asthma research, many demonstrations have been made of suppression of asthmatic responses that cannot be accounted for fully by the bronchodilator actions of the drugs and is assumed to represent an anti-inflammatory effect (Persson *et al.*, 1988; Banner and Page, 1995a,b).

In the following sections, the pharmacology of alkylxanthine PDE inhibitors in cells of the immune system is described. Where appropriate, the relative contributions of adenosine receptor antagonism and PDE inhibition to these actions is discussed.

3.1.1 Lymphocytes

Lymphocytes are the pivotal cells of immune responses, being the only cells capable of specifically recognizing and distinguishing antigens. B lymphocytes, as antibody-producing cells, are critical for humoural immunity and contribute, through such processes as opsonization, to cellular immune responses. B cells produce antibodies of distinct immunoglobulin classes which mediate different types of response. The relative proportions of B cells producing immunoglobulins of each class is governed by a cytokine-dependent process of class switching, so that manipulation of cytokine production by other immune cells can determine the extent of production of, for example, IgE antibodies, which mediate allergic responses. T lymphocytes recognize antigenic peptide fragments on the surface of antigen-presenting cells such as macrophages. Some (known as T4 or $CD4^+$) function as "helper" cells, producing cytokines which promote proliferation and differentiation of other immune cells and recruitment of inflammatory leucocytes, whereas others (T8 or $CD8^+$) function as cytotoxic cells, lysing cells that bear foreign antigens. In addition, a subpopulation of $CD8^+$ T cells appears to exert a suppressor function upon other T-cell responses. A third class, the large granular lymphocytes or natural killer (NK) cells, lyse tumour and virus-infected cells without an apparent antigenic stimulation (Abbas *et al.*, 1991).

Immunoglobulin secretion from B cells *in vitro* can be reduced by high concentrations of theophylline and IBMX (Strannegard and Strannegard, 1984; Shearer *et al.*, 1988) and the secretion of IgE from lymphocytes of patients with atopic dermatitis has specifically been demonstrated to be affected. The mechanism for these actions is unclear and, in fact, lower concentrations of theophylline – as well as other cAMP-elevating agents – may enhance IgE production in mixed lymphocyte preparations, possibly through an inhibition of suppressor T-cell function (Strannegard and Strannegard, 1984; Shah *et al.*, 1991).

Since immune responses to antigens depend on the proliferation of antigen-specific lymphocytes, one mechanism through which responses can be suppressed is the inhibition of this proliferation. Methylxanthines possess the capability to inhibit cell cycle progression and, thereby, proliferation of murine B cells induced by certain stimuli, including antigen, but enhances progression in response to other stimuli (Cohen and Rothstein, 1989; Muthusamy *et al.*, 1991). In T cells, whose function is clearly regulated by intracellular cAMP (Mary *et al.*, 1987; Giembycz *et al.*, 1994), the situation is clearer. Antigen-induced proliferation of purified mouse and rat T cells (Rosenthal *et al.*, 1992; Rott *et al.*, 1993), as well as human peripheral blood lymphocytes (Tilg *et al.*, 1993) is suppressed by pentoxifylline and proliferation of hamster lymph node cells has been shown to be inhibited by theophylline (Hart, 1988). Mimicking of antigenic stimulation with monoclonal antibodies directed against the T-cell antigen receptor (CD3) similarly induces cell proliferation and this response can be suppressed by theophylline at concentrations within the normal therapeutic range and by pentoxifylline (Scordamaglia *et al.*, 1988; Singer *et al.*, 1992). Proliferation of lymphocytes can also be stimulated by mitogenic plant lectins, such as phytohaemagglutinin, concanavalin A (Con A) and pokeweed mitogen, as well as tumour-promoting phorbol esters. Proliferation of T lymphocytes from a number of species induced by these agents can be inhibited by theophylline (Novogrodsky *et al.*, 1983; Crosti *et al.*, 1986; Scordamaglia *et al.*, 1988; Kotecki *et al.*, 1989), IBMX (Goto *et al.*, 1988), pentoxifylline (Bessler *et al.*, 1987; Rao *et al.*, 1991; Rosenthal *et al.*, 1992; Singer *et al.*, 1992; Rott *et al.*, 1993; Tilg *et al.*, 1993; Hecht *et al.*, 1995) and papaverine (Nokta *et al.*, 1993). In contrast, theophylline and IBMX fail to inhibit interleukin-2 (IL-2)-induced proliferation of mouse thymocytes and cultured T cells, even though the proliferation of these cells in response to phorbol esters is profoundly suppressed by the drugs (Novogrodsky *et al.*, 1983; Goto *et al.*, 1988). The signalling pathways through which the different mitogens exert their action have not been fully elucidated, so the reason for the stimulus-specificity of PDE inhibitors against T-cell proliferation remains unclear.

Although many studies have utilized very high concentrations of methylxanthines, concentrations in the range of 50–500 μM (equivalent to 10–100 μg/ml) can be shown to inhibit colony formation of human peripheral blood mononuclear cells and of purified T cells (Ghio *et al.*, 1988; Scordamaglia *et al.*, 1988). Furthermore, these actions appear to be largely or exclusively cAMP-mediated and dependent on cyclic AMP-dependent protein kinase (PKA) (Rosenthal *et al.*, 1992).

Cytokine expression and secretion by T lymphocytes are also sensitive to suppression by cAMP (Mary *et al.*, 1987; Giembycz *et al.*, 1994) and methylxanthines mimic these effects. Theophylline inhibits phorbol ester-induced IL-2 release from mouse thymocytes and mitogen-induced IL-2 and TNF-β (lymphotoxin) release from human peripheral blood lymphocytes (Prieur and Granger, 1975; Novogrodsky *et al.*, 1983; Scordamaglia *et al.*, 1988) whereas pentoxifylline has also been shown to suppress mitogen or anti-CD3-induced release of TNF-α (Fig. 3.3a) and interferon γ (IFNγ) but not of IFNα or IL-1β (Schandené *et al.*, 1992; Singer *et al.*, 1992; Rott *et al.*, 1993; Tilg *et al.*, 1993). The release of TNF-α into the systemic circulation of mice in response to anti-CD3 *in vivo* is also reduced by pentoxifylline (Fig. 3.3b). As observed in monocytes (see section 3.1.4), IL-6 release from mixed human lymphocyte cultures is enhanced by pentoxifylline (Tilg *et al.*, 1993), although the response of purified T cells is inhibited (Schandené *et al.*, 1992).

Since TNF-α appears to be involved in the demyelination of nerve fibres occurring in experimental autoimmune encephalomyelitis (EAE, a model of multiple sclerosis) in Lewis rats (Ruddle *et al.*, 1990), it has been suggested that suppression of TNF-α expression or secretion might be beneficial in this condition (Rott *et al.*, 1993; Sommer *et al.*, 1995; see also Chapter 7). The inhibition of T-cell TNF-α production by pentoxifylline is associated with an abrogation of EAE in myelin basic protein-sensitized rats and a suppression of the specific *ex vivo* reactivity of their T cells to the protein (Rott *et al.*, 1993). Since concentrations of pentoxifylline that suppressed T-cell TNF-α production *in vitro* were ineffective against spontaneous or induced IL-4 release, it was suggested that methylxanthine inhibition of cytokine elaboration is more marked in helper cells of the T_H1 than the T_H2 type (Rott *et al.*, 1993), in keeping with the apparent preference of cAMP for suppression of T_H1 cell function (Novak and Rothenberg, 1990). T_H2 cells are regarded as being important in the pathophysiology of allergic diseases, owing to their production of IL-4, which promotes IgE class-switching (Anderson and Coyle, 1994), and methylxanthines may, therefore, be less effective in allergy than in T_H1-mediated reactions.

The adherence of T lymphocytes to endothelial and epithelial cells is an essential stage in their accumulation at sites of infection or inflammation. The phorbol ester-stimulated adherence of T-cells to keratinocytes stimulated with IFNγ or TNF-α is inhibited by pentoxifylline, which also suppresses TNF-α-induced expression of intercellular adhesion molecule 1 (ICAM-1) in human skin biopsies (Bruynzeel *et al.*, 1995). It remains unclear whether pentoxifylline inhibits adhesion molecule expression on lymphocytes, endothelial cells, or both.

In asthmatic patients, a reduction in the number of suppressor T cells in the circulation is observed that is reversed after a period of theophylline treatment (Shohat *et al.*, 1983) or that can be provoked by withdrawal of theophylline from the patients' therapy (Fink *et al.*, 1987), reflecting the ability of theophylline to enhance the suppression of autologous cell proliferation in mixed lymphocytes *in vitro* (Zocchi *et al.*,

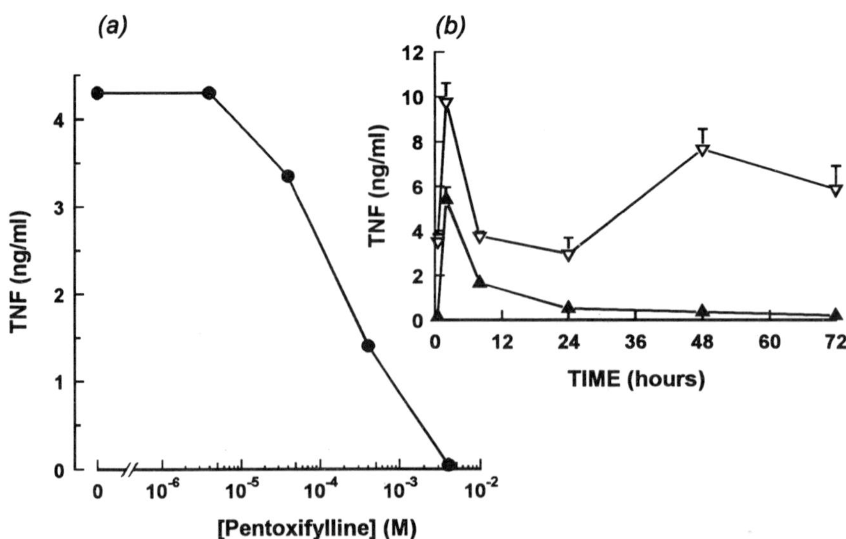

Figure 3.3 Pentoxifylline inhibits anti-CD3-stimulated TNF-α secretion *in vitro* and *in vivo*. (a) TNF-α release from purified human T lymphocytes stimulated with CLB-T3/3 anti-CD3 in the presence of increasing concentrations of pentoxifylline. Data taken from Schandené *et al.* (1992) (b) TNF-α concentrations in the serum of mice injected with 145–2C11 anti-CD3 without pre-treatment (▽) or after pre-treatment with 2 mg pentoxifylline (▲). Reproduced, with permission, from Bemelmans *et al.* (1994), © 1994, The American Association of Immunologists.

1985). This stimulation of suppressor T-cell activity is diminished in the lymphocytes of patients with insulin-dependent diabetes, suggesting a possible connection with autoimmune processes (Crosti *et al.*, 1986). Recently, withdrawal of theophylline from asthmatic patients has been shown to increase the numbers of T cells observed in bronchial biopsies in parallel with a decrease in T cell numbers in the peripheral circulation, suggesting that theophylline may suppress a process of lymphocyte trafficking between the circulation and the airways (Kidney *et al.*, 1995). These effects were observed in patients whose mean serum theophylline concentrations prior to withdrawal were relatively low (between 5 and 10 μg/ml).

Theophylline and IBMX elevate cAMP levels in NK cells and suppress their cytotoxic activity (Goto *et al.*, 1983; Takayama *et al.*,1988). IBMX inhibits anti-CD3-induced degranulation of human NK cells and also inhibits their response to the combination of phorbol ester and calcium ionophore (Takayama *et al.*, 1988). Possibly via the same mechanism, the ability of NK cells to kill tumour cells *in vitro* is reduced in the presence of moderate concentrations of theophylline (55–100 μM or 10–20 μg/ml). This action of theophylline is mimicked by other PDE inhibitors, but not by adenosine antagonists, and correlates with the degree of elevation of intracellular cAMP (Coskey *et al.*, 1993). The adherence of IL-2-activated killer cells to cultured tumour cells, as well as to fibronectin and cultured umbilical cord endothelial cells, is inhibited by pentoxifylline (Kovach *et al.*, 1994), indicating a further possible mechanism for suppression of cytotoxicity.

3.1.2 Neutrophils

Neutrophils are effector cells of natural immunity, responsible for phagocytosis of invading microbes, and form the major cell population in the acute inflammatory response (Abbas *et al.*, 1991). Through the release of neutral proteases and acid hydrolases from their granules and the generation of reactive oxygen species, neutrophils also contribute to tissue destruction associated with chronic inflammation and are implicated in the pathology of conditions such as emphysema, the adult respiratory distress syndrome, rheumatoid arthritis, inflammatory bowel disease, gout, neutrophilic dermatoses (including Sweet's syndrome, Beçhet's disease and psoriasis) and ischaemia/reperfusion-mediated injury in the heart, kidney and skeletal muscle (Leff and Repine, 1993).

Theophylline and IBMX exert actions on neutrophils that are broadly inhibitory. IBMX, for example, inhibits respiratory burst and lysosomal enzyme release by human neutrophils stimulated with the chemotactic tripeptide, *N*-formylmethionyl-L-leucyl-L-phenylalanine (FMLP), or the complement fragment C5a, although it is very poorly effective against phagocytosis-induced respiratory burst (Wright *et al.*, 1990). At concentrations of theophylline within the commonly quoted range of therapeutic serum concentrations (20–100 μM), intracellular concentrations of cAMP are elevated by 200% within 45 s of addition of the drug. This parallels an inhibition of reactive oxygen species generation (measured as lucigenin-enhanced chemiluminescence, CL) and leukotriene B_4 (LTB_4) release induced by FMLP or calcium ionophores (ionomycin or A23187), but not by synthetic diacylglycerol, in the first six minutes after theophylline addition (Nielson *et al.*, 1988). Confirming the importance of cAMP in these effects, the inhibition of the CL response by a β-adrenoceptor agonist, isoprenaline, is enhanced by both theophylline and enprofylline (Nielson *et al.*, 1986, 1988).

Enprofylline, which is more potent than theophylline as a PDE inhibitor but a much weaker adenosine antagonist (Persson and Karlsson, 1987), is marginally more potent than theophylline both in suppressing neutrophil respiratory burst and in augmenting the inhibitory action of isoprenaline. Furthermore, theophylline antagonizes the inhibition of A23187-induced respiratory burst by adenosine whereas enprofylline causes a slight augmentation of adenosine's inhibitory action presumably due to amplification by PDE inhibition of the cAMP elevation resulting from A_2 adenosine receptor activation (Nielson *et al.*, 1986). In contrast, another study showed FMLP-induced aggregation, superoxide anion (O_2^-) generation and degranulation to be inhibited only at higher concentrations of theophylline and IBMX (greater than 100 μM), whereas lower concentrations enhance the responses (Fig. 3.4a,b). The latter effect probably results from antagonism at adenosine A_2 receptors since the enhancement is mimicked by adenosine deaminase or 8-phenyltheophylline, which antagonizes adenosine receptors but does not inhibit PDE (Dianzani *et al.*, 1994), and is reversed by exogenous adenosine (Schmeichel and Thomas, 1987). These data are supported by the ability of enprofylline, in the concentration range 1–100 μM, to inhibit FMLP-induced O_2^- generation whereas theophylline, at the same concentrations, enhances the response; in the presence of adenosine deaminase, however, both drugs inhibit the response (Kaneko *et al.*, 1990). Theophylline, in the form of the water-soluble salt, aminophylline (theophylline ethylenediamine), has also been shown to cause a slight enhancement of FMLP-induced neutrophil chemotaxis and O_2^- generation at concentrations below 250 μM while inhibiting these functions at a millimolar concentration (Llewellyn-Jones and Stockley, 1994).

In vivo, intraperitoneal administration of aminophylline to mice causes a marked suppression of pulmonary bactericidal defences: killing of *Staphylococcus aureus*, given by aerosol inhalation, is significantly attenuated by a dose of 80 mg/kg aminophylline whereas *Proteus mirabilis* actually proliferates in the lungs of mice treated

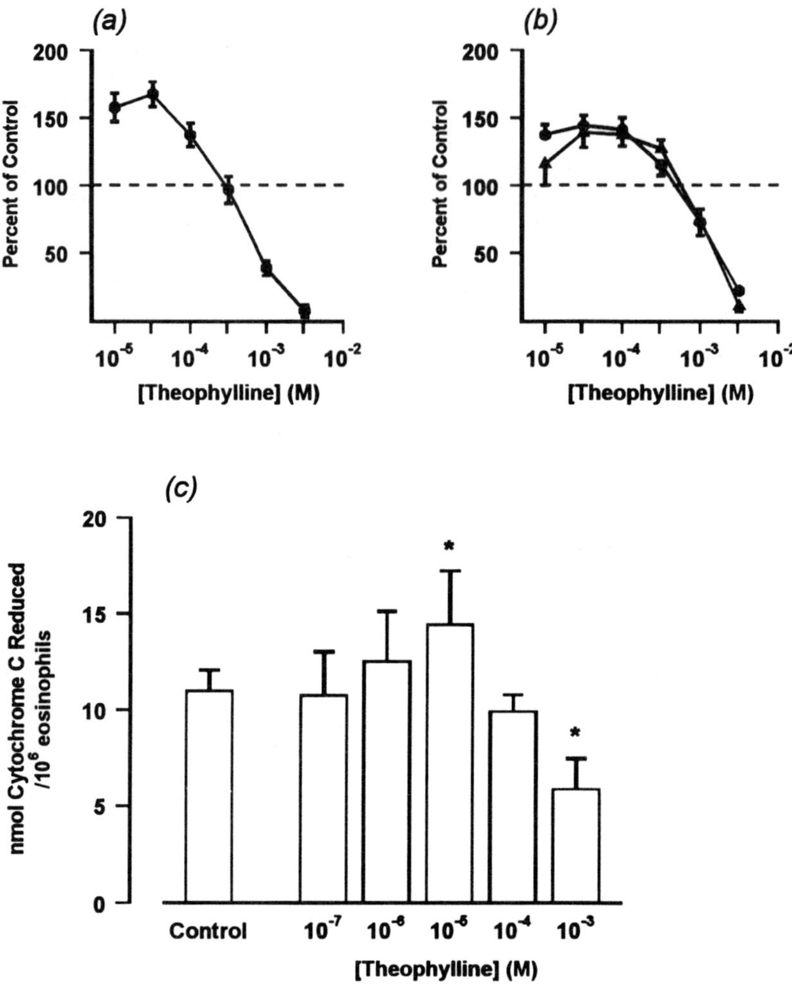

Figure 3.4 Theophylline exerts biphasic effects on human neutrophil and eosinophil functions. Release of (a) superoxide anion (O_2^-) and (b) the lysosomal enzymes β-glucuronidase (●) and lysozyme (▲) from cytochalasin B-treated neutrophils stimulated with FMLP following pre-treatment with increasing concentrations of theophylline. Reproduced, with permission, from Schmeichel and Thomas (1987) © 1987/1995 – The American Association of Immunologists. (c) Release of O_2^- from eosinophils stimulated with opsonized zymosan following pre-treatment with increasing concentrations of theophylline. Reproduced, with permission, from Yukawa *et al.* (1989).

with 40 or 80 mg/kg aminophylline. In parallel with this suppression, the numbers of neutrophils recovered from bronchoalveolar lavage of aminophylline (80 mg/kg)-treated mice at 4 hours after challenge with *P. mirabilis*, are reduced by approximately 65%, compared to control (Nelson *et al.*, 1985). The surmise that aminophylline is exerting a suppressive action on neutrophils is supported by the finding that bronchoalveolar neutrophils from *P. mirabilis*-challenged rats pre-treated with aminophylline display diminished bactericidal activity *ex vivo* when compared with cells from animals that have not been treated with the methylxanthine (Nelson *et al.*, 1985).

Inhibitory effects of theophylline on neutrophil function have also been observed *ex vivo* in cells obtained from the blood of human patients treated with theophylline. Chemotactic responsiveness of neutrophils (and monocytes) from chronic asthmatic children receiving regular oral theophylline is impaired compared to cells obtained from patients whose theophylline had been withdrawn 7 days before experimentation; the recovery of chemotactic responses after theophylline withdrawal was associated in this study with a mean 30% decrease in basal intracellular cAMP levels (Condino-Neto *et al.*, 1991). One week's treatment with oral theophylline, leading to a mean serum concentration of 9.4 μg/ml (approximately 50 μM), causes an increase in basal intracellular cAMP and enhances the ability of isoprenaline to stimulate cAMP accumulation and to inhibit A23187-induced CL (Nielson *et al.*, 1988). The correlation of cAMP levels with suppression of neutrophil function remains unclear, however.

Although IBMX enhances the inhibition of FMLP-induced neutrophil chemotaxis by isoprenaline, prostaglandin E_1 (PGE_1) or the AC activator, forskolin, the degree of elevation in intracellular cAMP following these treatments does not correlate with the degree of functional inhibition. Furthermore, isoprenaline or PGE_1, in the presence of IBMX, inhibits neutrophil chemotaxis in response to LTB_4 whereas forskolin, which causes a substantially greater elevation of cAMP in the presence of IBMX, has no effect on this response (Harvath *et al.*, 1991). IBMX has also been shown to cause a suppression of TNF-α receptor gene expression in neutrophil precursor cells that can not be mimicked by forskolin or by the stable cAMP analogue, N^6,2′-*O*-dibutyryladenosine 3′:5′-cyclic monophosphate (diBu-cAMP) (Lindvall *et al.*, 1990).

Numerous studies have been performed in human neutrophils of the actions of pentoxifylline. Pentoxifylline (Fig. 3.2) is a theobromine derivative that is used therapeutically in the treatment of claudication due to chronic occlusive arterial disease and is effective at doses that do not affect vascular smooth muscle tone, heart rate or cardiac output. Its therapeutic effectiveness is thought to result from haemorheological actions, primarily the increased elasticity of erythrocytes (Rall, 1990). Pentoxifylline is a more potent inhibitor of PDE than theophylline (Cortijo *et al.*, 1993) but is clinically effective at concentrations much lower than those routinely investigated in experiments *in vitro*. This appears to be due to the drug's major metabolites – 1-[5-hydroxyhexyl]-3,7-dimethylxanthine, 1-[4-carboxybutyl]-3,7-dimethylxanthine and 1-[3-carboxypropyl]-3,7-dimethylxanthine – being more potent and effective than pentoxifylline itself in many assays of cell function (Sullivan *et al.*, 1988; Hand *et al.*, 1989; Crouch and Fletcher, 1992).

Pentoxifylline inhibits FMLP-induced polymerization of G-actin to the filamentous F-actin form, although it does not affect the incorporation of actin into the neutrophil cytoskeleton (Rao *et al.*, 1988; Freyburger *et al.*, 1990), inhibits the formation of pseudopodia and decreases the rigidity – and, thereby, the viscosity – of human neutrophils *in vitro* (Armstrong *et al.*, 1990; Wong and Schmid-Schonbein, 1991). These effects of pentoxifylline on microfilament assembly may underlie both the clinical effectiveness of the drug in reducing blood viscosity and its ability to inhibit the phagocytosis by neutrophils of latex or zymosan particles and bacteria (Bessler *et al.*, 1986; Hand *et al.*, 1989). Concentrations of the drug that are effective against these cell functions increase intracellular cAMP but also inhibit adenosine uptake, so that the increase in cAMP cannot be said with certainty to result from PDE inhibition.

Pentoxifylline exhibits a similar range of actions to theophylline and IBMX, inhibiting respiratory burst and degranulation of neutrophils *in vitro* (Hammerschmidt *et al.*, 1988; Boogaerts *et al.*, 1990; Currie *et al.*, 1990; Hoffmann *et al.*, 1991; Oka *et al.*, 1991; Thiel *et al.*, 1991; Zheng *et al.*, 1990, 1991) and *ex vivo* (Crouch and Fletcher, 1992) but has been demonstrated, at high concentrations, to increase the chemotactic responsiveness of neutrophils to FMLP and C5a and to restore the responsiveness of cells from donors with impaired neutrophil chemotaxis resulting from disease (Boogaerts *et al.*, 1990). Perhaps significantly, low concentrations of pentoxifylline suppress priming of neutrophils by platelet activating factor (PAF) to subsequent stimulation by other mediators (Hammerschmidt *et al.*, 1988), so that amplification of inhibitory effects may occur through reduction of both the degree of priming and the resultant response to a subsequent stimulus.

In vivo, pentoxifylline inhibits neutrophil-mediated lung, liver and gastrointestinal injury associated with septicaemia or ischaemia/reperfusion in several species (Welsh *et al.*, 1988; Lilly *et al.*, 1989; Hoffmann *et al.*, 1991; Hewett *et al.*, 1993; Reignier *et al.*, 1994; Santucci *et al.*, 1994) and inhibits the increases in plasma levels of neutrophil elastase, lactoferrin, TNF-α and IL-6, and in bronchoalveolar lavage (BAL) levels of TNF-α and lysozyme, occurring after intravenous bolus injection of *Escherichia coli* endotoxin in chimpanzees or aerosol administration of *Streptococcus pneumoniae* in rabbits, respectively (van Leenen *et al.*, 1993; Mah *et al.*, 1993), without having significant effects on neutrophil numbers in the general, hepatic or gastric circulation or in the BAL (Hewett *et al.*, 1993; van Leenen *et al.*, 1993; Mah *et al.*, 1993; Santucci *et al.*, 1994). Surprisingly, proliferation of *S. pneumoniae* is reduced in the airways of pentoxifylline-treated rabbits, despite the drug's lack of effect on neutrophil migration and its inhibition of some mechanisms thought to be involved in the cells' bactericidal action (Mah *et al.*, 1993; Andres *et al.*, 1995).

3.1.3 Eosinophils

Eosinophils are immune effector cells that are particularly prominent in IgE-dependent reactions, such as parasite killing and allergy (Spry, 1988). Eosinophils occur in elevated numbers in the blood of atopic individuals and are prominent in the late phase inflammatory infiltrates associated with immediate type hypersensitivity reactions, for example following allergen exposure in the skin or lung (Abbas *et al.*, 1991). A large body of evidence exists implicating eosinophils in the pathophysiology of bronchial asthma, a chronic inflammatory disease of the airways (Kroegel, 1990; Moqbel, 1994); this phenomenon is germane to the immunopharmacology of methylxanthines, since theophylline and related drugs have been used in the therapy of asthma for several generations (Rall, 1990; Sullivan *et al.*, 1994b).

As described above for neutrophils, theophylline and IBMX have been demonstrated to exert a range of actions on eosinophils both *in vitro* and *in vivo*. At millimolar concentrations, theophylline decreases the survival time of eosinophils cultured in the presence of IL-5 (Hossain *et al.*, 1994), apparently by inducing apoptosis (Ohta *et al.*, 1994). Such high concentrations, however, are unlikely to be achieved *in vivo* during theophylline therapy, so that the pharmacological significance of these observations is questionable. At concentrations at the upper end of the therapeutic serum concentration range, theophylline has been shown to inhibit partially the chemotaxis of eosinophils in response to a variety of stimuli *in vitro* (Numao *et al.*, 1991) but, within this concentration range, theophylline enhances the generation of O_2^- in response to opsonized zymosan (OZ) particles (Fig. 3.4c). The latter effect, like similar phenomena observed in neutrophils and platelets, is assumed to result from the antagonist action of theophylline at A_2 adenosine receptors, since it can be mimicked by addition of adenosine deaminase or reversed by exogenous adenosine or A_2 receptor agonists (Yukawa *et al.*, 1989). At 1 mM, theophylline exerts an inhibitory action on the OZ-induced O_2^- generation. When oxygen radical generation is measured using luminol-enhanced chemiluminescence (CL), the response induced by the anaphylotoxin C5a is inhibited in a concentration dependent manner by theophylline with an IC_{50} of 525 μM (Hatzelmann *et al.*, 1995); concentrations within the therapeutic range, therefore, cause somewhat less than 50% inhibition of this response. Similarly, inhibition of C5a-induced eosinophil degranulation, measured as the release of eosinophil-derived neurotoxin (EDN) or eosinophil cationic protein (ECP), reaches 50% only at concentrations above 300 μM (Hatzelmann *et al.*, 1995). Leukotriene C_4 release induced by FMLP, however, is inhibited at lower concentrations of theophylline with an IC_{50} of only 50 μM, 6-fold lower than the IC_{50} for PDE inhibition in the same study (Tenor *et al.*, 1995). In contrast to the cell responses listed above, antibody-dependent killing of schistosomes of the parastic trematode helminth *Schistosoma mansoni* by human eosinophils is unaffected by theophylline at concentrations up to 550 μM (Thorne *et al.*, 1988).

IBMX exhibits a similar spectrum of actions to theophylline *in vitro*, inhibiting OZ-stimulated oxygen radical generation in guinea-pig peritoneal eosinophils (IC_{50} = 36 μM) (Dent *et al.*, 1991) and FMLP-stimulated lucigenin-dependent CL and C5a-induced degranulation in human peripheral blood eosinophils (IC_{50} = 16 μM and 50 μM, respectively) (Bray and Mueller, 1994; Hatzelmann *et al.*, 1995). Curiously, much higher concentrations of IBMX – almost identical to the necessary concentrations of theophylline – are required to suppress C5a-stimulated CL (IC_{50} = 524 μM) (Hatzelmann *et al.*, 1995).

Both theophylline and IBMX – as well as other cyclic AMP-elevating agents such as cholera toxin, β-adrenoceptor agonists and prostaglandin E_2 (PGE_2) – inhibit degranulation of eosinophils stimulated with IgG or secretory IgA immunoglobulins (Kita *et al.*, 1991). This suppressive effect is related to the magnitude of increases in intracellular cyclic AMP levels and the co-administration of IBMX with a β-agonist or PGE_2 leads to enhancement of both the rise in cyclic AMP and the inhibition of degranulation. Similarly, theophylline reduces the release of GM-CSF from eosinophils stimulated with IgA-coated sepharose beads (Shute *et al.*, 1995).

In vivo, systemic administration of theophylline reduces the influx of eosinophils to inflammatory sites in the skin and lungs of experimental animals and humans. Substantial reduction in the accumulation of ^{111}In-labelled eosinophils in the skin of sensitized guinea pigs injected intradermally with zymosan-activated plasma (ZAP, a source of C5a), PAF or antigen is observed following treatment with theophylline (Texeira *et al.*, 1994). Eosinophil influx into the bronchoalveolar space of sensitized guinea pigs or rabbits following antigen inhalation is also reduced after theophylline pre-treatment (Sanjar *et al.*, 1990b; Gozzard *et al.*, 1996) although administration of theophylline up to 4 hours after allergen inhalation does not prevent the late phase eosinophil accumulation (Tarayre *et al.*, 1991; Chand *et al.*, 1993). A newer xanthine derivative, isbufylline (Fig. 3.2), which inhibits PDE but has negligible adenosine antagonistic potency (Manzini *et al.*, 1990), also reduces eosinophil recruitment into the airways of sensitized guinea pigs following antigen inhalation (Manzini *et al.*, 1993). Six weeks' treatment with oral theophylline leads to decreased bronchial sub-epithelial eosinophil numbers observed following allergen inhalation in asthmatic humans (Sullivan *et al.*, 1994a); this effect occurs at serum theophylline concentrations below the normal therapeutic range (mean concentration = 6.6 μg/ml or 37 μM) and has been proposed to indicate a significant anti-inflammatory action of theophylline dissociated from the drug's bronchodilator action (Banner and Page, 1995a). It must be noted, however, that suppression by theophylline of airway eosinophil influx following antigen or PAF exposure in animals is associated with a reduction in neither acute bronchoconstriction nor bronchial hyperreactivity (Sanjar *et al.*, 1989, 1990a,b; Gozzard *et al.*, 1996), although isbufylline has been shown to suppress PAF-induced bronchial responsiveness to intravenous histamine in guinea pigs (Manzini *et al.*, 1993).

3.1.4 Monocytes and Macrophages

Mononuclear phagocytes, like neutrophils, effect natural immunity through phagocytosis and intracellular degradation of foreign particles and dead or injured "self" tissues. Monocytes are incompletely differentiated cells, released from the bone marrow into the circulation, that settle in organs or connective tissues and differentiate under the influence of local stimuli, including T lymphocyte-derived chemokines, into macrophages of various types, including central nervous system microglia, hepatic Kupffer cells and alveolar macrophages (Stein and Keshav, 1992; Abbas *et al.*, 1991). Macrophages process and present foreign antigens, leading to stimulation of antigen-specific T lymphocytes, which, in turn, secrete cytokines that activate macrophages for phagocytosis and release of inflammatory mediators and connective tissue-degrading enzymes. Macrophages and monocytes themselves secrete a range of cytokines that may participate in amplification of inflammatory responses at sites of infection, and in angiogenesis and wound healing (Stein and Keshav, 1992). Methylxanthines exert effects on most of these functions.

Theophylline and IBMX have both been demonstrated to increase intracellular cAMP concentrations in human peripheral blood monocytes (Kassis *et al.*, 1989; Nokta and Pollard, 1992) and alveolar macrophages (Hjemdahl *et al.*, 1990; Bachelet *et al.*, 1991), whereas IBMX also enhances isoprenaline-induced increases in macrophage cAMP levels (Hjemdahl *et al.*, 1990). Interestingly, the elevation of cAMP by IBMX in alveolar macrophages *in vitro* is lower in cells obtained from asthmatic patients (Bachelet *et al.*, 1991), suggesting that these cells may have a lower basal AC activity or altered PDE enzymes.

Both the phagocytosis and the presentation of antigenic particles are inhibited by theophylline – which reduces the expression of the class II major histocompatability complex (MHC) molecule, HLA-DR, induced by bacterial lipopolysaccharide (LPS) in human monocytes (McLeish *et al.*, 1987) – and IBMX and pentoxifylline, which reduce the transcription of MHC II genes and expression of MHC II molecules in murine macrophages (Figueiredo *et al.*, 1990; Hecht *et al.*, 1995), whose phagocytic function is also inhibited by theophylline (Hisadome *et al.*, 1989). The attachment of human monocytes to antibody-coated target cells is reduced by theophylline (Herlin and Kragballe, 1982) and antibody-dependent cellular cytotoxicity (ADCC) of human monocytes is also reduced by theophylline and IBMX, although the magnitude of this inhibition does not correlate with the elevation in intracellular cAMP, implying that theophylline may exert an additional action independent of PDE inhibition (Herlin and Kragballe, 1982, 1983). The phagocytosis of protozoan parasites *in vitro* by mouse peritoneal macrophages is also inhibited by theophylline (Wirth and Kierszenbaum, 1982), while bacterial phagocytosis and intracellular killing *ex vivo* are reduced in alveolar macrophages obtained from human subjects treated with oral theophylline for 14 days (O'Neill *et al.*, 1986). The actions of non-selective PDE inhibitors on anti-viral defences are unclear: although IBMX has been shown to accelerate HIV-3 replication in a human monocyte cell line by an apparently cAMP-related mechanism (Nokta and Pollard, 1992), papaverine causes a significant inhibition of the replication of the same virus in the same cell (Nokta *et al.*, 1993) and pentoxifylline decreases HIV-1 replication in peripheral blood mononuclear cells (Fazely *et al.*, 1991). Since TNF-α enhances HIV replication, it is thought that suppression of viral replication by methylxanthines results from the drugs' ability to suppress TNF-α production by monocytes and T lymphocytes (Fazely *et al.*, 1991).

The migration of monocytes to tissues, where they differentiate to macrophages, also appears to be under the influence of cyclic nucleotides and is inhibited by theophylline (Stephens and Snyderman, 1982). This inhibition seems to represent interference with an early stage of cell activation as theophylline reduces chemoattractant-induced polarization of human monocytes and their chemotaxis *in vitro* (Stephens and Snyderman, 1982). Guinea-pig peritoneal macrophages exhibit a similar response, with the increased Ca^{2+} efflux and actin polymerization induced by FMLP being suppressed by theophylline and papaverine (Hamachi *et al.*, 1984). The effect of these drugs on microfilaments appears to be to promote depolymerization of F-actin, since the level of monomeric actin in the cells increases whereas total actin content is unchanged (Hamachi *et al.*, 1984). The expression of the integrin α^M subunit, CD11b (also known as CR3, LeuCAMb or MAC-1), in response to LPS is reduced by theophylline in human monocytes (McLeish *et al.*, 1987), possibly interfering in the migration of monocytes to tissues. *In vivo*, the influx of macrophages to the airways of sensitized guinea pigs challenged by allergen aerosol is also inhibited by pre-treatment of the animals with theophylline (Santing *et al*, 1995), although it is unclear whether this represents a direct effect on monocytes/macrophages or a diminished production of chemotactic factors at the site of challenge.

Whereas theophylline and IBMX display a clear inhibition of the production of reactive oxygen species and arachidonic acid metabolites by human monocytes and alveolar macrophages in response to opsonized particles (Godfrey *et al.*, 1987; Wiik, 1989; Calhoun *et al.*, 1991; Baker and Fuller, 1992; Dent *et al.*, 1994a), theophylline has no effect on granule enzyme release from either murine or human macrophages (Fig. 3.5b) (Ackerman and Beebe, 1975; Baker and Fuller, 1992). Since both arachidonic acid mobilization and oxygen radical generation are mediated by phospholipase A_2 (PLA_2) in other leucocytes (Henderson *et al.*, 1989; White *et al.*, 1993) and PLA_2 activation is suppressed in methylxanthine-treated platelets (Rossignol *et al.*, 1988a,b), it was conceivable that an action upon this enzyme might account for the pharmacological actions of theophylline in mononuclear phagocytes. Theophylline failed to inhibit PLA_2 in monocyte homogenates, however (Godfrey *et al.*, 1987), indicating that the drug must exert its action either through a different

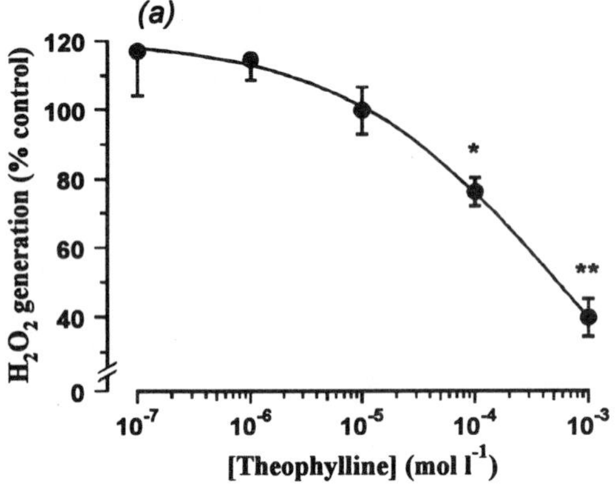

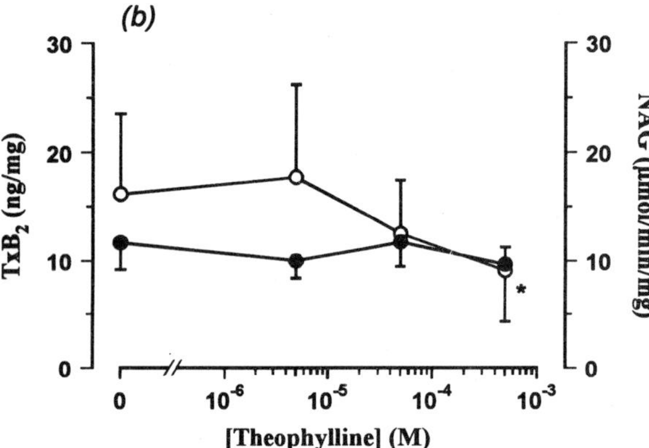

Figure 3.5 Effects of theophylline on human alveolar macrophage function. (a) Theophylline inhibits opsonized zymosan (OZ)-induced respiratory burst in the same range of concentrations that inhibit PDE activity. Reproduced, with permission, from Dent *et al.* (1994a). (b) Theophylline inhibits OZ-induced thromboxane generation (○) at high concentrations but does not affect granule enzyme release (●). Data taken from Baker and Fuller (1992). * $p < 0.05$, ** $p < 0.01$, compared to control.

pathway or at an earlier or later stage of the cell activation process.

The suppression of human alveolar macrophage respiratory burst appears to be primarily – if not exclusively – mediated by PDE inhibition since the inhibition of the cell function by theophylline is unaffected by addition of adenosine deaminase and is largely blocked by a selective inhibitor of PKA (Dent *et al.*, 1994a). The participation of inhibitory A_2 adenosine receptors in regulation of opsonized zymosan-stimulated generation of reactive oxygen species by macrophages seems to be less pronounced than in regulation of the same response in neutrophils and eosinophils, since elevation of alveolar macrophage H_2O_2 production by lower concentrations of theophylline is small and not statistically significant (Fig. 3.5a).

Cytokine production by monocytes and macrophages in response to bacterial LPS has been widely reported to be inhibited by methylxanthines. Release of TNF-α, as well as transcription of its gene, is suppressed by theophylline and IBMX in human and rat monocytes and in rat alveolar macrophages (Bailly *et al.*, 1990; Spatafora *et al.*, 1994) whereas pentoxifylline also reduces LPS-induced TNF-α production by cultured monocytes and decreases circulating TNF-α levels in humans during endotoxaemia (Zabel *et al.*, 1993). IBMX has been reported to inhibit IL-1 release from human monocytes and a mouse peritoneal macrophage cell line (Kassis *et al.*, 1989; Brandwein, 1986), although other workers have found no effect of the drug on release of IL-1α or IL-1β from monocytes (Bailly *et al.*, 1990). In contrast, a small but significant stimulation of IL-6 production has been demonstrated in human monocytes treated with IBMX or pentoxifylline at concentrations that inhibit LPS-induced TNF-α production (Bailly *et al.*, 1990; Schandené *et al.*, 1992), a finding that concurs with reports that production of IL-6 – which is induced by the AC-activating receptor agonist, PGE_1 – is increased by theophylline in a murine monocyte cell line (Dendorfer *et al.*, 1994).

That theophylline, at concentrations as low as 10 μM, should be effective in suppressing LPS-induced TNF-α production by human blood monocytes (Spatafora *et al.*, 1994) is interesting since this response is also inhibited potently by A_2 adenosine receptor agonists (Prabhakar *et al.*, 1995). It appears likely, therefore, that endogenous adenosine does not play a major role in regulating the response and that theophylline exerts its actions through PDE inhibition.

In addition to cytokines and connective tissue-degrading enzymes, monocytes also express tissue factor (TF), a transmembrane factor that enables activated monocytes to initiate blood coagulation cascades (Ollivier *et al.*, 1993). The induction of TF expression in human monocytes by LPS is inhibited by pentoxifylline at concentrations which elevate intracellular cAMP levels (de Prost *et al.*, 1990, 1992) whereas IBMX has been shown to suppress the LPS-induced increase in TF mRNA in the same cells (Ollivier *et al.*, 1993).

Finally, monocyte-dependent angiogenesis is also sensitive to methylxanthines: the *in vitro* neovascularization induced by monocytes obtained from diabetic patients with proliferative retinopathy is suppressed by both theophylline and theobromine (Skopinski *et al.*, 1993).

3.1.5 Mast Cells and Basophils

Mast cells and basophils bear high affinity receptors for the Fc portion of IgE and, therefore, are involved in immediate allergic responses. Mast cells are resident in tissues and occur in high numbers at surfaces, such as the skin and the respiratory and gastrointestinal mucosa, where exposure to environmental antigens occurs. A distinct population of "connective tissue" mast cells, differing in the profile of inflammatory mediators they produce and the predominant serine proteases and acid proteoglycans contained in their granules, is found in lung parenchyma and the serosa of body cavities. Basophils, which appear to derive from the same precursor cells as eosinophils, are end-stage circulating leucocytes and are found in tissues only when recruited into inflammatory sites (Abbas *et al.*, 1991).

The majority of studies of the pharmacology of mast cells have employed cells from the abdominal cavity of rats. Two pharmacologically distinct populations of mast cells appear to be represented: "connective tissue" cells from the peritoneal cavity exhibit responses – usually histamine release – upon receptor-mediated immunological stimulation with anti-IgE antibodies, antigen or lectins such as Con A, as well as non-immunological stimulation with basic peptides (e.g. compound 48/80) and calcium ionophores; cells from the intestinal mucosa respond strongly to immunological stimulation, weakly to ionophores and not at all to basic peptides (Befus *et al.*, 1982). This heterogeneity extends to the effectiveness of drugs in inhibiting mast cell responses. Theophylline inhibits IgE-mediated responses in peritoneal (Holgate *et al.*, 1981; Goto *et al.*, 1982; Pearce *et al.*, 1982; Sydbom and Fredholm, 1982; Ennis *et al.*, 1983; White *et al.*, 1984; Chand *et al.*, 1985; Schick and Austen, 1985; Shanahan *et al.*, 1986; Tomioka *et al.*, 1989) but not in mucosal mast cells (Pearce *et al.*, 1982; Shanahan *et al.*, 1986; Tomioka *et al.*, 1989). Histamine release from rat peritoneal mast cells stimulated with 48/80 (Sullivan *et al.*, 1975; Ishizaka *et al.*, 1981), calcium ionophore (Truneh and Pearce, 1984) or orthovanadate (al-Laith and Pearce, 1989) is also inhibited by theophylline, which additionally suppresses 48/80- or substance P-induced Ca^{2+} mobilization in these cells (Tasaka *et al.*, 1986a,b), as well as the mobilization induced by antigen (White *et al.*, 1984). Antigen-induced histamine release from guinea-pig lung and mesenteric mast cells is inhibited by theophylline (Pearce *et al.*, 1978; Lau *et al.*, 1994) whereas ionophore responses require higher theophylline concentrations for inhibition (Pearce *et al.*, 1978).

That the suppressive actions of theophylline on cell function are mediated by PDE inhibition is suggested by the elevation of intracellular cAMP levels and the consequent activation of PKA evoked by concentrations of theophylline that inhibit anti-IgE- or 48/80-induced histamine release from rat peritoneal mast cells (Sullivan *et al.*, 1975; Holgate *et al.*, 1981; Ishizaka *et al.*, 1981). Theophylline appears to exert an additional effect via adenosine receptor antagonism, since adenosine enhances histamine release in response to anti-IgE and this enhancement is blocked by lower concentrations of theophylline than those which inhibit the anti-IgE response itself (Sydbom and Fredholm, 1982).

In contrast to those from rodents, mucosal mast cells from primates appear to be sensitive to theophylline, which inhibits anti-IgE- or Con A-induced degranulation of mast cells from the intestinal mucosa of monkeys (Barrett and Metcalfe, 1985) and humans (Fox *et al.*, 1988; Liu *et al.*, 1991).

Mast cells recovered from parenchymal tissues of rodents are of the "connective tissue" type, resembling rat peritoneal mast cells in their pharmacology. Immunologically evoked mediator release from rat or guinea-pig lung and mouse spleen mast cells can be inhibited by theophylline or IBMX (Ennis *et al.*, 1983; Barrett *et al.*, 1984; Tomioka *et al.*, 1989; Undem *et al.*, 1990). Both IgG-mediated (actively sensitized guinea-pig lung) and IgE-mediated (actively sensitized rat peritoneal mast cells) mediator release are sensitive to inhibition by IBMX (Ennis *et al.*, 1983).

Most investigations of human mast cell pharmacology have been undertaken using chopped skin or lung fragments. These preparations exhibit histamine release in response to anti-IgE that is suppressed by theophylline and IBMX (Peters *et al.*, 1982; Fox *et al.*, 1988; Peachell *et al.*, 1988; Louis and Radermecker, 1990; Louis *et al.*, 1992). In contrast to rodent "connective tissue" mast cells, the cells from human lung are poorly responsive to peptide secretagogues, whereas skin cells respond strongly to 48/80 and substance P (Louis *et al.*, 1992; Schulman *et al.*, 1988). Human lung mast cells display heterogenous density when isolated by centrifugation over Percoll density gradients however, corresponding to varying granule histamine content, and the responsiveness of the cells to immunological stimuli and secretagogues varies among the subpopulations. Inhibition of anti-IgE-induced histamine release also varies among these separate populations, ranging from 30% inhibition by 300 μM IBMX to 60% (Schulman *et al.*, 1988).

Theophylline was demonstrated, more than twenty years ago, to elevate intracellular cAMP levels and to suppress antigen-induced release of histamine and PAF from passively sensitized rabbit basophils (Bussolino and Benveniste, 1980) and similar effects have been observed in human cells in the intervening decades. Theophylline inhibits histamine release from human basophils stimulated by anti-IgE (Spirer *et al.*, 1979; Louis and Radermecker, 1990; Louis *et al.*, 1992; Peachell *et al.*, 1992), allergen or complement factors (Grant *et al.*, 1975, 1976), and substance P or 48/80 (Louis and Radermecker, 1990), whereas IBMX has also been observed to suppress histamine and LTC_4 release induced by anti-IgE or allergen (Marone *et al.*, 1981; Warner *et al.*, 1988; Peachell *et al.*, 1992). Both theophylline and enprofylline inhibit human basophil PDE activity and suppress immunologically evoked histamine release, with enprofylline exhibiting higher potency for both actions (Toll and Andersson, 1984). The concentration dependence of intracellular cAMP elevation by IBMX or theophylline is very similar to that for inhibition of histamine release (Peachell *et al.*, 1992).

Basophils obtained from the peripheral blood of untreated asthmatic patients exhibit a spontaneous release of histamine *in vitro* that is higher than that observed with basophils from healthy or chronically medicated asthmatic subjects; this spontaneous degranulation is also inhibited by theophylline (Akagi *et al.*, 1989).

The eosinophil granule protein MBP, which is released into the extracellular medium at sites of eosinophilic inflammation, induces a Ca^{2+}/CaM-dependent degranulation of human basophils that can be inhibited by theophylline (Thomas *et al.*, 1989). Similarly, histamine release from basophils stimulated with PAF is inhibited by IBMX (Columbo *et al.*, 1993). The ability of these drugs both to inhibit mediator release from inflammatory cells and to reduce their subsequent actions on other cells may be important in the study of the modulation of inflammatory conditions by PDE inhibitors. The influence of PDE inhibitors on cytokine secretion from mast cells has not been studied in detail although IBMX has been demonstrated to enhance potently the release of IL-3 from anti-IgE-stimulated PB-3c murine mast cells, apparently by increasing the half-life of the cytokine's mRNA (Hahn and Moroni, 1994).

3.1.6 Platelets

The contribution of platelets to immune responses remains unclear, but their expression of the low affinity receptor for IgE (FcεRII or CD23) implies a potential for involvement in both parasite resistance and allergy (Capron *et al.*, 1986). Platelets are a source of several factors that can affect other immune cells (Van Damme, 1991), as well as of PAF and other lipid mediators of inflammation.

Platelets differ from most immune cells in having little PDE4 activity; the predominant isoenzymes in platelets are of the PDE3 and PDE5 families (see Chapter 2), although their precursor megakaryocytes do contain high levels of PDE4 (Akaike *et al.*, 1993). The non-selective methylxanthine PDE inhibitors exhibit similar suppressive actions on platelets to those observed in other immune cells, although some of these actions may be mediated by cGMP – the preferred substrate for PDE5 – rather than cAMP (Williams *et al.*, 1988). IBMX, for example, inhibits human platelet functions to a greater extent than the selective PDE3 inhibitor, SKF 94120, at concentrations of the two drugs that cause similar increases in cyclic AMP levels (Simpson *et al.*, 1988) whereas IBMX also enhances the inhibitory actions of the nitric oxide donors, sodium nitroprusside and 3-morpholinosydnonimine (SIN-1), which act through activation of soluble GC leading to elevation of cGMP levels (Salvemini *et al.*, 1990).

High concentrations of theophylline and IBMX inhibit aggregation of human, bovine, canine, rabbit, rat and guinea-pig platelets in response to several stimuli, including thrombin, collagen, ADP, arachidonic acid, thromboxanes and PAF (Tsien *et al.*, 1982; Cox *et al.*, 1984; Kitagawa *et al.*, 1984; Vinge *et al.*, 1984; Simpson *et al.*, 1988; Salvemini *et al.*, 1990; Misso and Thompson, 1992). Enprofylline and pentoxifylline have also been demonstrated to inhibit thrombin- and PAF-stimulated aggregation of human platelets (Misso and Thompson, 1992; Rossignol *et al.*, 1988a,b).

The involvement of adenosine in the regulation of platelet function is similar to that in neutrophils, eosinophils and monocytes (Vinge *et al.*, 1984). PAF-induced aggregation of human platelets in platelet-rich plasma (PRP) could be enhanced by the elimination of plasma adenosine (Agarwal *et al.*, 1994) whereas dipyridamole, which – in addition to inhibiting PDE5 – increases extracellular adenosine concentrations by blocking cellular reuptake of the nucleoside, causes an inhibition of platelet aggregation that is reversed by adenosine deaminase (Gresele *et al.*, 1986). The adenosine-mediated inhibition of aggregation is partially blocked by theophylline so that, in the presence of dipyridamole, theophylline enhances platelet aggregation (Gresele *et al.*, 1986). In fact, lower concentrations of theophylline, such as those occurring in the plasma of patients receiving therapeutic doses of the drug, also enhance PAF-induced platelet aggregation in PRP containing adenosine (Agarwal *et al.*, 1994). Enprofylline, whose potency as an antagonist at adenosine A_2 receptors is low relative to that of theophylline, is approximately five to tenfold more potent than theophylline in inhibiting ADP- and PAF-induced platelet aggregation. Furthermore, enprofylline increases the inhibition of aggregation by dipyridamole or adenosine under conditions where theophylline reverses the inhibition (Vinge *et al.*, 1984; Misso and Thompson, 1992).

In addition to aggregation, methylxanthine PDE inhibitors also suppress platelet ATP and 5-hydroxytryptamine (5-HT) secretion (Friedman and Detwiler, 1975; Rossignol *et al.*, 1988a,b) and adherence of platelets to vascular endothelium (Hammerschmidt *et al.*, 1988). The drugs' actions on secretion appear to involve inhibition of an early stage in signal transduction since responses to receptor-mediated stimuli are inhibited by theophylline whereas those to Ca^{2+} ionophores are not (Friedman and Detwiler, 1975). Pentoxifylline has been demonstrated to inhibit activation of phospholipases A_2 and C in response to thrombin (Rossignol *et al.*, 1988a,b), suggesting the coupling of receptors to these intracellular effectors as possible targets for cyclic nucleotide-mediated suppression of platelet function.

3.2 OTHER CELLS

Alkylxanthines exert actions in cells apart from those of the immune system. Although some of these actions, such as bronchodilation and increased cardiac contractility, may be of therapeutic benefit, others may be involved in the undesired effects of drugs such as theophylline.

3.2.1 Smooth Muscle

Methylxanthines relax airway and vascular smooth muscle and, as mentioned earlier (section 2.1), a relationship can be observed between PDE inhibition by methylxanthines and the relaxation of airways smooth muscle. When applied to human bronchial rings *in vitro*, theophylline, IBMX and pentoxifylline cause concentration-dependent relaxations in the same range of concentrations that inhibits PDE-catalysed cAMP hydrolysis. In this preparation no role can be demonstrated for adenosine receptor antagonism in the relaxation induced by the methylxanthines. A similar pattern of PDE inhibition and smooth muscle relaxation is observed in human pulmonary arteries (reviewed in Dent *et al.*, 1994b; Rabe *et al.*, 1995). Theophylline can relax vascular smooth muscle in the absence of endothelium, suggesting a participation of cAMP as well as the stimulated cGMP resulting from endothelial NO release (Marukawa *et al.*, 1994).

Proliferation of rat vascular smooth muscle cells is also inhibited by IBMX and papaverine in the same concentration range in which the drugs inhibit PDE (Pan *et al.*, 1994). This action is mimicked by selective PDE3 and PDE4 inhibitors, suggesting that PDE inhibition is likely to be the mechanism through which theophylline exerts its effect.

3.2.2 Cardiac Muscle

Elevation of cAMP levels in cardiac muscle increases the force of muscle contraction (see Chapter 6). As described in Chapter 5, however, cAMP and cGMP exert opposite actions in the heart, largely through the stimulation of cardiac PDE2 by cGMP; probably as a result of this phenomenon, non-selective PDE inhibitors have variable and weak effects on the heart *in vivo*. The combination of a small net increase in force of contraction combined with a decreased pre-load resulting from vasodilation leads to a modest increase in cardiac output of patients treated with intravenous aminophylline (Ogilvie *et al.*, 1977).

Theophylline also exhibits chronotropic effects, particularly sinus tachycardia (Sullivan *et al.*, 1994b), which may result from the elevation of cytosolic cAMP in cardiac muscle and the resulting calcium overload (Lubbe *et al.*, 1992) or as a reflex reaction to peripheral vasodilation (Esquivel *et al.*, 1986). Since adenosine has a negative chronotropic action, however, the adenosine antagonistic action of theophylline may contribute to the effects of theophylline on heart rate (Sullivan *et al.*, 1994b).

3.2.3 Renal Tubules

Theophylline has long been recognized as a diuretic, an action that results at least in part from increased renal blood flow and glomerular filtration rate. However, theophylline also exerts direct actions on the renal tubules, leading to increased excretion of Na^+ and Cl^- without any significant change in urinary H^+ concentration (Weiner, 1990), apparently through an inhibition of solute resorption (Brater *et al.*, 1983). *In vivo*, theophylline treatment can cause electrolyte disturbances, with hypokalaemia being a particular problem (Sigurd and Olesen, 1978). The diuretic actions of theophylline have been associated with the drug's antagonism of adenosine receptors (Fredholm, 1984), although the substitution of drugs displaying weaker adenosine antagonism may not reduce the incidence of electrolyte disturbances (Persson, 1986).

3.2.4 Central Nervous System

Theophylline and caffeine stimulate the CNS at relatively low doses *in vivo*, since the drugs are concentrated in the cerebrospinal fluid (Sullivan *et al.*, 1994b). Actions of methylxanthines in the CNS include a beneficial increase in ventilatory drive and a general increase in alertness and lessening of fatigue but also account for the risk of tremor and seizures during theophylline therapy. Some of the CNS actions of theophylline, such as the reversal of benzodiazepine sedation, appear to be mediated by adenosine receptor antagonism (Sullivan *et al.*, 1994b) whereas others, such as the induction of nausea and vomiting, are mimicked by selective PDE4 inhibitors and apparently result from the inhibition of PDE in the emesis centres of the brain (Palfreyman, 1995).

3.2.5 Pancreas

Phosphodiesterases of the PDE3 type in the liver are subject to stimulation by insulin (Houslay and Kilgour, 1990; see also Chapter 6). It is noteworthy that the pancreas also contains a cGMP-inhibited PDE activity which is inhibited by theophylline (Lemon and Bhoola, 1975) and the inhibition of which enhances insulin release from the islets of Langerhans (Montague and Howell, 1975). Thus, theophylline can both stimulate release of insulin and inhibit its action in hepatocytes. The significance of these antagonistic actions remains unclear.

The exocrine function of the pancreas is also affected by methylxanthine PDE inhibitors: IBMX, as well as selective PDE4 inhibitors, causes a concentration-dependent, cyclic AMP-mediated increase in pancreatic juice secretion in dogs (Iwatsuki *et al.*, 1991).

4. Summary and Directions for Future Research

Theophylline and other methylxanthines exert inhibitory actions on a variety of immune cells, particularly on those functions that are involved in inflammatory processes. Most of these actions can be shown to be related to the elevation of intracellular cyclic nucleotide concentrations resulting from the inhibition of PDE. Theophylline, IBMX and pentoxifylline inhibit different PDE isoenzymes with similar potencies in a variety of cells (Table 3.1), so that the actions of the drugs on cell functions do not reflect the isoenzyme profile in the cell.

Recent demonstrations of decreased inflammation in the airways of asthmatic patients treated chronically with low doses of theophylline have extended the *in vitro* findings of suppression of immune cell function to a disease state and interest has been re-awakened in PDE inhibitors as therapeutic agents in the treatment of inflammatory conditions. Although effects can be observed with doses of theophylline that produce low serum concentrations, it is likely that higher concentrations and a greater degree of inhibition of the target PDE isoenzymes will lead to greater suppression of inflammation. On these grounds, selective inhibitors, which can be used at doses causing a large degree of inhibition of a single isoenzyme family without significant non-specific actions (such as those listed in section 3.2), are viewed as being of greater potential therapeutic benefit. Since some, if not all, side-effects of theophylline are mediated at least in part by PDE inhibition, the search for alkylxanthines with lower potency as adenosine receptor antagonists in an attempt to minimize side-effects may prove fruitless. Thus, isoenzyme-selective drugs appear to present a more desirable target.

The PDE isoenzyme complements of many tissues and cells have been characterized, as described in detail in the other chapters of this volume. The identification of restricted PDE expression in specific cells has raised the possibility of developing drugs that can be directed against those cells and, thereby, exert suppressive effects where desired without causing undesired effects in other cells. Although the ubiquity of most isoenzyme families (see Chapters 1 and 2) may preclude highly specific therapeutic actions of inhibitors of individual families, the recent identification of drugs that selectively inhibit the enzyme activity of a single PDE protein or a particular state of a PDE (see Chapter 11) might indicate that drugs developed in the future will allow highly specific targeting of PDE inhibitor actions, so that the actions of PDE inhibitors on immune cells can be of significant value in the treatment of inflammatory diseases.

5. *References*

Abbas, A.K., Lichtman, A.H. and Pober, J.S. (1991). Cellular and Molecular Immunology. W.B. Saunders, Philadelphia.

Ackerman, N.R. and Beebe, J.R. (1975). Effects of pharmacologic agents on release of lysosomal enzymes from alveolar mononuclear cells. J. Pharmacol. Exp. Ther. 193, 603–613.

Agarwal, K.C., Clarke, E., Rounds, S., Parks, R.E., Jr and Huzoor-Akbar. (1994). Platelet-activating factor (PAF)-induced platelet aggregation: modulation by plasma adenosine and methylxanthines. Biochem. Pharmacol. 48, 1909–1916.

Akagi, K., Kohi, F., Trivedi, R. and Townley, R.G. (1989). Pharmacologic modulation of spontaneous histamine release. Ann. Allergy 63, 39–46.

Akaike, N., Uneyama, H., Kawa, K. and Yamashita, Y. (1993). Existence of rolipram-sensitive phosphodiesterase in rat megakaryocyte. Br. J. Pharmacol. 109, 1020–1023.

al-Laith, M. and Pearce, F.L. (1989). Some further characteristics of histamine secretion from rat mast cells stimulated with sodium orthovanadate. Agents Actions 27, 65–67.

Anderson, G.P. and Coyle, A.J. (1994). T_H2 and "T_H2-like" cells in allergy and asthma: pharmacological perspectives. Trends Pharmacol. Sci. 15, 324–332.

Andres, D.W., Kutkoski, G J., Quinlan, W.M., Doyle, N.A. and Doerschuk, C.M. (1995). Effect of pentoxifylline on changes in neutrophil sequestration and emigration in the lungs. Am. J. Physiol. 268, L27-L32.

Armstrong, M., Jr, Needham, D., Hatchell, D.L. and Nunn, R.S. (1990). Effect of pentoxifylline on the flow of polymorphonuclear leukocytes through a model capillary. Angiology 41, 253–262.

Bachelet, M., Vincent, D., Havet, N., Marrash-Chahla, R., Pradalier, A., Dry, J. and Vargaftig, B.B. (1991). Reduced responsiveness of adenylate cyclase in alveolar macrophages from patients with asthma. J. Allergy Clin. Immunol. 88, 322–328.

Bailly, S., Ferrua, B., Fay, M. and Gougerot-Pocidalo, M.A. (1990). Differential regulation of IL 6, IL 1 A, IL 1 beta and TNF alpha production in LPS-stimulated human monocytes: role of cyclic AMP. Cytokine 2, 205–210.

Baker, A.J. and Fuller, R.W. (1992). Effect of cyclic adenosine monophosphate, 5′-(N-ethylcarboxamido)-adenosine and methylxanthines on the release of thromboxane and lysosomal enzymes from human alveolar macrophages and peripheral blood monocytes in vitro. Eur. J. Pharmacol. 211, 157–161.

Banner, K.H. and Page, C.P. (1995a). Theophylline and selective phosphodiesterase inhibitors as anti-inflammatory drugs in the treatment of bronchial asthma. Eur. Respir. J. 8, 996–1000.

Banner, K.H. and Page, C.P. (1995b). Immunomodulatory actions of xanthines and isoenzyme selective phosphodiesterase inhibitors. Monaldi Arch. Chest Dis. 50, 286–292.

Barrett, K.E. and Metcalfe, D.D. (1985). The histologic and functional characterization of enzymatically dispersed intestinal mast cells of nonhuman primates: effects of secretagogues and anti-allergic drugs on histamine secretion. J. Immunol. 135, 2020–2026.

Barrett, K.E., Pluznik, D.H. and Metcalfe, D.D. (1984). Histamine release from the cultured mouse mast cell line PT18 in response to immunologic and non-immunologic stimuli. Agents Actions 14, 488–493.

Beavo, J.A. (1988). Multiple isozymes of cyclic nucleotide phosphodiesterase. Adv. Second Messenger Phosphoprotein Res. 22, 1–38.

Befus, A.D., Pearce, F.L., Goodacre, R. and Bienenstock, J. (1982). Unique functional characteristics of mucosal mast cells. Adv. Exp. Med. Biol. 149, 521–527.

Bemelmans, M.H.A., Abramowicz, D., Gouma, D.J., Goldman, M. and Buurman, W.A. (1994). In vivo T cell activation by anti-CD3 monoclonal antibody induces soluble TNF receptor release in mice: effects of pentoxifylline, methylprednisolone, anti-TNF, and anti-IFN-γ antibodies. J. Immunol. 153, 499–506.

Bergstrand, H. (1980). Phosphodiesterase inhibition and theophylline. Eur. J. Respir. Dis. 61 (Suppl. 109), 37–44.

Bessler, H., Gilgal, R., Djaldetti, M. and Zahavi, I. (1986). Effect of pentoxifylline on the phagocytic activity, cAMP levels, and superoxide anion production by monocytes and polymorphonuclear cells. J. Leukoc. Biol. 40, 747–754.

Bessler, H., Gilgal, R., Zahavi, I. and Djaldetti, M. (1987). Effect of pentoxifylline on E-rosette formation and on the mitogenic response of human mononuclear cells. Biomed. Pharmacother. 41, 439–441.

Boogaerts, M.A., Malbrain, S., Meeus, P., van Hove, L. and Verhoef, G.E. (1990). *In vitro* modulation of normal and diseased human neutrophil function by pentoxifylline. Blut 61, 60–65.

Brandwein, S.R. (1986). Regulation of interleukin 1 production by mouse peritoneal macrophages: effects of arachidonic acid metabolites, cyclic nucleotides, and interferons. J. Biol. Chem. 261, 8624–8632.

Brater, D.C., Kaojaren, S. and Chennavasin, P. (1983). Pharmacodynamics of the diuretic effects of aminophylline and acetazolamide alone and combined with furosemide in normal subjects. J. Pharmacol. Exp. Ther. 227, 92–97.

Bray, K.M. and Mueller, T. (1994). Inhibition of phosphodiesterase type IV attenuates the oxidative burst in human isolated eosinophils. Br. J. Pharmacol. 113, 88P. [Abstract]

Bruynzeel, I., van der Raaij, L.M., Stoof, T.J. and Willemze, R. (1995). Pentoxifylline inhibits T-cell adherence to keratinocytes. J. Invest. Dermatol. 104, 1004–1007.

Buckle, D.R., Arch, J.R.S., Connolly, B.J., Fenwick, A.E., Foster, K.A., Murray, K.J., Readshaw, S.A., Smallridge, M. and Smith, D.G. (1994). Inhibition of cyclic nucleotide phosphodiesterase by derivatives of 1,3-bis(cyclopropylmethyl)xanthine. J. Med. Chem. 37, 476–485.

Bussolino, F. and Benveniste, J. (1980). Pharmacological modulation of platelet-activating factor (PAF) release from rabbit leucocytes. I. Role of cAMP. Immunology 40, 367–376.

Butcher, R.W. and Sutherland, E.W. (1962). Adenosine-3′,5′-phosphate in biological materials. I. Purification and properties of cyclic 3′,5′- nucleotide phosphodiesterase and use of this enzyme to characterize adenosine 3′,5′-phosphate in human urine. J. Biol. Chem. 237, 1244–1250.

Calhoun, W.J., Stevens, C.A. and Lambert, S.B. (1991). Modulation of superoxide production of alveolar macrophages and peripheral blood mononuclear cells by β-agonists and theophylline. J. Lab. Clin. Med. 117, 514–522.

Capron, M., Jounalt, T., Prin, L., Joseph, M., Ameisen, J., Butterworth, A.E., Papin, J., Kunsnierz, J. and Capron, A. (1986). Functional study of a monoclonal antibody to IgE Fc receptor (FcεR2) of eosinophils, platelets and macrophages. J. Exp. Med. 164, 72–89.

Chand, N., Pillar, J., Diamantis, W. and Sofia, R.D. (1985). Inhibition of IgE-mediated allergic histamine release from rat peritoneal mast cells by azelastine and selected antiallergic drugs. Agents Actions 16, 318–322.

Chand, N., Harrison, J.E., Rooney, S.M., Nolan, K.W., DeVine, C.L., Jakubicki, R.G., Pillar, J., Diamantis, W. and Sofia, R.D. (1993). Allergic bronchial eosinophilia: a therapeutic approach for the selection of potential bronchial anti-inflammatory drugs. Allergy 48, 624–626.

Cohen, D.P. and Rothstein, T.L. (1989). Adenosine 3′,5′-cyclic monophosphate modulates the mitogenic responses of murine B lymphocytes. Cell Immunol. 121, 113—124.

Columbo, M., Horowitz, E.M., McKenzie-White, J., Kagey-Sobotka, A. and Lichtenstein, L.M. (1993). Pharmacologic control of histamine release from human basophils induced by platelet-activating factor. Int. Arch. Allergy Immunol. 102, 383–390.

Condino-Neto, A., Vilela, M.M., Cambiucci, E.C., Ribeiro, J.D., Guglielmi, A.A., Magna, L.A. and De Nucci, G. (1991). Theophylline therapy inhibits neutrophil and mononuclear cell chemotaxis from chronic asthmatic children. Br. J. Clin. Pharmacol. 32, 557–561.

Cortijo, J., Bou, J., Beleta, J., Cardelús, I., Llenas, J., Morcillo, E. and Gristwood, R.W. (1993). Investigation into the role of phosphodiesterase IV in bronchorelaxation, including studies with human bronchus. Br. J. Pharmacol. 108, 562–568.

Coskey, L.A., Bitting, J. and Roth, M.D. (1993). Inhibition of natural killer cell activity by therapeutic levels of theophylline. Am. J. Respir. Cell Mol. Biol. 9, 659–665.

Cox, C.P., Linden, J. and Said, S.I. (1984). VIP elevates platelet cyclic AMP (cAMP) levels and inhibits *in vitro* platelet activation induced by platelet-activating factor (PAF). Peptides 5, 325–328.

Crosti, F., Secchi, A., Ferrero, E., Falqui, L., Inverardi, L., Pontiroli, A.E., Ciboddo, G.F., Pavoni, D., Protti, P. and Rugarli, C. (1986). Impairment of lymphocyte-suppressive system in recent-onset insulin-dependent diabetes mellitus: correlation with metabolic control. Diabetes 35, 1053–1057.

Crouch, S.P. and Fletcher, J. (1992). Effect of ingested pentoxifylline on neutrophil superoxide anion production. Infect. Immun. 60, 4504–4509.

Currie, M.S., Rao, K.M., Padmanabhan, J., Jones, A., Crawford, J. and Cohen, H.J. (1990). Stimulus-specific effects of pentoxifylline on neutrophil CR3 expression, degranulation, and superoxide production. J. Leukoc. Biol. 47, 244–250.

Dendorfer, U., Oettgen, P. and Libermann, T.A. (1994). Multiple regulatory elements in the interleukin-6 gene mediate induction by prostaglandins, cyclic AMP, and lipopolysaccharide. Mol. Cell Biol. 14, 4443–4454.

Dent, G., Giembycz, M.A., Rabe, K.F. and Barnes, P.J. (1991). Inhibition of eosinophil cyclic nucleotide PDE activity and opsonised zymosan-stimulated respiratory burst by "type IV"-selective PDE inhibitors. Br. J. Pharmacol. 103, 1339–1346.

Dent, G., Giembycz, M.A., Rabe, K.F., Wolf, B., Barnes, P.J. and Magnussen, H. (1994a). Theophylline suppresses human alveolar macrophage respiratory burst through phosphodiesterase inhibition. Am. J. Respir. Cell Mol. Biol. 10, 565–572.

Dent, G., Magnussen, H. and Rabe, K.F. (1994b). Cyclic nucleotide phosphodiesterases in the human lung. Lung 172, 129–146.

de Prost, D., Ollivier, V. and Hakim, J. (1990). Pentoxifylline inhibition of procoagulant activity generated by activated mononuclear phagocytes. Mol. Pharmacol. 38, 562–566.

de Prost, D., Ollivier, V. and Hakim, J. (1992). Inhibition par la pentoxifylline de l'activité procoagulante produite par les monocytes activés par l'endotoxine. C.R. Soc. Biol. (Paris) 186, 186–192.

Dianzani, C., Brunelleschi, S., Viano, I. and Fantozzi, R. (1994). Adenosine modulation of primed human neutrophils. Eur. J. Pharmacol. 263, 223–226.

Ennis, M., Robinson, C. and Dollery, C.T. (1983). Action of 3-isobutyl-1-methylxanthine and prostaglandins D_2 and E_1 on histamine release from rat and guinea pig mast cells. Int. Arch. Allergy Appl. Immunol. 72, 289–293.

Esquivel, M., Burns, R.J. and Ogilvie, R.I. (1986). Cardiovascular effects of enprofylline and theophylline. Clin. Pharmacol. Ther. 39, 395–402.

Fazely, F., Dezube, B.J., Allen-Ryan, J., Pardee, A.B. and Ruprecht, R.M. (1991). Pentoxifylline (Trental) decreases the replication of the human immunodeficiency virus type 1 in human peripheral blood mononuclear cells and in cultured T cells. Blood 77, 1653–1656.

Figueiredo, F., Uhing, R.J., Okonogi, K., Gettys, T.W., Johnson, S.P., Adams, D.O. and Prpic, V. (1990). Activation of the cAMP cascade inhibits an early event involved in murine macrophage Ia expression. J. Biol. Chem. 265, 12317–12323.

Fink, G., Mittelman, M., Shohat, B. and Spitzer, S.A. (1987). Theophylline-induced alterations in cellular immunity in asthmatic patients. Clin. Allergy 17, 313–316.

Fox, C.C., Wolf, E.J., Kagey-Sobotka, A. and Lichtenstein, L.M. (1988). Comparison of human lung and intestinal mast cells. J. Allergy Clin. Immunol. 81, 89–94.

Fredholm, B.B. (1980). Are methylxanthine effects due to antagonism of endogenous adenosine? Trends Pharmacol. Sci. 2, 129–133.

Fredholm, B.B. (1984). Cardiovascular and renal actions of methylxanthines. In "The Methylxanthine Beverages and Foods: Chemistry, Consumption and Health Effects" (ed. G.A. Spiller), pp. 303–330. Liss, New York.

Freyburger, G., Belloc, F. and Boisseau, M.R. (1990). Pentoxifylline inhibits actin polymerization in human neutrophils after stimulation by chemoattractant factor. Agents Actions 31, 72–78.

Friedman, F. and Detwiler, T.C. (1975). Stimulus-secretion coupling in platelets: effects of drugs on secretion of adenosine 5′-triphosphate. Biochemistry 14, 1315–1320.

Galvan, M. and Schudt, C. (1990). Actions of the phosphodiesterase inhibitor zardaverine on guinea-pig ventricular muscle. Naunyn-Schmiedebergs Arch. Pharmacol. 342, 221–227.

Ghio, R., Scordamaglia, A., D'Elia, P., Pizzorno, G., Romagnoli, M., Ciprandi, G. and Canonica, G.W. (1988). Inhibiton of the colony-forming capacity of human T-lymphocytes exerted by theophylline. Int. J. Immunopharmacol. 10, 299–302.

Giembycz, M.A., Corrigan, C.J., Kay, A.B., Barnes, P.J. (1994). Inhibition of CD4 and CD8 T-lymphocytes (T-LC) proliferation and cytokine secretion by isoenzyme-selective phosphodiesterase (PDE) inhibitors: correlation with intracellular cyclic AMP (cAMP) concentrations. J. Allergy Clin. Immunol. 93, 167. [Abstract]

Godfrey, R.W., Manzi, R.M., Gennaro, D.E. and Hoffstein, S.T. (1987). Phospholipid and arachidonic acid metabolism in zymosan-stimulated human monocytes: modulation by cAMP. J. Cell Physiol. 131, 384–392.

Goto, K., Hisadome, M. and Terasawa, M. (1982). Inhibitory effect of traxanox sodium on IgE-mediated histamine release from passively-sensitized mast cells of the rat *in vitro*. Int. Arch. Allergy Appl. Immunol. 68, 332–337.

Goto, T., Herberman, R.B., Maluish, A. and Strong, D.M. (1983). Cyclic AMP as a mediator of prostaglandin E-induced suppression of human natural killer cell activity. J. Immunol. 130, 1350–1355.

Goto, Y., Takeshita, T. and Sugamura, K. (1988). Adenosine 3′,5′-cyclic monophosphate (cAMP) inhibits phorbol ester-induced growth of an IL-2-dependent T cell line. FEBS Lett. 239, 165–168.

Gozzard, N., Herd, C.M., Blake, S.M., Holbrook, M., Hughes, B., Higgs, G.A. and Page, C.P. (1996). Effects of theophylline and rolipram on antigen-induced airway responses in neonatally immunised rabbits. Br. J. Pharmacol. 117, 1405–1412.

Grant, J.A., Dupree, E., Goldman, A.S., Schultz, D.R. and Jackson, A.L. (1975). Complement-mediated release of histamine from human leukocytes. J. Immunol. 114, 1101–1106.

Grant, J.A., Settle, L., Whorton, E.B. and Dupree, E. (1976). Complement-mediated release of histamine from human basophils. II. Biochemical characterization of the reaction. J. Immunol. 117, 450–456.

Gresele, P., Arnout, J., Deckmyn, H. and Vermylen, J. (1986). Mechanism of the antiplatelet action of dipyridamole in whole blood: modulation of adenosine concentration and activity. Thromb. Haemost. 55, 12–18.

Hahn, S. and Moroni, C. (1994). Modulation of cytokine expression in PB-3c mastocytes by IBMX and PMA. Lymphokine Cytokine Res. 13, 247–252.

Hamachi, T., Hirata, M. and Koga, T. (1984). Effect of cAMP-elevating drugs on Ca^{2+} efflux and actin polymerization in peritoneal macrophages stimulated with *N*-formyl chemotactic peptide. Biochim. Biophys. Acta 804, 230–236.

Hamet, P., Franks, D.J., Tremblay, J. and Coquil, J.F. (1983). Rapid activation of cAMP phosphodiesterase in rat platelets. Can. J. Biochem. Cell Biol. 61, 1158–1165.

Hammerschmidt, D.E., Kotasek, D., McCarthy, T., Huh, P.W., Freyburger, G. and Vercellotti, G.M. (1988). Pentoxifylline inhibits granulocyte and platelet function, including granulocyte priming by platelet activating factor. J. Lab. Clin. Med. 112, 254–263.

Hand, W.L., Butera, M.L., King-Thompson, N.L. and Hand, D.L. (1989). Pentoxifylline modulation of plasma membrane functions in human polymorphonuclear leukocytes. Infect. Immun. 57, 3520–3526.

Hart, D.A. (1988). Lithium potentiates antigen-dependent stimulation of lymphocytes only under suboptimal conditions. Int. J. Immunopharmacol. 10, 153–160.

Harvath, L., Robbins, J.D., Russell, A.A. and Seamon, K.B. (1991). cAMP and human neutrophil chemotaxis: elevation of cAMP differentially affects chemotactic responsiveness. J. Immunol. 146, 224–232.

Hatzelmann, A., Tenor, H. and Schudt, C. (1995). Differential effects of non-selective and selective phosphodiesterase inhibitors on human eosinophil functions. Br. J. Pharmacol. 114, 821–831.

Hecht, M., Muller, M., Lohmann-Matthes, M.L. and Emmendörffer, A. (1995). *In vitro* and *in vivo* effects of pentoxifylline on macrophages and lymphocytes derived from autoimmune MRL-1pr/1pr mice. J. Leukoc. Biol. 57, 242–249.

Henderson, L.M., Chappell, J.B. and Jones, O.T.G. (1989). Superoxide generation is inhibited by phospholipase A_2 inhibitors: role for phospholipase A_2 in the activation of the NADPH oxidase. Biochem. J. 264, 249–255.

Herlin, T. and Kragballe, K. (1982). Divergent effects of methylxanthines and adenylate cyclase agonists on monocyte cytotoxicity and cyclic AMP levels. Eur. J. Clin. Invest. 12, 293–299.

Herlin, T. and Kragballe, K. (1983). Facilitation of cAMP increments during ADCC mediated by monocytes pretreated with cAMP-elevating agents. Int. Arch. Allergy Appl. Immunol. 72, 1–5.

Hewett, J.A., Jean, P.A., Kunkel, S.L. and Roth, R.A. (1993). Relationship between tumor necrosis factor-alpha and neutrophils in endotoxin-induced liver injury. Am. J. Physiol. 265, G1011–G1015.

Hisadome, M., Terasawa, M., Goto, K., Kawazoe, Y. and Okumoto, T. (1989). Stimulation of phagocytic activity of leukocytes and macrophages by traxanox sodium in mice and rats. Jpn. J. Pharmacol. 49, 275–284.

Hjemdahl, P., Larsson, K., Johansson, M.-C., Zetterlund, A. and Eklund, A. (1990). β-Adrenoceptors in human alveolar macrophages isolated by elutriation. Br. J. Clin. Pharmacol. 30, 673–682.

Hoffmann, H., Hatherill, J.R., Crowley, J., Harada, H., Yonemaru, M., Zheng, H., Ishizaka, A. and Raffin, T.A. (1991). Early post-treatment with pentoxifylline or dibutyryl cAMP attenuates *Escherichia coli*-induced acute lung injury in guinea pigs. Am. Rev. Respir. Dis. 143, 289–293.

Holgate, S.T., Winslow, C.M., Lewis, R.A. and Austen, K.F. (1981). Effects of prostaglandin D_2 and theophylline on rat serosal mast cells: discordance between increased cellular levels of cyclic AMP and activation of cyclic AMP-dependent protein kinase. J. Immunol. 127, 1530–1533.

Hossain, M., Okubo, Y. and Sekiguchi, M. (1994). Effects of various drugs (staurosporine, herbimycin A, ketotifen, theophylline, FK506 and cyclosporin A) on eosinophil viability. Arerugi 43, 711–717.

Houslay, M.D. and Kilgour, E.(1990). Cyclic nucleotide phosphodiesterases in liver: a review of their characterisation, regulation by insulin and glucagon and their role in controlling intracellular cyclic AMP concentrations. In "Cyclic Nucleotide Phosphodiesterases: Structure, regulation and Drug Action" (eds J. Beavo and M.D. Houslay), pp. 185–224. Wiley, Chichester.

Ishizaka, T., Hirata, F., Sterk, A.R., Ishizaka, K. and Axelrod, J.A. (1981). Bridging of IgE receptors activates phospholipid methylation and adenylate cyclase in mast cell plasma membranes. Proc. Natl Acad. Sci. USA 78, 6812–6816.

Iwatsuki, K., Horiuchi, A., Ren, L.M. and Chiba, S. (1991). Effects of the cyclic nucleotide phosphodiesterase inhibitors, rolipram, 3-isobutyl-1-methylxanthine, amrinone and zaprinast, on pancreatic exocrine secretion in dogs. Eur. J. Pharmacol. 209, 63–68.

Kaneko, M., Suzuki, K., Furui, H., Takagi, K. and Satake, T. (1990). Comparison of theophylline and enprofylline effects on human neutrophil superoxide production. Clin. Exp. Pharmacol. Physiol. 17, 849–859.

Kassis, S., Lee, J.C. and Hanna, N. (1989). Effects of prostaglandins and cAMP levels on monocyte IL-1 production. Agents Actions 27, 274–276.

Kidney, J.C., Dominguez, M., Taylor, P.M., Rose, M., Chung, K.F. and Barnes, P.J. (1995). Immunomodulation by theophylline: demonstration by withdrawal of therapy. Am. J. Respir. Crit. Care Med. 151, 1907–1914.

Kita, H., Abu-Ghazaleh, R.I., Gleich, G.J. and Abraham, R.T. (1991). Regulation of Ig-induced eosinophil degranulation by adenosine 3′,5′-cyclic monophosphate. J. Immunol. 146, 2712–2718.

Kitagawa, S., Endo, J. and Kametani, F. (1984). Effects of four types of reagents on ADP-induced aggregation and Ca^{2+} mobilization of bovine blood platelets. Biochim. Biophys. Acta 798, 210–215.

Kotecki, M., Pawlak, A.L. and Wiktorowicz, K.E. (1989). The inhibitory effect of theophylline on cell cycle kinetics of human lymphocytes *in vitro*. Arch. Immunol. Ther. Exp. (Warsaw) 37, 725–733.

Kovach, N.L., Lindgren, C.G., Fefer, A., Thompson, J.A., Yednock, T. and Harlan, J.M. (1994). Pentoxifylline inhibits integrin-mediated adherence of interleukin-2-activated human peripheral blood lymphocytes to human umbilical vein endothelial cells, matrix components, and cultured tumor cells. Blood 87, 2234–2242.

Kroegel, C. (1990). The role of eosinophils in asthma. Lung 168 (Suppl.), 5–17.

Lahat, N., Nir, E., Horenstien, L. and Colin, A.A. (1985). Effect of theophylline on the proportion and function of T-suppressor cells in asthmatic children. Allergy 40, 453–457.

Lau, H.Y., Wong, P.L. and Lai, C.K. (1994). Effects of β_2-adrenergic agonists on isolated guinea pig lung mast cells. Agents Actions 42, 92–94.

Leff, J.A. and Repine, J.E. (1993). Neutrophil-mediated tissue injury. In "The Neutrophil" (eds. J.S. Abramson and J.G. Wheeler), pp. 229–262. IRL Press (Oxford University Press), New York.

Lemon, M.J.C. and Bhoola, K.D. (1975). Excitation-secretion coupling in exocrine glands: properties of cyclic AMP phosphodiesterase and adenylate cyclase from the submaxillary gland and pancreas. Biochim. Biophys. Acta 385, 101–113.

Lilly, C.M., Sandhu, J.S., Ishizaka, A., Harada, H., Yonemaru, M., Larrick, J.W., Shi, T.X., O'Hanley, P.T. and Raffin, T.A. (1989). Pentoxifylline prevents tumor necrosis factor-induced lung injury. Am. Rev. Respir. Dis. 139, 1361–1368.

Lindvall, L., Lantz, M., Gullberg, U. and Olsson, I. (1990). Modulation of the constitutive gene expression of the 55 kD tumor necrosis factor receptor in hematopoietic cells. Biochem. Biophys. Res. Commun. 172, 557–563.

Liu, W.L., Boulos, P.B., Lau, H.Y. and Pearce, F.L. (1991). Mast cells from human gastric mucosa: a comparative study with lung and colonic mast cells. Agents Actions 33, 13–15.

Llewellyn-Jones, C.G. and Stockley, R.A. (1994). The effects of β_2-agonists and methylxanthines on neutrophil function *in vitro*. Eur. Respir. J. 7, 1460–1466.

Louis, R.E. and Radermecker, M.F. (1990). Substance P-induced histamine release from human basophils, skin and lung fragments: effect of nedocromil sodium and theophylline. Int. Arch. Allergy Appl. Immunol. 92, 329–333.

Louis, R., Bury, T., Corhay, J.L. and Radermecker, M. (1992). LY 186655, a phosphodiesterase inhibitor, inhibits histamine release from human basophils, lung and skin fragments. Int. J. Immunopharmacol. 14, 191–194.

Lubbe, W.F., Podzuweit, T. and Opie, L.H. (1992). Potential arrhythmogenic role of cyclic adenosine monophosphate (AMP) and cytosolic calcium overload: implications for prophylactic effects of beta-blockers in myocardial infarction and proarrhythmic effects of phosphodiesterase inhibitors. J. Am. Coll. Cardiol. 19, 1622–1633.

Mah, M.P., Aeberhard, E.E., Gilliam, M.B. and Sherman, M.P. (1993). Effects of pentoxifylline on *in vivo* leukocyte function and clearance of group B streptococci from preterm rabbit lungs. Crit. Care Med. 21, 712–720.

Manzini, S., Meini, S., Giachetti, A., Beani, L., Borea, P.A., Antonelli, T., Ballati, L. and Bacciarelli, C. (1990). Pharmacodynamic profile of isbufylline, a new antibronchospastic xanthine devoid of central excitatory actions. Arzneimittelforschung 40, 1205–1213.

Manzini, S., Perretti, F., Abelli, L., Evangelista, S., Seeds, E.A. and Page, C.P. (1993). Isbufylline, a new xanthine derivative, inhibits airway responsiveness and airway inflammation in guinea pigs. Eur. J. Pharmacol. 249, 251–257.

Marone, G., Kagey-Sobotka, A. and Lichtenstein, L.M. (1981). IgE-mediated histamine release from human basophils: differences between antigen and anti-IgE-induced secretion. Int. Arch. Allergy Appl. Immunol. 65, 339–348.

Marukawa, S., Hatake, K., Wakabayashi, I. and Hishida, S. (1994). Vasorelaxant effects of oxpentifylline and theophylline on rat isolated aorta. J. Pharm. Pharmacol. 46, 342–345.

Mary, D., Aussel, C., Ferrua, B. and Fehlmann, M. (1987). Regulation of interleukin 2 synthesis by cAMP in human T cells. J. Immunol. 139, 1179–1184.

McLeish, K.R., Wellhausen, S.R. and Dean, W.L. (1987). Biochemical basis of HLA-DR and CR3 modulation on human peripheral blood monocytes by lipopolysaccharide. Cell Immunol. 108, 242–248.

Misso, N.L. and Thompson, P.J. (1992). Xanthines inhibit human platelet aggregation induced by platelet-activating factor. Clin. Exp. Pharmacol. Physiol. 19, 599–602.

Miyamoto, K.-i., Kurita, M., Ohmae, S., Sakai, R., Sanae, F. and Takagi, K. (1994). Selective tracheal relaxation and phosphodiesterase-IV inhibition by xanthine derivatives. Eur. J. Pharmacol. 267, 317–322.

Montague, W. and Howell, S.L. (1975). Cyclic AMP and the physiology of the islets of Langerhans. Adv. Cyclic Nucleotide Res. 6, 201–243.

Moqbel, R. (1994). Eosinophils, cytokines, and allergic inflammation. Ann. NY Acad. Sci. 725, 223–233.

Muthusamy, N., Baluyut, A.R. and Subbarao, B. (1991). Differential regulation of surface Ig- and Lyb2-mediated B cell activation by cyclic AMP. I. Evidence for alternative regulation of signaling through two different receptors linked to phosphatidylinositol hydrolysis in murine B cells. J. Immunol. 147, 2483–2492.

Nelson, S., Summer, W.R. and Jakab, G.J. (1985). Aminophylline-induced suppression of pulmonary antibacterial defenses. Am. Rev. Respir. Dis. 131, 923–927.

Newman, D.J., Colella, D.F., Spainhour, C.B., Jr, Braun, E.G., Zabko-Potapovich, B. and Wardelt J.R. Jr (1978). cAMP-phosphodiesterase inhibitors and tracheal smooth muscle relaxation. Biochem. Pharmacol. 27, 729–732.

Nielson, C.P., Crowley, J.J., Cusack, B.J. and Vestal, R.E. (1986). Therapeutic concentrations of theophylline and enprofylline potentiate catecholamine effects and inhibit leukocyte activation. J. Allergy Clin. Immunol. 78, 660–667.

Nielson, C.P., Crowley, J.J., Morgan, M.E. and Vestal, R.E. (1988). Polymorphonuclear leukocyte inhibition by therapeutic concentrations of theophylline is mediated by cyclic-3′,5′-adenosine monophosphate. Am. Rev. Respir. Dis. 137, 25–30.

Nokta, M.A. and Pollard, R.B. (1992). Human immunodeficiency virus replication: modulation by cellular levels of cAMP. AIDS Res. Hum. Retroviruses 8, 1255–1261.

Nokta, M., Albrecht, T. and Pollard, R. (1993). Papaverine hydrochloride: effects on HIV replication and T-lymphocyte cell function. Immunopharmacology 26, 181–185.

Novak, T.J. and Rothenberg, E.V. (1990). Cyclic AMP inhibits induction of interleukin 2 but not of interleukin 4 in T cells. Proc. Natl Acad. Sci. USA 87, 9353–9357.

Novogrodsky, A., Patya, M., Rubin, A.L. and Stenzel, K.H. (1983). Agents that increase cellular cAMP inhibit production of interleukin-2, but not its activity. Biochem. Biophys. Res. Commun. 114, 93–98.

Numao, T., Fukuda, T., Akutsu, I. and Makino, S. (1991). Effects of anti-asthmatic drugs on human eosinophil chemotaxis. Nippon Kyobu Shikkan Gakkai Zasshi 29, 65–71.

Ogilvie, R.I., Fernandez, P.G. and Winsberg, F. (1977). Cardiovascular response to increasing theophylline concentrations. Eur. J. Clin. Pharmacol. 12, 409–414.

Ohta, K., Sawamoto, S., Nakajima, M., Nagai, A., Tanaka, Y., Hirai, K., Mano, K. and Miyashita, H. (1994). Theophylline induces apoptosis in eosinophils surviving with interleukin-5 *in vitro*. J. Allergy Clin. Immunol. 93, 200. [Abstract]

Oka, Y., Murata, A., Nishijima, J., Hiraoka, N., Yasuda, T., Kitagawa, K., Ogawa, M. and Mori, T. (1991). Inhibitory effect of pentoxifylline and prostaglandin E_1 on the release of neutrophil elastase from FMLP-stimulated neutrophils. J. Med. 22, 371–382.

Ollivier, V., Houssaye, S., Ternisien, C., Leon, A., de Verneuil, H., Elbim, C., Mackman, N., Edgington, T.S. and de Prost, D. (1993). Endotoxin-induced tissue factor messenger RNA in human monocytes is negatively regulated by a cyclic AMP-dependent mechanism. Blood 81, 973–979.

O'Neill, S.J., Sitar, D.S., Klass, D.J., Taraska, V.A., Kepron, W. and Mitenko, P.A. (1986). The pulmonary disposition of theophylline and its influence on human alveolar macrophage bactericidal function. Am. Rev. Respir. Dis. 134, 1225–1228.

Palfreyman, M.N. (1995). Phosphodiesterase type IV inhibitors as antiinflammatory agents. Drugs Fut. 20, 793–804.

Pan, X., Arauz, E., Krzanowski, J.J., Fitzpatrick, D.F. and Polson, J.B. (1994). Synergistic interactions between selective phosphodiesterase isozyme families PDE III and PDE IV to attenuate proliferation of rat vascular smooth muscle cells. Biochem. Pharmacol. 48, 827–835.

Peachell, P.T., MacGlashan, D.W., Jr, Lichtenstein, L.M. and Schleimer, R.P. (1988). Regulation of human basophil and lung mast cell function by cyclic adenosine monophosphate. J. Immunol. 140, 571–579.

Peachell, P.T., Undem, B.J., Schleimer, R.P., MacGlashan, D.W., Lichtenstein, L.M., Cieslinski, L.B. and Torphy, T.J. (1992). Preliminary identification and role of phosphodiesterase isozymes in human basophils. J. Immunol. 148, 2503–2510.

Pearce, F.L., Blum, U., Poblete-Freundt, G. and Schmutzler, W. (1978). Studies on the release of histamine from isolated guinea pig mast cells stimulated by ionophore A23187 or by the anaphylactic reaction. Naunyn Schmiedebergs Arch. Pharmacol. 302, 165–172.

Pearce, F.L., Befus, A.D., Gauldie, J. and Bienenstock, J. (1982). Mucosal mast cells. II. effects of anti-allergic compounds on histamine secretion by isolated intestinal mast cells. J. Immunol. 128, 2481–2486.

Persson, C.G.A. (1986). Development of safer xanthine drugs for treatment of obstructive airways disease. J. Allergy Clin. Immunol. 78, 817–824.

Persson, C.G.A. and Karlsson, J.-A. (1987). In vitro responses to bronchodilator drugs. In "Drug Therapy for Asthma: Research and Clinical Practice" (eds. J.W. Jenne and S. Murphy), pp. 129–176. Marcel Dekker, New York.

Persson, C.G.A., Erjefält, I. and Gustafsson, B. (1988). Xanthines – symptomatic or prophylactic in asthma? Agents Actions Suppl. 23, 137–155.

Peters, S.P., Schulman, E.S., Schleimer, R.P., MacGlashan, D.W., Jr, Newball, H.H. and Lichtenstein, L.M. (1982). Dispersed human lung mast cells: pharmacologic aspects and comparison with human lung tissue fragments. Am. Rev. Respir. Dis. 126, 1034–1039.

Polson, J.B., Krzanowski, J.J., Fitzpatrick, D.F. and Szentivanyi, A. (1978). Studies on the inhibition of phosphodiesterase-catalyzed cyclic AMP and cyclic GMP breakdown and relaxation of canine smooth muscle. Biochem. Pharmacol. 27, 254–256.

Polson, J.B., Krzanowski, J.J., Anderson, W.H., Fitzpatrick, D.F., Hwang, D.P. and Szentivanyi, A. (1979). Analysis of the relationship between pharmacological inhibition of cyclic nucleotide phosphodiesterase and relaxation of canine tracheal smooth muscle. Biochem. Pharmacol. 28, 1391–1395.

Prabhakar, U., Brooks, D.P., Lipshlitz, D. and Esser, K.M. (1995). Inhibition of LPS-induced TNFα production in human monocytes by adenosine (A_2) receptor selective agonists. Int. J. Immunopharmacol. 17, 221–224.

Prieur, A.M. and Granger, G.A. (1975). The effect of agents which modulate the levels of cyclic nucleotides on human lymphotoxin secretion and activity in vitro. Transplantation 20, 331–337.

Rabe, K.F., Magnussen, H. and Dent, G. (1995). Theophylline and selective PDE inhibitors as bronchodilators and smooth muscle relaxants. Eur. Respir. J. 8, 637–642.

Rall, T.W. (1990). Drugs used in the treatment of asthma: the methylxanthines, cromolyn sodium, and other agents. In "Goodman and Gilman's The Pharmacological Basis of Therapeutics" 8th edition (eds. A.G. Gilman, T.W. Rall, A.S. Nies and P. Taylor), pp. 618–637. Pergamon, New York.

Rao, K.M., Crawford, J. Currie, M.S. and Cohen, H.J. (1988). Actin depolymerization and inhibition of capping induced by pentoxifylline in human lymphocytes and neutrophils. J. Cell Physiol. 137, 577–582.

Rao, K.M., Currie, M.S., McCachren, S.S. and Cohen, H.J. (1991). Pentoxifylline and other methylxanthines inhibit interleukin-2 receptor expression in human lymphocytes. Cell Immunol. 135, 314–325.

Reignier, J., Mazmanian, M., Detruit, H., Chapelier, A., Weiss, M., Libert, J.M., Herve, P. and Paris-Sud University Lung Transplantation Group. (1994). Reduction of ischemia-reperfusion injury by pentoxifylline in the isolated rat lung. Am. J. Respir. Crit. Care Med. 150, 342–347.

Rosenthal, L.A., Taub, D.D., Moors, M.A. and Blank, K.J. (1992). Methylxanthine-induced inhibition of the antigen- and superantigen-specific activation of T and B lymphocytes. Immunopharmacology 24, 203–217.

Rossignol, L., Plantavid, M., Chap, H. and Douste-Blazy, L. (1988a). Effets de la pentoxifylline et de la propentofylline sur le renouvellement des polyphosphoinositides de la membrane erythrocytaire et sur le metabolisme de l'acide arachidonique dans les plaquettes. Pathol. Biol. (Paris) 36, 1073–1075.

Rossignol, L., Plantavid, M., Chap, H. and Douste-Blazy, L. (1988b). Effects of two methylxanthines, pentoxifylline and propentofylline, on arachidonic acid metabolism in platelets stimulated by thrombin. Biochem. Pharmacol. 37, 3229–3236.

Rott, O., Cash, E. and Fleischer, B. (1993). Phosphodiesterase inhibitor pentoxyfylline, a selective suppressor of T-helper type 1- but not type 2-associated lymphokine production, prevents induction of experimental autoimmune encephalomyelitis in Lewis rats. Eur. J. Immunol. 23, 1745–1751.

Ruddle, N.H., Bergman, C.M., McGrath, K.M., Lingenheld, E.G., Grunnet, M.L., Padula, S.J. and Clark, R.B. (1990). An antibody to lymphotoxin and tumor necrosis factor prevents transfer of experimental allergic encephalomyelitis. J. Exp. Med. 172, 1193–1200.

Salvemini, D., Radziszewski, W., Korbut, R. and Vane, J. (1990). The use of oxyhaemoglobin to explore the events underlying inhibition of platelet aggregation induced by NO or NO-donors. Br. J. Pharmacol. 101, 991–995.

Sanjar, S., Aoki, S., Boubekeur, K., Burrows, L., Colditz, I., Chapman, I. and Morley, J. (1989). Inhibition of PAF-induced eosinophil accumulation in pulmonary airways of guinea pigs by anti-asthma drugs. Jpn. J. Pharmacol. 51, 167–172.

Sanjar, S., Aoki, S., Boubekeur, K., Chapman, I.D., Smith, D., Kings, M.A. and Morley, J. (1990a). Eosinophil accumulation in pulmonary airways of guinea-pigs induced by exposure to an aerosol of platelet-activating factor: effect of anti-asthma drugs. Br. J. Pharmacol. 99, 267–272.

Sanjar, S., Aoki, S., Kristersson, A., Smith, D. and Morley, J. (1990b). Antigen challenge induces pulmonary airway eosinophil accumulation and airway hyperreactivity in sensitized guinea-pigs: the effect of anti-asthma drugs. Br. J. Pharmacol. 99, 679–686.

Santing, R.E., Olymulder, C.G., Van der Molen, K., Meurs, H. and Zaagsma, J. (1995). Phosphodiesterase inhibitors reduce bronchial hyperreactivity and airway inflammation in unrestrained guinea pigs. Eur. J. Pharmacol. 275, 75–82.

Santucci, L., Fiorucci, S., Giansanti, M., Brunori, P.M., Di Matteo, F.M. and Morelli, A. (1994). Pentoxifylline prevents indomethacin induced acute gastric mucosal damage in rats: role of tumour necrosis factor alpha. Gut 35, 909–915.

Schandené, L., Vandenbussche, P., Crusiaux, A., Alègre, M.L., Abramowicz, D., Dupont, E., Content, J. and Goldman, M. (1992). Differential effects of pentoxifylline on the production of tumour necrosis factor-alpha (TNF-α) and interleukin-6 (IL-6) by monocytes and T cells. Immunology 76, 30–34.

Schick, B. and Austen, K.F. (1985). Pharmacological modulation of activation-secretion of rat serosal mast cells by chymase, an endogenous secretory granule protease. Immunology 56, 513–522.

Schmeichel, C.J. and Thomas, L.L. (1987). Methylxanthine bronchodilators potentiate multiple human neutrophil functions. J. Immunol. 138, 1896–1903.

Schudt, C., Winder, S., Eltze, M., Kilian, U. and Beume, R. (1991a). Zardaverine: a cyclic AMP specific PDE III/IV inhibitor. Agents Actions Suppl. 34, 379–402.

Schudt, C., Winder, S., Forderkunz, S., Hatzelmann, A. and Ullrich, V. (1991b). Influence of selective phosphodiesterase inhibitors on human neutrophil functions and levels of cAMP and Ca_i. Naunyn-Schmiedebergs Arch. Pharmacol. 344, 682–690.

Schudt, C., Winder, S., Müller, B. and Ukena, D. (1991c). Zardaverine as a selective inhibitor of phosphodiesterase isozymes. Biochem. Pharmacol. 42, 153–162.

Schulman, E.S., Post, T.J. and Vigderman, R.J. (1988). Density heterogeneity of human lung mast cells. J. Allergy Clin. Immunol. 82, 78–86.

Scordamaglia, A., Ciprandi, G., Ruffoni, S., Caria, M., Paolieri, F., Venuti, D. and Canonica, G.W. (1988). Theophylline and the immune response: *in vitro* and *in vivo* effects. Clin. Immunol. Immunopathol. 48, 238–246.

Shah, T.P., Lichtenstein, L.M., Undem, B. and MacDonald, S.M. (1991). Effects of cyclic-AMP on human IgE synthesis *in vitro*. J. Allergy Clin. Immunol. 89, 173. [Abstract]

Shahid, M., Van Amsterdam, R.G.M., de Boer, J., ten Berge, R.E., Nicholson, C.D. and Zaagsma, J. (1991). The presence of five cyclic nucleotide phosphodiesterase isoenzyme activities in bovine tracheal smooth muscle and the functional effects of selective inhibitors. Br. J. Pharmacol. 104, 471–477.

Shanahan, F., Lee, T.D., Bienenstock, J. and Befus, A.D. (1986). Mast cell heterogeneity: effect of anti-allergic compounds on neuropeptide-induced histamine release. Int. Arch. Allergy Appl. Immunol. 80, 424–426.

Shearer, W.T., Patke, C.L., Gilliam, E.B., Rosenblatt, H.M., Barron, K.S. and Orson, F.M. (1988). Modulation of a human lymphoblastoid B cell line by cyclic AMP: Ig secretion and phosphatidylcholine metabolism. J. Immunol. 141, 1678–1686.

Shohat, B., Volovitz, B. and Varsano, I. (1983). Induction of suppressor T cells in asthmatic children by theophylline treatment. Clin. Allergy 13, 487–493.

Shute, J.K., Tenor, H., Church, M.K. and Schudt, C. (1995). Theophylline inhibits GM-CSF release from eosinophils. Eur. Respir. J. 8 (Suppl. 19), 9s. [Abstract]

Sigurd, B. and Olesen, K.H. (1978). Comparative natiuretic and diuretic efficacy of theophylline ethylenediamine and of bendroflumethiazide during long-term treatment with the potent diuretic bumetanide. Acta Med. Scand. 203, 113–119.

Simpson, A.W., Reeves, M.L. and Rink, T.J. (1988). Effects of SK&F 94120, an inhibitor of cyclic nucleotide phosphodiesterase type III, on human platelets. Biochem. Pharmacol. 37, 2315–2320.

Singer, J.W., Bianco, J.A., Takahashi, G., Simrell, C., Petersen, J. and Andrews, D.F. III. (1992). Effect of methylxanthine derivatives on T cell activation. Bone Marrow Transplant. 10, 19–25.

Skopinski, P., Zukowska, M., Malkowska-Zwierz, W., Radomska, D., Retelska, D., Kecik, D., Domoslawski, M., Sobczynska, A., Pawinska, M. and Skopinska-Rozewska, E. (1993). The effect of TPP, theophylline and theobromine on the angiogenic activity of mononuclear leucocytes obtained from diabetic patients with proliferative retinopathy. Acta Pol. Pharm. 50, 409–411.

Sommer, N., Löschmann, P.-A., Northoff, G.H., Weller, M., Steinbrecher, A., Steinbach, J.P., Lichtenfels, R., Meyermann, R., Riethmüller, A., Fontana, A., Dichgans, J. and Martin, R. (1995). The antidepressant rolipram suppresses cytokine production and prevents autoimmune encephalomyelitis. Nat. Med. 1, 244–248.

Spatafora, M., Chiappara, G., Merendino, A.M., D'Amico, D., Bellia, V. and Bonsignore, G. (1994). Theophylline suppresses the release of tumour necrosis factor-α by blood monocytes and alveolar macrophages. Eur. Respir. J. 7, 223–228.

Spirer, Z., Zakuth, V., Diamant, S., Stabinsky, Y. and Fridkin, M. (1979). Studies on the activity of phorbol myristate acetate on the human polymorphonuclear leukocytes. Experientia 35, 830–831.

Spry, C.J.F. (1988). Eosinophils: a Comprehensive Review and Guide to the Scientific and Medical Literature. Oxford University Press, Oxford.

Stein, M. and Keshav, S. (1992). The versatility of macrophages. Clin. Exp. Allergy 22, 19–27.

Stephens, C.G. and Snyderman, R. (1982). Cyclic nucleotides regulate the morphologic alterations required for chemotaxis in monocytes. J. Immunol. 128, 1192–1197.

Strannegard, O. and Strannegard, I.L. (1984). Effect of cyclic AMP-elevating agents on human spontaneous IgE synthesis *in vitro*. Int. Arch. Allergy Appl. Immunol. 74, 9–14.

Sullivan, T.J., Parker, K.L., Eisen, S.A. and Parker, C.W. (1975). Modulation of cyclic AMP in purified rat mast cells. II. Studies on the relationship between intracellular cyclic AMP concentrations and histamine release. J. Immunol. 114, 1480–1485.

Sullivan, G.W., Carper, H.T., Novick, W.J., Jr and Mandell, G.L. (1988). Inhibition of the inflammatory action of interleukin-1 and tumor necrosis factor (alpha) on neutrophil function by pentoxifylline. Infect. Immun. 56, 1722–1729.

Sullivan, P., Bekir, S., Jaffar, Z., Page, C., Jeffery, P. and Costello, J. (1994a). Anti-inflammatory effects of low-dose oral theophylline in atopic asthma. Lancet 343, 1006–1008.

Sullivan, P., Page, C.P. and Costello, J.F. (1994b). Xanthines. In "Drugs and the Lung" (eds. C.P. Page and W.J. Metzger), pp. 69–99. Raven Press, New York.

Sydbom, A. and Fredholm, B.B. (1982). On the mechanism by which theophylline inhibits histamine release from rat mast cells. Acta Physiol. Scand. 114, 243–251.

Takayama, H., Trenn, G. and Sitkovsky, M.V. (1988). Locus of inhibitory action of cAMP-dependent protein kinase in the antigen receptor-triggered cytotoxic T lymphocyte activation pathway. J. Biol. Chem. 263, 2330–2336.

Tarayre, J.P., Aliaga, M., Barbara, M., Malfetes, N., Vieu, S. and Tisne-Versailles, J. (1991). Theophylline reduces pulmonary eosinophilia after various types of active anaphylactic shock in guinea-pigs. J. Pharm. Pharmacol. 43, 877–879.

Tasaka, K., Mio, M. and Okamoto, M. (1986a). Intracellular calcium release induced by histamine releasers and its inhibition by some antiallergic drugs. Ann. Allergy 56, 464–469.

Tasaka, K., Mio, M. and Okamoto, M. (1986b). Changes in intracellular Ca^{2+} distribution of rat peritoneal mast cells before and after histamine release. Agents Actions 18, 61–64.

Tenor, H., Shute, J.K., Church, M.K., Hatzelmann, A. and Schudt, C. (1995). Inhibition of human peripheral blood eosinophil LTC_4 production by PDE inhibitors. Eur. Respir. J. 8 (Suppl. 19), 9s. [Abstract]

Texeira, M.M., Rossi, A.G., Williams, T.J. and Hellewell, P.G. (1994). Effects of phosphodiesterase isoenzyme inhibitors on cutaneous inflammation in the guinea-pig. Br. J. Pharmacol. 112, 332–340.

Thiel, M., Bardenheuer, H., Poch, G., Madel, C. and Peter, K. (1991). Pentoxifylline does not act via adenosine receptors in the inhibition of the superoxide anion production of human polymorphonuclear leukocytes. Biochem. Biophys. Res. Commun. 180, 53–58.

Thomas, L.L., Zheutlin, L.M. and Gleich, G.J. (1989). Pharmacological control of human basophil histamine release stimulated by eosinophil granule major basic protein. Immunology 66, 611–615.

Thompson, W.J., Ross, C.P., Hersh, E.M., Epstein, P.M. and Strada, S.J. (1980). Activation of human lymphocyte high affinity cyclic AMP phosphodiesterase by culture with 1-methyl-3-isobutylxanthine. J. Cyclic Nucleotide Res. 6, 25–36.

Thorne, K.J., Richardson, B.A., Butterworth, A.E., Hay, I. and Higenbottam, T.W. (1988). Effect of drugs used in the treatment of asthma on the production of eosinophil-activating factor by monocytes. Int. Arch. Allergy Appl. Immunol. 85, 257–259.

Tilg, H., Eibl, B., Pichl, M., Gachter, A., Herold, M., Brankova, J., Huber, C. and Niederwieser, D. (1993). Immune response modulation by pentoxifylline *in vitro*. Transplantation 56, 196–201.

Toll, J.B.C. and Andersson, R.G.G. (1984). Effects of enprofylline and theophylline on purified human basophils. Allergy 39, 515–520.

Tomioka, M., Goto, T., Lee, T.D., Bienenstock, J. and Befus, A.D. (1989). Isolation and characterization of lung mast cells from rats with bleomycin-induced pulmonary fibrosis. Immunology 66, 439–444.

Tremblay, J., Lachance, B. and Hamet, P. (1985). Activation of cyclic GMP-binding and cyclic AMP-specific phosphodiesterases of rat platelets by a mechanism involving cyclic AMP-dependent phosphorylation. J. Cyclic Nucleotide Protein Phosphor. Res. 10, 397–411.

Triner, L., Vulliemoz, Y. and Veronsky, M. (1977). Cyclic 3′,5′-adenosine monophosphate and bronchial tone. Eur. J. Pharmacol. 41, 37–46.

Truneh, A. and Pearce, F.L. (1984). Effect of cyclic AMP, disodium cromoglycate and other anti-allergic drugs on histamine secretion from rat mast cells stimulated with the calcium ionophore ionomycin. Agents Actions 14, 179–184.

Tsien, W.H., Sass, S.P. and Sheppard, H. (1982). The lack of correlation between inhibition of aggregation and cAMP levels with canine platelets. Thromb. Res. 28, 509–519.

Ukena, D., Schudt, C. and Sybrecht, G.W. (1993). Adenosine receptor-blocking xanthines as inhibitors of phosphodiesterase isozymes. Biochem. Pharmacol. 45, 847–851.

Undem, B.J., Torphy, T.J., Goldman, D. and Chilton, F.H. (1990). Inhibition by adenosine 3′:5′-monophosphate of eicosanoid and platelet-activating factor biosynthesis in the mouse PT-18 mast cell. J. Biol. Chem. 265, 6750–6758.

Van Damme, J. (1991). IL-8 and related molecules. In "The Cytokine Handbook" 1st edition (ed. A. Thomson), pp. 201–213. Academic Press, London.

van Leenen, D., van der Poll, T., Levi, M., ten Cate, H., van Deventer, S.J.H., Hack, C.E., Aarden, L.A. and ten Cate, J.W. (1993). Pentoxifylline attenuates neutrophil activation in experimental endotoxemia in chimpanzees. J. Immunol. 151, 2318–2325.

Vinge, E., Andersson, K.E. and Persson, C.G. (1984). Effects on aggregation of human platelets of two xanthines and their interactions with adenosine. Acta Physiol. Scand. 120, 117–121.

Warner, J.A., MacGlashan, D.W., Peters, S.P., Kagey-Sobotka, A. and Lichtenstein, L.M. (1988). The pharmacologic modulation of mediator release from human basophils. J. Allergy Clin. Immunol. 82, 432–438.

Weiner, I.M. (1990). Diuretics and other agents employed in the mobilization of edema fluid. In "Goodman & Gilman's Pharmacological Basis of Therapeutics", eighth edition (eds. A.G. Gilman, T.W. Rall, A.S. Nies and P. Taylor), pp. 713–731. Pergamon, New York.

Welsh, C.H., Lien, D., Worthen, G.S. and Weil, J.V. (1988). Pentoxifylline decreases endotoxin-induced pulmonary neutrophil sequestration and extravascular protein accumulation in the dog. Am. Rev. Respir. Dis. 138, 1106–1114.

White, J.R., Ishizaka, T., Ishizaka, K. and Sha'afi, R. (1984). Direct demonstration of increased intracellular concentration of free calcium as measured by quin-2 in stimulated rat peritoneal mast cell. Proc. Natl Acad. Sci. USA 81, 3978–3982.

White, S.R., Strek, M.E., Kulp, G.V.P., Spaethe, S.M., Burch, R.A., Neeley, S.P. and Leff, A.R. (1993). Regulation of human eosinophil degranulation and activation by endogenous phospholipase A_2. J. Clin. Invest. 91, 2118–2125.

Wiik, P. (1989). Vasoactive intestinal peptide inhibits the respiratory burst in human monocytes by a cyclic AMP-mediated mechanism. Regul. Pept. 25, 187–197.

Williams, W.R., Kagamimori, S., Shimizu, T., Kakiuchi, H. and Naruse, Y. (1988). Effects of theophylline, beta-adrenergic stimulants and prednisolone on platelet and urine cyclic guanosine monophosphate (cGMP). Tohoku J. Exp. Med. 156, 331–340.

Wirth, J.J. and Kierszenbaum, F. (1982). Inhibitory action of elevated levels of adenosine-3′:5′ cyclic monophosphate on phagocytosis: effects on macrophage-*Trypanosoma cruzi* interaction. J. Immunol. 129, 2759–2762.

Wong, P.M. and Schmid-Schonbein, G.W. (1991). Attenuation of spontaneous pseudopod formation in human neutrophils by pentoxifylline. Cell Biophys. 18, 203–215.

Wright, C.D., Kuipers, P.J., Kobylarz-Singer, D., Devall, L.J., Klinkefus, B.A. and Weishaar, RE. (1990). Differential inhibition of human neutrophil functions: role of cyclic AMP-specific, cyclic GMP-insensitive phosphodiesterase. Biochem. Pharmacol. 40, 699–707.

Yukawa, T., Kroegel, C., Chanez, P., Dent, G., Ukena, D., Chung, K.F. and Barnes, P.J. (1989). Effect of theophylline and adenosine on eosinophil function. Am. Rev. Respir. Dis. 140, 327–333.

Zabel, P., Schade, F.U. and Schlaak, M. (1993). Inhibition of endogenous TNF formation by pentoxifylline. Immunobiology 187, 447–463.

Zheng, H., Crowley, J.J., Chan, J.C., Hoffmann, H., Hatherill, J.R., Ishizaka, A. and Raffin, T.A. (1990). Attenuation of tumor necrosis factor-induced endothelial cell cytotoxicity and neutrophil chemiluminescence. Am. Rev. Respir. Dis. 142, 1073–1078.

Zheng, H., Crowley, J.J., Chan, J.C. and Raffin, T.A. (1991). Attenuation of LPS-induced neutrophil thromboxane B_2 release and chemiluminescence. J. Cell Physiol. 146, 264–269.

Zocchi, M.R., Pardi, R., Gromo, G., Ferreno, E., Ferreno, M.E., Besana, C. and Rugarli, C. (1985). Theophylline induced nonspecific suppressor activity in human peripheral blood lymphocytes. J. Immunopharmacol. 7, 217–234.

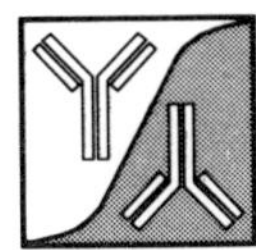

4. Ca^{2+}/Calmodulin-Dependent Cyclic Nucleotide Phosphodiesterase (PDE1)

Rajendra K. Sharma *and* Robert A. Hickie

1. Introduction 65
2. Purification and Characterization 66
3. Isoenzymes of CaM-PDE 66
   3.1 Demonstration of Isoenzymes 66
   3.2 Kinetic Properties 68
   3.3 Differential Activation by Calmodulin and Ca^{2+} 68
   3.4 Regulation by Phosphorylation 69
   3.5 Inhibitors 73
4. Activity in Cancer Cells 74
5. Conclusions 74
6. Acknowledgements 74
7. References 75

1. Introduction

The purpose of this chapter is to summarize some of the more significant advances in research on Ca^{2+}/calmodulin-dependent cyclic nucleotide phosphodiesterase (CaM-PDE, PDE1) during the past five years. Excellent reviews on this subject have been published previously (see Beavo, 1988, 1990; Sharma *et al.*, 1988; Wang *et al.*, 1990).

Kakiuchi and Yamazaki (1970) were the first to demonstrate the existence of a Ca^{2+}-dependent cyclic nucleotide phosphodiesterase (PDE) in rat brain. In addition, Kakiuchi *et al.* (1970) discovered an endogenous brain protein factor which could enhance the Ca^{2+} sensitivity of this enzyme. Independently, Cheung (1970,1971) established the existence of a protein activator in bovine brain which stimulated the PDE activity severalfold; Cheung coined the term "calmodulin" (CaM) for this protein activator. In 1973, Teo and Wang demonstrated that Kakiuchi's protein factor was identical to the protein activator of PDE discovered by Cheung. Stimulation of this PDE activity required the presence of both Ca^{2+} and CaM (Teo and Wang, 1973; Kakiuchi *et al.* 1973).

CaM-PDE has been demonstrated in most mammalian tissues (Kakiuchi *et al.*, 1975; Wells and Hardman, 1977; Beavo, 1988; Sharma *et al.*, 1988; Wang *et al.*, 1990). Mammalian brain contains the highest CaM-PDE activities, i.e. approximately 10 mg/kg in bovine brain (Sharma *et al.*, 1980). Bovine heart, the second most abundant source of the enzyme, contains about 10–20% of the activity found in brain (La Porte *et al.*, 1979). In addition to mammalian brain and heart, CaM-PDE activity has been detected in many other tissues, including pancreas (Rutten *et al.*, 1973), spermatozoa (Bhatnagar and Anand, 1982), chicken gizzard (Birnbaum and Head, 1983), quail oviduct (Dumas *et al.*, 1988), frog ovary (Jedlicki *et al.*, 1985), *Drosophila* head (Yamanaka and Kelly, 1981; Solti *et al.*, 1983; Walter and Kiger, 1984), mollusc nervous system (Calhoon and Gillette, 1983) and *Neurospora crassa* (Perez *et al.*, 1983). This enzyme has also been demonstrated in various immune cell types, including rat splenic lymphocytes (Hait and Weiss, 1977) and human peripheral blood lymphocytes (Thompson *et al.*, 1976) and monocytes (Thompson *et al.*, 1980). These lists indicate a widespread distribution of the enzyme in eukaryotes. The relative amount of CaM-PDE is very low compared to other PDEs; consequently it is difficult to determine CaM-PDE activity by measuring the existence of the enzyme level in the presence or absence of Ca^{2+} or a CaM antagonist. Many compounds (e.g.

Phosphodiesterase Inhibitors
ISBN 0–12–210720–9

phospholipids, fatty acids, gangliosides, proteases and an *Escherichia coli* glycolipid) can stimulate CaM-PDE in the absence of Ca^{2+} (Wolff and Brostrom, 1976; Hidaka *et al.*, 1977, 1978; Davis and Daly, 1980; Gietzen *et al.*, 1982; Walters and Jirsa, 1988). In view of this, it is difficult to detect Ca^{2+}/CaM activation in CaM-PDE which has been previously activated by one or more of these compounds. In a number of cells and tissues, such as pancreas (Vandermeers *et al.*, 1983), lung (Sharma and Wirch, 1979; Sharma and Wang, 1986a), spermatozoa (Wasco and Orr, 1984) and neutrophils (Engerson *et al.*, 1986; Grady and Thomas, 1986), early studies failed to detect CaM-PDE activity. More recent studies have, however, demonstrated that these tissues contain this form of PDE. In order to demonstrate the presence of CaM-PDE activity, it is sometimes necessary to separate the enzyme from the other multiple forms of PDE by column chromatography procedures. In most tissues, CaM-PDE activity is found predominantly in the cytosolic fraction with only a small amount of enzyme activity (10–20%) associated with particulate – particularly microsomal – fractions (Kakiuchi *et al.*, 1978; Grab *et al.*, 1981). In rat sperm, however, CaM-PDE is reported to exist exclusively in the particulate fraction (Wasco and Orr, 1984; Chaudhry and Casillas, 1988). It is not known whether the membrane-bound CaM-PDE(s) is(are) different from the soluble enzymes.

2. *Purification and Characterization*

Several groups of investigators have purified a number of CaM-PDE isoenzyme forms from bovine brain and heart to apparent homogeneity; these isoenzyme forms have different molecular weights as well as distinct catalytic and regulatory properties (Morrill *et al.*, 1979; La Porte *et al.*, 1979; Sharma *et al.*, 1980; Hansen and Beavo, 1982; Kincaid and Vaughan, 1983; Kincaid *et al.*, 1984; Shenolikar *et al.*, 1985; Sharma, 1991). The most efficient purification procedure for this enzyme is CaM affinity Sepharose 4B column chromatography. This column is commonly used to isolate proteins capable of undergoing Ca^{2+}-dependent association with CaM. The purified enzyme is activated severalfold by CaM in the presence of Ca^{2+} and this activation is completely inhibited when EGTA is present to chelate free Ca^{2+}. All pure CaM-PDE preparations essentially contain a single polypeptide band on SDS-PAGE and the reported subunit molecular masses fall within a narrow range of 58–63 kD (Klee *et al.*, 1979; Morrill *et al.*, 1979; Sharma *et al.*, 1980; Hansen and Beavo, 1982; Kincaid and Vaughan, 1983; Kincaid *et al.*, 1984). This variation was initially thought to arise from differences in SDS-PAGE conditions and/or from partial proteolysis and these investigators suggested that CaM-PDE consists of a single species since no significant differences in enzymatic and molecular properties were noted in enzyme preparations from different tissues. We now know that these discrepancies were due to the fact that there are distinct, but closely related, isoenzyme forms of CaM-PDE in different mammalian tissues, with each form having different properties. The availability of monoclonal antibodies (mAb) has greatly facilitated the direct demonstration of CaM-PDE isoenzymes and also improved their purification. Hansen and Beavo (1982), using mAb ACAP-1 immunoaffinity chromatography, have purified to homogeneity CaM-PDE from both bovine brain and heart. In 1991, Sharma purified CaM-PDE to apparent homogeneity from the total CaM-binding fraction of bovine heart in a single step by using C1 mAb immunoaffinity chromatography. This rapid purification method not only minimizes the possibility of post-homogenization proteolysis but also separates out other CaM-binding proteins and decreases the possibility of contamination with different CaM-PDE isoenzymes. Under these conditions, bovine heart was found to contain only one type of CaM-PDE.

3. *Isoenzymes of CaM-PDE*

3.1 DEMONSTRATION OF ISOENZYMES

Sharma *et al.* (1984) provided the first direct evidence for the existence of different forms of bovine brain CaM-PDE. A preparation of this enzyme was initially considered to be homogeneous (Sharma *et al.*, 1980) when examined by sodium dodecyl sulphate – polyacrylamide gel electrophoresis (SDS-PAGE) (Fig. 4.1A; lane 1), using the procedure of Weber and Osborn (1969). However, when SDS-PAGE was carried out using the procedure of Laemmli (1970), two major polypeptides were observed with different molecular weights of 60 and 63 kD (Fig. 4.1A; lane 2). Furthermore, peptide mapping by partial proteolysis (Sharma *et al.*, 1984) suggested that the two polypeptides were distinctly different (Fig. 4.1B; lanes 1 and 2). Western immunoblot analysis subsequently showed that some of the mAbs react with both 60 kD and 63 kD polypeptides, whereas others react with only the 60 kD polypeptide (Sharma *et al.*, 1984), further suggesting that the two polypeptides, though related, are immunologically different (Table 4.1). A 60 kD polypeptide-specific mAb, C1, has been used to study the nature of the two polypeptides (Sharma *et al.*, 1984). When the immunocomplex of CaM-PDE and C1 mAb was subjected to sucrose density centrifugation, three PDE activity peaks were found (Fig. 4.2A). Analysis of the protein in the activity peak fractions by SDS-PAGE showed that the first peak contained only the 63 kD polypeptide, whereas the second peak contained both the 63 kD and 60 kD polypeptides and the third peak contained only

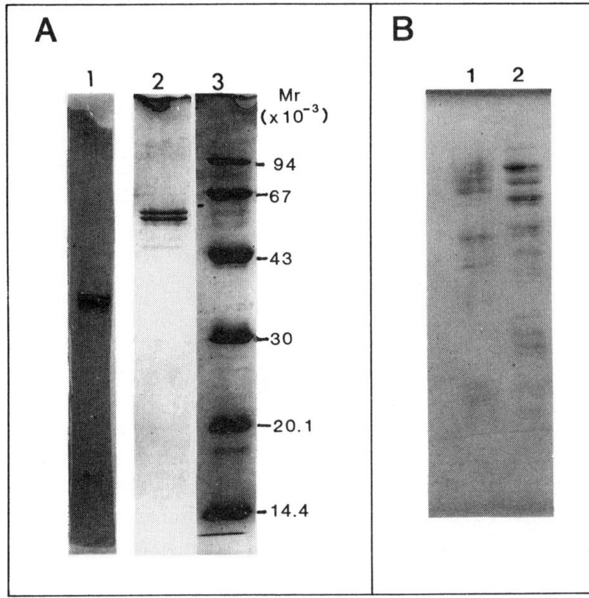

Figure 4.1 A. SDS-PAGE of CaM-PDE. Purified CaM-PDE was analysed according to the methods of Weber and Osborn (1969), lane 1, or Laemmli (1970), lane 2. Molecular weight standards are shown in lane 3. B. Peptide maps of CaM-PDE: 63 kD, lane 1; 60 kD, lane 2. For details of experimental conditions, see Sharma *et al.*, (1984).

Table 4.1 Cross-reaction of monoclonal antibodies with bovine CaM-PDE isoenzymes

	Monoclonal antibody			
Isoenzyme	*C1*[a]	*C2*[a]	*A6*[a]	*ACAP-16*[b]
Brain 60 kD isoenzyme	+	+	+	+
Brain 63 kD isoenzyme	–	–	+	–
Heart isoenzyme	+	+	+	+
Lung isoenzyme	+	+	+	+

[a] Antibody raised against purified bovine brain CaM-PDE containing both 60 kD and 63 kD isoenzymes (Sharma *et al.*, 1984; Sharma and Wang, 1986; Sharma, 1991).
[b] Antibody raised against purified bovine heart CaM-PDE (Hansen and Beavo, 1986).

the 60 kD polypeptide. The results of this study suggest that bovine brain contains three isoenzymes of CaM-PDE: two homodimeric proteins containing two subunits each of 63 kD and 60 kD, respectively, and one heterodimeric protein form containing a 63 kD and 60 kD subunit (Fig. 4.2B). Immunoaffinity chromatography was used to obtain individual isoenzymes (Fig. 4.3). The first peak, which did not interact with Cl mAb, appeared in the flow-through fractions and contained only the 63 kD homodimeric isoenzyme. The second peak, which was eluted with $MgCl_2$, was greatly enriched in 60 kD homodimeric isoenzyme; however, this peak always contained 5–7% 63 kD isoenzyme contamination. Since the brain is a highly heterogeneous organ, containing many different cell types, it is possible that the two different major isoenzymes (60 kD and

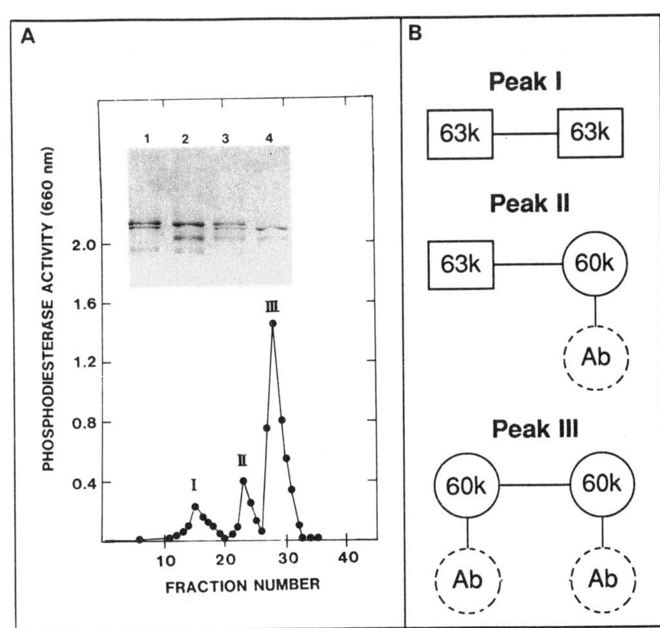

Figure 4.2 A. Sucrose density gradient centrifugation profile of immunocomplexes of bovine brain CaM-PDE with C1 monoclonal antibody. For details of experimental conditions, see Sharma *et al.*, (1984). B. Schematic illustration of peaks I, II and III from the density gradient.

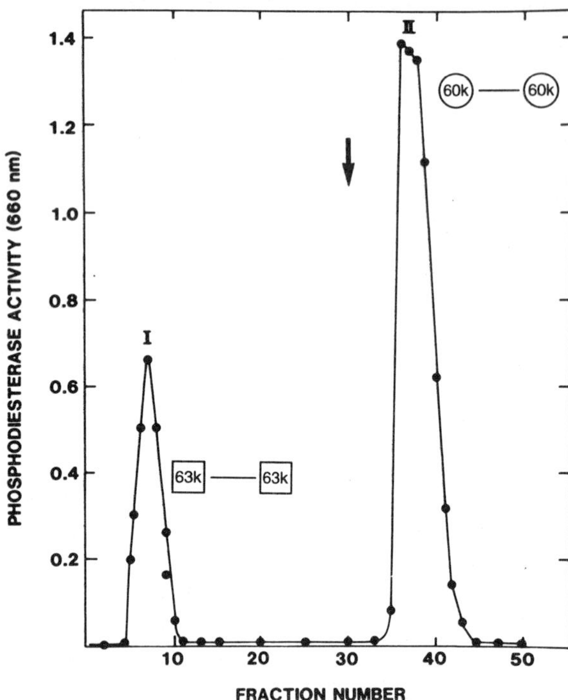

Figure 4.3 Separation of CaM-PDE isoenzymes by C1 monoclonal antibody/Sepharose 4B column chromatography. For details of experimental conditions, see Sharma *et al.* (1984).

63 kD) may have originated from different cell types. The existence in the pure enzyme sample of the heterodimer seems to suggest that the different subunit types are not totally cell-type specific. On the other hand, the existence of heterodimeric isoenzyme forms may arise during purification or during storage of the purified CaM-PDE (Sharma *et al.*, 1984). These results suggest that the two homodimeric isoenzymes, 63 kD and 60 kD, are the predominant forms. The monoclonal antibody C1, which reacts with 60 kD CaM-PDE from bovine brain, also cross-reacts with bovine heart (Sharma, 1991) and bovine lung (Sharma and Wang, 1986a). The bovine lung CaM-PDE isoenzyme is different from other CaM-PDEs in that the lung isoenzyme contains CaM as a subunit (Sharma and Wang, 1986a). In addition, the C1 monoclonal antibody also cross-reacts with several tissues examined, such as eye, liver, kidney, spleen and uterus (Sharma *et al.*, unpublished observations), suggesting the presence of this isoenzyme form in these tissues as well. Furthermore, a bovine heart CaM-PDE mAb, ACAP–1, produced by Hansen and Beavo (1986) was also found to cross-react with the 60 kD but not the 63 kD isoenzyme from bovine brain. These results are summarized in Table 4.1.

Another isoenzyme has been purified from bovine brain (Shenolikar *et al.*, 1985) which has a higher molecular mass (150 kD with subunit mass of 75 kD) than any of the other known CaM-PDE isoenzymes from bovine brain. This novel isoenzyme is strictly specific for cGMP (Shenolikar *et al.*,1985). It has also been reported that two forms of CaM-PDE isoenzyme can be detected in brain extracts by isoelectric focusing (Kincaid *et al.*, 1984). Another CaM-PDE isoenzyme from bovine testis has been purified, which exists as a monomeric protein with a molecular mass of 70 kD; however, it dimerizes in the presence of Ca^{2+} and CaM to form a molecular mass of 180 kD (Rossi *et al.*, 1988).

3.2 KINETIC PROPERTIES

Table 4.2 shows kinetic data of more recent studies of CaM-PDE obtained from several mammalian tissues by various laboratories. The results of early kinetic studies of CaM-PDE showed considerable discrepancies in K_m and/or *V*max. These discrepancies may have been due to the purity of the enzyme. The kinetic data presented in the table were obtained from either homogeneous or highly purified preparations of CaM-PDE. Table 4.2 also shows that isoenzymes from bovine brain, heart and lung and from porcine coronary arteries generally have a higher affinity for cGMP than cAMP. The 60 kD isoenzymes from these tissues have similar kinetic properties whereas the bovine brain 63 kD CaM-PDE isoenzyme and rat pancreas Peak II have a 2- to 3-fold higher affinity for both substrates, cAMP and cGMP. Furthermore, the novel bovine brain CaM-PDE (Shenolikar *et al.*, 1985) and that from rat testis have similar affinities for both cAMP and cGMP.

3.3 DIFFERENTIAL ACTIVATION BY CALMODULIN AND Ca^{2+}

Detailed studies on the activation of CaM-PDE by CaM have been carried out using bovine brain and heart CaM-PDE isoenzymes. The general mechanism of action has been reviewed by Sharma *et al.* (1988) and Wang *et al.* (1990).

As discussed above, the brain 63 kD CaM-PDE isoenzyme is kinetically different from the brain 60 kD, heart and lung CaM-PDE isoenzymes (Table 4.2). Although the latter isoenzymes are almost identical in terms of immunological properties (Sharma and Wang, 1986a; Sharma, 1991; Sharma and Kalra, 1994a), the heart CaM-PDE isoenzyme has a higher affinity for CaM (Mutus *et al.*, 1985; Hansen and Beavo, 1986; Sharma, 1991) than the bovine brain isoenzymes. The difference in CaM affinity exhibited by the heart and brain enzymes may be related to the relative concentrations of CaM in these two tissues (Klee and Vanaman, 1982). The lung CaM-PDE isoenzyme has the highest apparent affinity for CaM, since it contains CaM as a subunit (Sharma and Wang, 1986a). Activation of the lung isoenzyme cannot be inhibited by common CaM

Table 4.2 Kinetic properties of CaM-PDE enzymes from mammalian tissues

Source of enzyme	K_m *(μM)* *cAMP*	*cGMP*	*Vmax (ratio) (cAMP/cGMP)*	*Reference*
Bovine brain				Sharma & Kalra (1994a)
60 kD Isoenzyme	12	1.2	0.3	
63 kD Isoenzyme	35	2.7	1.8	
Bovine brain	2.9	2.7	0.07	Shenolikar *et al.* (1985)
Bovine spermatozoa	7.5, 95	–	–	Wasco and Orr (1984)
Bovine coronary artery	2.5, 140	3.0,17	0.18	Weishaar *et al.* (1986)
Bovine lung	42	2.8	1.7	Sharma and Kalra (1994a)
Bovine heart	215	9.0	3.3	Ho *et al.* (1976)
	40	3.2	3.0	Sharma and Kalra (1994)
Guinea-pig left ventricle	0.8	0.9, 53	1.1	Reeves *et al.* (1987)
Human cardiac ventricle	0.75	1.0	1.0	Keravis *et al.* (1987)
Pig coronary artery	70	3.0	0.1	Keravis *et al.* (1987)
Rat brain				Hansen and Beavo (1986)
Low molecular weight isoenzyme	–	–	1.2	
High molecular weight isoenzyme	–	–	0.68	
Rat pancreas				Vandermeers *et al.* (1983)
Peak I	0.35	0.16	1.4	
Peak II	9.0	1.7	1.1	
Rat testis				Purvis *et al.* (1981)
Peak I	1.0	1.0	–	
Peak II	30	3.0	–	
Peak III	–	1.5	–	

antagonists, such as compound 48/80, or by a CaM-binding protein, calcineurin (Sharma and Wang, 1986a). At present, the significance of CaM as a tightly bound subunit is not known, although results suggest that the CaM subunit in the lung isoenzyme is not subject to competition by other CaM-binding proteins after an increase in intracellular free Ca^{2+} concentrations in a stimulated cell. Similar results have been also observed in porcine brain, showing that this CaM-PDE has a lower affinity for CaM than the isoenzyme from pig artery (Keravis *et al.*, 1986).

The interaction of CaM and Ca^{2+} with CaM-PDE in a CaM-stimulated reaction depends on the binding of Ca^{2+} to CaM and the association of Ca^{2+}/CaM with the enzyme. It has been observed (Sharma and Kalra, 1994a) that, at an identical CaM concentration, the bovine heart CaM-PDE isoenzyme is stimulated at much lower Ca^{2+} concentrations than is the bovine brain isoenzyme (Table 4.3). Although the physiological significance of the observed differential Ca^{2+} sensitivity of the CaM-PDE isoenzymes is not known, these results suggest that the differential Ca^{2+} affinity of the tissue-specific isoenzymes may be a mechanism by which the CaM-regulatory reactions are adapted in the respective tissues. Activation of the bovine lung CaM-PDE isoenzyme by Ca^{2+} is unaffected by the addition of exogenous CaM (Table 4.3), suggesting that this isoenzyme does not undergo Ca^{2+}-dependent reversible association with CaM. Since CaM-PDE isoenzymes respond differently to CaM and Ca^{2+} stimulation, this differential regulation may represent a fine-tuning mechanism for CaM action.

3.4 REGULATION BY PHOSPHORYLATION

The main difference in the regulation of CaM-PDE isoenzymes appears to be their response to phosphorylation. The phosphorylation of CaM-PDE by cAMP-dependent protein kinase (PKA) (Sharma *et al.*, 1980) and CaM-dependent protein kinase II (Fukunaga *et al.*, 1984) was reported before the discovery of the CaM-PDE isoenzymes (Sharma *et al.*, 1984). These studies were extended using purified CaM-PDE isoenzymes and suggested that these isoenzymes are differentially regulated by protein phosphorylation mechanisms (Sharma and Wang, 1985, 1986b,c; Sharma, 1991). The main difference is that heart and brain 60 kD CaM-PDE isoenzymes are substrates for PKA (Sharma and Wang, 1984; Sharma, 1991; Florio *et al.*, 1994), whereas the brain 63 kD isoenzyme is phosphorylated by CaM-dependent protein kinases in a Ca^{2+}/CaM-dependent

Table 4.3 Ca^{2+} activation of CaM-PDE isoenzymes at various concentrations of CaM

[Calmodulin] (μM)	[Ca^{2+}] (μM) Required for half-maximal activation			
	Heart CaM-PDE isoenzyme	Brain 60 kD CaM-PDE isoenzyme	Brain 63 kD CaM-PDE isoenzyme	Lung CaM-PDE isoenzyme
0.04	0.50	ND	ND	ND
1.00	0.08	0.090	0.70	0.15
10.0	0.01	0.35	0.30	0.15

ND, not determined.

reaction (Sharma and Wang, 1986b; Hashimoto *et al.*, 1989; Zhang *et al.*, 1993a). Therefore, these phosphorylation reactions are highly specific for the respective isoenzyme forms (Table 4.4).

A number of studies have revealed highly complex mechanisms for the regulation of the CaM-PDE isoenzymes by protein kinases and protein phosphatase (Sharma and Wang, 1985, 1986b,c; Sharma, 1991, 1995; Sharma and Kalra, 1994b). Various second messenger effects on the CaM-PDE isoenzymes are summarized in Table 4.4. To achieve meaningful regulation, the multiple regulatory activities of each isoenzyme have to interact in a co-ordinated manner. Working hypotheses have been proposed to integrate the multiple second messenger-dependent regulatory activities directed towards the individual isoenzymes during the surge in cell Ca^{2+} and cAMP (Sharma and Wang, 1986b,c, Sharma *et al.* 1988; Wang *et al.*, 1990; Sharma and Kalra, 1994b; Sharma, 1995). These hypotheses can work only if the multiple CaM actions are temporally separated during the surges of cAMP and Ca^{2+} in the cell.

Although these isoenzymes are phosphorylated by different protein kinases, they can be dephosphorylated by a single CaM-dependent protein phosphatase (Table 4.4). Phosphorylation results in a decrease in the affinity of the isoenzyme for CaM but does not abolish CaM activation (Sharma and Wang, 1985, 1986b; Sharma, 1991). Furthermore, the phosphorylated CaM-PDE isoenzymes require a higher concentration of Ca^{2+} than the non-phosphorylated isoenzymes. In all cases, phosphorylation can be reversed by CaM-dependent protein phosphatase and this dephosphorylation is accompanied by an increase in the affinity of CaM-PDE for CaM. For each of the CaM-PDE isoenzymes, a working hypothesis has been presented to indicate how the multiple regulatory reactions may be organized in the cells to achieve advantageous control of the cAMP concentration (Fig. 4.4). In most cases, cell activation involves transitory increases in both cAMP and cell Ca^{2+}. The two signal systems interact with each other through many regulatory reactions and a change in concentration of one second messenger will affect the other. Therefore, the two signal fluxes are closely coupled in a dynamic fashion. The operation of the different regulatory mechanisms on the CaM-PDE may be temporally separated during the signal fluxes. Consequently, an initial increase in cAMP concentration during cell activation may bring about the phosphorylation of heart and 60 kD bovine brain CaM-PDE isoenzymes with concomitant phosphodiesterase inhibition

Table 4.4 Characteristics of the regulation of CaM-PDE isoenzymes by phosphorylation and dephosphorylation mechanisms

Regulatory factor/ characteristic	Isoenzyme		
	Bovine heart CaM-PDE isoenzyme	Bovine brain 60 kD CaM-PDE isoenzyme	Bovine brain 63 kD CaM-PDE isoenzyme
Protein kinase	cAMP-dependent protein kinase	cAMP-dependent protein kinase	CaM-dependent protein kinase II
Stoichiometry	1 mol phosphate/mol subunit	1 mol phosphate/mol subunit	1 mol phosphate/mol subunit
Effect on PDE	Decreases CaM affinity, increases K_a of Ca^{2+} activation	Decreases CaM affinity, increases K_a of Ca^{2+} activation	Decreases CaM affinity, increases K_a of Ca^{2+} activation
Effectors	Ca^{2+}/CaM inhibits phosphorylation by binding to PDE	Ca^{2+}/CaM inhibits phosphorylation by binding to PDE	Ca^{2+}/CaM does not inhibit phosphorylation
Protein phosphatase	CaM-dependent protein phosphatase	CaM-dependent protein phosphatase	CaM-dependent protein phosphatase

at the low concentrations of Ca^{2+} existing at the early stage of cell activation. As a result, the intracellular concentrations of cAMP are increased (Fig. 4.4A). At the later stage of cell activation (when the cell Ca^{2+} concentration is increased) the CaM-dependent protein phosphatase is activated to dephosphorylate the heart and 60 kD CaM-PDE isoenzymes; the dephosphorylated isoenzymes are then fully activated by Ca^{2+} and CaM. Ca^{2+}/CaM also inhibits the phosphorylation of heart and 60 kD CaM-PDE isoenzymes; thus, the dephosphorylated form of the isoenzyme will be maintained even when the cAMP concentration is high in the cells. The concerted actions of these regulatory mechanisms on the heart and 60 kD CaM-PDE isoenzymes bring about a rapid decline in cAMP concentrations in the cell (Fig. 4.4A).

In contrast to heart and 60 kD CaM-PDE isoenzymes, the 63 kD CaM-PDE isoenzyme has been shown to undergo a Ca^{2+} and CaM-dependent phosphorylation that is accompanied by an increase in the Ca^{2+} concentration required for the enzyme activation by CaM (Sharma and Wang, 1986b). The 63 kD CaM-PDE is under multiple, and often antagonistic, CaM regulatory actions (Table 4.4). Earlier we proposed a working hypothesis (Fig. 4.4B) suggesting that the CaM actions are temporally separated, with the activation of CaM-dependent protein kinase(s) occurring at an early stage and activation of the 63 kD CaM-PDE isoenzyme and the phosphatase occurring at the later stage during a Ca^{2+} surge (Sharma and Wang, 1986b,c). We postulated that, in order to achieve such temporal separation of the CaM actions, CaM activation of the CaM-dependent protein kinase(s) should require a much lower Ca^{2+} concentration than the CaM activation of CaM-dependent protein phosphatase and the 63 kD CaM-PDE isoenzyme (Sharma and Wang, 1986b,c). To test the validity of this working hypothesis, we purified and characterized one of the CaM-dependent protein kinases which was shown to be a member of the CaM-dependent protein kinase II family (Goldering *et al.*, 1983; Kennedy *et al.*, 1983; Kuret and Schulman, 1984). On the basis of its molecular mass, subunit size, protein substrate specificity and mode of autophosphorylation, the purified bovine brain CaM-dependent protein kinase is considered to belong to the CaM-dependent protein kinase II family (Fukunaga *et al.*, 1984; Hashimoto *et al.*, 1989). Phosphorylation of the

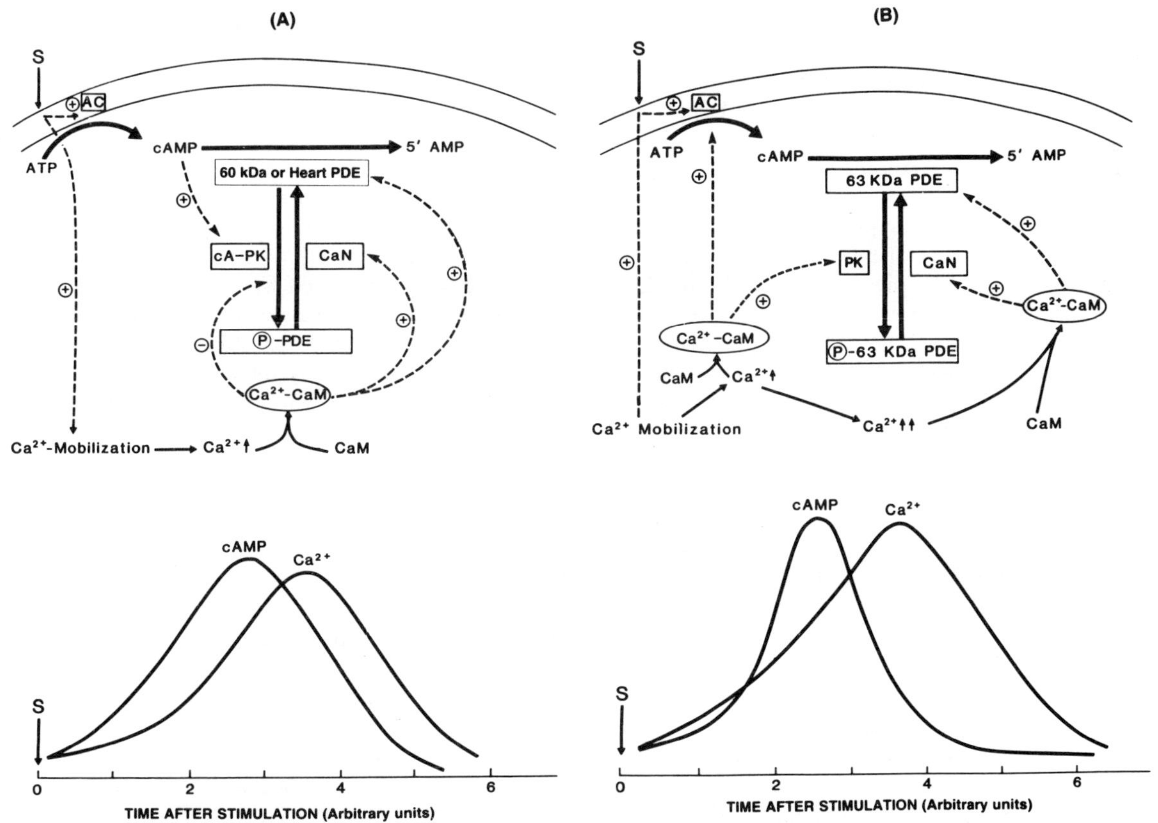

Figure 4.4 Hypotheses for the temporally separated regulation of 60 kD (A) and 63 kD (B) CaM-PDE isoenzymes by Ca^{2+} and cAMP. AC, adenylate cyclase; PDE, CaM-PDE; CaN, CaM-dependent protein phosphatase; cA-PK, cAMP-dependent protein kinase (PKA); PK, CaM-dependent protein kinase; Ⓟ, phosphorylated; +, activation; –, inhibition. Upper panels, organization of regulatory reactions; lower panels, stimulated Ca^{2+} and cAMP fluxes.

63 kD CaM-PDE isoenzyme by CaM-dependent protein kinase II resulted in a maximal incorporation of one mole of phosphate per mole of subunit (Zhang *et al.*, 1993a), instead of two moles of phosphate per mole of subunit observed when the total CaM-binding protein fraction was used as a source of protein kinase (Sharma and Wang, 1986b). Preliminary results suggested that the total CaM-binding protein fraction contains two CaM-dependent protein kinases, one of which has been purified to near homogeneity (Zhang *et al.*, 1993a). The second CaM-dependent protein kinase has an apparent molecular weight of 180 kD and has not yet been purified (Zhang *et al.*, 1993a); this CaM-dependent protein kinase is responsible for phosphorylation of this distinct site.

The phosphorylation of 63 kD CaM-PDE by CaM-dependent protein kinase II depends absolutely on the presence of Ca^{2+} and CaM. After phosphorylation, further increases in Ca^{2+} concentrations are required for enzyme activation by CaM (Zhang *et al.*, 1993a). We previously postulated this CaM-dependent protein kinase to be activated by CaM at much lower concentrations of Ca^{2+} than the CaM-dependent protein phosphatase and the 63 kD CaM-PDE isoenzyme (Sharma and Wang, 1986b). However, this hypothesis was not supported when the concentration-dependence of activation by Ca^{2+} of the 63 kD CaM-PDE isoenzyme was compared with CaM-dependent protein kinase II at identical concentrations of CaM (R.K. Sharma, unpublished observations). These results suggest that CaM-dependent protein kinase II and the 63 kD CaM-PDE isoenzyme have similar Ca^{2+} concentration-dependences at identical concentrations of CaM.

Like other members of the protein kinase family, this CaM-dependent protein kinase II is also autophosphorylated rapidly in the presence of Ca^{2+} and CaM (Zhang *et al.*, 1993a,b); however, it is converted into a Ca^{2+}-independent activity, i.e. the phosphorylation of 63 kD CaM-PDE isoenzyme by autophosphorylated protein kinase becomes Ca^{2+}-independent (Zhang *et al.*, 1993a), suggesting that binding of CaM does not change the substrate activity of this PDE isoenzyme. Similar results have been reported which indicate that the CaM-dependent protein kinase II from a variety of different sources can be converted into a Ca^{2+}-independent form after autophosphorylation (Lai *et al.*, 1986; Hashimoto *et al.*, 1987, 1989; Colbran, 1992, 1993; Zhang *et al.*, 1993a; Bronstein *et al.*, 1993).

We conclude that the autophosphorylation reaction can be used to achieve the required temporal separation of the activation of protein kinase from that of phosphatase and/or 63 kD CaM-PDE isoenzyme. Therefore, upon very brief exposure to high concentrations of Ca^{2+}, the CaM-dependent protein kinase II becomes active and insensitive to subsequent increases in Ca^{2+} concentration, whereas the activation of 63 kD CaM-PDE requires the continued presence of high concentrations of Ca^{2+} (Zhang *et al.*, 1993a).

There are several possible ways by which such a brief exposure of CaM-dependent protein kinase II to high concentration of Ca^{2+} can occur at the onset of Ca^{2+} flux. Recent studies of agonist-induced Ca^{2+} flux in single cells have suggested that the overall Ca^{2+} surge may be composed of a series of rapid Ca^{2+} transients (Woods *et al.*, 1986, 1987; Monck *et al.*, 1988). Such Ca^{2+} transients may therefore serve to trigger the autophosphorylation of the protein kinase at the onset of the Ca^{2+} surge. Alternatively, it is possible that CaM-dependent protein kinase II may be localized proximally to the sites of Ca^{2+} entry and therefore autophosphorylated rapidly at the onset of a Ca^{2+} flux. Immunocytochemical studies have shown that CaM-dependent protein kinase II is localized at the inner surface of plasma membranes, as well as at the outer surface of mitochondria and at synaptic vesicles and microtubules (Ouimet *et al.*, 1984). Therefore, autophosphorylation of CaM-dependent protein kinase II provides an additional mechanism to be incorporated into a revised hypothesis of the regulation of the 63 kD CaM-PDE isoenzyme; the revised hypothesis is schematically presented in Fig. 4.5.

In addition to the temporal separation, a hypothesis is required to include a number of other regulation possibilities. For example, autophosphorylation of CaM-dependent protein kinase II can be reversed by protein phosphatase I (Saitoh *et al.*, 1987; Colbran, 1992; Bronstein *et al.*, 1993), which is subjected to the protein phosphatase inhibitor I. When cAMP levels rise in the cell, PKA phosphorylates inhibitor I to activate it; phosphorylated phosphatase inhibitor I can then inhibit protein phosphatase I (Huang and Glinsmann, 1976; Hemmings *et al.*, 1984). When inhibitor I is dephosphorylated and inactivated by CaM-dependent protein phosphatase, protein phosphatase I is reactivated. As a result, cAMP may exert an inhibitory effect on the 63 kD CaM-PDE isoenzyme through a regulatory cascade involving protein phosphatase inhibitor I, protein phosphatase I and CaM-dependent protein kinase II. This complex regulatory interaction is in agreement with the previously suggested role for the 63 kD CaM-PDE isoenzyme in the dynamic coupling of cAMP and Ca^{2+} fluxes in the cell (Sharma and Wang, 1986b).

In summary, during the early stage of cell activation, the initial increases in cAMP and Ca^{2+} cause a temporary suppression of 63 kD CaM-PDE isoenzyme activity to maintain the rise in cAMP concentration. As the Ca^{2+} concentration in the cell is subsequently elevated, the CaM-dependent protein phosphatase is activated to reverse phosphorylation of – and, thereby, to reactivate – the 63 kD CaM-PDE isoenzyme. Since CaM-dependent protein phosphatase also dephosphorylates protein phosphatase inhibitor I to cause the reactivation of protein phosphatase I, autophosphorylation of CaM-dependent protein kinase II is also

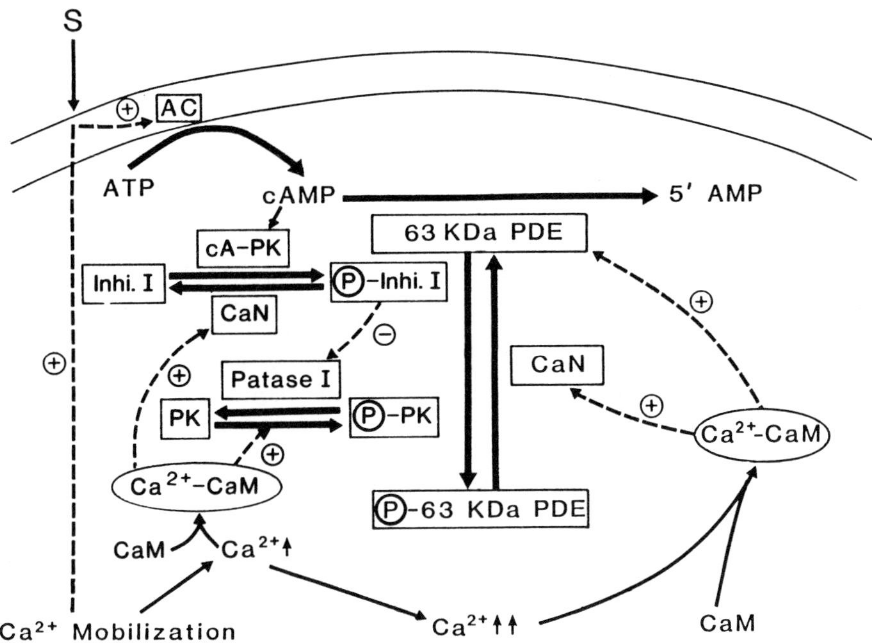

Figure 4.5 Schematic diagram of the regulation of 63 kD CaM-PDE isoenzyme by Ca^{2+} and cAMP as mediated by the autophosphorylation mechanism of CaM-dependent protein kinase II. The scheme depicts the complex interactions among: cA-PK, cAMP-dependent protein kinase (PKA); Patase I, protein phosphatase I; ℗-Inhi. I, phosphorylated protein inhibitor I; CaN, CaM-dependent protein phosphatase. Broken arrow, early events; solid arrow, late events.

reversed, so that rephosphorylation of the 63 kD CaM-PDE isoenzyme will no longer occur as Ca^{2+} concentrations subside in the cell.

3.5 INHIBITORS

To address the question of the function of individual forms of phosphodiesterase, a pharmacological approach is also used. A large number of pharmacological agents are capable of inhibiting CaM-PDE (Wang *et al.*, 1990). Many of the inhibition studies were carried out by using either purified or partially purified CaM-PDE; however, it is not clear from previous studies which of the specific CaM-PDE isoenzymes were used. Very recently, we have demonstrated that isoenzymes of CaM-PDE may be distinguished by the kinetics of inhibition by some pharmacological agents (Sharma and Kalra, 1993). We reported that ginsenosides can inhibit only heart CaM-PDE and 60 kD CaM-PDE isoenzymes but not the brain 63 kD isoenzyme (Sharma and Kalra, 1993). These results suggest that ginsenosides compete with CaM to prevent activation of the enzyme and that both the heart and 60 kD CaM-PDE isoenzymes have similar affinities for the ginsenosides (Table 4.5). Inhibition of the heart CaM-PDE and 60 kD CaM-PDE isoenzymes can be overcome by the addition of excess CaM, suggesting that ginsenosides are specific and act by simple competition with free CaM (Sharma and Kalra, 1993).

In addition, Wu *et al.* (1992) have reported that 3-isobutyl-1-methylxanthine (IBMX) exhibits marked differences in its inhibition potency for the 60 kD and 63 kD CaM-PDE isoenzymes: the 60 kD CaM-PDE isoenzyme displays significantly higher affinity (≈40-fold) for IBMX than does the 63 kD CaM-PDE isoenzyme. On the other hand, nicardipine does not show any significant difference in the inhibition of 60 kD and 63 kD CaM-PDE isoenzymes; moreover, inhibition of these isoenzymes never approaches 100%. Therefore, the differential inhibition of 60 kD and 63 kD CaM-PDE isoenzymes by IBMX gives further evidence for differences in active regions of these isoenzymes.

Sharma *et al.* (1991) have reported that certain peptide antagonists (sarilesin and sarmesin), as well as some peptide agonists (angiotensin II and angiotensin III), inhibit the 60 kD CaM-PDE isoenzyme. However, sarilesin and angiotensin III are more potent inhibitors of the 60 kD CaM-PDE isoenzyme than are sarmesin and angiotensin II (Sharma *et al.*, 1991). Nakanishi *et al.* (1992) demonstrated that compound KS-505a, isolated from *Streptomyces argenteolus*, inhibits CaM-PDE from bovine brain and bovine heart. Bovine brain CaM-PDE has a 20-fold higher affinity than the bovine heart enzyme for KS-505a. It is not clear from this study which isoenzyme from bovine brain was inhibited, since partially purified enzyme was used.

Table 4.5 Effects of ginsenosides Rb, Rc and Re on CaM-PDE isoenzymes

	IC_{50} (μg/ml) *for Ginsenoside*		
CaM-PDE isoenzyme	*Rb*	*Rc*	*Re*
Bovine brain 60 kD isoenzyme	7.0	4.0	12.0
Bovine heart isoenzyme	7.5	4.0	14.0

IC_{50} is the concentration of ginsenoside required to produce 50% inhibition of CaM-PDE activity (Sharma and Kalra, 1993).

4. *Activity in Cancer Cells*

It has been evident for some time that cyclic nucleotides play a role in cell growth regulation and that this role may be altered in diseases characterized by unregulated cell growth, such as cancer. This altered role appears to be related to changes in cyclic nucleotide levels resulting from corresponding changes in activity of synthetic enzymes (cyclases) and/or hydrolytic enzymes (PDE) (Hickie *et al.*, 1974, 1975a,b; Hickie 1978; Helfman and Kuo, 1982a; Whitfield, 1990). The cyclic nucleotide phosphodiesterases have been studied in a variety of tumours, for example leukaemic lymphocytes and lymphoblasts (Hait and Weiss, 1977, 1979; Epstein and Hachisu, 1984; Onali *et al.*, 1985), Hodgkin's disease lymphocytes (Aleksijevic *et al.*, 1987), hepatomas (Hickie *et al.*, 1975b, 1977; Hickie, 1978; Wei and Hickie, 1983; Turnbull and Hickie, 1984), neuroblastoma (Kumar *et al.*, 1975) and breast cancer (Singer *et al.*, 1976). Taken together, the results suggest that the ratio of cGMP to cAMP tends to be higher in tumour cells, primarily due to relative changes in PDE activity.

High levels of cellular cAMP (or its stable analogues) can inhibit the growth of tumour cells and promote differentiation (Hickie, 1978; Puck 1987; Cho-Chung, 1990; Van Lookeren *et al.*, 1991; Rohlff *et al.*, 1993) by modulating transcription and gene expression (Nagamine and Reich, 1985; Najam *et al.*, 1986; Puck 1987; Ally *et al.*, 1989; Mirossay *et al.*, 1992). Alternatively, PDEs can also be considered as potential targets for tumour growth inhibition (Weiss and Hait, 1977; Bertram *et al.*, 1982; Helfman and Kuo, 1982b; Drees *et al.*, 1993). There is, however, very little known at present about the role of specific PDE isoenzymes in growth regulation or their change in activity in tumours. One isoenzyme form whose activity is probably altered in tumours is CaM-PDE, since it has been shown that both Ca^{2+} levels (Hickie and Kalant, 1967; Wei *et al.*, 1982; Hickie *et al.*, 1983) and CaM levels (Wei and Hickie, 1981; Wei *et al.*, 1982; Hickie *et al.*, 1983) are significantly elevated (2–3-fold). It has been further demonstrated (Hickie *et al.*, 1983, 1984; Wei *et al.*, 1983; Gehrig *et al.*, 1984; Borsa *et al.*, 1986; Vandonselaar *et al.*, 1994) that drugs which inhibit CaM activity – either by binding directly to CaM (e.g. trifluoperazine) or by reducing the availability of Ca^{2+} to CaM (e.g. verapamil) – can inhibit the growth of human tumours. There is a good correlation between anti-CaM activity of the drug and its tumour-inhibiting efficacy (Wei *et al.*, 1983; Hickie *et al.*, 1984). Clinical studies are currently under way in patients with refractory tumours using combinations of CaM antagonists with conventional anti-cancer drugs; the results obtained so far seem promising (Hickie *et al.*, unpublished observations). Concurrent studies are also under way to investigate the expression of CaM genes in human melanocytes and melanomas (Hickie *et al.*, 1992). These tissues have been found to possess three CaM genes, which have been mapped to three different chromosomes (McPherson *et al.*, 1991). These genes have recently been sublocalized to chromosomes 14 q24-q31, 2 p21.1-p21.3 and 19 q13.2-q13.3 (Berchtold *et al.*, 1993). The expression of one of these genes appears to be linked to cell growth regulation whereas the other two genes are constitutively expressed under conditions that modify growth (Hickie *et al.*, 1992).

In summary, it is evident that cyclic nucleotides (particularly cAMP) can influence cell growth and are altered in malignancy. These changes are likely to be related to altered activities of the phosphodiesterases. In view of the changes in Ca^{2+} and CaM levels in tumours, there is a need for future studies to examine, in more detail, the role of CaM-PDE in cell growth regulation and cancer.

5. *Conclusions*

CaM-PDE is one of the key enzymes involved in the complex interaction between the cyclic nucleotide and Ca^{2+} second messenger systems. The activity of CaM-PDE is found to be widely distributed. CaM-PDE exists in different isoforms which exhibit distinct molecular and/or catalytic properties. Immunological, kinetic, activation and regulatory characterizations have revealed subtle differences between these isoenzymes. Accumulating evidence suggests that the activity of CaM-PDE is selectively regulated by cross-talk between Ca^{2+} and cAMP signalling pathways. However, the exact physiological functions of CaM-PDE isoenzymes are still not clear since most studies undertaken so far have been on *in vitro* systems. In view of the importance of these isoenzymes *in vivo*, it is essential that the emphasis of further studies on CaM-PDE isoenzymes be directed to *in vivo* systems of both normal cells and abnormal cells (such as tumours).

6. *Acknowledgements*

This work was supported by grants from the Heart and Stroke Foundation of Saskatchewan to R.K. Sharma and from the Saskatchewan Cancer Foundation to R.A. Hickie. We thank Ms Suniti Saini and Ms Linda Cronin for typing the manuscript.

7. *References*

Aleksijevic, A., Lugnier, C., Giron, C., Mayer, S., Stoclet, J.C. and Lang, J.M. (1987). Cyclic AMP and cyclic GMP phosphodiesterase activities in Hodgkin's disease lymphocytes. Int. J. Immunopharmacol. 9, 525–531.

Ally, S., Clair, T., Katsaros, D., Tortora, G., Yokozaki, H., Finch, R.A., Avery, T.L. and Cho-Chung, Y.S. (1989). Inhibition of growth and modulation of gene expression in human lung carcinoma in athymic mice by site-selective 8-Cl-cyclic adenosine monophosphate. Cancer Res. 49, 5650–5655.

Beavo, J.A. (1988). Multiple isozymes of cyclic nucleotide phosphodiesterase. Adv. Second Messenger Phosphoprotein Res. 22, 1–39.

Beavo, J.A. (1990). Multiple phosphodiesterase isozymes: background, nomenclature and implications. In "Cyclic Nucleotide Phosphodiesterases: Structure, Regulation and Drug Action" (eds. J. Beavo and M.D. Houslay), pp. 3–15. John Wiley, Chichester.

Berchtold, M.W., Egli, R., Rhyner, J.A., Hameister, H. and Strehler, E.E. (1993). Localization of the human *bona fide* calmodulin genes CALM 1, CALM 2, and CALM 3 to chromosomes 14 q24-q31, 2 p21.1-p21.3, and 19 q13.2-q13.3. Genomics 16, 461–465.

Bertram, J.S., Bertram, B.B. and Janik P. (1982). Inhibition of neoplastic cell growth by quiescent cells is mediated by serum concentration and cAMP phosphodiesterase inhibitors. J. Cell. Biochem. 18, 515–538.

Bhatnagar, S.K. and Anand, S.R. (1982). Cyclic nucleotide phosphodiesterase activity in midpiece and tail of buffalo spermatozoa and its role in sperm motility. Biochim. Biophys. Acta 716, 133–139.

Birnbaum, R.J. and Head, J.E. (1983). Studies on the soluble phosphodiesterases of chicken gizzard smooth muscle. Biochem. J. 215, 627–636.

Borsa, J., Enspenner, M., Sargent, M.D. and Hickie, R.A. (1986). Selective cytotoxicity of calmidazolium and trifluoperazine for cycling versus noncycling C3H 10 T 1/2 cells *in vitro*. Cancer Res. 46, 133–136.

Bronstein, J.M., Farber, D.B. and Wasterlain, C.G. (1993). Regulation of type-II calmodulin kinase: functional implications. Brain Res. Brain Res. Rev. 18, 135–147.

Calhoon, R.D. and Gillette, R. (1983). Ca^{2+} activated and pH sensitive cyclic AMP phosphodiesterase in the nervous system of the mollusc *Pleurobranchaea*. Brain Res. 271, 371–374.

Chaudhry, P.S. and Casillas, E.R. (1988). Calmodulin-stimulated cyclic nucleotide phosphodiesterase in plasma membranes of bovine epididymal spermatozoa. Arch. Biochem. Biophys. 262, 439–444.

Cheung, W.Y. (1970). Cyclic 3′:5′-nucleotide phosphodiesterase: determination of an activator. Biochem. Biophys. Res. Commun. 38, 533–538.

Cheung, W.Y. (1971). Cyclic 3′:5′-nucleotide phosphodiesterase. J. Biol. Chem. 246, 2859–2869.

Cho-Chung, Y.S. (1990). Role of cyclic AMP receptor proteins in growth, differentiation and suppression of malignancy: new approaches to therapy. Cancer Res. 50, 7093 -7100.

Colbran, R.J. (1992). Regulation and role of brain calcium/calmodulin-dependent protein kinase II. Neurochem. Int. 21, 469–497.

Colbran, R.J. (1993). Inactivation of Ca^{2+}/calmodulin-dependent protein kinase II by basal autophosphorylation. J. Biol. Chem. 268, 7163–7170.

Davis, C.W. and Daly, J.W. (1980). Activation of rat cerebral cortical 3′:5′-cyclic nucleotide phosphodiesterase activity by gangliosides. Mol. Pharmacol. 17, 206–211.

Drees, M., Zimmerman, R. and Eisenbrand, G. (1993). 3′:5′-cyclic nucleotide phosphodiesterase in tumor cells as potential target for tumor growth inhibition. Cancer Res. 53, 3058–3061.

Dumas, M.Y., Fanidi, A., Pageaux, J.F., Courion, C., Némoz, G., Prigent, A.F., Pacheco, H. and Laugier, C. (1988). Cyclic nucleotide phosphodiesterase activity in the quail oviducts: hormonal regulation and involvement in estrogen-induced growth. Endocrinology 122, 165–172.

Engerson, T., Legendre, J.L. and Johnes, H.P. (1986). Calmodulin dependency of human neutrophil phosphodiesterase. Inflammation 10, 31–35.

Epstein, P.M. and Hachisu, R. (1984). Cyclic nucleotide phosphodiesterase in normal and leukemic human lymphocytes and lymphoblasts. Adv. Cyclic Nucleotide Protein Phosphorylation Res. 16, 303–324.

Florio, V.A., Sonnenburg, W.K., Johnson, R., Kwak, K.S., Jensen, G. S., Walsh, K.A. and Beavo, J.A. (1994). Phosphorylation of the 61-kDa calmodulin-stimulated cyclic nucleotide phosphodiesterase at serine 120 reduces its affinity for calmodulin. Biochemistry 33, 8948–8954.

Fukunaga, K., Yamamoto, H., Tanaka, E., Iwasa, T. and Miyamoto, E. (1984). Phosphorylation and activation of calmodulin-sensitive cyclic nucleotide phosphodiesterase by a brain Ca^{2+}, calmodulin-dependent protein kinase. Life Sci. 35, 493–499.

Gehrig, L.M.B., Delbaere, L.T.J. and Hickie, R.A. (1984). Preliminary X-ray data for the calmodulin/trifluoperazine complex. J. Mol. Biol. 177, 559–561.

Gietzen, K., Xu, Y.H., Galla, H.J. and Bader, H. (1982). Multimers of anionic amphiphiles mimic calmodulin stimulation of cyclic nucleotide phosphodiesterase. Biochem. J. 207, 637–640.

Goldering, J.R., Gonzalez, B., McGuire, J.S. and Delorenzo, R.J. (1983). Purification and characterization of a calmodulin-dependent protein kinase from rat brain cytosol able to phosphorylate tubulin and microtubule-associated proteins. J. Biol. Chem. 258, 12632–12640.

Grab, D.J., Carlin, R.K. and Siekevitz, P. (1981). Function of calmodulin in post synaptic densities. I. Presence of calmodulin-activatable cyclic nucleotide phosphodiesterase activity. J. Cell Biol. 89, 433–439.

Grady, P.G. and Thomas, L.L. (1986). Characterization of cyclic nucleotide phosphodiesterase activities in resting and *N*-formylmethionylleucylphenylalanine-stimulated human neutrophils. Biochim. Biophys. Acta 885, 282–293.

Hait, W.N. and Weiss, B. (1977). Characteristics of the cyclic nucleotide phosphodiesterases of normal and leukemic lymphocytes. Biochim. Biophys. Acta 497, 86–100.

Hait, W.N. and Weiss, B. (1979). Cyclic nucleotide phosphodiesterase of normal and leukemic lymphocytes. Mol. Pharmacol. 16, 851–864.

Hansen, R.S. and Beavo, J.A. (1982). Purification of two calcium calmodulin-dependent forms of cyclic nucleotide phosphodiesterase by using conformation-specific monoclonal antibody chromatography. Proc. Natl Acad. Sci. USA 79, 2788–2792.

Hansen, R.S. and Beavo, J.A. (1986). Differential recognition of calmodulin–enzyme complexes by a conformation-specific anticalmodulin monoclonal antibody. J. Biol. Chem. 261, 14636–14645.

Hashimoto, Y., Schworer, C.M., Colbran, R.J. and Soderling, T.R. (1987). Autophosphorylation of Ca^{2+}/calmodulin protein kinase II: effects on total and Ca^{2+} independent activities and kinetic parameters. J. Biol. Chem. 262, 8051–8055.

Hashimoto, Y., Sharma, R.K. and Soderling, T.R. (1989). Regulation of Ca^{2+}/calmodulin-dependent cyclic nucleotide phosphodiesterase by the autophosphorylated form of Ca^{2+}/calmodulin-dependent protein kinase II. J. Biol. Chem. 264, 10884–10887.

Helfman, D.M. and Kuo, J.F. (1982a). A homogeneous cyclic CMP phosphodiesterase hydrolyzes both pyrimidine and purine cyclic 2′:3′- and 3′:5′-nucleotides. J. Biol. Chem. 257, 1044–1047.

Helfman, D.M. and Kuo, J.F. (1982b). Differential effects of various phosphodiesterase inhibitors, pyrimidine and purine compounds and inorganic phosphates on cyclic CMP, cyclic AMP and cyclic GMP phosphodiesterases. Biochem. Pharmacol. 31, 43–47.

Hemmings, H.C., Greengard, P., Tung, H.Y.L. and Cohen, P. (1984). DARPP-32, a dopamine-regulated neuronal phosphoprotein, is a potent inhibitor of protein phosphatase-I. Nature (London) 310, 503–505.

Hickie, R.A. (1978). Regulation of cyclic AMP and cyclic GMP. In "Morris Hepatomas: Mechanisms of Regulation" (eds. H.P. Morris and W.E. Criss), pp. 451–488. Plenum, New York.

Hickie, R.A. and Kalant, H. (1967). Calcium and magnesium content of rat liver and Morris hepatoma 5123 t.c. Cancer Res. 27, 1053–1057.

Hickie, R.A., Walker, C.M. and Croll, G.A. (1974). Decreased cyclic adenosine 3′:5′-monophosphate levels in Morris hepatoma 5123 t.c.(h). Biochem. Biophys. Res. Commun. 59, 167–173.

Hickie, R.A., Jan, S.H. and Datta, A. (1975a). Comparative adenylate cyclase activities in homogenate and plasma membrane fractions in Morris hepatoma 5123 t.c.(h). Cancer Res. 35, 596–600.

Hickie, R.A., Walker, C.M. and Datta, A. (1975b). Increased activity of low K_m cyclic adenosine 3′:5′-monophosphate phosphodiesterases in Morris hepatoma 5123 t.c.(h). Cancer Res. 35, 601–605.

Hickie, R.A., Thompson, W.J., Strada, S.J., Couture-Murillo, B., Morris, H.P. and Robison, G.A. (1977). Comparison of cyclic adenosine 3′:5′-monophosphate and cyclic guanosine 3′:5′-monophosphate levels, cyclases, and phosphodiesterases in Morris hepatomas and liver. Cancer Res. 37, 3599–3606.

Hickie, R.A., Wei, J.-W., Blyth, L.M., Wong, M.M. and Klaassen, D.J. (1983). Cations and calmodulin in normal and neoplastic cell growth regulation. Can. J. Biochem. Cell Biol. 61, 934–941.

Hickie, R.A., Klaassen, D.J., Carl, G.Z., Meyskens, F.L., Jr, Kreutzfeld, K.L. and Thomson, S.P. (1984). Anticalmodulin agents as inhibitors of tumor cell clonogenicity. In "Human Tumor Cloning" (eds. S.E. Salmon and J.M. Trent), pp. 409–424. Grune and Stratton, New York.

Hickie, R.A., Graham, M.J., Buckmeier, J.A. and Meyskens, F.L., Jr (1992). Comparison of calmodulin gene expression in human neonatal melanocytes and metastatic melanoma cell lines. J. Invest. Dermatol. 99, 764–773.

Hidaka, H., Yamaki, T., Ochiai, Y., Asano, T. and Yamabe, H. (1977). Cyclic 3′:5′-nucleotide phosphodiesterase determined in various human tissues by DEAE-cellulose chromatography. Biochim. Biophys. Acta 484, 398–407.

Hidaka, H., Yamaki, T. and Yamabe, H. (1978). Two forms of Ca^{2+}-dependent cyclic 3′:5′-nucleotide phosphodiesterase from human aorta and effect of free fatty acids. Arch. Biochem. Biophys. 187, 315–321.

Ho, H.C., Teo, T.S., Desai, R. and Wang, J.H. (1976). Catalytic and regulatory properties of two forms of bovine heart cyclic nucleotide phosphodiesterase. Biochim. Biophys. Acta 429, 461–473.

Huang, F.L. and Glinsmann, W.H. (1976). Separation and characterization of two phosphorylase phosphatase inhibitors from rabbit skeletal muscle. Eur. J. Biochem. 70, 419–426.

Jedlicki, E., Orellane, O., Allende, C.C. and Allende, J.E. (1985). A protein inhibitor of calmodulin-regulated cyclic nucleotide phosphodiesterase in amphibian ovaries. Arch. Biochem. Biophys. 241, 215–224.

Kakiuchi, S. and Yamazaki, R. (1970). Calcium dependent phosphodiesterase activity and its activating factor (PAF) from brain studies on cyclic 3′:5′-nucleotide phosphodiesterase (3). Biochem. Biophys. Res. Commun. 41, 1104–1110.

Kakiuchi, S., Yamazaki, R. and Nakajima, H. (1970). Properties of a heat-stable phosphodiesterase activating factor isolated from brain extract. Proc. Japan Acad. 46, 587–592.

Kakiuchi, S., Yamazaki, R., Teshima, Y. and Uenishi, K. (1973). Regulation of nucleoside cyclic 3′:5′-monophosphate phosphodiesterase activity from rat brain by a modulator protein and Ca^{2+}. Proc. Natl Acad. Sci. USA 70, 3526–3530.

Kakiuchi, S., Yamazaki, R., Teshima, Y., Uenishi, K. and Miyamato, E. (1975). Multiple cyclic nucleotide phosphodiesterase activities from rat tissues and occurrence of a calcium-plus-magnesium-ion-dependent phosphodiesterase and its protein activator. Biochem. J. 146, 109–120.

Kakiuchi, S., Yamazaki, R., Teshima, Y., Uenishi, K., Yasuda, S., Kashiba, A., Sobue, K., Ohshima, M. and Nakajima, T. (1978). Membrane-bound protein modulator and phosphodiesterase. Adv. Cyclic Nucleotide Res. 9, 253–263.

Kennedy, M.B., McGuinness, T. and Greengard, P. (1983). A calcium/calmodulin-dependent protein kinase from mammalian brain that phosphorylates synapsin I: partial purification and characterization. J. Neurosci. 3, 818–831.

Keravis, T.M., Duemler, B.H. and Wells, J.M. (1987). Calmodulin-sensitive phosphodiesterase of porcine cerebral cortex: kinetic behaviour, calmodulin activation, and stability. J. Cyclic Nucleotide Protein Phosphorylation Res. 11, 365–372.

Kincaid, R.L. and Vaughan, M. (1983). Affinity chromatography of brain cyclic nucleotide phosphodiesterase using 3-(2-pyridyldithiol)propionyl-substituted calmodulin linked to thiol-sepharose. Biochemistry 22, 826–830.

Kincaid, R.L., Manganiello, V.C., Odya, C.E., Osborne, J.C., Smith-Coleman, I.E., Danello, M.A. and Vaughan, M. (1984). Purification and properties of calmodulin-stimulated phosphodiesterase from mammalian brains. J. Biol. Chem. 259, 5158–5166.

Klee, C.B. and Vanaman, T.C. (1982). Calmodulin. Adv. Protein Chem. 35, 213–321.

Klee, C.B., Crouch, T.H. and Krinks, M.H. (1979). Subunit structure and catalytic properties of bovine brain Ca^{2+}-dependent cyclic nucleotide phosphodiesterase. Biochemistry 18, 722–729.

Kumar, S., Becker, G. and Prasad, K.N. (1975). Cyclic adenosine 3′:5′-monophosphate phosphodiesterase activity in malignant and cyclic adenosine 3′:5′-monophosphate-induced "differentiated" neuroblastoma cells. Cancer Res. 35, 82–87.

Kuret, J.A. and Schulman, H. (1984). Purification and characterization of a Ca^{2+}/calmodulin-dependent protein kinase from rat brain. Biochemistry 23, 5459–5504.

Laemmli, U.K. (1970). Cleavage of structural proteins during the assembly of the head of bacteriophage T4. Nature (London) 227, 680–685.

Lai, Y., Nairn, A.C. and Greengard, P. (1986). Autophosphorylation reversibly regulates the Ca^{2+}/calmodulin dependence of the Ca^{2+}/calmodulin-dependent protein kinase II. Proc. Natl Acad. Sci. USA 83, 4253–4257.

La Porte, D.C., Toscano, W.A. and Strom, D.R. (1979). Cross-linking of iodine-125-labeled, calcium-dependent regulatory protein to the Ca^{2+}-sensitive phosphodiesterase purified from bovine heart. Biochemistry 18, 2820–2825.

McPherson, J.D., Hickie, R.A., Wasmuth, J.J., Meyskens, F.L., Jr, Perham, R.N., Strehler, E.E. and Graham, M.J. (1991). Chromosomal localization of multiple genes encoding calmodulin. Cytogenet. Cell Genet. 58, 1951. [Abstract]

Mirossay, L., Chastre, E., Empereur, S. and Gespach, C. (1992). Cyclic AMP-responsive gene transcription in cellular proliferation and transformation. Int. J. Oncol. 1, 1–13.

Monck, J.R., Raynolds, E.E., Thomas, A.P. and Williamson, J.R. (1988). Novel kinetics of single cell Ca^{2+} transients in stimulated hepatocytes and A10 cells measured using fura-2 and fluorescent videomicroscopy. J. Biol. Chem. 263, 4569–4575.

Morrill, M.E., Thompson, S.T. and Stellwagen, E. (1979). Purification of a cyclic nucleotide phosphodiesterase from bovine brain using blue dextran-Sepharose chromatography. J. Biol. Chem. 254, 4371–4374.

Mutus, B., Karuppiah, N., Sharma, R.K. and MacManus, J.P. (1985). The differential stimulation of brain and heart cyclic-AMP phosphodiesterase by oncomodulin. Biochem. Biophys. Res. Commun. 131, 500–506.

Nagamine, Y. and Reich, E. (1985). Gene expression and cAMP. Proc. Natl Acad. Sci. USA 82, 4606–4610.

Najam, N., Clair, T., Bassin, R.H. and Cho-Chung, Y.S. (1986). Cyclic AMP suppresses expression of v-rasH oncogene linked to the mouse mammary tumor virus promoter. Biochem. Biophys. Res. Commun. 134, 436–442.

Nakanishi, S., Osawa, K., Saito, Y., Kawamoto, I., Kuroda, K. and Kase, H. (1992). KS-505a, a novel inhibitor of bovine brain Ca^{2+} and calmodulin-dependent cyclic nucleotide phosphodiesterase from *Streptomyces argenteolus*. J. Antibiot. 45, 341–347.

Onali, P., Strada, S.J., Chang, L., Epstein, P.M., Hersh, E.M. and Thompson, W.J. (1985). Purification and characterization of high-affinity cyclic adenosine 5′-monophosphate phosphodiesterases from human acute myelogenous leukemic cells. Cancer Res. 45, 1384–1391.

Ouimet, C.C., McGuiness, T.L. and Greengard, P. (1984). Immunocytochemical localization of calcium/calmodulin-dependent protein kinase II in rat brain. Proc. Natl Acad. Sci. USA 81, 5604–5608.

Perez, R.O., Tuinen, D.V., Marme, D. and Turian, G. (1983). Calmodulin-stimulated cyclic nucleotide phosphodiesterase from *Neurospora crassa*. Biochim. Biophys. Acta 758, 84–87.

Puck, T.T. (1987). Genetic regulation of growth control: role of cyclic AMP and cell cytoskeleton. Somat. Cell Mol. Genet. 13, 451–457.

Purvis, K., Olsen, A. and Hansson, V. (1981). Calmodulin-dependent cyclic nucleotide phosphodiesterase in the immature rat testis. J. Biol. Chem. 256, 11434–11441.

Reeves, M.L., Leigh, B.K. and England, P.J. (1987). The identification of a new cyclic nucleotide phosphodiesterase activity in human and guinea-pig cardiac ventricle: implications for the mechanism of action of selective phosphodiesterase inhibitors. Biochem. J. 241, 535–541.

Rohlff, C., Clair, T. and Cho-Chung, Y.S. (1993). 8-Cl-cAMP induces truncation and down-regulation of the RIa subunit and up-regulation of the RIIb subunit of cAMP-dependent protein kinase leading to type II holoenzyme-dependent growth inhibition and differentiation of HL-60 leukemia cells. J. Biol. Chem. 268, 5774–5782.

Rossi, P., Giorgi, M., Geremia, R. and Kincaid, R.L. (1988). Testis-specific calmodulin-dependent phosphodiesterase: a distinct high affinity cAMP isozyme immunologically related to brain calmodulin-dependent cGMP phosphodiesterase. J. Biol. Chem. 263, 15521–15527.

Rutten, W.J., Schoot, B.M., De Pont, J.J.H.M. and Bonting, S.L. (1973). Adenosine 3′:5′-monophosphate phosphodiesterase in rat pancreas. Biochim. Biophys. Acta 315, 384–393.

Saitoh, Y., Yamamato, H., Fukunaga, K., Matsukado, Y. and Miyamoto, E. (1987). Inactivation and reactivation of the multifunctional calmodulin-dependent protein kinase from brain by autophosphorylation and dephosphorylation: involvement of protein phosphatases from brain. J. Neurochem. 99, 1286–1292.

Sharma, R.K. (1991). Phosphorylation and characterization of bovine heart calmodulin-dependent phosphodiesterase. Biochemistry 30, 5964–5968.

Sharma, R.K. (1995). Signal transduction: regulation of cAMP concentrations in cardiac muscle by calmodulin-dependent cyclic nucleotide phosphodiesterase. Mol. Cell. Biochem. 149/150, 241–247.

Sharma, R.K. and Kalra, J. (1993). Ginsenosides are potent and selective inhibitors of some calmodulin-dependent phosphodiesterase isozymes. Biochemistry 32, 4975–4978.

Sharma, R.K. and Kalra, J. (1994a). Characterization of calmodulin-dependent cyclic nucleotide phosphodiesterase isoenzymes. Biochem. J. 299, 97–100.

Sharma, R.K. and Kalra, J. (1994b). Molecular interaction between cAMP and calcium in calmodulin-dependent cyclic nucleotide phosphodiesterase system. Clin. Invest. Med. 17, 374–382.

Sharma, R.K. and Wang, J.H. (1985). Differential regulation of bovine brain calmodulin-dependent cyclic nucleotide phosphodiesterase isozyme by cyclic AMP-dependent protein kinase and calmodulin-dependent phosphatase. Proc. Natl Acad. Sci. USA 82, 2603–2607.

Sharma, R.K. and Wang, J.H. (1986a). Purification and characterization of bovine lung calmodulin-dependent cyclic nucleotide phosphodiesterase: an enzyme containing calmodulin as a subunit. J. Biol. Chem. 261, 14160–14166.

Sharma, R.K. and Wang, J.H. (1986b). Regulation of 63 kDa subunit containing isozyme of bovine brain calmodulin-dependent cyclic nucleotide phosphodiesterase by a calmodulin-dependent protein kinase. J. Biol. Chem. 261, 1322–1328.

Sharma, R.K. and Wang, J.H. (1986c). Regulation of cAMP concentration by calmodulin-dependent cyclic nucleotide phosphodiesterase. Biochem. Cell Biol. 64, 1072–1080.

Sharma, R.K. and Wirch, E. (1979). Ca^{2+}-dependent cyclic nucleotide phosphodiesterase from rabbit lung. Biochem. Biophys. Res. Commun. 91, 338–344.

Sharma, R.K., Wang, T.H., Wirch, E. and Wang, J.H. (1980). Purification and properties of bovine brain calmodulin-dependent cyclic nucleotide phosphodiesterase. J. Biol. Chem. 255, 5916–5923.

Sharma, R.K., Adachi, A.M., Adachi, K. and Wang, J.H. (1984). Demonstration of bovine brain calmodulin-dependent cyclic nucleotide phosphodiesterase isozymes by monoclonal antibodies. J. Biol. Chem. 259, 9248–9254.

Sharma, R.K., Mooibroek, M. and Wang, J.H. (1988). Calmodulin-stimulated cyclic nucleotide phosphodiesterase isozymes. In "Molecular Aspects of Cellular Regulation" (eds. P. Cohen and C.B. Klee), vol. 5, pp. 265–295. Elsevier, Amsterdam.

Sharma, R.K., Smith, J.R. and Moore, G.J. (1991). Inhibition of bovine brain calmodulin-dependent cGMP phosphodiesterase by peptide and non-peptide angiotensin receptor ligands. Biochem. Biophys. Res. Commun. 179, 85–89.

Shenolikar, S., Thompson, W.J. and Strada, S.J. (1985). Characterization of Ca^{2+}-calmodulin-stimulated cyclic GMP phosphodiesterase from bovine brain. Biochemistry 24, 672–678.

Singer, A.L., Sherwin, R.P., Dunn, A.S. and Appleman, M.M. (1976). Cyclic nucleotide phosphodiesterases in neoplastic and non-neoplastic human mammary tissues. Cancer Res. 36, 60–66.

Solti, M., Davey, P., Kiss, I., Londesborough, J. and Friedrich, P. (1983). Cyclic nucleotide phosphodiesterase in larval brain of wild type and dunce mutant strains of *Drosophila melanogaster*: isozyme pattern and activation by Ca^{2+}-calmodulin. Biochem. Biophys. Res. Commun. 111, 652–658.

Teo, T.S. and Wang, J.H. (1973). Mechanism of activation of cyclic adenosine 3′:5′-monophosphate phosphodiesterase from bovine heart by calcium ions. J. Biol. Chem. 248, 5950–5955.

Thompson, W.J., Ross, C.P., Pledger, W.J., Strada, S.J., Banner, R.L. and Hersh, E.M. (1976). Cyclic adenosine 3′:5′-monophosphate phosphodiesterase: distinct forms in human lymphocytes and monocytes. J. Biol. Chem. 251, 4922–4929.

Thompson, W.J., Ross, C.P., Strada, S.J., Hersh, E.M. and Lavis, V.R. (1980). Comparative analyses of cyclic adenosine 3′:5′-monophosphate phosphodiesterase of human peripheral blood monocytes and cultured P388D cells. Cancer Res. 40, 1955–1960.

Turnbull, J.L. and Hickie, R.A. (1984). The isolation and characterization of cyclic nucleotide phosphodiesterases from Morris hepatoma 5123 t.c. (h) and rat liver. Int. J. Biochem. 16, 19–29.

Vandermeers, A., Vandermeers-Piret, M.C., Rathe, J. and Christophe, J. (1983). Purification and kinetic properties of two soluble forms of calmodulin-dependent cyclic nucleotide phosphodiesterase from rat pancreas. Biochem. J. 211, 341–347.

Vandonselaar, M., Hickie, R.A., Quail, J.W. and Delbaere, L.T.J. (1994). Trifluoperazine-induced conformational change in Ca^{2+}-calmodulin. Nature (Structural Biology) 1, 795–801.

Van Lookeren Campagne, M.M., Diaz, F.V., Jastorff, B. and Kessin, R.H. (1991). 8-Chloroadenosine 3′:5′-monophosphate inhibits the growth of chinese hamster ovary and molt-4 cells through its adenosine metabolite. Cancer Res. 51, 1600–1605.

Walter, M.F. and Kiger, J.A. (1984). The dunce gene of *Drosophila*: roles of Ca^{2+} and calmodulin in adenosine 3′:5′-cyclic monophosphate-specific phosphodiesterase activity. J. Neurosci. 4, 495–501.

Walters, J.D. and Jirsa, R.C. (1988). Activation of cyclic nucleotide phosphodiesterase by a monosaccharide precursor of *Escherichia coli* lipid A. FEBS Lett. 236, 312–314.

Wang, J.H., Sharma, R.K. and Mooibroek, M.J. (1990). Calmodulin-stimulated cyclic nucleotide phosphodiesterase. Mol. Pharm. Cell. Reg. 2, 19–59.

Wasco, W.M. and Orr, G.A. (1984). Function of calmodulin in mammalian sperm: presence of a calmodulin-dependent cyclic nucleotide phosphodiesterase associated with demembranated rat caudal epididymal sperm. Biochem. Biophys. Res. Commun. 118, 636–642.

Weber, K. and Osborn, M. (1969). The reliability of molecular weight determinations by dodecyl sulfate-polyacrylamide gel electrophoresis. J. Biol. Chem. 244, 4406–4412.

Wei, J.-W. and Hickie, R.A. (1981). Increased content of calmodulin in Morris hepatoma 5123 t.c. (h). Biochem. Biophys. Res. Commun. 100, 1562–1568.

Wei, J.-W. and Hickie, R.A. (1983). Decreased activities of cyclic cytidine 3′:5′-monophosphate phosphodiesterase in Morris hepatomas having varying growth rates. Int. J. Biochem. 15, 789–795.

Wei, J.-W., Morris, H.P. and Hickie, R.A. (1982). Positive correlation between calmodulin content and hepatoma growth rates. Cancer Res. 42, 2571–2574.

Wei, J.-W., Hickie, R.A. and Klaassen, D.J. (1983). Inhibition of human breast cancer colony formation by anticalmodulin agents: trifluoperazine, W-7 and W-13. Cancer Chemother. Pharmacol. 11, 86–90.

Weishaar, R.E., Burrows, S.D., Kobylarz, D.C., Quade, M.M. and Evans, D.B. (1986). Multiple molecular forms of cyclic nucleotide phosphodiesterase in cardiac and smooth muscle and in platelets: isolation, characterization, and effects of various reference phosphodiesterase inhibitors and cardiotonic agents. Biochem. Pharmacol. 35, 787–800.

Weiss, B. and Hait, W.N. (1977). Selective cyclic nucleotide phosphodiesterase inhibitors as potential therapeutic agents. Ann. Rev. Pharmacol. Toxicol. 17, 441–477.

Wells, J.M. and Hardman, J.G. (1977). Cyclic nucleotide phosphodiesterases. Adv. Cyclic Nucleotide Res. 8, 119–143.

Whitfield, J.F. (1990). Calcium, Cell Cycles and Cancer. CRC Press Inc., Boca Raton.

Wolff, D.J. and Brostrom, C.O. (1976). Calcium-dependent cyclic nucleotide phosphodiesterase from brain: identification of phospholipids as calcium-independent activators. Arch. Biochem. Biophys. 173, 720–731.

Woods, N.M., Cutherbertson, K.S.R. and Cobbold, P.H. (1986). Repetitive transient rises in cytoplasmic free calcium in hormone-stimulated hepatocytes. Nature 319, 600–602.

Woods, N.M., Cutherbertson, K.S.R. and Cobbold, P.H. (1987). Agonist-induced oscillations in cytoplasmic free calcium concentration in single rat hepatocytes. Cell Calcium 8, 79–100.

Wu, Z., Sharma, R.K. and Wang, J.H. (1992). Catalytic and regulatory properties of CaM-stimulated phosphodiesterase isozyme. Adv. Cyclic Nucleotide Protein Phosphorylation Res. 25, 29–43.

Yamanaka, M.K. and Kelly, L.E. (1981). A calcium/calmodulin-dependent cyclic adenosine monophosphate phosphodiesterase from *Drosophila* heads. Biochim. Biophys. Acta 674, 277–286.

Zhang, G.Y., Wang, J.H. and Sharma, R.K. (1993a). Purification and characterization of bovine brain calmodulin-dependent protein kinase II: the significance of autophosphorylation in the regulation of 63 kDa calmodulin-dependent cyclic nucleotide phosphodiesterase isozyme. Mol. Cell. Biochem. 122, 159–169.

Zhang, G.Y., Wang, J.H. and Sharma, R.K. (1993b). Bovine brain calmodulin-dependent protein kinase II: molecular mechanisms of autophosphorylation. Biochem. Biophys. Res. Commun. 191, 669–674.

5. EHNA as an Inhibitor of PDE2: A Pharmacological and Biochemical Study in Cardiac Myocytes

Pierre-François Méry, Catherine Pavoine, Françoise Pecker *and* Rodolphe Fischmeister

1. Introduction 81
   1.1 Cardiac Ca^{2+} Current is Inhibited by cGMP via Activation of PDE2 81
   1.2 Use of the Cardiac I_{Ca} in Determining the Effects of EHNA on PDE2 82
   1.3 Use of Purified Cardiac PDE Isoforms to Determine the Selectivity of Action 82
2. Methods 82
   2.1 Electrophysiology 82
   2.2 PDE Assays for cAMP 83
3. Results 83
   3.1 EHNA has no Effect on I_{Ca} in the Absence of cGMP 83
   3.2 EHNA Antagonizes the Inhibitory Effect of cGMP on I_{Ca} 84
   3.3 EHNA Antagonizes the Inhibitory Effect of Nitric Oxide Donors on I_{Ca} 84
   3.4 EHNA Inhibits a cGMP-Stimulated PDE in the Crude Particulate Fraction 85
   3.5 EHNA Selectively Inhibits the Purified Soluble PDE2 86
   3.6 Participation of Adenosine Deaminase in the Effects of EHNA? 86
4. Discussion 86
   4.1 EHNA Acts as a Selective Inhibitor of PDE2 in Cardiac Myocytes 86
   4.2 EHNA Should be Useful in Evaluating the Role of PDE2 in Various Tissues 87
5. Acknowledgements 87
6. References 87

1. Introduction

1.1 Cardiac Ca^{2+} Current is Inhibited by cGMP via Activation of PDE2

By controlling the activity of various cardiac proteins, cyclic nucleotides are well-recognized to contribute to the performances of the normal and the diseased heart (Hartzell, 1988). The fine tuning of L-type Ca^{2+} channel activity by cyclic AMP (cAMP) and cyclic GMP (cGMP) receives careful attention since these channels are responsible for the triggering of cardiac contraction. Cyclic AMP activation of cAMP-dependent protein kinase leads to the phosphorylation of cardiac L-type Ca^{2+} channels (or a closely associated protein), resulting in an increase in the mean probability of channel opening and stimulation of macroscopic calcium current (I_{Ca}) (Hartzell, 1988). Cyclic GMP has often been shown to produce contractile effects opposite to those of

Phosphodiesterase Inhibitors
ISBN 0-12-210720-9

cAMP in the heart (Hartzell, 1988; Lohmann *et al.*, 1991). In isolated cardiomyocytes from different species, exogenous or endogenous cGMP can strongly inhibit I_{Ca} (reviewed in Lohmann *et al.*, 1991; see also Méry *et al.*, 1993; Levi *et al.*, 1994; Whaler and Dollinger, 1995).

In frog ventricular myocytes, the inhibitory effect of cGMP has been attributed to the stimulation of a specific cAMP-phosphodiesterase, PDE2 (Hartzell and Fischmeister, 1986, 1987). This conclusion was suggested by the finding that cAMP levels are reduced when the cGMP level is increased in the frog heart (Flitney and Singh, 1981). It is now strengthened by pharmacological and biochemical findings. In the isolated frog myocyte, cGMP inhibited cAMP stimulated I_{Ca} but did not affect I_{Ca} that had been increased by the hydrolysis-resistant cAMP analogue, 8-bromoadenosine 3':5'-cyclic monophosphate (8-Br-cAMP) (Fischmeister and Hartzell 1986, 1987). The sensitivity of the cAMP-stimulated I_{Ca} ($IC_{50} \approx 0.6\ \mu M$) to cGMP correlates well with the ability of cGMP to activate the PDE2 ($K_a \approx 1.1\ \mu M$) (Simmons and Hartzell, 1988). Moreover, in this preparation, the inhibitory effect of cGMP on cAMP-stimulated I_{Ca} is largely antagonized by 3-isobutyl-1-methylxanthine (IBMX), a non-selective phosphodiesterase inhibitor (Hartzell and Fischmeister, 1986; Fischmeister and Hartzell, 1987). The effects of cGMP on I_{Ca} and PDE2 are sensitive to the same range of (high) concentrations of IBMX (Simmons and Hartzell, 1988). Unlike in mammalian myocytes, the cGMP-dependent protein kinase is practically excluded from the inhibitory effect of cGMP on the cAMP-stimulated I_{Ca} in the amphibian cells. Indeed, while being a potent stimulator of this kinase, the weak stimulator of PDE2, 8-bromoguanosine 3':5'-cyclic monophosphate (8-Br-cGMP), does not mimic the inhibitory effect of cGMP on cAMP-stimulated I_{Ca} (Hartzell and Fischmeister, 1986; Fischmeister and Hartzell, 1987). Therefore, the inhibition of cardiac I_{Ca} by cGMP in the frog myocyte can be viewed as a sensitive *in vivo* assay for PDE2 activity.

1.2 USE OF THE CARDIAC I_{Ca} IN DETERMINING THE EFFECTS OF EHNA ON PDE2

Podzuweit *et al.* (1992, 1993) have established, under *in vitro* conditions, that erythro-9-(2-hydroxyl-3-nonyl)-adenine (EHNA, which they called MEP1) acted as an inhibitor of a cGMP-stimulated peak of PDE activity purified from both human and porcine hearts. In the micromolar range of concentrations, this compound was reported to have no effect on three other purified peaks of cardiac PDE (PDE1, PDE3 and PDE4) from the same preparations (Podzuweit *et al.*, 1992, 1993). However, EHNA is most commonly used as a specific inhibitor of adenosine deaminase (K_i = 7 nM) (Cristalli *et al.*, 1994) and as a potential preservative agent against ischaemia-reperfusion injury (Zhu *et al.*, 1994, and references therein). Therefore, it remained to be verified whether EHNA can act as a selective PDE2 inhibitor when applied to intact cells. For this reason, we have studied the effects of EHNA on the regulation of I_{Ca} by exogenously dialysed and endogenously synthesized cGMP. The study was performed in frog ventricular cells because of the body of evidence listed above for the participation of PDE2 in the regulation of I_{Ca} in this preparation.

1.3 USE OF PURIFIED CARDIAC PDE ISOFORMS TO DETERMINE THE SELECTIVITY OF ACTION

The effect of EHNA on cardiac PDE activity was measured in the particulate fraction of frog ventricle and on the different purified PDEs of the soluble fraction. The crude particulate fraction of frog myocytes is a useful assay to study the effect of EHNA since it contains a high specific activity of PDE2, sensitive to cGMP (Simmons and Hartzell, 1988). Moreover, the putative regulatory proteins that modulate PDE2 activity in intact cells probably remain present in this preparation. For instance, it was found suitable to study the regulation of PDE3 by glucagon involving the guanine nucleotide-binding protein G_i (Méry *et al.*, 1990; Brechler *et al.*, 1992). However, the presence of the other PDEs in this type of assay will preclude the study of the precise kinetics of PDE2. Although purification may perturb the pharmacological sensitivity of cardiac PDEs (Pang, 1992), it is required to determine the site of action of EHNA on PDE2. In addition, there is no better means than the purification of each different PDE to study the selectivity of EHNA action.

2. *Methods*

2.1 ELECTROPHYSIOLOGY

The whole-cell configuration of the patch-clamp technique was used to record L-type Ca^{2+} current (I_{Ca}) on Ca^{2+}-tolerant myocytes isolated from frog cardiac ventricle, as described by Fischmeister and Hartzell (1986, 1987).

Several compounds were generously supplied as follows: milrinone was gift from Sterling-Winthrop, Ro 20-1724 from Hoffman LaRoche, and 3-morpholinosydnonimine (SIN–1) from Dr J. Winicki (Hoechst Laboratories, France). EHNA was kindly provided by Dr T. Podzuweit (Max Planck Institute, Bad Nauheim, Germany) or purchased from Sigma Chemical Co. (St Louis MO, USA), with no change in the results. EHNA was either dissolved immediately before application, or prepared as a 10 mM stock solution in distilled water and stored at −20°C in small aliquots until use.

The results are expressed as mean ± SEM. In the text, the "basal" current refers to the activity of non-phosphorylated Ca^{2+} channels in the absence of either isoprenaline or cAMP. In the case of single applications, the effect of a compound is referred to as the percentage variation from the basal level. Since EHNA has no effect on basal I_{Ca}, its effects are expressed as the percentage variation over the cAMP-dependent stimulation of I_{Ca}, i.e.:

$$\frac{\text{test } I_{Ca} - \text{reference } I_{Ca}}{\text{reference } I_{Ca} - \text{basal } I_{Ca}} \times 100$$

2.2 PDE Assays for cAMP

The preparation of the particulate fraction from frog cardiac ventricle has been described by Brechler *et al.* (1992) and Méry *et al.* (1993). The separation of the soluble PDE isoforms was performed according to the method of Bethke *et al.* (1992). PDE activity in frog ventricle particulate fraction has been characterized by Méry *et al.* (1990) and Brechler *et al.* (1992) and was determined according to the two-step assay procedure of Thompson *et al.* (1979). The assay medium (0.4 ml) consisted of: 20 mM HEPES, pH 7.6, 120 mM CsCl, 5 mM EGTA, 4 mM $MgCl_2$, 2 μM [^{3}H]cAMP (10^5 c.p.m.) with or without 5 μM cGMP, unless indicated. Incubation was initiated by the addition of 50 μg protein and was terminated after 10 min at 30°C by boiling for 45 s. Data are the mean of triplicate determinations. Results are expressed as nmol cAMP hydrolysed/mg protein/10 min.

3. *Results*

3.1 EHNA has no Effect on I_{Ca} in the Absence of cGMP

Because PDE2 hydrolyses cAMP so efficiently, the effect of EHNA on I_{Ca} was first studied after stimulation of cAMP production with isoprenaline, a β-adrenoceptor agonist. In the experiment shown in Fig. 5.1A, I_{Ca} was stimulated by 1 nM isoprenaline. This concentration of the β-agonist produced a submaximal stimulation of I_{Ca} (187 ± 32%, $n = 21$). As reported earlier (Fischmeister and Hartzell, 1990), inhibitors of PDE3 and PDE4 – milrinone (10 μM) and Ro 20-1724 (10 μM), respectively – or a high concentration of IBMX (200 μM) produced a substantial further increase in I_{Ca} when not maximally stimulated by cAMP phosphorylation. Under these conditions, however, EHNA (30 μM) was found to have no effect on I_{Ca} (Fig. 5.1A). The results of several similar experiments are summarized in Fig. 5.1B. Whereas milrinone (10 μM), Ro 20-1724 (10 μM) and IBMX (200 μM) produced substantial and significant increases of I_{Ca} on top of a stimulation with 1 nM isoprenaline, the effect of EHNA (30 μM) was not significant.

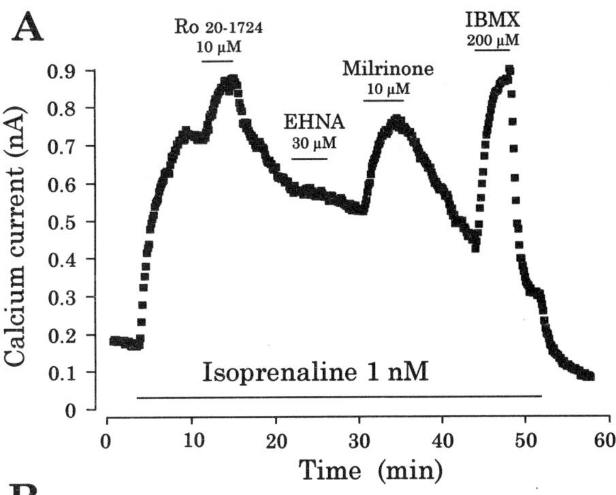

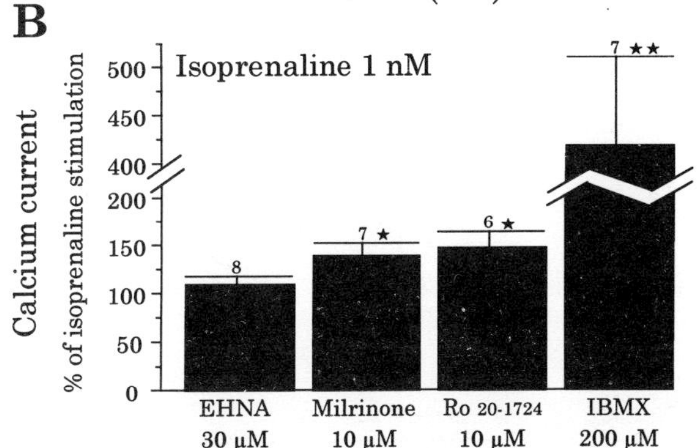

Figure 5.1 Effects of EHNA and other PDE inhibitors on isoprenaline-stimulated I_{Ca}. (A) A frog ventricular cell was initially superfused with control Cs Ringer solution. During the periods indicated, the cell was successively exposed to isoprenaline (1 nM) alone and in the presence of Ro 20-1724 (10 μM), EHNA (30 μM), milrinone (10 μM) or IBMX (200 μM). (B) Summary of the effects of milrinone (10 μM), Ro 20-1724 (10 μM), IBMX (200 μM) and EHNA (30 μM) on isoprenaline (1 nM)-stimulated I_{Ca}. The bars indicate the mean and SEM of the number of experiments indicated. Significant statistical differences from isoprenaline-stimulated level (100%) are indicated as *$P<0.05$, **$P<0.005$. Reproduced with permission from Méry *et al.* (1995). © Williams and Wilkins.

Another means of elevating intracellular cAMP level is to dialyse exogenous cAMP into the myocyte. On average, perfusion of 10 μM cAMP in frog myocytes induced a 695 ± 44% increase of I_{Ca} over its basal amplitude ($n = 34$). EHNA, at concentrations ranging from 0.3 to 30 μM, exerts no significant effects on cAMP-elevated I_{Ca} (Fig. 5.2B). Thus EHNA does not alter the coupling of the cAMP pathway to the cardiac L-type Ca^{2+} channels in the absence of cGMP. Furthermore, since EHNA (0.1–30 μM) does not affect the basal I_{Ca} in the absence of cAMP, it is unlikely to bind directly to the Ca^{2+} channels (Fig. 5.2B).

3.2 EHNA ANTAGONIZES THE INHIBITORY EFFECT OF cGMP ON I_{Ca}

Intracellular perfusion of a frog ventricular cell with cGMP strongly antagonizes the stimulatory action of cAMP (Hartzell and Fischmeister, 1986; Fischmeister and Hartzell, 1987). Figure 5.2A shows such an experiment

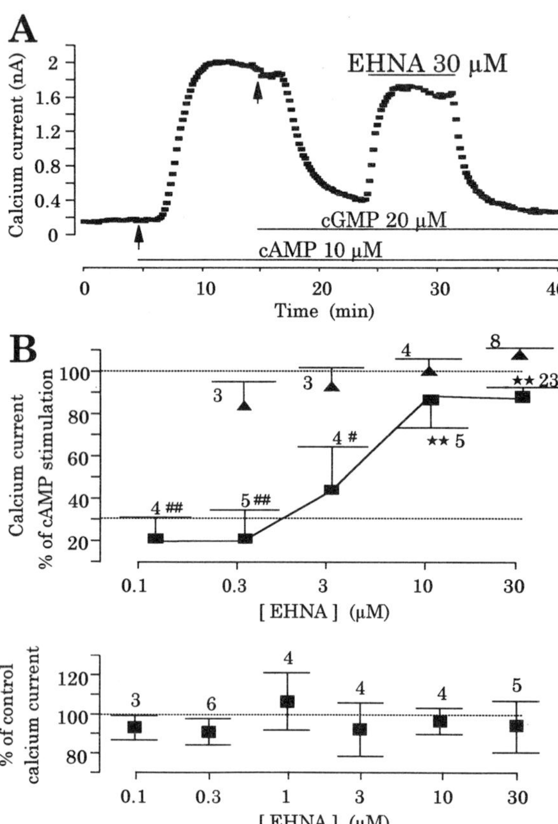

Figure 5.2 Effects of EHNA on cGMP-inhibited I_{Ca}. (A) A frog ventricular cell was initially superfused with control Cs Ringer solution and internally dialysed with control intracellular Cs solution. At the *first arrow*, 10 µM cAMP was added to the intracellular solution, which then perfused the cell throughout the rest of the experiment. At the *second arrow*, 20 µM cGMP was added to the cAMP-containing intracellular solution. During the period indicated, the cell was exposed to 30 µM EHNA. (B) Summary of the effects of EHNA on I_{Ca} in the presence of 10 µM cAMP (upper panel, ▲) or 10 µM cAMP plus 20 µM cGMP (upper panel, ■) in the intracellular solution, or in the absence of cyclic nucleotides (lower panel). The points and bars indicate the mean and SEM of the number of experiments indicated. Addition of cGMP reduced the cAMP-induced stimulation (100%) to the percentage level indicated by the dotted line in (B). Significant statistical differences from cAMP-stimulated level (# and ##) or cAMP + cGMP level (**) are indicated as #$P < 0.05$, and ** or ##$P < 0.005$. Reproduced with permission from Méry *et al.* (1995). © Williams and Wilkins.

where I_{Ca} had been first stimulated by 10 µM cAMP and then 20 µM cGMP was added to the intracellular medium containing cAMP. Intracellular perfusion with cGMP antagonized by approximately 80% the stimulatory effect of cAMP. When EHNA (30 µM) was superfused onto the cell, it fully antagonized the inhibitory effect of cGMP. Thus, unlike the lack of effect on basal and isoprenaline- or cAMP-stimulated I_{Ca}, EHNA produces a strong stimulatory effect on I_{Ca} when cGMP is dialysed into the cell.

Figure 5.2B shows the concentration–response curve for the stimulatory effect of EHNA on I_{Ca}. In the presence of 20 µM cGMP, the stimulation of I_{Ca} was on average only 31 ± 13% ($n = 27$) of its value in the presence of cAMP alone (100%). At concentrations above 0.3 µM, EHNA significantly increased I_{Ca}. At a concentration of 3 µM, EHNA reversed by approximately 50% the inhibitory effect of cGMP. Increasing the concentration further induced a larger stimulation of I_{Ca} until the inhibitory effect of cGMP was totally reversed, which occurred at 30 µM concentration. This effect occurs in a voltage-independent manner, since addition of 30 µM EHNA only scales up the current–voltage relationship recorded in the presence of cAMP plus cGMP. Increasing the concentration to 100 µM produced no additional effect on I_{Ca} (not shown). Thus EHNA is a total antagonist of the inhibitory effect of cGMP on I_{Ca} in frog ventricular cells.

Since the non-selective PDE inhibitor, IBMX, could antagonize the inhibitory effect of cGMP on I_{Ca} (Hartzell and Fischmeister, 1986, 1990; Fischmeister and Hartzell, 1987; Simmons and Hartzell, 1988), we compared the effects of EHNA and IBMX in the same cells. In nine cells where 20 µM cGMP decreased cAMP (10 µM)-stimulated I_{Ca} from 778 ± 84% to 132 ± 24% of the basal level, the effects of EHNA (30 µM) and IBMX (500 µM) were similar (74 ± 10% and 84 ± 6% recovery from cGMP inhibition, respectively) and not additive. Although not statistically significant, the effect of IBMX appears somewhat larger than that of EHNA. Since IBMX is likely to antagonize PDE3 and PDE4 (Hartzell and Fischmeister, 1990), together with the effect of cGMP on PDE2, we investigated the effects of Ro 20-1724 and milrinone in the presence of EHNA. Ro 20-1724 (10 µM) potentiated the effect of 30 µM EHNA (to 98 ± 13% recovery from cGMP inhibition, $n = 5$). In contrast, milrinone (10 µM) did not change the effect of 30 µM EHNA ($n = 3$). Thus, it is likely that PDE3 was inhibited by the dialysis of cGMP. Neither milrinone ($n = 3$) nor Ro 20-1724 ($n = 5$) had any effect on I_{Ca} in the presence of cAMP + cGMP alone, i.e. in the absence of EHNA (Hartzell and Fischmeister, 1990).

3.3 EHNA ANTAGONIZES THE INHIBITORY EFFECT OF NITRIC OXIDE DONORS ON I_{Ca}

The endogenous production of cGMP can be enhanced by nitric oxide (NO) donors such as SIN-1 and sodium nitroprusside (SNP), which stimulate guanylate cyclase activity (Hartzell, 1988; Lohmann *et al.*, 1991). In frog myocytes, these compounds can mimic the inhibitory effect of exogenous cGMP on isoprenaline- or cAMP-stimulated I_{Ca} but have no effect on 8-Br-cAMP-stimulated I_{Ca} (Méry *et al.*, 1993). In a first set of experiments, 1 μM isoprenaline increased I_{Ca} by 1330 ± 209% ($n = 5$), and the addition of 30 μM SIN-1 inhibited 38 ± 4% of the isoprenaline-stimulated I_{Ca} ($n = 6$). The inhibitory effect of SIN-1 was abolished by 30 μM EHNA (to 99 ± 3% of the initial isoprenaline stimulation). In similar experiments, SNP (1 mM) reduced by 77 ± 7% the amplitude of the isoprenaline (0.1 μM)-stimulated I_{Ca} ($n = 5$). Further addition of 10 μM EHNA partially antagonized the effect of SNP (to 67 ± 7% of the isoprenaline stimulation, $n = 4$), but 30 μM EHNA fully suppressed the effect of the NO donor on I_{Ca} (to 84 ± 6% of the isoprenaline stimulation, $n = 4$). Again, in these experiments, the antagonistic effects of EHNA occurred in a voltage-independent manner.

3.4 EHNA INHIBITS A CGMP-STIMULATED PDE IN THE CRUDE PARTICULATE FRACTION

We examined the effects of EHNA on PDE activity in frog ventricle particulate fraction, a condition which approximates to those used in electrophysiological

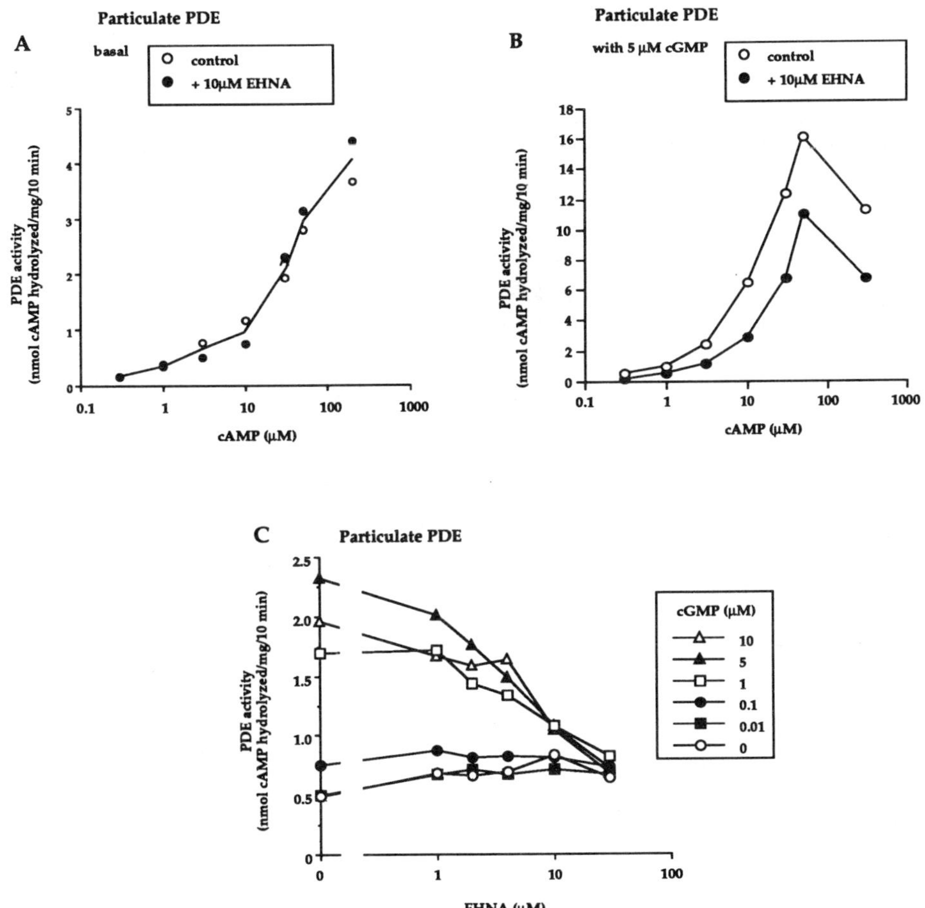

Figure 5.3 Effect of EHNA on the PDE activity of frog ventricle particulate fraction. The effect of EHNA on basal (A) and cGMP-stimulated (B) PDE activities was measured at varying cAMP concentrations. (A) Basal PDE activity was measured in frog ventricle particulate fraction, in the absence (control) and in the presence of 10 μM EHNA, as indicated in section 3.4. (B) PDE activity was measured in the presence of 5 μM cGMP, either in the absence (control) or in the presence of 10 μM EHNA. Note that EHNA inhibited cGMP stimulation of PDE activity to the same degree, irrespective of the concentration of cAMP. (C) Concentration-dependent inhibition of cGMP-stimulated PDE activity by EHNA in frog ventricle particulate fraction. PDE activity was measured in the presence of 2 μM cAMP, at different cGMP concentrations, and with varying EHNA concentrations as described in section 3.4. (C) reproduced with permission from Méry *et al.* (1995). © Williams & Wilkins.

recordings. In the absence of cGMP, the total PDE activity of the preparation was unaltered by 10 μM EHNA when measured at concentrations of cAMP ranging from 0.5 to 100 μM (Fig. 5.3A). In contrast, the PDE activity, which increased four-fold upon addition of 5 μM cGMP to the preparation, was clearly reduced by 10 μM EHNA (Fig. 5.3B). A 40% reduction of the apparent cGMP-stimulated PDE was observed at all concentrations of cAMP tested. The antagonistic effect of EHNA was concentration-dependent (Fig. 5.3C) and attributed to a decrease in the maximal velocity (Vmax). In the presence of 5 μM cGMP, a maximal 75% reduction in Vmax was observed at 30 μM EHNA, with half-maximal inhibition occurring at 4 μM EHNA. Overall, these data are in good agreement with the results of patch-clamp experiments.

3.5 EHNA SELECTIVELY INHIBITS THE PURIFIED SOLUBLE PDE2

To assess the selectivity of the action of EHNA, each PDE was purified from the soluble fraction of frog cardiac ventricle. Under our conditions, PDE1 activity was twice as high as PDE3 activity, whereas that of PDE4 was much lower. In the presence of 5 μM cGMP, PDE2 activity became comparable to that of PDE1. Irrespective of the concentration of cGMP, the activity of the purified PDE2 was suppressed in a concentration-dependent manner by EHNA (Fig. 5.4A). As for the particulate PDE, the effect of EHNA on the soluble PDE2 resulted from a decrease in Vmax, with a comparable potency, half maximal inhibition occurring at 5 μM EHNA. Interestingly, inhibition by EHNA of soluble PDE2 occurred in a non-competitive manner with respect to cGMP activation of the enzyme (K_a = 40 nM). This suggests that the drug binds at a site other than the allosteric cGMP regulator site. In contrast, the other PDEs were not significantly affected by EHNA (Fig. 5.4B), even at the highest concentration (30 μM) shown to overcome fully the inhibitory effect of cGMP on cAMP-stimulated I_{Ca}. These experiments demonstrate that EHNA is a selective inhibitor of PDE2, among the different cardiac isoforms of PDE.

3.6 PARTICIPATION OF ADENOSINE DEAMINASE IN THE EFFECTS OF EHNA?

Because EHNA is an inhibitor of adenosine deaminase, an indirect effect of EHNA due to adenosine accumulation needs to be ruled out. Adenosine (10 μM) had no effect on PDE activity in the frog ventricle particulate fraction, measured under basal conditions or in the presence of 5 μM cGMP to stimulate the PDE2 activity (not shown). At the same concentration, adenosine induced an inhibitory effect on isoprenaline-stimulated I_{Ca} in two out of four cells (not shown; see Alvarez *et al.*, 1990). Therefore, it seems unlikely that the effects of EHNA could be due to some contamination by adenosine, because, as shown above, EHNA produces stimulatory, not inhibitory, effects on I_{Ca}.

4. *Discussion*

4.1 EHNA ACTS AS A SELECTIVE INHIBITOR OF PDE2 IN CARDIAC MYOCYTES

We examined the effects of EHNA on Ca^{2+} current and PDE activity in frog ventricular cardiomyocytes. We conclude that EHNA acts primarily to inhibit PDE2 in this preparation.

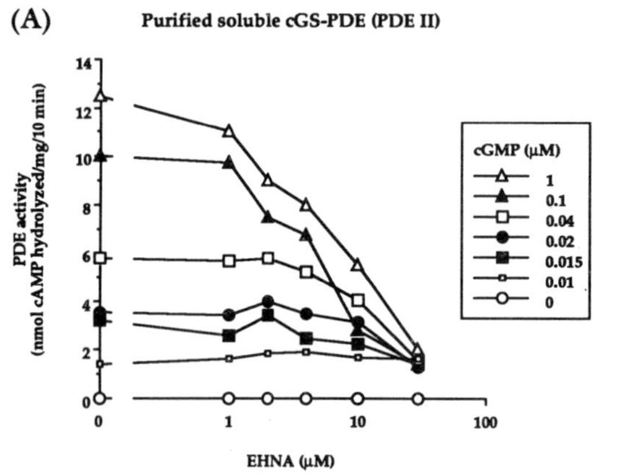

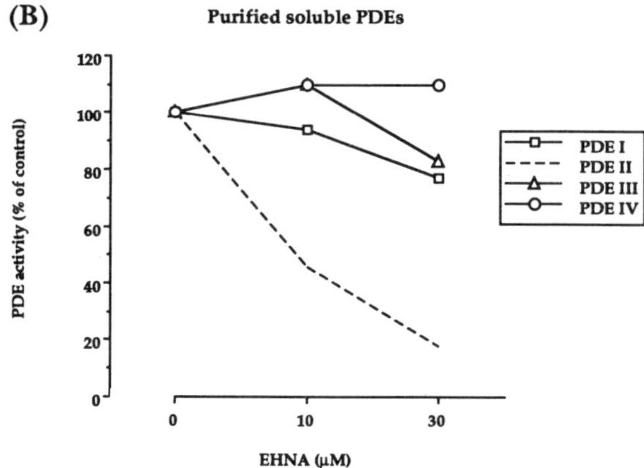

Figure 5.4 (A) Concentration-dependent inhibition of purified soluble PDE2 activity by EHNA examined under the same conditions as for the particulate fraction. (B) Specificity of EHNA effect. The effects of 10 μM and 30 μM EHNA were examined on the purified soluble PDE 1 to 4 isoforms under the same conditions as in Figure 5.3. The data are normalized with respect to the PDE activities in the absence of EHNA. These were: PDE1, 14.2 ± 2.5; PDE3, 7.0 ± 0.6; PDE4, 1.0 ± 0.1 nmol cAMP/mg/10 min. The dotted line represents the effect of EHNA on PDE2 activity in the presence of 0.1 μM cGMP and is taken from A. Reproduced with permission from Méry *et al.* (1995). © Williams & Wilkins.

An important feature of the present study is the correspondence between biochemical and electrophysiological data. For example, EHNA had no effect on isoprenaline- or cAMP-elevated I_{Ca}, in contrast to other selective PDE inhibitors such as milrinone, Ro 20-1724 or the non-selective PDE inhibitor IBMX. This suggests that, whereas different PDEs – more specifically PDE3 and PDE4 isoforms (Fischmeister and Hartzell, 1990) – were active under our experimental conditions, EHNA did not modify their activity. The data obtained with purified soluble PDE isoforms supported this observation, since EHNA (up to 30 μM) was found to have little or no effect on PDE1, PDE3 and PDE4. Moreover, both in intact myocytes and in particulate fractions, the effect of EHNA requires prior elevation of the cGMP level. Also, the concentration–response curve for the effects of EHNA on cGMP-inhibited I_{Ca} was superimposable upon that obtained for the effects of EHNA on cGMP-stimulated PDE activity in particulate fraction or on purified soluble cGMP-stimulated PDE. Finally, both biochemical (this report) and electrophysiological data (Méry *et al.*, 1995) support the hypothesis that EHNA inhibits PDE2 in a non-competitive manner with respect to the effect of cGMP on the enzyme. This may indicate that EHNA does not bind to the cGMP allosteric regulator site of the PDE2.

4.2 EHNA Should be Useful in Evaluating the Role of PDE2 in Various Tissues

Although PDE2 is expressed in numerous tissues, relatively little is known about its function. One reason for this is that no agent has been described as a truly selective inhibitor of PDE2 (Weishaar *et al.*, 1985; Nicholson *et al.*, 1991). Various compounds, such as dipyridamole or the isoquinoline derivatives HL-725 (trequensin) and papaverine, have been shown to exert a somewhat greater inhibition of PDE2 than of PDE1 and PDE3 (Weishaar *et al.*, 1985; Nicholson *et al.*, 1991; Whalin *et al.*, 1991). However, in platelets dipyridamole inhibits PDE1 and PDE2 activities to a comparable degree (Weishaar *et al.*, 1985) and in frog heart dipyridamole and papaverine were shown to inhibit PDE4 with a lower K_i than PDE2 (Lugnier *et al.*, 1992). For these reasons, the function of PDE2 in various cell types, such as human fibroblasts (Lee *et al.*, 1988), bovine adrenal glomerulosa cells (MacFarland *et al.*, 1991), rat phaeochromocytoma cells (Whalin *et al.*, 1991) and frog cardiomyocytes (Hartzell and Fischmeister, 1986; Fischmeister and Hartzell, 1987, 1990; Simmons and Hartzell, 1988) was identified by means other than direct and selective inhibition of PDE2.

We suggest EHNA to be a suitable pharmacological agent to identify PDE2 among other PDE activities. As such, the effect of EHNA should be opposite to that of reasonable concentrations of cGMP on cAMP hydrolysis. A major drawback in the use of EHNA to inhibit PDE2 is that EHNA is a potent inhibitor of adenosine deaminase (Cristalli *et al.*, 1994). An accumulation of adenosine – in the extracellular or the intracellular compartment – may, therefore, participate in the effects of EHNA, particularly in complex preparations. For instance, binding of adenosine to purinoceptors can strongly modify the activity of second messenger pathways (Tucker and Linden, 1993). In addition, binding of adenosine to the intracellular P-site of adenylate cyclase can reduce cAMP production. However, adenosine deaminase is not expressed in every cell type. For instance, it could not be detected in cardiac myocytes while present in other cardiac cell types (Schrader and West, 1991). Also the K_m of adenosine deaminase for adenosine is in the range of 20–50 μM (Schrader and West, 1991). Thus, unless adenosine deaminase activity occurs in a compartment not readily accessible to internal perfusion, e.g. in the close vicinity of the membrane, the continuous dialysis of the cell in whole cell patch-clamp experiments will prevent such a high accumulation of adenosine.

Nevertheless, the effect of EHNA can be compared to that of adenosine. In frog myocytes, adenosine does not mimic the effects of EHNA on either PDE or I_{Ca}. Superfusion of adenosine on frog myocytes has either no effect or induces an inhibition of isoprenaline stimulated I_{Ca}, likely mediated by the activation of A_1 adenosine receptors, which are negatively coupled to adenylate cyclase (Alvarez *et al.*, 1990; Tucker and Linden, 1993). In addition, adenosine deaminase is also inhibited by pentostatin (2′-deoxycoformycin; Parke-Davis, Ann Arbor MI, USA), the chemical structure of which is totally different from that of EHNA. Lacking the adenine ring, this compound is unlikely to mimic the effect of EHNA on PDE2 and may also be used to discriminate between PDE2 and adenosine deaminase activities.

5. *Acknowledgements*

We thank Patrick Lechêne for skilful technical assistance, Florence Lefèvre for preparation of the cells, Dr Thomas Podzuweit for the generous gift of MEP-1 and for confirming its identity with EHNA, and Dr Jacques Hanoune for permanent support. This work was supported by grants from the Fondation pour la Recherche Médicale, the Association Française contre les Myopathies, and Hoechst Pharma (France).

6. *References*

Alvarez, J.L., Mongo, K., Scamps, F. and Vassort, G. (1990). Effects of purinergic stimulation on the Ca current in single frog cardiac cells. Pflügers Arch. 416, 189–195.

Bethke, T., Meyer, W., Schmitz, W., Scholz, H., Stein, B., Thomas, K. and Wenzlaff, H. (1992). Phosphodiesterase inhibition in ventricular cardiomyocytes from guinea-pig hearts. Br. J. Pharmacol. 107, 127–133.

Brechler, V., Pavoine, C., Hanf, R., Garbarz, E., Fischmeister, R. and Pecker, F. (1992). Inhibition by glucagon of the cGMP-inhibited low-K_m cAMP phosphodiesterase in heart is mediated by a pertussis toxin-sensitive G-protein. J. Biol. Chem. 267, 15496–15501.

Cristalli, G., Eleuteri, A., Volpini, R., Vittori, S., Camaioni, E. and Lupidi, G. (1994). Adenosine deaminase inhibitors: synthesis and structure–activity relationships of 2-hydroxy-3-nonyl derivatives of azoles. J. Med. Chem. 37, 201–205.

Fischmeister, R. and Hartzell, H.C. (1986). Mechanism of action of acetylcholine on calcium current in single cells from frog ventricle. J. Physiol. (London) 376, 183–202.

Fischmeister, R. and Hartzell, H.C. (1987). Cyclic guanosine 3′,5′-monophosphate regulates the calcium current in single cells from frog ventricle. J. Physiol. (London) 387, 453–472.

Fischmeister, R. and Hartzell, H.C. (1990). Regulation of calcium current by low-K_m cyclic AMP phosphodiesterases in cardiac cells. Mol. Pharmacol. 38, 426–433.

Flitney, F.W. and Singh, J. (1981). Evidence that cyclic GMP may regulate cyclic AMP metabolism in isolated frog ventricle. J. Mol. Cell. Cardiol. 13, 963–979.

Hartzell, H.C. (1988). Regulation of cardiac ion channels by catecholamines, acetylcholine and 2nd messenger systems. Prog. Biophys. Mol. Biol. 52, 165–247.

Hartzell, H.C. and Fischmeister, R. (1986). Opposite effects of cyclic GMP and cyclic AMP on Ca^{2+} current in single heart cells. Nature 323, 273–275.

Lee, M.A, West, R.E., Jr and Moss, J. (1988). Atrial natriuretic factor reduces cyclic adenosine monophosphate content of human fibroblasts by enhancing phosphodiesterase activity. J. Clin. Invest. 82, 388–393.

Levi, R.C., Alloatti, G., Penna, C. and Gallo, M.P. (1994). Guanylate-cyclase-mediated inhibition of cardiac I_{Ca} by carbachol and sodium nitroprusside. Pflügers Arch. 426, 419–426.

Lohmann, S.M., Fischmeister, R. and Walter, U. (1991). Signal transduction by cGMP in heart. Basic Res. Cardiol. 86, 503–514.

Lugnier, C., Gauthier, C., Le Bec, A. and Soustre, H. (1992). Cyclic nucleotide phosphodiesterases from frog atrial fibers: isolation and drug sensitivities. Am. J. Physiol. 262, H654–H660.

MacFarland, R. T., Zelus, B.D. and Beavo, J.A. (1991). High concentrations of a cGMP-stimulated phosphodiesterase mediate ANP-induced decreases in cAMP and steroidogenesis in adrenal glomerulosa cells. J. Biol. Chem. 266, 136–142.

Méry, P.-F., Brechler, V., Pavoine, C., Pecker, F. and Fischmeister, R. (1990). Glucagon stimulates the cardiac Ca^{2+} current by activation of adenylyl cyclase and inhibition of phosphodiesterase. Nature 345, 158–161.

Méry, P.-F., Pavoine, C., Belhassen, L., Pecker, F. and Fischmeister, R. (1993). Nitric oxide regulates cardiac Ca^{2+} current: involvement of cGMP-inhibited and cGMP-stimulated phosphodiesterases through guanylyl cyclase activation. J. Biol. Chem. 268, 26286–26295.

Méry, P.-F., Pavoine, C., Pecker, F. and Fischmeister, R. (1995). Erythro-9-(2-hydroxy-3-nonyl)adenine inhibits cGMP-stimulated phosphodiesterase in isolated cardiac myocytes. Mol. Pharmacol. 48, 121–130.

Nicholson, C.D., Challis, R.A.J. and Shahid, M. (1991). Differential modulation of tissue function and therapeutic potential of selective inhibitors of cyclic nucleotide phosphodiesterase isoenzymes. Trends Pharmacol. Sci. 12, 19–27.

Pang, D.C. (1992). Tissue and species specificity of cardiac cAMP-phosphodiesterase inhibitors. Adv. Second Messenger Phosphoprotein Res. 25, 207–320.

Podzuweit, T., Müller, A. and Nennstiel, P. (1992). Selective inhibition of the cGMP-stimulated cyclic nucleotide phosphodiesterase from pig and human myocardium. J. Mol. Cell. Cardiol. 24 (Suppl. V), 102. [Abstract]

Podzuweit, T., Müller, A and Opie, L.H. (1993). Anti-arrhythmic effects of selective inhibition of myocardial phosphodiesterase-II. Lancet 341, 760.

Schrader, W.P. and West, C.A. (1990). Localization of adenosine deaminase and adenosine deaminase complexing protein in rabbit heart: implications for adenosine metabolism. Circ. Res. 66, 754–762.

Simmons, M.A. and Hartzell, H.C. (1988). Role of phosphodiesterase in regulation of calcium current in isolated cardiac myocytes. Mol. Pharmacol. 33, 664–671.

Thompson, W.J., Terasaki, W.L., Epstein, P.M. and Strada, J.S. (1979). Assay of cyclic nucleotide phosphodiesterase and resolution of multiple molecular forms of the enzyme. Adv. Cyclic Nucleotide Res. 10, 69–92.

Tucker, A.L. and Linden, J. (1993). Cloned receptors and cardiovascular responses to adenosine. Cardiovasc. Res. 27, 62–67.

Weishaar, R.E., Cain, M.H. and Bristol, J.A. (1985). A new generation of phosphodiesterase inhibitors: multiple molecular forms of phosphodiesterase and the potential for drug selectivity. J. Med. Chem. 28, 537–545.

Whaler, G.M. and Dollinger, S.J. (1995). The nitric oxide donor SIN-1 inhibits the mammalian cardiac calcium current through cGMP-dependent protein kinase. Am. J. Physiol. 37, C45–C54.

Whalin, M.E., Scammel, J.G., Strada, S.J. and Thompson, W.J. (1991). Phosphodiesterase II, the cGMP-activatable cyclic nucleotide phosphodiesterase, regulates cyclic AMP metabolism in PC12 cells. Mol. Pharmacol. 39, 711–717.

Zhu, Q., Yang, X., Claydon, M.A., Hicks, G.L., Jr and Wang, T. (1994). Adenosine deaminase inhibitor in cardioplegia enhanced function preservation of the hypothermically stored rat heart. Transplantation 57, 35–40.

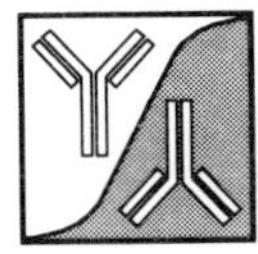

6. cGMP-Inhibited Phosphodiesterases (PDE3)

Narcisse Komas, Matthew Movsesian, Sasko Kedev, Eva Degerman, Per Belfrage *and* Vincent C. Manganiello

1. Introduction 89
2. Purification and Characterization 89
3. Molecular Cloning and Domain Organization 92
4. Structure/Function Relationships 93
   4.1 Catalytic Domain 93
   4.2 Membrane-Association Domain 95
   4.3 Regulatory Domain Phosphorylation/Activation 95
5. Pharmacology and Potential Therapeutic Usage of PDE3 Inhibitors 96
   5.1 Inotropic Agents 97
   5.2 Vasodilators 98
   5.3 Relaxation of Airway Smooth Muscle 99
   5.4 Antithrombotic Agents 100
   5.5 Anti-Inflammatory Agents 100
6. Therapeutic Use of PDE3 Inhibitors 101
7. Acknowledgements 101
8. References 101

1. Introduction

After Sutherland and co-workers initially described cyclic nucleotide hydrolytic activity in tissue extracts (Sutherland and Rall, 1958), Butcher and Sutherland (1962) partially purified what was thought to be an ubiquitous enzyme responsible for the specific hydrolysis of the 3′-bond of cyclic nucleotides. In the early 1970s, however, it became clear that multiple forms of cyclic nucleotide phosphodiesterases (PDEs) were present in various tissues (Thompson and Appleman, 1971a,b; Appleman and Terasaki, 1975). These multiple forms were initially classified as three major types, one of which exhibited high affinity for cAMP and was designated as the "low K_m" cAMP PDE. This "low K_m" cAMP PDE was later discovered to consist of two distinct isoenzymes, each with high affinity for cAMP but with distinct physical properties, kinetic characteristics and inhibitor specificities. One was very sensitive to inhibition by cilostamide and cGMP, and is now known as the cGMP-inhibited cyclic nucleotide phosphodiesterase (cGI-PDE) or PDE3, whereas the other, which was very sensitive to inhibition by Ro 20-1724, is now classified as PDE4 (Yamamoto *et al.*, 1984; Harrison *et al.*, 1986; Reeves *et al.*, 1987b; Kariya and Dage, 1988). With the development of selective PDE inhibitors and the use of molecular cloning strategies seven PDE families (PDE1–7) have been identified, with almost every family containing at least two subfamilies (Beavo and Reifsnyder, 1990; Conti *et al.*, 1991; Thompson, 1991; Michaeli *et al.*, 1991; Beavo *et al.*, 1994; Manganiello *et al.*, 1995a; see also Chapters 1 and 2). This chapter focuses on the PDE3 or cGI-PDE family.

2. Purification and Characterization

Purification of PDE3 isoenzymes to homogeneity has been difficult owing to their low abundance and sensitivity to proteolysis. We used an affinity matrix, composed of the *N*-(2-isothiocyanato)ethyl derivative of cilostamide, a specific PDE3 inhibitor (Weishaar *et al.*, 1985a; Alvarez *et al.*, 1986; Harrison *et al.*,

Phosphodiesterase Inhibitors
ISBN 0-12-210720-9

1986; Degerman *et al.*, 1987; Reeves *et al.*, 1987b; Kariya and Dage, 1988; Manganiello *et al.*, 1988, 1990, 1995a; Beavo and Reifsnyder, 1990; Conti *et al.*, 1991; Thompson, 1991; Beavo *et al.*, 1994), to purify these enzymes from several tissues, including rat and bovine adipose tissue (Degerman *et al.*, 1987, 1988), bovine aortic smooth muscle (Rascon *et al.*, 1992), human platelets (Degerman *et al.*, 1993) and human placenta (Le Bon *et al.*, 1992). Other procedures have yielded homogeneous preparations from bovine heart, rat liver and human platelets (Harrison *et al.*, 1986; Pyne *et al.*, 1987; Boyes and Loten, 1988; Grant and Colman, 1984). Due to proteolysis during purification, even in the presence of protease inhibitors, the final enzyme preparations in most cases contain several immunologically related polypeptides with molecular weights of 60–135 kD (SDS-PAGE). PDE3 activities have been detected in vascular and airway smooth muscle preparations (Weishaar *et al.*, 1986; Prigent *et al.*, 1988; Silver *et al.*, 1988a; Torphy and Cieslinski, 1990; Komas *et al.*, 1991a; de Boer *et al.*, 1992; Rabe *et al.*, 1993, 1994) and inflammatory cells, including macrophages (Tenor *et al.*, 1995a) and T cells (Robicsek *et al.*, 1991; Tenor *et al.*, 1995b) but these enzymes have not yet been characterized at the molecular level.

Immunoprecipitation or Western blotting of PDE3 isoenzymes from different tissues has demonstrated subunit molecular weights of 105–135 kD (determined by SDS-PAGE), one exception being the rat liver PDE3, often referred to as "dense vesicle" PDE, with a molecular weight of approximately 57 kD (Pyne *et al.*, 1987). Rat liver may contain two PDE3 isoforms, the "dense vesicle" PDE and a ~73 kD enzyme. The native subunit molecular weight of the latter is probably higher since the enzyme was solubilized using chymotrypsin (Boyes and Loten, 1988). Thus, in most cases, polypeptides obtained after purification probably originate from larger native forms that are proteolytically truncated during purification.

PDE3 isoenzymes can be distinguished from PDEs of other families by their high affinities for both cAMP and cGMP (K_m values of 0.1–0.8 μM); *V*max for cAMP is higher (approximately 4–10-fold) than for cGMP (Manganiello *et al.*, 1990). The properties of several purified PDE3 preparations are summarized in Table 6.1. In view of the K_m values for cAMP and cGMP, it is not surprising that cGMP is a competitive inhibitor of cAMP hydrolysis.

A second defining characteristic of PDE3 isoenzymes relative to other PDE families is their sensitivity to a number of drugs that augment myocardial contractility, inhibit platelet aggregation, relax smooth muscle, inhibit smooth muscle and lymphocyte proliferation, and inhibit the anti-lipolytic action of insulin (Weishaar *et al.*, 1985a, 1992; Alvarez *et al.*, 1986; Colucci *et al.*, 1986; Harrison *et al.*, 1986; Reeves *et al.*, 1987b; Kariya and Dage, 1988; Manganiello *et al.*, 1988, 1990, 1995a,b; Beavo and Reifsnyder, 1990; Houslay and Kilgour, 1990; Conti *et al.*, 1991; Nicholson *et al.*, 1991; Thompson, 1991; Torphy and Undem, 1991; Giembycz and Dent, 1992; Arnold, 1993; Beltman *et al.*, 1993; Endoh and Hori, 1993; Hall, 1993; Beavo *et al.*, 1994; Raeburn *et al.*, 1994; Degerman *et al.*, 1995). These compounds include cilostamide (OPC

Table 6.1 Properties of PDE3 isoforms

	Rat adipose tissue	*Bovine aortic smooth muscle*	*Bovine heart*	*Human platelets*	*Rat liver*		*Human placenta*
Catalytic properties							
K_m (μM)							
cAMP	0.4	0.16	0.15	0.2	0.24	0.3/29	0.57
cGMP	0.3	0.09	0.10	0.3	0.17	10	15
*V*max (μmol/min/mg)							
cAMP	8.5	3.1	6.0	6.1	6.2	0.114/0.633	0.86
cGMP	2.0	0.3	0.6	0.9	2.1	0.0041	0.47
Inhibitors							
IC_{50} (μM)[a]							
OPC-3911	0.04	0.054	0.005[b]	—	—	—	0.22
Milrinone	0.6	0.40	0.26[b]	0.71	—	1	—
CI-930	0.4	0.40	—	0.2	—	—	—
Ro 20-1724	190	>30[c]	62[b]	316	—	50	120
cGMP	0.2	0.25	0.06[b]	0.32	0.18	2.0	0.12

[a] IC_{50} is the concentration of drug causing 50% inhibition of PDE activity using 1 μM cAMP as substrate.
[b] Values for the bovine heart enzyme are given as K_i.
[c] Inhibition at 30 μM was less than 20%.
Taken from Degerman *et al.* (1995).

3689) and related OPC (Otsuka Pharmaceutical) derivatives, milrinone, enoximone, imazodan (CI-914), indolidan (LY 195115), Y-590, anegralide, lixazinone, SKF 94120 and ICI 1233188 (Fig. 6.1, Table 6.2). Some of the compounds listed in Table 6.2 were developed as cardiotonic/vasodilator drugs in the quest (ultimately not realized) to replace the cardiac glycoside digitalis in treatment of cardiovascular disease, especially congestive heart failure (Erhardt, 1987). Evaluation of these drugs involved structure/function studies which related inhibition of partially purified PDE3 preparations (usually contaminated with PDE4) to inotropic, anti-platelet or vasodilator effects. Many are relatively specific for PDE3, with submicromolar IC_{50} and K_i values. None apparently is selective for any specific tissue PDE3 isoform.

Many PDE3 inhibitors belong to several chemically and structurally related classes, with similarity to cAMP based on three dimensional structural models (Moos *et al.*, 1987; Weishaar and Bristol, 1989; Erhardt, 1990; Pang, 1992; Robertson and Boyd, 1992). The bipyridines, amrinone and milrinone, were among those initially developed, with milrinone undergoing the most extensive clinical trials for potential utility in the treatment of heart failure. The methyl substitution in the pyridine ring of milrinone was responsible for its greater potency than amrinone and its relative selectivity for PDE3 (Fig. 6.1) (Robertson and Boyd, 1992). Enoximone and piroximone are imidazolone derivatives of the bipyridines (Fig. 6.1) (Kariya *et al.*, 1982, 1984). A number of 4,5-dihydropyridazinone derivatives, including imazodan (CI 914) and indolidan (LY 195115), proved to be selective and potent PDE3 inhibitors (Fig. 6.1). In the imazodan group, the imidazole ring was important for specificity and the 5′-methyl and 4,5-dihydropyridazinone ring for potency (Fig. 6.1) (Bristol *et al.*, 1984; Sircar *et al.*, 1985, 1987; Moos *et al.*, 1987). Replacement of the imidazolyphenyl with an indolone moiety led to 4,5-dihydropyridazinones of the indolidan type, which were more potent than imazodan (Fig. 6.1) (Kauffman *et al.*, 1987a,b). Cilostamide, cilostazol and other OPC compounds are dihydroquinolinone derivatives with butoxy side-chains that are responsible for inhibitory potency (Fig. 6.1) (Hidaka *et al.*, 1979; Hidaka and Endo, 1984; Tani *et al.*,1992). Lixazinone, a very potent PDE3 inhibitor was developed by combining structural features of the two unrelated inhibitors cilostamide and anagrelide (Jones *et al.*, 1987; Venuti *et al.*, 1987, 1988).

These selective PDE3 inhibitors have also been useful in dissecting physiological processes regulated by PDE3 enzymes in intact cells and tissues. In isolated rat adipocytes PDE3 inhibitors or non-hydrolysable cAMP analogues block the antilipolytic action of insulin or of a putative phosphatidylinositol glycan insulin mediator (Misek and Saltiel, 1992; Schmitz-Peiffer *et al.*, 1992; Eriksson *et al.*, 1994; Beebe *et al.*, 1985). In isolated hepatocytes the non-hydrolysable analogues also block insulin-mediated inhibition of phosphorylase A and insulin-induced activation of gluconeogenesis. In frog oocytes stimulation of meiosis by insulin is associated with inhibition of adenylate cyclase (AC) and activation of a cilostamide-inhibited PDE; PDE3 inhibitors, but not PDE4 inhibitors such as rolipram, block the meiotic response to insulin (Sadler, 1991).

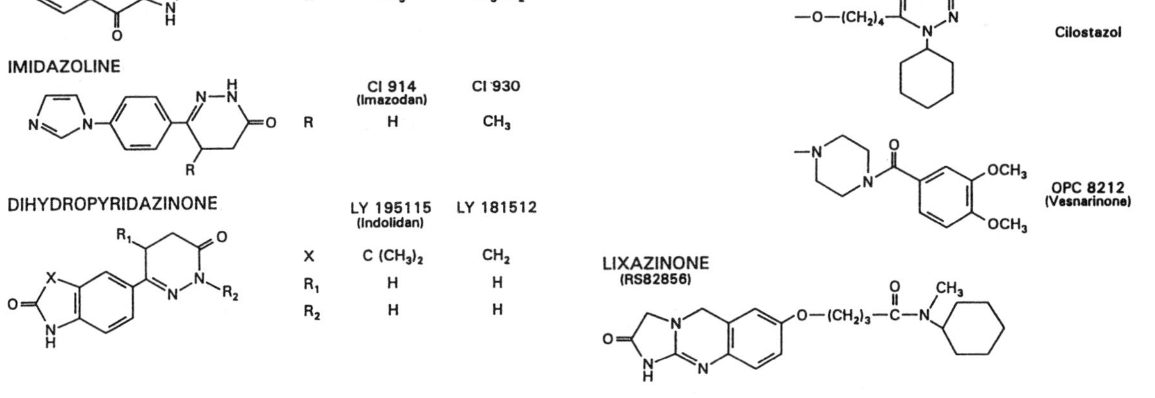

Figure 6.1 Chemical structures of some selective PDE3 inhibitors.

Table 6.2 Selective inhibitors of PDE3

Amrinone	Milrinone
Anagrelide	Motapizone
CI-930	OPC 3911
Cilostamide (OPC 3689)	Pimobendan (UD-CG115)[a]
Cilostazol	Piroximone
Enoximone	R 80122
ICI 1233188	Siguazodan
Imazodan (CI-914)	SKF 94120
Indolidan (LY 195115)	Vesnarinone (OPC 8212)[a]
Lixazinone	Y-590

[a] Therapeutic effects of these drugs may be related to actions other than or in addition to PDE3 inhibition.

In some instances PDE3 enzymes apparently function in concert with other PDE isoenzymes. For example, in T lymphocytes (Robicsek *et al.*, 1991; Marcoz *et al.*, 1993) and in isolated cultured vascular smooth muscle cells (Souness *et al.*, 1992; Pan *et al.*, 1994) and renal mesangial cells (Matousovic *et al.*, 1995), inhibition of PDE3 and 4 is associated with inhibition of DNA synthesis. In T lymphocytes, PDE3 inhibitors could be replaced by nitric oxide (NO)-generating agents, suggesting that, in conjunction with inhibition of PDE4, inhibition of PDE3 by either selective inhibitors or endogenous cGMP (produced via the stimulation by endothelium-derived NO of guanylate cyclase (GC)) is important for effective inhibition of DNA synthesis (Marcoz *et al.*, 1993). In rabbit platelets, inhibition of aggregation by prostacyclin-induced increases in cAMP was potentiated by inhibition of PDE3 through endogenous cGMP produced by nitrovasodilator stimulation of GC (Maurice and Haslam, 1990a). In rat aorta, PDE3 inhibitors alone relaxed pre-contracted aortic rings but the PDE4 inhibitor rolipram did not. Combination of PDE3 and 4 inhibitors did, however, produce synergic effects on aortic ring relaxation (Lindgren *et al.*, 1991).

A third important general characteristic of PDE3 enzymes involves their short-term regulation by hormones. Incubation of intact rat adipocytes (Pawlson *et al.*, 1974; Zinman and Hollenberg, 1974; Makino and Kono, 1980; Boyes and Loten, 1989), rat hepatocytes (Allan and Sneyd, 1975; Heyworth *et al.*, 1983; Loten *et al.*, 1987) and human platelets (Grant *et al.*, 1988; MacPhee *et al.*, 1988) with hormones that increase cAMP leads to increased PDE3 activity. This increased activity is thought to be important in "feedback" regulation of cAMP content and biological processes initiated by hormonal activation of AC. In adipocytes and hepatocytes, insulin reduces hormone-stimulated cAMP accumulation and cAMP-dependent protein kinase (PKA) activities by activating PDE3 (Manganiello and Vaughan, 1973; Pawlson *et al.*, 1974; Sakai *et al.*, 1974; Zinman and Hollenberg, 1974; Allan and Sneyd, 1975; Kono *et al.*, 1975; Loten *et al.*, 1978; Makino and Kono, 1980; Weber and Appleman, 1982; Heyworth *et al.*, 1983; Benelli *et al.*, 1986; Boyes and Loten, 1989; Lindgren *et al.*, 1991), resulting in inhibition of lipolysis and glycogenolysis, respectively. The role of PDE3 activation in insulin action has been discussed in several recent reviews (Belfrage *et al.*, 1986; Houslay and Kilgour, 1990; Loten, 1991; Beltman *et al.*, 1993; Degerman *et al.*, 1995; Manganiello *et al.*, 1995b).

3. Molecular Cloning and Domain Organization

Complementary DNAs encoding two PDE3 subfamilies, PDE3A and PDE3B (previously known as cGIP2 and cGIP1, respectively), have been cloned from rat and human adipose tissue and human cardiac muscle cDNA libraries (Meacci *et al.*, 1992; Taira *et al.*, 1993) and rat and human genomic libraries (Fig. 6.2). We have recently cloned a mouse (MM)PDE3B cDNA from a mouse adipocyte library; it is very similar to rat (RN)PDE3B (unpublished observations). PDE3A and PDE3B are products of distinct but related genes. Southern blot hybridizations of human (HS)PDE3A and HSPDE3B cDNAs with genomic digests of human–hamster somatic cell hybrids and fluorescent *in situ* hybridization of human metaphase chromosomes with HSPDE3A and HSPDE3B genomic clones indicate that the gene for HSPDE3B is located on human chromosome 11 and that for HSPDE3A on

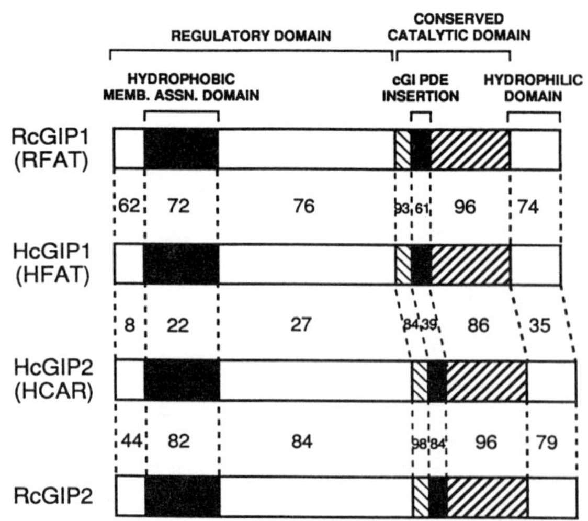

Figure 6.2 Domain organization and deduced amino acid identities of four PDE3 isoenzymes: RcGIP1 (RNPDE3B) and RcGIP2 (RNPDE3A) from rat adipose tissue, HcGIP1 (HSPDE3B) from human adipose tissue and HcGIP2 (HSPDE3A) from human cardiac muscle.

human chromosome 12 (unpublished observations; see also Chapter 1). RNPDE3B cDNA hybridizes weakly, if at all, with cardiac mRNA. RNPDE3B mRNA is relatively abundant in rat adipocytes and increases dramatically during differentiation of cultured murine 3T3-L1 adipocytes (Taira *et al.*, 1993). Earlier studies had indicated that particulate 3T3-L1 adipocyte PDE3 activity and responsiveness to insulin also appear or dramatically increase during differentiation of 3T3-L1 adipocytes (Murray and Russell, 1980; Elks *et al.*, 1983; see also Chapter 1). On the other hand, RNPDE3A cDNA hybridizes strongly with heart and weakly, if at all, with rat and 3T3-L1 adipocyte mRNAs (Taira *et al.*, 1993). Northern blot hybridizations and RNase protection assays reveal ~4.4 kb and ~7.6 kb transcripts in mRNA from HeLa cells and from human placental and cardiac tissue (Meacci *et al.*, 1992; Kasuya *et al.*, 1995). These two mRNA species, encoding ~80 and ~125 kD PDE3 isoforms, respectively, are thought to be transcribed from different initiation sites in the HSPDE3A gene in a tissue-specific manner (Kasuya *et al.*, 1995). A PDE3A ~4.4 kb transcript has also been detected in human erythroleukaemia cells (Cheung *et al.*, 1994) and T84 human colon carcinoma cells (Meacci *et al.*, 1992). Multiple species of RNPDE3A and HSPDE3B mRNAs have been detected on Northern blots of rat heart, lung, and other rat tissues, and human adipose tissue, respectively (Taira *et al.*, 1993, and unpublished observations).

PDE3A and PDE3B cDNAs encode proteins with molecular weights of 122–125 kD, consistent with those of PDE3 enzymes from intact rat adipocytes and human cardiac muscle microsomal preparations (Degerman *et al.*, 1990; Smith *et al.*, 1991, 1993). The entire deduced amino acid sequence of RNPDE3B is more closely related to that of HSPDE3B than RNPDE3A, which is similar to HSPDE3A. Mammalian PDE gene families, including PDE3, possess similar domain organizations with conserved catalytic domains in C-terminal regions and N-terminal regulatory domains (see Chapter 1). The domain organization and hydropathy plots of all PDE3A and PDE3B isoforms are quite similar, consistent with the notion of related structural and functional domains (Fig. 6.2). Within the conserved domain of PDE3 isoenzymes is found a sequence of 44 amino acids that does not align with sequences of other PDE families (Meacci *et al.*, 1992; Taira *et al.*, 1993). Deduced sequences of the conserved domains of all four compared PDE3 enzymes are very similar except for the 44 amino acid insertion, the sequence of which is similar in HSPDE3B and RNPDE3B and differs from those in HPDE3A and RNPDE3A, which are similar to each other. Thus, this 44 amino acid insertion, which is unique to the PDE3 gene family, may be important in identifying PDE3 subfamilies.

Available amino acid sequence data of a PDE3 purified from human platelets indicate it to be a HSPDE3A isoform (Degerman *et al.*, 1993; Taira *et al.*, 1993). Realizing that hormone-sensitive PDE3 activity is present in liver, and assuming that different PDE3s might have different 44 amino acid insertions within the conserved catalytic domain, an ~300 bp fragment containing the PDE3 insertion was cloned by reverse transcriptase/polymerase chain reaction (RT-PCR) from human hepatoma (HepG2 cells) mRNA. The sequence of this fragment, however, was essentially identical to that of the analogous region in HSPDE3B, consistent with the relatively strong hybridization of HepG2 poly(A)$^+$RNA with HSPDE3B cDNA and little, if any, with HSPDE3A cDNA (T. Murata *et al.*, unpublished observations). The regulatory domain contains hydrophobic putative membrane-association domains and several consensus sequences (-RRXS-) for a PKA substrate. The N-terminal portions of the deduced sequences of RNPDE3B and HSPDE3B, including those in the hydrophobic domains and those adjacent to the consensus substrate sequences for PKA phosphorylation, are very similar and differ from the analogous sequences of RNPDE3A and HSPDE3A, which are similar to each other.

4. *Structure/Function Relationships*

4.1 CATALYTIC DOMAIN

The conserved catalytic domains of different PDE families presumably include family-specific features accounting for distinct substrate affinities, catalytic rates, inhibitor sensitivities and structural features common to all PDEs, such as histidine-containing, Zn^{2+}-binding domains, which may participate in cyclic nucleotide bond hydrolysis and account for overlapping substrate specificities (Francis *et al.*, 1994). From the use of histidine- and sulfhydryl-modifying agents, Omburo *et al.* (1995) recently concluded that essential histidines and cysteines were present near or at the active site of the platelet PDE3 catalytic domain. To define essential interactions of cAMP and cGMP with the catalytic sites of PDEs and to begin to map the topology of these sites a series of cAMP and cGMP analogues have been used as competitive inhibitors (Beltman *et al.*, 1995; Butt *et al.*, 1995). Comparisons of IC_{50} values for different analogues relative to those for cAMP and cGMP were utilized to predict the functional groups on cAMP and cGMP that interact with each PDE family. It was found that each PDE family had a unique profile of binding interactions and that both cAMP and cGMP interact with the different PDE families in specific and unique fashions. The structural determinants involved in interactions of cGMP and cAMP with PDE catalytic sites were different from the interactions with analogous sites in other cyclic nucleotide binding molecules, such as PKA and cGMP-dependent protein kinase (PKG). PDE3, which exhibits the highest affinity for cGMP among PDEs, was found to interact with cGMP via multiple sites, since almost every modification of the guanine ring resulted in altered potency for cGMP inhibition of PDE3. PDE3 was found to bind to the anti-conformer of cAMP and cGMP (Beltman *et al.*, 1995; Butt *et al.*, 1995).

Studies of structure–activity relationships of a number of PDE3 inhibitors have also provided insights into the topography of PDE3 catalytic sites. It has been suggested that PDE3 inhibitors of the extended type (dihydpyridazinones, cilostamide, anagrelide, lixazinone, etc.) conform to a "H-P-I" model: they are composed of three basic subunits of heterocyclic, phenyl and imidazole groups with small lipophilic substitutions and assume an overall planar or nearly flat structure that resembles the anti-conformation of cAMP (Weishaar and Bristol, 1989; Erhardt, 1990; Pang, 1992; Robertson and Boyd, 1992) (Fig. 6.1). From the three-dimensional structure of cAMP in the anti-conformation, the N1 and $6NH_2$ atoms of the adenine moiety can align with the imidazole moiety of the inhibitors and the cyclic phosphate with amidic atoms of the heterocyclic substituents. The electronic structure of the ionized cyclic monophosphate is similar to that of the neutral heterocyclic moiety (Weishaar and Bristol, 1989; Erhardt, 1990; Pang, 1992; Robertson and Boyd, 1992). A five-point pharmacophore model has also been developed, which incorporates many of the features of the H-P-I model, with (i) a dipole moiety (carbonyl group) at one end, (ii) an adjacent amidic hydrogen, (iii) a small lipophilic substituent on the heterocyclic moiety, (iv) a hydrogen bonding moiety (imidazole ring) at the other end, and (v) an overall nearly planar or flat topography (Bristol *et al.*, 1984; Sircar *et al.*, 1985, 1987; Moos *et al.*, 1987; Weishaar and Bristol, 1989; Pang, 1992; Robertson and Boyd, 1992). The heterocyclic moiety is thought to account for inhibitory potency and the imidazole ring for potency and PDE3-specificity. Small alkyl substitutions in the heterocyclic moiety that occupy space corresponding to portions of the ribose moiety of cAMP affect inhibitor potency. The phenyl moiety probably serves other than a space function and may regulate structural planarity, since in cAMP there is a slight twist from planarity established by the adenine and cyclic phosphate moieties (Bristol *et al.*, 1984; Sircar *et al.*, 1985, 1987; Moos *et al.*, 1987; Weishaar and Bristol, 1989; Erhardt, 1990; Pang, 1992; Robertson and Boyd, 1992). Compounds of the milrinone and enoximone series do not adopt a planar conformation; other interactions with structural determinants on PDE3 presumably compensate for the binding energy lost in deviating from a planar conformation (Robertson and Boyd, 1992).

Studies with lixazinone, one of the most potent PDE3 inhibitors, added additional important features to the pharmacophore model for the PDE3 receptor site. Lixazinone was synthesized by combining features of two structurally unrelated PDE3 inhibitors, cilostamide and anagrelide (Jones *et al.*, 1987; Venuti *et al.*, 1987; Robertson and Boyd, 1992) (Fig. 6.1). The lipophilic oxybutyramide side-chain of cilostamide, responsible for the potency of cilostamide and other OPC derivatives, was incorporated into the heterocyclic ring structure of anagrelide. Structure–activity studies with a series of lixazinone derivatives indicated that the lipophilic oxybutyramide side-chain was responsible for inhibitory potency. Molecular modelling demonstrated that lixazinone could assume an overall planar conformation and structural features similar to imazodan, milrinone and enoximone. In lixazinone, however, the lipophilic oxybutyramide side-chain with its electron-donating ether moiety extended beyond the other inhibitors (Jones *et al.*, 1987; Venuti *et al.*, 1987; Robertson and Boyd, 1992). It has been suggested that this long side-chain reaches a secondary hydrophobic binding domain in PDE3, which perhaps stabilizes the enzyme–inhibitor complex at the active site, and that the ether moiety is involved in hydrogen bonding analogous to pyridyl or imidazolyl moieties in milrinone or imazodan.

The catalytic core of PDE3 has been identified in studies with N- and C-terminal deletion recombinant PDE3 enzymes (Komas *et al.*, 1994; Pillai *et al.*, 1994). Truncated recombinant RNPDE3B and HSPDE3A isoforms (expressed in *E. coli* and Sf9 cells) encoding both full-length and truncated proteins which included the entire catalytic domain exhibited the same characteristics as native PDE3, e.g. high affinity for cAMP and cGMP and sensitivity to selective PDE3 inhibitors. These findings suggest that the considerable differences in deduced amino acid sequence of RNPDE3B and HSPDE3A N-terminal regulatory domains are not reflected in dramatic alterations in substrate affinity or inhibitor sensitivities. Studies with truncated recombinant proteins indicated that the PDE3 catalytic core included the conserved catalytic domain (including the 44 amino acid insertion) plus some additional upstream and downstream sequences (Komas *et al.*, 1994; Pillai *et al.*, 1994). Whether the N-terminal domains contain autoinhibitory domains that affect *V*max is unknown. HSPDE3A recombinant proteins were, however, more sensitive than RNPDE3B recombinants to inhibition by cGMP (unpublished observations and Komas *et al.*, 1994). Despite the overall similarity in deduced sequences of RNPDE3B and HSPDE3A catalytic domains, the sequences of these two PDE3s do differ in the 44 amino acid insertion which is unique to PDE3 catalytic domains. It is possible, therefore, that these inserts not only distinguish the catalytic domains of PDE3 from other PDEs but may identify catalytic domains of different PDE3 subfamilies and be involved in determining their sensitivities to cGMP.

4.2 MEMBRANE-ASSOCIATION DOMAIN

In Sf9 cells, full-length recombinant RNPDE3B and HSPDE3A were predominantly found in particulate fractions, whereas the truncated forms were predominantly soluble (T. Murata *et al.*, unpublished observations). These results suggest that determinants for PDE3 association with intracellular membranes may well be located in the N-terminal portion, perhaps within the N-terminal hydrophobic domains. Whether cytosolic PDE3 enzymes lack this sequence has not been clarified. Based on activity measurements and immunoprecipitation studies, the human platelet cGI-PDE (the native form of which exhibits monomeric molecular weight of 105–110 kD) is thought to be almost entirely cytosolic (Grant and Colman, 1984; Alvarez *et al.*, 1986; Grant *et al.*, 1988; MacPhee *et al.*, 1988; Degerman *et al.*, 1993). Rat adipocyte PDE3, which is predominantly microsomal, exhibits a monomeric molecular weight of ~135 kD (Degerman *et al.*, 1990; Smith *et al.*, 1991). [^{32}P]PDE3 enzymes with molecular weights of ~135 kD were immuno-isolated from solubilized dog, rabbit, human and guinea-pig cardiac sarcoplasmic reticulum (SaR) microsomal fractions (Smith *et al.*, 1993). On the other hand, [^{32}P]PDE3s with lower molecular weights (~116 kD and ~90 kD) were found in both myocardial SaR and cytosolic fractions (Smith *et al.*, 1993). Although it is very likely that at least some of the lower molecular weight forms and cytosolic PDE3 enzymes represent proteolytic fragments, it cannot be ruled out that some of them are products of related PDE3 genes or arise by alternative mRNA splicing or from use of alternative transcription initiation sites (Kasuya *et al.*, 1995), especially since multiple mRNA species have been observed on Northern blots of rat tissue RNAs hybridized with RNPDE3A cDNA (Taira *et al.*, 1993 and unpublished observations).

4.3 REGULATORY DOMAIN PHOSPHORYLATION/ACTIVATION

The PDE3 regulatory domain is thought to contain phosphorylation sites that regulate enzyme activity, including site(s) phosphorylated in response to insulin and cAMP-elevating agents. Phosphorylation of PDE3 has been demonstrated in intact adipocytes, hepatocytes and platelets, and in broken cell preparations.

4.3.1 Adipocyte PDE3

Incubation of rat adipocytes with insulin, agents that increase cAMP or with cAMP analogues, results in serine-phosphorylation and activation of a PDE3 that is associated with adipocyte microsomes (Smith and Manganiello, 1988; Degerman *et al.*, 1990; Smith *et al.*, 1991). Phosphorylation and activation of the PDE3 induced by effectors that increase cAMP results in feedback regulation of cAMP. This, however, is not merely a response to excess cAMP, since in adipocytes activation occurs over virtually the entire range of isoprenaline-induced activation of AC and lipolysis, including conditions in which PKA is not saturated with cAMP (Smith and Manganiello, 1988). It may seem paradoxical that the PDE3 is rapidly activated in response to two opposing effectors such as insulin and isoprenaline. This dual regulation of PDE3 has been studied in detail in rat adipocytes, where, in the presence of both insulin and lipolytic effectors (a "physiological" condition in which insulin can reduce isoprenaline-induced increases in PKA activity and lipolysis), synergic phosphorylation (activation) of the adipocyte PDE3 has been demonstrated (Smith and Manganiello, 1988; Degerman *et al.*, 1990; Smith *et al.*, 1991). These results suggest "crosstalk" between the cAMP and insulin pathways, insofar that cAMP may sensitize the insulin pathway. In intact adipocytes incubated with insulin, isoprenaline or both, phosphorylation occurs – based on the deduced amino acid sequence of rat PDE3 (Taira *et al.*, 1993) – at a single site (Ser^{302}) localized within a PKA consensus sequence in the adipocyte enzyme (T. Rahn *et al.*, unpublished observations). Although Ser^{302} is phosphorylated to a similar extent in intact cells by hormones or PKA, Ser^{427} is the major site phosphorylated by PKA in solubilized cells (Rascon *et al.*, 1994). The finding that the same site on PDE3 is phosphorylated in intact cells in response to insulin, isoprenaline and the combination of the two agents suggests that the synergic phosphorylation and activation of PDE3 is due to crosstalk between the two pathways upstream of PDE3. An insulin-stimulated PDE3 kinase (PDE3 IK) has been partially characterized but not yet identified (Gettys *et al.*, 1988; Shibata and Kono, 1990a,b; Rahn *et al.*, 1994). This kinase or another upstream component in the insulin signal transduction pathway, such as phosphatidyl-inositol-3-kinase (PI-3-Kinase) (Rahn *et al.*, 1994), could be a converging point for the insulin and cAMP pathways in the adipocyte. It will be of considerable interest to identify signal transduction components between the PI-3-Kinase and PDE3 and to identify at which level the cAMP and insulin pathways crosstalk in regulating phosphorylation/activation of PDE3 and control of lipolysis.

4.3.2 Liver PDE3

Like adipocytes, isolated hepatocytes respond to insulin and glucagon with increases in PDE3 activity (Allan and Sneyd, 1975; Loten *et al.*, 1978; Heyworth *et al.*, 1983). In liver, however, insulin alone was reported to stimulate a peripheral plasma membrane enzyme with characteristics of PDE4 isoforms (Heyworth *et al.*, 1983; Houslay and Kilgour, 1990), whereas both insulin and glucagon activated a PDE3 associated with the Golgi network or an incompletely characterized microsomal fraction designated as "dense vesicles" (Allan and Sneyd, 1975; Loten *et al.*, 1978; Heyworth *et al.*, 1983; Benelli *et al.*, 1986). Activation of the hepatocyte PDE3 by PKA seemed to require prior dephosphorylation of a "basal" site (Kilgour *et al.*, 1989). Incubation of liver "dense vesicles" with PKA was found to result in a time-dependent phosphorylation and activation of the PDE3, provided the membranes had been previously incubated with Mg^{2+}. Since [^{32}P]PDE3 isolated from unstimulated ^{32}P-labelled hepatocytes was dephosphorylated in a Mg^{2+}-dependent manner (Kilgour *et al.*, 1989), and since phosphatase inhibitors prevented the PKA-dependent phosphorylation and activation of the PDE3, it was suggested that dephosphorylation of a specific phosphorylation site by a Mg^{2+}-dependent phosphatase (with no effect on enzyme activity) was required before PKA could act on the PDE3 (Kilgour *et al.*, 1989).

4.3.3 Platelet PDE3

More recently it was found that stimulation of platelets with insulin (Lopez-Aparicio *et al.*, 1992), as well as effectors that increase cAMP (Grant *et al.*, 1988; MacPhee *et al.*, 1988), results in activation of platelet PDE3. The physiological role for the insulin-mediated activation of the platelet PDE3 is not certain. As is the case in adipocytes, activation of the platelet PDE3 by insulin is associated with serine-phosphorylation of the enzyme (Lopez-Aparicio *et al.*, 1993). PKA (Grant *et al.*, 1988; MacPhee *et al.*, 1988) and a partially purified PDE3 IK (Lopez-Aparicio *et al.*, 1993) were demonstrated to activate and phosphorylate the platelet PDE3 *in vitro*. Specific inhibition of PDE3 in platelets prevented platelet aggregation, suggesting an important role for this enzyme in platelet function. It has been reported that in the diabetic state platelets exhibit greater than normal sensitivity towards aggregating agents and greater resistance to anti-aggregatory agents, perhaps indicative of altered cAMP metabolism in diabetic platelets (Hendra and Betteridge, 1989). In other studies, however, insulin apparently reduced platelet aggregation (Trovati *et al.*, 1984).

4.3.4 Other PDE3 Enzymes

PDE3s from bovine aorta and myocardial preparations from several species have been shown to be phosphorylated *in vitro* by PKA (Rascon *et al.*, 1992; Smith *et al.*, 1993). Incubation of cultured bovine aortic smooth muscle cells with forskolin resulted in phosphorylation and activation of PDE3 (D. Ekholm *et al.*, unpublished observations).

Molecular cloning and characterization of expressed PDE3A and PDE3B, together with data from intact cells and tissues, are consistent with the notion that the PDE3A and PDE3B isoforms are products of different genes, are differentially expressed and regulated in different cells and tissues, and may be involved in regulation of different physiological processes, e.g. PDE3B in hormonal regulation of lipolysis and glycogenolysis and PDE3A in myocardial and smooth muscle contractility and platelet aggregation. Further evidence regarding the possible role of specific PDE3 isoforms in different tissues comes from *in situ* hybridization studies in rats. These studies demonstrated distinct tissue-specific patterns of mRNAs for the two PDE3 isoenzymes (Reinhart *et al.*, 1995; see Chapter 1). PDE3A mRNA was present in the cardiovascular system, including myocardium and arterial and venous smooth muscle, throughout development. It was found also in bronchial, gastrointestinal and genito-urinary smooth muscle and epithelium, megakaryocytes, and oocytes. PDE3A mRNA demonstrated a complex, developmentally regulated pattern of gene expression in the CNS. The cellular distribution of PDE3B mRNA clearly differed from that of PDE3A. PDE3B mRNA was found in adipocytes (white and brown) during embryogenesis and throughout adult life. PDE3B mRNA was also present in liver throughout development (although at lower levels). PDE3B mRNA was found in spermatocytes, renal collecting duct epithelium, and neuroepithelium, including the developing neural retina.

5. *Pharmacology and Potential Therapeutic Usage of PDE3 Inhibitors*

The following sections are focused on the potential therapeutic use of PDE3 inhibitors as inotropic/vasodilator agents, anti-thrombotic and anti-inflammatory agents.

There has been long-standing interest in discovering and developing specific and selective inhibitors of the different PDE families, with the goal of providing potent therapeutic agents without the side-effects of non-selective PDE inhibitors. Although selective inhibitors of several PDE families are available, particularly of PDE3 and PDE4, none distinguishes between individual isoforms (e.g. PDE3A and PDE3B) within the same PDE family, an issue of paramount importance if inhibition of a single PDE isoform in a specific cell or tissue is needed for therapeutic benefit. Whether the lack of selectivity among members of the same PDE family or non-specific effects of PDE inhibitors (or both) are responsible for the side-effects (toxicity) observed in clinical trials of PDE3 and PDE4 inhibitors is not known. In addition, the possibility that different isoforms within the same family (e.g. PDE3A and PDE3B) have discrete, cell-specific functions has serious implications for development of therapeutic agents, in terms of drug delivery to specific target cell types and the development of inhibitors for different isoforms within the same family.

5.1 INOTROPIC AGENTS

Historically, the discovery of PDE3 inhibitors was associated temporally with a concerted effort by the pharmaceutical industry to develop alternative therapies applicable to cardiovascular disease, especially in the search for novel positive inotropic agents that could replace drugs such as digitalis (with its narrow therapeutic window) for the treatment of congestive heart failure. Although many of these PDE3 inhibitors, which served as inotropic agents, produced favourable haemodynamic effects, none has provided long-term improvement in clinical symptoms. They are, therefore, of only limited clinical utility. Nonetheless, these potent and relatively specific PDE3 inhibitors have provided considerable information and insight into functions of cGI-PDEs.

Since cAMP is an important intracellular second messenger in regulation of myocardial contractility, vascular and airway smooth muscle relaxation and platelet aggregation, PDEs have been logical targets in the development of drugs designed to increase cAMP concentrations in these tissues. Increases in cAMP in cardiac myocytes are accompanied by enhancement of myocardial contraction, thought to be mediated by PKA-catalysed phosphorylation of L-type Ca^{2+} channels (which increases the probability of their being open and enhances the inward Ca^{2+} current during contraction) (Klockner *et al.*, 1992; Sculptoreanu *et al.*, 1993) and of phospholamban (which increases the rate of Ca^{2+} sequestration by the SaR during relaxation) (Sham *et al.*, 1991; Movsesian, 1993; Luo *et al.*, 1994). Fischmeister and associates have recently suggested that NO may regulate L-type calcium current in frog ventricle and human atrial cells by activation of GC leading to increased cGMP, inhibition of PDE3, and subsequently increased cAMP (Méry *et al.*, 1993; Kirstein *et al.*, 1995; see also Chapter 5). However, increased myocardial cAMP can also induce tachycardia and arrhythmias (Taniguchi *et al.*, 1977; Tsien, 1977).

In the late 1970s and early 1980s it became clear that inhibitors of myocardial cAMP PDE such as amrinone and enoximone increased cAMP, enhanced contractility in isolated canine and guinea-pig ventricular myocardial strips or papillary muscle preparations and produced cardiotonic and vasodilator effects *in vivo* in animals and in patients with severe heart failure (Benotti *et al.*, 1978; Alousi *et al.*, 1979; Honerjäger *et al.*, 1981; Dage *et al.*, 1982; Endoh *et al.*, 1982; Roebel *et al.*, 1984). These drugs apparently inhibited myocardial low K_m cAMP PDE activity without affecting Na^+/K^+- or Ca^{2+}-ATPase, AC, or SaR- or mitochondrial Ca^{2+} uptake (Kariya *et al.*, 1982; Dage *et al.*, 1982; Roebel *et al.*, 1984; Weishaar *et al.*, 1985b; Pang, 1988). Early structure/function studies of these drugs, however, were carried out with relatively crude or, at best, partially purified preparations of "low K_m" cAMP PDE which contained – depending on the species and tissue source – variable proportions of PDE3 and 4 isoenzymes (Pang 1988). Yamamoto *et al.* (1984) first demonstrated that PDE3 and PDE4 isoforms could be distinguished by their inhibitor specificities, the former inhibited by cilostamide and the latter by Ro 20-1724. Harrison *et al.* (1986) demonstrated that a highly purified PDE3 preparation from bovine heart was inhibited by a number of cardiotonic drugs and suggested that PDE3 served as a "receptor" for these agents. Other workers later demonstrated a high correlation between the ability of dihydropyridazinone cardiotonics of the indolidan (LY 195115) series and other drugs, including milrinone, imazodan, enoximone and piroximone, to inhibit SaR-associated PDE3 and enhance myocardial contractility in dogs (Kauffman *et al.*, 1987a) and in isolated rabbit ventricular papillary muscle or guinea-pig atrial preparations (Kithas *et al.*, 1988; Ahn *et al.*, 1986). In myocardial SaR preparations a radiolabelled derivative of indolidan ([^{3}H]LY 186126), a potent cGI-PDE inhibitor, exhibited reversible, high affinity ($K_d \approx 6$ nM) binding to a single class of SaR binding sites. Ligand competition and displacement correlated with inhibition of PDE3 activity (Kithas *et al.*, 1988; Kauffman *et al.*, 1989a,b; Lugnier *et al.*, 1993). Although it is possible that a SaR-associated PDE3 regulates cAMP-dependent phosphorylation of phospholamban and Ca^{2+} movement into the SaR, cardiotonic activity can not be related solely to selective inhibition of SaR PDE3 since PDE3 enzymes are present in myocardial cytosol as well as SaR (Harrison *et al.*, 1986; Weishaar *et al.*, 1987, 1992; Kauffman *et al.*, 1989a,b; Silver *et al.*, 1990; Movsesian *et al.*, 1991; Lugnier *et al.*, 1993; Smith *et al.*, 1993).

Several studies have demonstrated species differences in inotropic responses to PDE3 inhibitors (Weishaar and Bristol, 1989; Weishaar *et al.*, 1987, 1992). Imadozan, for example, increased myocardial contractility in Rhesus monkey > dog > guinea pig > hamster > rat (Weishaar *et al.*, 1987). Species specificity of effects of inhibitors on isolated PDE3 enzymes was apparently not responsible for these differences in cardiotonic effects. Structure/function studies with imazodan derivatives, for example, correlated inhibition of guinea-pig cardiac PDE3 with inotropic responses in anaesthetized dogs (Sircar *et al.*, 1987) whereas effects of lixazinone derivatives on inhibition of human platelet aggregation were related to inhibition of platelet, rat heart and dog heart PDE3s (Jones *et al.*, 1987; Venuti *et al.*, 1987, 1988). It is not known whether the species differences in inotropic responses relate to physical or functional compartmentation of PDE3s, to differences in total amounts of PDEs and/or relative amounts and proportions of different PDEs, or to other factors, including functional compartmentation of cAMP in heart (Brunton *et al.*, 1981; Bode and Brunton, 1988). In isolated perfused rat heart and rabbit ventricular myocytes, although both PGE_1 and isoprenaline increased cAMP and activated PKA, only isoprenaline activated phosphorylase and inhibited glycogen synthase (Brunton *et al.*, 1981; Bode and Brunton, 1988). A PDE (or PDEs) activated by α_1-adrenoceptor agonists may somehow participate in this functional compartmentation of cAMP and PKA (Brunton *et al.*, 1981). In the guinea pig, however, PDE3 and PDE4 may regulate these different compartments or pools of cAMP (Gristwood and Owen, 1986; Murray *et al.*, 1987; Reeves *et al.*, 1987a; Reeves and England, 1990). Although both rolipram (a PDE4 inhibitor) and SKF 94120 (a PDE3 inhibitor) increased cAMP and activated PKA, only SKF 94120 exhibited inotropic activity (Murray *et al.*, 1987; Reeves *et al.*, 1987a; Reeves and England, 1990). In these studies both PGE_1 and isoprenaline increased cAMP and PKA but only isoprenaline activated phosphorylase A and increased myocardial contractility, suggesting that isoprenaline and PDE3 regulate the same pool or compartment of cAMP (Reeves *et al.*, 1987a). Rolipram acted synergically with SKF 94120 in increasing cAMP and contractility in isolated guinea-pig ventricle strips (Gristwood and Owen, 1986). The PDE3 inhibitor milrinone exerted small inotropic effects on isolated rabbit papillary muscle but not on rat ventricle preparations. Rolipram, ineffective alone, augmented responses of both preparations to milrinone and the mixed PDE3/4 inhibitor Org 30029 produced effects similar to those of the combination of milrinone and rolipram (Shahid and Nicholson, 1990). Furthermore, in the rat Ro 20-1724, ineffective alone, enhanced the effect of isoprenaline on both cAMP accumulation and force of cardiac contraction (Katano and Endoh, 1990). Taken together, these data suggest that inhibition of a fraction of the total cAMP PDE activity (i.e. PDE3 and 4 isoenzymes), perhaps at specific intracellular locations, may be sufficient to induce positive inotropic effects.

Partially purified PDE3 has been prepared from canine sino-atrial node regions and ventricular muscle (Komas *et al.*, 1989). Whereas the preparations exhibited similar K_m values for cAMP and cGMP they differed in sensitivity to some PDE3 inhibitors. These findings suggested that with selective PDE3 inhibitors it might be possible to dissociate toxic arrhythmogenic effects of these drugs (mediated perhaps by cAMP) from therapeutically desirable inotropic responses.

5.2 VASODILATORS

The mechanisms by which cAMP regulates relaxation of smooth muscle are less well understood than are its effects on myocardial contractility. In vascular smooth muscle, activation of protein kinase G (PKG) may be the predominant mechanism for inducing relaxation. There is an excellent correlation between the ability of cyclic nucleotide analogues to activate PKG and to relax pig coronary arteries and guinea-pig tracheal smooth muscle (Francis *et al.*, 1988; Sekhar *et al.*, 1992). NO, nitrovasodilators and endothelium-derived relaxing factor (EDRF: NO or an NO derivative generated in endothelial cells in response to several vasodilatory substances, e.g. bradykinin, acetylcholine, substance P) activate soluble GC, increasing cGMP and activating PKG in smooth muscle myocytes (Knowles and Moncada, 1992). Whereas activation of PKA might decrease smooth muscle Ca^{2+} and induce smooth muscle relaxation by mechanisms similar to those in cardiac myocytes, recent evidence has suggested that in vascular smooth muscle cAMP and cAMP analogues can activate PKG (Jiang *et al.*, 1992), resulting in phosphorylation of K^+ channels (increasing their open probability) leading to K^+ efflux, hyperpolarization of the plasma membrane and reduced Ca^{2+} efflux (Lincoln *et al.*, 1990; Archer *et al.*, 1994). PDE3 enzymes, which hydrolyse both cAMP and cGMP, might be important regulators of this "cross-talk" between cAMP and cGMP signalling.

Some of the interactions between EDRF/NO, nitrovasodilators such as 3-morpholinosydnonimine (SIN-1), PDE inhibitors and activators of AC that lead to relaxation of vascular smooth muscle may be mediated via increases in cAMP brought about by inhibition of PDE3 by endogenous cGMP generated in response to EDRF/NO (Komas *et al.*, 1991b; Maurice and Haslam, 1990b; Maurice *et al.*, 1991; Lugnier and Komas, 1993). Lugnier and Komas (1993) also found that, whereas PDE3 inhibitors such as milrinone, imazodan, indolidan, and SKF 94120 relaxed rat aorta in an endothelium-independent fashion, effects of PDE4 inhibitors such as rolipram were endothelium-dependent and potentiated by agents that increased cAMP or by PDE3 inhibitors. They suggested that participation of PDE4 in regulation of vascular smooth muscle relaxation is endothelium-dependent and may involve increases in cAMP brought about in part either by inhibition of smooth muscle PDE3 by cGMP – produced in response to EDRF/NO – or by PDE3 inhibitors.

The haemodynamic responses to PDE3 inhibitors are related to their inotropic actions as well as to effects on peripheral vasculature, where they are potent vasodilators. PDE3 inhibitors promote coronary artery relaxation (Ludmer *et al.*, 1986), reduce pulmonary artery pressure (Harris *et al.*, 1992) and reduce systemic arterial pressure and resistance in human patients and animals (Jaski *et al.*, 1985; Feneck, 1990; Silver *et al.*, 1992). There was an excellent correlation between the ability of a series of indolidan derivatives and other PDE3 inhibitors, including milrinone, enoximone, imazodan and CI-930, to inhibit a PDE3 preparation from canine ventricle SaR and to relax rat aortic strips that had been contracted with 5-hydroxytryptamine (5-HT) (Kauffman *et al.*, 1987b). Relaxation of phenylephrine-contracted guinea-pig aortic rings by milrinone, imazodan, and CI–930 correlated with their inhibition of guinea-pig aorta PDE3 (Silver *et al.*, 1988b). The presence of PDE3 in rat mesenteric artery and relaxation of human mesenteric arteries by milrinone and OPC 3911 are consistent with a role for PDE3s in dilation of resistance/conductance arteries (Komas *et al.*, 1991a; Lindgren *et al.*, 1989). PDE1, 3, 4 and 5 isoenzymes have been detected in isolated human pulmonary artery preparations; pulmonary artery rings contracted with $PGF_{2\alpha}$ were relaxed with inhibitors of PDE3 (motapizone), PDE3/4 (zardaverine) or PDE5 (zaprinast) but not PDE4 (rolipram) (Rabe *et al.*, 1994).

In rat aorta preparations, hydrolysis of cAMP (0.5 μM substrate) was inhibited ~50% by OPC 3911, ~20% by rolipram and ~90% with both inhibitors (Lindgren *et al.*, 1991). In this study the effects of OPC 3911 and milrinone were endothelium-independent whereas those of CI-930 were not. In other studies, however, effects of PDE3 inhibitors, including CI-930, were endothelium-independent (Komas *et al.*, 1991b; Kauffman *et al.*, 1987b). The relaxant effects of indolidan and milrinone on rat aortic strips were, for the most part, endothelium-independent (Kauffman *et al.*, 1987b) but methylene blue (an inhibitor of GC) shifted the concentration–response curves for indolidan and milrinone rightwards, suggesting a possible role for cGMP (perhaps NO) in their vasorelaxant actions (Kauffman *et al.*, 1987b). Although OPC 3911, CI-930 and milrinone relaxed precontracted aortic rings, rolipram alone was ineffective. When rolipram was combined with OPC 3911 or milrinone, however, supra-additive effects on cAMP content and relaxation were observed. These results suggested that regulation of vascular tone by PDE4 became important when PDE3 was inhibited and that inhibition of both PDE3 and 4 produced synergic effects on vasorelaxation. These studies also pointed out potential problems of interpreting some studies with milrinone, since some vasorelaxant effects in rat aorta (Lindgren *et al.*, 1991) and effects on cAMP content (Silver *et al.*, 1988b) were observed at concentrations of milrinone that could inhibit PDE4 as well as PDE3. Synergy between the PDE3 inhibitor SKF 94120 and rolipram to increase contractility of isolated guinea-pig ventricle (Gristwood and Owen, 1986) and rat and rabbit papillary muscle preparations (Shahid and Nicholson, 1990) has also been reported.

Vasorelaxant effects of PDE3 inhibitors may also be related to signalling pathways utilized by different vasoconstrictors. In rat aorta, OPC 3911 and CI–930 were more effective in relaxing contraction induced by serotonin than by phenylephrine (Lindgren *et al.*, 1991). In the presence of rolipram, however, effects of OPC 3911 and CI-930 on phenylephrine-induced contraction correlated well with PDE3 inhibition. Similarly, it was found that synergic effects of PDE inhibitors or nitrovasodilators and isoprenaline were more pronounced when inhibition of contraction rather than vasorelaxation of contracted aorta was measured (Maurice and Haslam, 1990b; Maurice *et al.*, 1991).

5.3 RELAXATION OF AIRWAY SMOOTH MUSCLE

PDE isoforms 1–5 have been demonstrated in airway smooth muscle preparations from several species, including humans (Silver *et al.*, 1988a; Harris *et al.*, 1989; Torphy and Cieslinski, 1990; de Boer *et al.*, 1992; Rabe *et al.*, 1993; Tomkinson *et al.*, 1993; Dent *et al.*, 1994; Raeburn, 1994). The relaxant effects of different PDE inhibitors in different species indicate species differences in the contribution of PDE3 and other PDE isoforms (especially PDE4) to regulation of airway smooth muscle relaxation (Dent *et al.*, 1994; Raeburn, 1994). In guinea-pig trachealis contracted by exposure to histamine or methacholine the rank order of potency for relaxation induced by PDE inhibitors was PDE4 > PDE3 > PDE5 >PDE1 (Tomkinson *et al.*, 1993). In another study, PDE3 and 4 inhibitors were approximately equipotent in inducing relaxation of isolated guinea-pig tracheal muscle contracted with histamine and carbachol (Harris *et al.*, 1989). Among a series of PDE3 inhibitors, including anagrelide, CI-930, milrinone and imazodan, there was also strong correlation between inhibition (IC_{50}) of a tracheal PDE3 preparation and relaxation of histamine-induced bronchoconstriction in anaesthetized guinea-pigs (Harris *et al.*, 1989). In isolated canine tracheal preparations contracted with methacholine, inhibitors of PDE3 (SKF 94120) or PDE4 (Ro 20-1724) potentiated effects of isoprenaline on both cAMP accumulation and relaxation (Torphy *et al.*, 1991). In contracted pig bronchial preparations, a PDE3 inhibitor (siguazodan) was a more effective relaxant than a PDE4 inhibitor (rolipram) (Tomkinson *et al.*, 1993; Raeburn, 1994) whereas the converse was observed in bovine trachea (Shahid *et al.*, 1991).

As observed in rat aorta preparations (Lindgren *et al.*, 1991), inhibitors of PDE3 and 4 also produced synergic or additive effects on airway smooth muscle. In guinea-pig preparations, low concentrations of either a PDE3 (imazodan) or PDE4 inhibitor (rolipram) increased relaxation of airway smooth muscle produced by the other (Harris *et al.*, 1989). In canine preparations, effects of PDE3 and 4 inhibitors were additive (Torphy *et al.*, 1991). Treatment of methacholine-contracted pig bronchus with low concentrations of a PDE3 inhibitor (siguazodan) potentiated rolipram-induced relaxation (Raeburn, 1994).

As in animals, both PDE3 and PDE4 isoenzymes are apparently important in relaxation of human airway smooth muscle. In isolated human bronchial ring preparations, inhibitors of PDE3 (Org 9935), PDE4 (rolipram) or PDE3 and 4 (Org 30029) produced concentration-dependent relaxation of histamine- and methacholine-contracted bronchial segments (de Boer *et al.*, 1992).

Org 9935 was more effective in relaxing bronchial rings contracted by histamine than by methacholine; whereas concentration–response curves for Org 9935 and rolipram were biphasic, that for Org 30029 was monophasic, suggesting that more effective relaxation was produced with simultaneous inhibition of both PDE3 and PDE4 (de Boer *et al.*, 1992). In another study, although the PDE3 inhibitor SKF 94120 was more effective than rolipram (PDE4 inhibitor) in reducing basal tone of isolated human bronchial ring preparations, relaxation was greater with either the mixed PDE3/4 inhibitor zardaverine or the combination of SKF 94120 and rolipram than with SKF 94120 alone (Rabe *et al.*, 1993). In preparations from large (4–15 mm diameter) and small (<0.5–2 mm diameter) bronchi, Torphy *et al.* (1993) found that the PDE3 inhibitor siguazodan was much more effective than rolipram in relaxing large (but not small) bronchi contracted with carbachol or leukotriene D_4. Effects of siguazodan and rolipram on relaxation of contracted large and small bronchi were additive or synergic (Torphy *et al.*, 1993). Whereas in this study siguazodan, but not rolipram, potentiated the effect of isoprenaline on relaxation of bronchial rings, in another study the converse was observed (Qian *et al.*, 1993).

In the anaesthetized intact dog, PDE3 inhibitors were potent bronchodilator agents but produced cardiovascular responses as well; PDE4 inhibitors produced substantial bronchodilation with minimal cardiovascular effects (Heaslip *et al.*, 1991). In anaesthetized guinea-pigs, however, administration of siguazodan directly into the airway had minimal effects on heart rate or systemic blood pressure at doses that produced maximal effects on bronchodilation (Raeburn *et al.*, 1992). Recently, oral administration of cilostazol (PDE3 inhibitor) was reported to reduce bronchial responsiveness to a methacholine challenge with headache as the only complaint (Fujimura *et al.*, 1995). The results of larger studies that evaluate the bronchodilator effects of aerosol administration of PDE3, PDE4 and PDE3/4 inhibitors should be of interest.

5.4 ANTITHROMBOTIC AGENTS

Platelet PDE3 is thought to be important in the regulation of platelet cAMP (Grant and Colman, 1984; Alvarez *et al.*, 1986; Weishaar *et al.*, 1986; Grant *et al.*, 1988; MacPhee *et al.*, 1988; Degerman *et al.*, 1993). Many inotropic agents that inhibit cardiac PDE3 – including cilostamide and other OPC derivatives, anagrelide, lixazinone, SKE 94120 and siguazodan – inhibited platelet PDE3 and aggregation at concentrations similar to those required for their inhibition of cardiac PDE3 and vasodilator and inotropic effects (Hidaka *et al.*, 1979; Hidaka and Endo, 1984; Alvarez *et al.*, 1986; Jones *et al.*, 1987; Venuti *et al.*, 1987; Simpson *et al.*, 1988; Murray *et al.*, 1990; de Chaffoy de Courcelles *et al.*, 1992; Tani *et al.*, 1992). Siguazodan, milrinone, R80122 and SKF 94120 increased platelet cAMP (Simpson *et al.*, 1988; Murray *et al.*, 1990; Roevens and de Chaffoy de Courcelles, 1993), decreased platelet cytosolic Ca^{2+} (Murray *et al.*, 1990; Roevens and de Chaffoy de Courcelles, 1993) and increased Ca^{2+} in intracellular storage sites (Roevens and de Chaffoy de Courcelles, 1993). The latter effect is thought to be analogous to the effect of PDE3 inhibitors in myocardium, where these drugs – via increases in cAMP – stimulate removal of Ca^{2+} from the cytoplasm and increase the rate of relaxation of the myocardium (lusitropic effect of cAMP). Although the rank order for inhibition of human platelet PDE3 (trequinsin > lixazinone > cilostamide > milrinone > siguazodan) correlated with enhancement of iloprost-induced increases in cAMP in intact platelets, much higher concentrations were required to increase cAMP (Tang *et al.*, 1994). Similarly, much lower concentrations of the PDE3 inhibitors were required to inhibit photolabelling of cytosolic platelet PDE3 with [^{32}P]cGMP than to increase cAMP in intact platelets; penetration of these compounds through the platelet membrane may account for these discrepancies (Tang *et al.*, 1994). It has been suggested that in human and rabbit platelets sodium nitroprusside (SNP) and nitrovasodilators increase cAMP and potentiate effects of cAMP-elevating agents in inhibiting platelet aggregation by increasing cGMP, which in turn inhibits platelet PDE3 (Maurice and Haslam, 1990a; Andersson and Vinge, 1991).

Although from a therapeutic perspective a cardiotonic agent with vasodilatory and antithrombotic properties might be useful in preventing some complications of cardiovascular disease and heart failure, such benefits have not been studied in any controlled trials.

5.5 ANTI-INFLAMMATORY AGENTS

PDE3s have been identified in lymphocytes. In several studies, combinations of PDE3 and PDE4 inhibitors were more effective in inhibiting T cell (and cultured smooth muscle cell) thymidine incorporation and/or proliferation than either alone (Sadler, 1991; Souness *et al.*, 1992; Marcoz *et al.*, 1993; Pan *et al.*, 1994). In mouse T lymphocytes SNP could replace inhibitors of PDE3, but not of PDE4, suggesting that under these conditions endogenous cGMP (generated by NO activation of GC) increased cAMP via inhibition of lymphocyte PDE3 (Marcoz *et al.*, 1993). Whether there is a role for PDE3 or mixed PDE3/4 inhibitors in treatment of immune or inflammatory disorders remains to be established (see Chapter 10).

6. Therapeutic Use of PDE3 Inhibitors

The only disease in which the therapeutic benefit of PDE3 inhibitors has been systematically evaluated is congestive heart failure. The rationale for the use of PDE3 inhibitors in this condition is based on the augmentation of myocardial inotropy and lusitropy associated with increased cAMP in cardiac myocytes and the decrease in vascular smooth muscle contraction associated with increased cAMP and possibly cGMP in vascular smooth muscle cells. These effects of PDE3 inhibitors increase cardiac output and reduce pulmonary congestion in patients with heart failure.

The administration of PDE3 inhibitors to patients with heart failure resulted in favourable haemodynamic effects acutely and improvements in exercise tolerance in the short term (Le Jemtel *et al.*, 1980; Baim *et al.*, 1983; Uretsky *et al.*, 1983; Jaski *et al.*, 1985). Unfortunately, these benefits were not sustained during long-term administration of the drugs. Results of clinical trials with amrinone (Di Bianco *et al.*, 1984; Massie *et al.*, 1985), milrinone (Di Bianco *et al.*, 1989) and enoximone and its analogues (Shah *et al.*, 1985; Uretsky *et al.*, 1990; Narahara *et al.*, 1991) gave no evidence of decreased mortality in patients taking PDE3 inhibitors. In some cases, there was a significant increase in mortality in patients treated with these drugs. A meta-analysis of many of these studies revealed the likelihood of death increased by 41% in patients to whom these PDE3 inhibitors were administered (Nony *et al.*, 1994). The reasons for the lack of sustained benefit and the increased mortality have not been identified.

Based on these observations, PDE3 inhibitors would seem to represent a valuable addition to the short-term treatment of heart failure, for instance in patients in cardiogenic shock awaiting cardiac transplantation. Most of these drugs have, however, no apparent role in long-term therapy, although clinical studies with newer PDE3 inhibitors suggest that the latter conclusion may be premature. Pimobendan (Fig. 6.1, Table 6.2), a drug that both inhibits PDE3 and increases the sensitivity of cardiac myofilaments to Ca^{2+}, seemed to have sustained beneficial effects on haemodynamics and exercise tolerance in patients with heart failure (Katz *et al.*, 1992; Remme *et al.*, 1994). Whether these effects resulted from pimobendan's inhibition of PDE3 or its Ca^{2+}-sensitizing properties is not clear. The fact that pimobendan, in contrast to other PDE3 inhibitors, reduces myocardial oxygen consumption suggests that the latter mechanism may be important (Hasenfuss *et al.*, 1989). The effects of pimobendan on mortality have not been systematically evaluated in a randomized double-blind trial.

Vesnarinone (Fig. 6.1, Table 6.2), a PDE3 inhibitor that decreases outward K^+ currents and increases inward Na^+ current in cardiac myocytes, improved mortality in heart failure (Lathrop and Schwartz, 1985; Iijima and Taira, 1987; Feldman *et al.*, 1993). The beneficial effect of vesnarinone on mortality occurred only at low doses of the drug, at which improvement in haemodynamics was not noted; mortality was increased, however, in patients who received higher doses of vesnarinone, at which inotropic effects could be expected (Feldman *et al.*, 1993). It has been suggested that the beneficial effects of vesnarinone might be related to inhibition of production of cytokines such as tumour necrosis factor (Packer, 1993), which may be involved in pathophysiological mechanisms of cardiac failure. As with pimobendan, therefore, the extent to which mechanisms other than inhibition of myocardial PDE3 contribute to the beneficial effects of vesnarinone is not clear. Whether judicious choice of drug dosage or patient classification and selection would improve results with other PDE3 inhibitors/inotropes also remains to be determined (Colucci, 1993).

7. Acknowledgements

We especially thank Dr T. Rahn, Dr T. Murata and Dr D. Ekholm for allowing us to include unpublished information, and Dr Martha Vaughan for critical reading of the manuscript. Financial support to E.D. and P.B. was given by the Swedish Medical Research Council (Grant 3362), the Medical Faculty, Lund University, and the following foundations: Swedish Diabetes Association, Stockholm; Albert Pahlsson, Malmö; Novo Nordisk, Copenhagen; Lars Hierta, Stockholm; Crafoord, Lund; Swedish Society of Medicine, Stockholm (M.M.); and U.S. Department of Veterans' Affairs Medical Research Funds.

8. References

Ahn, H.-S., Eardley, D., Watkins, R. and Prioli, N. (1986). Effects of several newer cardiotonic drugs on cardiac cyclic AMP metabolism. Biochem. Pharmacol. 35, 1113–1121.

Allan, E.H. and Sneyd, J.G.T. (1975). An effect of glucagon on 3′,5′ cyclic AMP phosphodiesterase in isolated hepatocytes. Biochem. Biophys. Res. Commun. 62, 594–601.

Alousi, A., Farah, A., Lesher, G. and Opalka, J. (1979). Cardiotonic activity of amrinone-Win 40680 [5-amino,3′,4′-bipyredin-6(1H)-one]. Circ. Res. 45, 666–677.

Alvarez, R., Banerjee, G., Bruno, J.J., Jones, G.H., Littschwager, K., Strosberg, A.M. and Venuti, M.C. (1986). A potent and selective inhibitor of cyclic AMP phosphodiesterase with potential cardiotonic and anti-thrombotic properties. Mol. Pharmacol. 29, 554–560.

Andersson, T.L. and Vinge, E. (1991). Interactions between isoprenaline, sodium nitroprusside, and isozyme-selective phosphodiesterase inhibitors on ADP-induced aggregation and cyclic nucleotide levels in human platelets. J. Cardiovasc. Pharmacol. 18, 237–242.

Appleman, M.M. and Terasaki, W.L. (1975). Regulation of cyclic nucleotide phosphodiesterase. Adv. Cyclic Nucleotide Res. 5, 153–163.

Archer, S.L., Huang, J.M.C., Hampi, V., Nelson, D.P., Schultz, P.J. and Weir, E.K. (1994). Nitric oxide and cGMP cause vasorelaxation by activation of a charybdotoxin-sensitive K channel by cGMP-dependent protein kinase. Proc. Natl Acad. Sci. USA 91, 7583–7587.

Arnold, J.M. (1993). The role of phosphodiesterase inhibitors in heart failure. Pharmacol. Ther. 57, 161–170.

Baim, D.S., McDowell, A.V., Cherniles, J., Monrad, E.S., Parker, J.A., Edelson, J., Braunwald, E. and Grossman, W. (1983). Evaluation of a new bipyridine inotropic agent, milrinone, in patients with severe congestive heart failure. N. Engl. J. Med. 309, 748–756.

Beavo, J.A. and Reifsnyder, D.H. (1990). Primary sequences of phosphodiesterase isoenzymes and design of selective inhibitors. Trends Pharmacol. Sci. 11, 150–155.

Beavo, J.A., Conti, M. and Heaslip, R.J. (1994). Multiple cyclic nucleotide phosphodiesterases. Mol. Pharmacol. 46, 399–405.

Beebe, S.J., Redmon, J.B., Blackmore, P.F. and Corbin, J.D. (1985). Discriminative insulin antagonism of stimulatory effects of various cAMP analogs on adipocyte lipolysis and hepatocyte glycogenolysis. J. Biol. Chem. 260, 15781–15788.

Belfrage, P., Donner, J., Eriksson, H. and Stralfors, P. (1986). Mechanisms for control of lipolysis by insulin and growth hormone. In "Mechanisms of Insulin Action" (eds. P. Belfrage, J. Donner and P. Stalfors), pp. 323–340. Elsevier, Amsterdam.

Beltman, J., Sonnenburg, W.K. and Beavo, J.A. (1993). The role of protein phosphorylation in the regulation of cyclic nucleotide phosphodiesterases. Mol. Cell. Biochem. 127/128, 239–253.

Beltman, J., Becker, D.E., Butt, E., Jensen, G.S., Rybalkin, S.D., Jastorff, B. and Beavo, J. (1995). Characterization of cyclic nucleotide phosphodiesterases with cyclic GMP analogs: topology of the catalytic domains. Mol. Pharmacol. 47, 330–339.

Benelli, C., Desbuquois, S. and De Galle, B. (1986). Acute *in vivo* stimulation of low K_m cyclic AMP phosphodiesterase activity by insulin in rat liver Golgi fractions. Eur. J. Biochem. 156, 211–220.

Benotti, J., Grossman, W., Braunwald, E., Davolos, D. and Alousi, A. (1978). Hemodynamic assessment of amrinone: a new inotropic agent. N. Engl. J. Med. 299, 1373–1377.

Bode, D.C. and Brunton, L.L. (1988). Post-receptor modulation of the effects of cyclic AMP in isolated cardiac myocytes. Mol. Cell. Biochem. 82, 13–18.

Boyes, S. and Loten, E.G. (1988). Purification of an insulin-sensitive cyclic AMP phosphodiesterase from rat liver. Eur. J. Biochem. 174, 303–309.

Boyes, S. and Loten, E.G. (1989). Insulin and lipolytic hormones stimulate the same phosphodiesterase isoform in rat adipose tissue. Biochem. Biophys. Res. Commun. 162, 814–820.

Bristol, J.A., Sircar, I., Moos, W.H., Evans, D.B. and Weishaar, R.E. (1984). Cardiotonic agents. 1. 4,5-Dihydro-6-[4-(1H-imidazol-1-yl)phenyl]-3-(2H)-pyridazinones: novel positive inotropic agents for the treatment of congestive heart failure. J. Med. Chem. 27, 1099–1011.

Brunton, L.L., Hayes, J.S. and Mayer, S.E. (1981). Functional compartmentation of cyclic AMP and protein kinase in heart. Adv. Cycl. Nucleotide Res. 14, 391–397.

Butcher, R.W. and Sutherland, E.W. (1962). Adenosine-3′,5′-phosphate in biological material. I. Purification and properties of cyclic 3′,5′-nucleotide phosphodiesterase and use of this enzyme to characterize adenosine 3′,5′-phosphate in human urine. J. Biol. Chem. 237, 1244–1250.

Butt, E., Beltman, J., Becker, D.E., Jensen, G.S., Rybalkin, S.D., Jastorff, B. and Beavo, J.A. (1995). Characterization of cyclic nucleotide phosphodiesterases with cyclic AMP analogs: topology of the catalytic sites and comparison with other cyclic AMP-binding proteins. Mol. Pharmacol. 47, 340–347.

Cheung, P., Xu, H. and Colman, R. (1994). Molecular cloning and analysis of human platelet erythroleukemia (HEL) cell cGMP-inhibited phosphodiesterase. FASEB J. 8, 82. [Abstract]

Colucci, W.S. (1993). Search for an effective and safe orally active positive inotropic drug for treatment of heart failure. Cardiol. Rev. 1, 42–49.

Colucci, W.S., Wright, R.F. and Braunwald, E. (1986). New positive inotropic agents in the treatment of congestive heart failure: mechanisms of action and recent clinical developments. N. Engl. J. Med. 314, 290–299 and 349–358.

Conti, M., Jin, S.-C., Monaco, L., Repaske, D.R. and Swinnen, J.V. (1991). Hormonal regulation of cyclic nucleotide phosphodiesterases. Endocr. Rev. 12, 218–234.

Dage, R.C., Roebel, L.E., Hsieh, C.-P., Weiner, D.L. and Woodward, J.K. (1982). Cardiovascular properties of a new cardiotonic agent MDL 17,043 (1,3 dihydro-4-methyl-5-[4-(methylthio)-benzoyl]-2H-imidazol-2-one). J. Cardiovasc. Pharmacol. 4, 500–508.

de Boer, J., Philpott, A., van Amsterdam, R., Shahid, M., Zaagsma, J. and Nicholson, C.D. (1992). Human bronchial cyclic nucleotide phosphodiesterase isoenzymes: biochemical and pharmacological analysis using selective inhibitors. Br. J. Pharmacol. 106, 1028–1034.

de Chaffoy de Courcelles, D., de Loore, K., Fryne, E. and Janssen, P.A. (1992). Inhibition of human cardiac cAMP phosphodiesterases by R 80122, a new selective cAMP phosphodiesterase III inhibitor: comparison with other cardiotonic compounds. J. Pharmacol. Exp. Ther. 263, 6–14.

Degerman, E., Newman, A., Rice, K., Belfrage, P. and Manganiello, V.C. (1987). Purification of the putative hormone-sensitive cyclic AMP phosphodiesterase from rat adipose tissue using a derivative of cilostamide as a novel affinity ligand. J. Biol. Chem. 162, 5797–5807.

Degerman, E., Manganiello, V.C., Newman, A., Rice, K. and Belfrage, P. (1988). Purification, properties and polyclonal antibodies for the particulate cAMP phosphodiesterase from bovine adipose tissue. Second Messengers Phosphoproteins 12, 171–181.

Degerman, E., Smith, C.J., Tornqvist, H., Vasta, V., Manganiello, V.C. and Belfrage, P. (1990). Evidence that insulin and isoprenaline activate the cGMP inhibited low K_m cAMP-phosphodiesterase in fat cells by phosphorylation. Proc. Natl Acad. Sci. USA 87, 533–537.

Degerman, E., Moos, M., Jr, Rascon, A., Vasta, V., Meacci, E., Smith, C.J., Lindgren, S., Andersson, K.-E., Belfrage, P. and Manganiello, V.C. (1993). Single step purification, partial structure and properties of human platelet cGMP inhibited cAMP phosphodiesterase. Biochim. Biophys. Acta 1205, 189–198.

Degerman, E., Leroy, M.-J., Taira, M., Belfrage, P. and Manganiello, V.C. (1995). A role for insulin-mediated regulation of cGMP-inhibited phosphodiesterase in the antilipolytic action of insulin. In "Diabetes Mellitus, A Clinical and Fundamental Textbook" (eds. D. Le Roith, J. Olefsky and S. Taylor). Lippincort, Philadelphia. [in press]

Dent, G., Magnussen, H. and Rabe, K.F. (1994). Cyclic nucleotide phosphodiesterases in the human lung. Lung 172, 129–146.

Di Bianco, R., Shabetai, R., Silverman, B.D., Leier, C.V. and Benotti, J.R. (1984). Oral amrinone for the treatment of chronic congestive heart failure: results of a multicenter randomized double-blind and placebo-controlled withdrawal study. J. Am. Coll. Cardiol. 4, 855–866.

Di Bianco, R., Shabetai, R., Kostuk, W., Moran, J., Schlant, R.C. and Wright, R. (1989). A comparison of oral milrinone, digoxin, and their combination in the treatment of patients with chronic heart failure. N. Engl. J. Med. 320, 677–683.

Elks, M.L., Manganiello, V.C. and Vaughan, M. (1983). Hormone-sensitive particulate cAMP phosphodiesterase activity in 3T3-L1 adipocytes. J. Biol. Chem. 258, 8582–8587.

Endoh, M. and Hori, M. (1993). Basic pharmacology and clinical application of new positive inotropic agents. Drugs Today 29, 29–56.

Endoh, M., Yamashita, A. and Taira, N. (1982). Positive inotropic effects of amrinone in relation to cyclic nucleotide metabolism in canine ventricular muscle. J. Pharmacol. Exp. Ther. 221, 775–783.

Erhardt, P. (1987). In search of the digitalis replacement. J. Med. Chem. 30, 231–237.

Erhardt, P. (1990). Second-generation phosphodiesterase inhibitors: structure–activity relationships and receptor models. In "Cyclic Nucleotide Phosphodiesterases: Structure, Regulation and Drug Action" (eds. J. Beavo and M.D. Houslay), pp. 316–332. Wiley, Chichester.

Eriksson, H., Degerman, E., Olsson, H., Smith, C.J., Manganiello, V.C. and Belfrage, P. (1994). Evidence for a key role of the cGMP-inhibited cAMP phosphodiesterase in the antilipolytic action of insulin. Biochim. Biophys. Acta 1266, 101–107.

Feldman, A.M., Bristow, M.R., Parmley, W.W., Carson, P.E., Pepine, C.J., Gilbert, E.M., Strobeck, J.E., Hendrix, G.H., Powers, E.R., Bain, R.P. and White, B.G (1993). Effects of vesnarinone on morbidity and mortality in patients with heart failure. N. Engl. J. Med. 329, 149–155.

Feneck, R. (1990). Effects of variable dose milrinone in patients with low cardiac output after cardiac surgery. Am. Heart J. 121, 1995–2000.

Francis, S.H., Noblett, B., Todd, B.W., Wells, J.N. and Corbin, J.D. (1988). Relaxation of vascular and tracheal smooth muscle by cyclic nucleotide analogs that preferentially activate purified cGMP-dependent protein kinase. Mol. Pharmacol. 34, 506–517.

Francis, S.H., Colbran, J.L., McAllister-Lucas, L.M. and Corbin, J.D. (1994). Zinc interactions and conserved motifs of the cGMP-binding cGMP-specific phosphodiesterase suggest that it is a zinc hydrolase. J. Biol. Chem. 269, 22477–22480.

Fujimura, M., Kamio, Y., Saito, M., Hashimoto, T. and Matsuda, T. (1995). Bronchodilator and bronchoprotective effects of cilostazol in humans *in vivo*. Am. J. Respir. Crit. Care Med. 151, 222–225.

Giembycz, M.A. and Dent, G. (1992). Prospects for selective cyclic nucleotide phosphodiesterase inhibitors in the treatment of bronchial asthma. Clin. Exp. Allergy 22, 337–344.

Grant, P.G. and Colman, R.W. (1984). Purification and characterization of a human platelet cyclic nucleotide phosphodiesterase. Biochemistry 23, 1801–1807.

Grant, D.G., Mannarino, A.F. and Colman, R.W. (1988). cAMP-mediated phosphorylation of the low K_m cAMP phosphodiesterase markedly stimulates its catalytic activity. Proc. Natl Acad. Sci. USA 85, 9071–9075.

Gristwood, R.W. and Owen, D.A.A. (1986). Effects of rolipram on guinea-pig ventricles in vitro: evidence of an unexpected synergism with SK&F 94120. Br. J. Pharmacol. 89, 91P. [Abstract]

Hall, I.P. (1993). Isoenzyme selective phosphodiesterase inhibitors: potential clinical uses. Br. J. Clin. Pharmacol. 35, 1–7.

Harris, A.L., Connell, M.J., Ferguson, E.W., Wallace, A.M., Gordon, R.J, Pagani, E.D. and Silver, P.J. (1989). Role of low K_m cyclic AMP phosphodiesterase inhibition in tracheal relaxation and bronchodilation in the guinea pig. J. Pharmacol. Exp. Ther. 251, 199–206.

Harris, M.N., Daborn, A.K. and O'Dwyer, J.P. (1992). Milrinone and the pulmonary vascular system. Eur. J. Anaesthesiol. Suppl. 1, 27–30.

Harrison, S.A., Reifsnyder, D.H., Gallis, B., Cadd, G.G. and Beavo, J.A. (1986). Isolation and characterization of a bovine cardiac muscle cGMP-inhibited phosphodiesterase. Mol. Pharmacol. 29, 506–514.

Hasenfuss, G., Holubarsch, C., Heiss, H.W. and Just, H. (1989). Influence of UDCG–115 on hemodynamics and myocardial energetics in patients with idiopathic dilated cardiomyopathy. Am. Heart J. 118, 512–519.

Heaslip, R.J., Buckley, S.K., Sickels, B.D. and Grimes, D. (1991). Bronchial vs. cardiovascular activities of selective phosphodiesterase inhibitors in the anesthetized beta-blocked dog. J. Pharmacol. Exp. Ther. 264, 609–615.

Hendra, T. and Betteridge, D.J. (1989). Platelet function, platelet prostanoids and vascular prostacyclin in diabetes mellitus. Prostaglandins Leukot. Essent. Fatty Acids 35, 197–212.

Heyworth, C.M., Wallace, A.V. and Houslay, M.D. (1983). Insulin and glucagon regulate the activation of two distinct membrane-bound cyclic AMP phosphodiesterases in hepatocytes. Biochem. J. 214, 99–110.

Hidaka, H. and Endo, T. (1984). Selective inhibitors of three forms of cyclic nucleotide phosphodiesterase – basic and potential clinical applications. Adv. Cyclic Nucleotide Protein Phosphorylation Res. 16, 245–259.

Hidaka, H., Hayashi, H., Kohri, H., Kimura, Y., Hosokawa, T., Igawa, T. and Saitoh, Y. (1979). Selective inhibitor of platelet cyclic adenosine monophosphate phosphodiesterase, cilostamide, inhibits platelet aggregation. J. Pharmacol. Exp. Ther. 211, 26–30.

Honerjäger, P., Schafer-Korting, M. and Reiter, M. (1981). Involvement of cyclic AMP in the direct inotropic action of amrinone: biochemical and functional evidence. Naunyn-Schmiedebergs. Arch. Pharmacol. 318, 112–120.

Houslay, M.D. and Kilgour, E. (1990). Cyclic nucleotide phosphodiesterases in liver: a review of their characterization, regulation by insulin and glucagon and their role in controlling intracellular cyclic AMP concentrations. In "Cyclic Nucleotide Phosphodiesterases: Structure, Regulation and Drug Action" (eds. J. Beavo and M.D. Houslay), pp. 185–224. Wiley, Chichester.

Iijima, T. and Taira, N. (1987). Membrane current changes responsible for the positive inotropic effect of OPC-8212, a new positive inotropic agent, in single ventricular cells of the guinea pig heart. J. Pharmacol. Exp. Ther. 240, 657–662.

Jaski, B.E., Fifer, M.A., Wright, R.F., Braunwald, E. and Colucci, W.S. (1985). Positive inotropic and vasodilator actions of milrinone in patients with severe heart failure: dose–response relationships and comparison to nitroprusside. J. Clin. Invest. 75, 643–649.

Jiang, H., Colbran, J.L., Francis, S.H. and Corbin, J.D. (1992). Direct evidence for cross-activation of cGMP-dependent protein kinase in pig coronary arteries. J. Biol. Chem. 267, 1015–1019.

Jones, G.H., Venuti, M.C., Alvarez, R., Bruno, J.J., Berks, A.H. and Prince, A. (1987). Inhibitors of cAMP phosphodiesterase. 1. Analogues of cilostamide and anagrelide. J. Med. Chem. 30, 295–303.

Kariya, T. and Dage, R.C. (1988). Tissue distribution and selective inhibition of subtypes of high affinity cAMP phosphodiesterase. Biochem. Pharmacol. 37, 3267–3270.

Kariya, T., Wille, L.J. and Dage, R.C. (1982). Biochemical studies on the mechanism of cardiotonic activity of MDL 17,043. J. Cardiovasc. Pharmacol. 4, 509–514.

Kariya, T., Wille, L.J. and Dage, R.C. (1984). Studies on the mechanism of the cardiotonic activity of MDL 19205: effects on several biochemical systems. J. Cardiovasc. Pharmacol. 6, 50–55.

Kasuya, J., Goko, H. and Fujita Yamaguchi, Y. (1995). Multiple transcripts for the human cardiac form of the cGMP-inhibited phosphodiesterase. J. Biol. Chem. 16, 14305–14312.

Katano, Y. and Endoh, M. (1990). Differential effects of Ro 20-1724 and isobutylmethylxanthine on the basal force of contraction and β-adrenoreceptor mediated responses in the rat ventricular myocardium. Biochem. Biophys. Res. Commun. 167, 123–129.

Katz, S.D., Kubo, S.H., Jessup, M., Brozena, S., Troha, J.M., Wahl, J., Cohn, J.N., Sonnenblick, E.H. and Le Jemtel, T.H. (1992). A multicenter, randomized, double-blind, placebo-controlled trial of pimobendan, a new cardiotonic and vasodilator agent, in patients with severe congestive heart failure. Am. Heart J. 123, 95–103.

Kauffmann, R.F., Crowe, V.G., Utterback, B.G. and Robertson, R.W. (1987a). LY19155: a potent, selective inhibitor of cyclic nucleotide phosphodiesterase located in the sarcoplasmic reticulum. Mol. Pharmacol. 30,609–616.

Kauffmann, R.F., Schenck, K.W., Utterback, B.G., Crowe, V.G. and Cohen, M.L. (1987b). *In vitro* vascular relaxation by new inotropic agents: relationship to inhibition and cyclic nucleotides. J. Pharmacol. Exp. Ther. 242, 864–871.

Kauffmann, R.F., Utterback, B.G. and Robertson, D.W. (1989a). Specific binding of [^{3}H] LY186126, an analogue of indolidan (LY195115), to cardiac membranes enriched in sarcoplasmic reticulum vesicles. Circ. Res. 64, 1037–1040.

Kauffman, R.F., Utterback, B.G. and Robertson, D.W. (1989b). Characterization and pharmacological relevance of high affinity binding sites for [^{3}H]LY186126, a cardiotonic phosphodiesterase inhibitor, in canine cardiac membranes. Circ. Res. 65, 154–163.

Kilgour, E., Anderson, N.G. and Houslay, M.D. (1989). Activation and phosphorylation of the "dense-vesicle" high-affinity cyclic AMP phosphodiesterase by cyclic AMP-dependent protein kinase. Biochem. J. 260, 27–36.

Kirstein, M., Rivet-Bastide, M., Hatem, S., Benardeau, A., Mercadier, J.-J. and Fischmeister, R. (1995). Nitric oxide regulates calcium current in isolated human atrial myocytes. J. Clin. Invest. 95, 794–802.

Kithas, P.A., Artman, M., Thompson, W.J. and Strada, S.J. (1988). Subcellular distribution of high affinity type IV cyclic AMP phosphodiesterase activity in rabbit ventricular myocardium: relations to the effects of cardiotonic drugs. Circ. Res. 62, 782–789.

Klockner, U., Itagaki, K., Bodi, I. and Schwartz, A. (1992). Beta-subunit expression is required for cAMP-dependent increase of cloned cardiac and vascular calcium channel currents. Pflugers Arch. 420, 413–415.

Knowles, R. and Moncada, S. (1992). Nitric oxide as a signal in blood vessels. Trends Biochem. Sci. 17, 399–402.

Komas, N., Lugnier, C., Le Bec, A., Serradil-Le Gal, C., Barthelemy, G. and Stoclet, J.-C. (1989). Differential sensitivity to cardiotonic drugs of cyclic AMP phosphodiesterases isolated from canine ventricular and sinoatrial-enriched tissues. J. Cardiovasc. Pharmacol. 14, 213–220.

Komas, N., Lugnier, C., Andriantsitohaina, R. and Stoclet, J.-C. (1991a). Characterization of cyclic nucleotide phosphodiesterases from rat mesenteric artery. Eur. J. Pharmacol. 208, 85–87.

Komas, N., Lugnier, C. and Stoclet, J.-C. (1991b). Endothelial-dependent and independent relaxation of the rat aorta by cyclic nucleotide phosphodiesterase inhibitors. Br. J. Pharmacol. 105, 495–503.

Komas, N., Taira, M., Murata, T. and Manganiello, V. (1994). Pharmacological responses of two recombinant cGMP-inhibited cyclic nucleotide phosphodiesterases. FASEB J. 8, 82. [Abstract]

Kono, T., Robinson, F.W. and Sarver, J.A. (1975). Insulin-sensitive phosphodiesterase: its location, hormonal stimulation, and oxidative stabilization. J. Biol. Chem. 250, 7826–7835.

Lathrop, D.A. and Schwartz, A. (1985). Evidence for possible increase of sodium channel open time and involvement of Na/Ca exchange by a new positive inotropic drug: OPC-8212. Eur. J. Pharmacol. 117, 391–392.

Le Bon, T.R., Kasuya, J., Paxton, R.J., Belfrage, P., Hockman, S., Manganiello, V.C. and Fujita Yamaguchi, Y. (1992). Purification and characterization of guanosine 3′,5′-monophosphate phosphodiesterase from human placental cytosolic fractions. Endocrinology 130, 3265–3274.

Le Jemtel, T.H., Keung, E., Ribner, H.S., Wexler, J., Blaufox, J.D. and Sonnenblick, E.H. (1980). Sustained beneficial effects of oral amrinone on cardiac and renal function in patients with severe congestive heart failure. Am. J. Cardiol. 45, 123–129.

Lincoln, T.M., Cornwell, T.L. and Taylor, A.E. (1990). cGMP-dependent protein kinase mediates the reduction of Ca^{2+} by cAMP in vascular smooth muscle cells. Am. J. Physiol. 258, C399–C407.

Lindgren, S., Andersson, K.-E., Belfrage, P., Degerman, E. and Manganiello, V.C. (1989). Relaxant effects of selective phosphodiesterase inhibitors milrinone and OPC 3911 on isolated human mesenteric vessels. Pharmacol. Toxicol. 64, 440–445.

Lindgren, S., Rascon, A., Andersson, K.-E., Manganiello, V. and Degerman, E. (1991). Selective inhibition of cGMP-inhibited and cGMP-noninhibited cyclic nucleotide phosphodiesterases and relaxation of rat aorta. Biochem. Pharmacol. 42, 545–552.

Lopez-Aparicio, P., Rascon, A., Manganiello, V.C., Andersson, K.E., Belfrage, P. and Degerman, E. (1992). Insulin-induced phosphorylation and activation of the cGMP-inhibited cAMP phosphodiesterase in human platelets. Biochem. Biophys. Res. Commun. 186, 517–523.

Lopez-Aparicio, P., Belfrage, P., Manganiello, V.C., Kono, T. and Degerman, E. (1993). Stimulation by insulin of a serine kinase in human platelets that phosphorylates and activates the cGMP-inhibited phosphodiesterase. Biochem. Biophys. Res. Commun. 193, 1137–1144.

Loten, E.G. (1991). Hormone-sensitive phosphodiesterase of liver and adipocyte tissue. Int. J. Biochem. 23, 649–655.

Loten, E.G., Assimacopoulos-Jeannet, F.D., Exton, J.H. and Park, C.P. (1978). Stimulation of a low K_m phosphodiesterase from liver by insulin and glucagon. J. Biol. Chem. 253, 746–757.

Ludmer, P.L., Wright, R.F., Arnold, M.O., Ganz, P., Braunwald, E. and Collucci, W.S. (1986). Separation of the direct myocardial and vasodilator actions of milrinone administered by an intracoronary infusion technique. Circulation 73, 130–137.

Lugnier, C. and Komas, N. (1993). Modulation of vascular cyclic nucleotide phosphodiesterases by cyclic GMP: role in vasodilation. Eur. Heart J. 14 (Suppl. 1), 141–148.

Lugnier, C., Muller, B., Le Bec, A., Beaudry, C. and Rousseau, E. (1993). Characterization of indolidan- and rolipram-sensitive cyclic nucleotide phosphodiesterases in canine and human cardiac microsomal fractions. Mol. Pharmacol. 265, 1142–1150.

Luo, W., Grupp, I.L., Harrer, J., Ponniah, S., Grupp, G., Duffy, J.J., Doetschman, T. and Kranias, E.G. (1994). Targeted ablation of the phospholamban gene is associated with markedly enhanced myocardial contractility and loss of beta-agonist stimulation. Circ. Res. 75, 401–409.

MacPhee, H., Reifsnyder, D.H., Moore, T.A., Lerea, K.M. and Beavo, J.A. (1988). Phosphorylation results in activation of a cAMP phosphodiesterase in human platelets. J. Biol. Chem. 21, 10353–10358.

Makino, G. and Kono, T. (1980). Characterization of insulin-sensitive phosphodiesterase in fat cells. II. Comparison of enzyme activities stimulated by insulin and isoproterenol. J. Biol. Chem. 255, 7850–7854.

Manganiello, V. and Vaughan, M. (1973). An effect of insulin on cyclic adenosine 3′,5′-monophosphate activity in fat cells. J. Biol. Chem. 248, 7164–7170.

Manganiello, V.C., Degerman, E. and Elks, M. (1988). Selective inhibitors of specific phosphodiesterases in intact adipocytes. Methods Enzymol. 159, 504–520.

Manganiello, V.C., Smith, C.J., Degerman, E. and Belfrage, P. (1990). cGMP-inhibited cyclic nucleotide phosphodiesterases. In "Cyclic Nucleotide Phosphodiesterases: Structure, Regulation and Drug Action" (eds. J. Beavo and M.D. Houslay), pp. 87–116. Wiley, Chichester.

Manganiello, V.C., Murata, T., Taira, M., Belfrage, P. and Degerman, E. (1995a). Diversity in cyclic nucleotide phosphodiesterase families. Arch. Biochem. Biophys. 322, 1–13.

Manganiello, V.C., Taira, M., Degerman, E. and Belfrage, P. (1995b). Type III cGMP-inhibited cyclic nucleotide phosphodiesterase (PDE3 Gene Family). Cell. Signal. 7, 445–455.

Marcoz, P., Prigent, A.F., Lagarde, M. and Némoz, G. (1993). Modulation of the rat thymocyte proliferative response by means of selective inhibitors and cGMP-elevating agents. Mol. Pharmacol. 44, 1027–1035.

Massie, B., Bourassa, M., Di Bianco, R., Hess, M., Konstam, M., Likoff, M. and Packer, M. (1985). Long-term oral administration of amrinone for congestive heart failure: lack of efficacy in a multicenter controlled trial. Circulation 71, 963–971.

Matousovic, K., Grande, T., Chini, C.C., Chini, E.N. and Dousa, T.P. (1995). Inhibitors of cyclic nucleotide phosphodiesterase isoenzymes type III and type IV suppress mitogenesis of rat mesangial cells. J. Clin. Invest. 96, 401–410.

Maurice, D.H. and Haslam, R.J. (1990a). Molecular basis for the synergistic inhibition of platelet function by nitrovasodilators and activators of adenylate cyclase: inhibition of cyclic AMP breakdown by cyclic GMP. Mol. Pharmacol. 37, 671–681.

Maurice, D.H. and Haslam, R.J. (1990b). Nitroprusside enhances isoprenaline-induced increases in cAMP in rat aortic smooth muscle. Eur. J. Pharmacol. 191, 471–475.

Maurice, D.H., Crankshaw, D. and Haslam, R.J. (1991). Synergistic actions of nitrovasodilators and isoprenaline on rat aortic smooth muscle. Eur. J. Pharmacol. 192, 235–242.

Meacci, E., Taira, M., Moos, M., Jr, Smith, C.J., Movsesian, M.A., Degerman, E., Belfrage, P. and Manganiello, V. (1992). Molecular cloning and expression of human myocardial cGMP-inhibited cAMP phosphodiesterase. Proc. Natl Acad. Sci. USA 89, 3721–3725.

Méry, F.-F., Pavoine, C., Belhassen, L., Pecker, F. and Fischmeister, R. (1993). Nitric oxide regulates cardiac Ca^{2+} current. J. Biol. Chem. 35, 26286–26295.

Michaeli, T., Bloom, T.J., Martins, T., Loughney, K., Ferguson, K., Riggs, M., Rodgers, L., Beavo, J.A. and Wigler, M. (1993). Isolation and characterization of a previously undetected human cAMP phosphodiesterase by complementation of a cAMP phosphodiesterase-deficient *Saccharomyces cerevesiae*. J. Biol. Chem. 268,12925–12932.

Misek, D.E. and Saltiel, A.R. (1992). An inositol phosphate glycan derived from *Trypanosoma brucei* glycosyl-phosphatidylinositol mimics some of the metabolic actions of insulin. J. Biol. Chem. 267, 16266–16273.

Moos, W.H., Humblet, C.C., Sircar, I., Rithner, C., Weishaar, R.E., Bristol, J.A. and McPhail, A.T. (1987). Cardiotonic agents. 8. Selective inhibitors of adenosine 3′,5′-cyclic phosphate phosphodiesterase III. Elaboration of a five-point model for positive inotropic activity. J. Med. Chem. 30, 1963–1972.

Movsesian, M.A. (1993). Calcium uptake and release by cardiac sarcoplasmic reticulum. In "Inotropic Drugs: Basic Mechanisms and Clinical Practice" (eds .P.D. Allen, J.K.

Gwathmey and M. Briggs), pp. 101–120. Marcel Dekker, New York.

Movsesian, M.A., Smith, C.J., Krall, J., Bristow, M.R. and Manganiello, V.C. (1991). Sarcoplasmic reticulum-associated cyclic adenosine 5′-monophosphate phosphodiesterase in normal and failing human hearts. J. Clin. Invest. 88, 15–19.

Murray, K.J., England, P.J. and Reeves, M.L. (1987). Positive inotropic actions of SKF 94120 occur via a cyclic AMP-dependent mechanism: activation of a cAMP-dependent protein kinase. Br. J. Pharmacol. 92, 755P. [Abstract]

Murray, K.J., England, P.J., Hallam, T.J., Maguire, J., Moores, K., Reeves, M.L., Simpson, A.W. and Rink, T.J. (1990). The effects of siguazodan, a selective phosphodiesterase inhibitor, on human platelet function. Br. J. Pharmacol. 99, 612–616.

Murray, T. and Russell, T.R. (1980). Acquisition of insulin-sensitive activity of adenosine-3′,5′-monophosphate phosphodiesterase during adipose conversion of 3T3-L1 cells. Eur. J. Biochem. 107, 217–224.

Narahara, K.A. and the Western Enoximone Study Group (1991). Oral enoximone therapy in chronic heart failure: a placebo-controlled randomized trial. Am. Heart J. 121, 1471–1479.

Nicholson, C.D., Challiss, R.A.J. and Shahid, M. (1991). Differential modulation of tissue function and therapeutic potential of selective inhibitors of cyclic nucleotide phosphodiesterase isoenzymes. Trends Pharmacol. Sci. 12, 19–27.

Nony, P., Boissel, J.P., Lievre, M., Leizorovicz, A., Haugh, M.C., Fareh, S. and de Breyne, B. (1994). Evaluation of the effect of phosphodiesterase inhibitors on mortality in chronic heart failure patients. Eur. J. Clin. Pharmacol. 46, 191–196.

Omburo, G.A., Brickus, T., Ghazaleh, F.A. and Colman, R.W. (1995). Divalent metal requirement and possible classification of cyclic GMP-inhibited cyclic nucleotide phosphodiesterase as a metallohydrolase. Arch. Biochem. Biophys. 323, 1–5.

Packer, M. (1993). The search for the ideal positive inotropic agent. N. Engl. J. Med. 329. 201–202. [Editorial]

Pan, X., Arauz, E., Krzanowski, J.J., Fitzpatrick, D.F. and Polson, J.B. (1994). Synergistic interactions between selective pharmacological inhibitors of phosphodiesterase isozyme families PDE III and PDE IV to attenuate proliferation of rat vascular smooth muscle cells. Biochem. Pharmacol. 48, 827–835.

Pang, D. (1988). Cyclic AMP and cylic GMP phosphodiesterases: targets for drug development. Drug Dev. Res. 12, 85–92.

Pang, D. (1992). Tissue and species specificity of cardiac cAMP phosphodiesterase inhibitors. Adv. Second Messenger Phosphoprotein Res. 25, 307–321.

Pawlson, L.G., Lovell-Smith, C.J., Manganiello, V.C. and Vaughan, M. (1974). Effects of insulin and lipolytic agents on rat adipocyte low K_m cyclic adenosine 3′,5′-monophosphate phosphodiesterase activity in fat cells. Proc. Natl Acad. Sci. USA 71, 1639–1642.

Pillai, R., Staub, S.F. and Colicelli, J. (1994). Mutational mapping of kinetic and pharmacological properties of a human cardiac cAMP phosphodiesterase. J. Biol. Chem. 269, 30676–30681.

Prigent, A., Fougier, S., Némoz, G., Anker, G., Pacheco, H., Lugnier, C., Le Bec, A. and Stoclet, J.C. (1988). Comparison of cyclic nucleotide phosphodiesterase isoforms from rat heart and bovine aorta. Biochem. Pharmacol. 37, 3671–3678.

Pyne, N.J., Cooper, M.E. and Houslay, M.D. (1987). The insulin- and glucagon-stimulated "dense-vesicle" high-affinity cyclic AMP phosphodiesterase from rat liver. Biochem. J. 242, 33–36.

Qian, Y., Nalene, E., Karlsson, J.-A., Raeburn, D. and Advenier, C. (1993). Effects of rolipram and siguazodan on the human isolated bronchus and their interaction with isoprenaline and sodium nitroprusside. Br. J. Pharmacol. 109, 774–778.

Rabe, K.F., Tenor, H., Dent, G., Schudt, C., Liebig, S. and Magnusson, H. (1993). Phosphodiesterase isoenzymes modulating inherent tone in human airways: identification and characterization. Am. J. Physiol. 264, L458-464.

Rabe, K.F., Tenor, H., Dent, G., Schudt, C., Nakashima, M. and Magnussen, H. (1994). Identification of PDE isoenzymes in human pulmonary artery and effect of selective PDE inhibitors. Am. J. Physiol. 266, L536–543.

Raeburn, D. (1994). Phosphodiesterase inhibitors and airways muscle. In "Methylxanthines and Phosphodiesterase Inhibitors in the Treatment of Airway Disease" (eds. J. Costello and P.J. Piper), pp. 101–117. Parthenon, New York.

Raeburn, D., Sharma, S., Buckley, G.B., Underwood, S.L., Tomkinson, A. and Karlsson, J.-A. (1992). Comparison of isoenzyme-selective phosphodiesterase (PDE) inhibitors and theophylline on histamine-induced bronchospasm in the anaesthetized guinea pig. Eur. Respir. J. 5 (Suppl. 15), 214s. [Abstract]

Raeburn, D., Souness, J.E., Tomkinson, A., and Karlsson, J.-A. (1994). Isoenzyme-selective cyclic nucleotide phosphodiesterase inhibitors: biochemistry, pharmacology and therapeutic potential in asthma. Prog. Drug Res. 40, 9–32.

Rahn, T., Ridderstrale, M., Tornqvist, H., Manganiello, V., Fredrickkson, G., Belfrage, P. and Degerman, E. (1994). Essential role of phosphatidylinositol 3-kinase in insulin-induced activation and phosphorylation of the cGMP-inhibited cAMP phosphodiesterase in rat adipocytes. FEBS Lett. 350, 314–318.

Rascon, A., Lindgren, S., Stavenow, L., Belfrage, P., Andersson, K.-E., Manganiello, V.C. and Degerman, E. (1992). Purification and properties of the cGMP-inhibited cAMP phosphodiesterase from bovine aortic smooth muscle. Biochim. Biophys. Acta 1134, 149–156.

Rascon, A., Degerman, E., Taira, M., Meacci, E., Smith, C.J., Manganiello, V., Belfrage, P. and Tornqvist, H. (1994). Identification of the phosphorylation site *in vitro* for cAMP dependent protein kinase on the rat adipocyte cGMP-inhibited cAMP phosphodiesterase. J. Biol. Chem. 269, 11962–11966.

Reeves, M.L. and England, P.J. (1990). Cardiac phosphodiesterases and the functional effects of selective inhibition. In "Cyclic Nucleotide Phosphodiesterases: Structure, Regulation and Drug Action" (eds. J. Beavo and M.D. Houslay), pp. 299–316. Wiley, Chichester.

Reeves, M.L., England, P.J. and Murray, K.J. (1987a). Compartments of cyclic AMP-dependent protein kinase in perfused guinea pig hearts. Biochem. Soc. Trans. 15, 955–956.

Reeves, M.L., Leigh, B.M. and England, P.J. (1987b). Identification of a new cyclic nucleotide phosphodiesterase activity in human and guinea pig cardiac ventricle: implications for the mechanism of action of selective phosphodiesterase inhibitors. Biochem. J. 241, 535–541.

Reinhart, R.R., Chin, E., Zhou, J., Taira, M., Murata, T., Manganiello, V.C. and Bondy, C.A. (1995). Distinctive anatomical patterns of gene expression for cGMP-inhibited cyclic nucleotide phosphodiesterases. J. Clin. Invest. 95, 1528–1538.

Remme, W.J., Krayenbuhl, H.P., Baumann, G., Frick, M.H., Haehl, M., Nehmiz, G. and Baiker, W. (1994). Long-term efficacy and safety of pimobendan in moderate heart failure. Eur. Heart J. 15, 947–956.

Robertson, D.W. and Boyd, D.B. (1992). Structural requirements for potent and selective inhibition of low-K_M, cyclic-AMP-specific phosphodiesterases. Adv. Second Messenger Phosphoprotein Res. 25, 321–340.

Robicsek, S.A., Blanchard, D.K., Djeu, J.Y., Krzanowski, J.J., Szentivanyi, A. and Polson, J.B. (1991). Multiple high affinity cAMP phosphodiesterases in human lymphocytes. Biochem. Pharmacol. 42, 869–877.

Roebel, L.E., Dage, R.C., Cheng, H.C. and Woodward, J.K. (1984). *In vitro* and *in vivo* assessment of the cardiovascular effects of the cardiotonic drug MDL 19205. J. Cardiovasc. Pharmacol. 6, 43–49.

Roevens, P. and de Chaffoy de Courcelles, D. (1993). Cyclic AMP phosphodiesterase IIIA1 inhibitors decrease cytosolic Ca^{2+} concentration and increase the Ca^{2+} content of intracellular storage sites in human platelets. Biochem. Pharmacol. 45, 2279–2282.

Sadler, S.E. (1991). Type III phosphodiesterase plays a necessary role in the growth promoting actions of insulin, insulin-like growth factor 1, and Ha21ras in *Xenopus laevis* oocytes. Mol. Endocrinol. 5, 1939–1946.

Sakai, T., Thompson, W.J., Lavis, V.R. and Williams, R.H. (1974). Cyclic nucleotide phosphodiesterase activities from isolated fat cells: correlation of subcellular distribution with effects of nucleotides and insulin. Arch. Biochem. Biophys. 142, 331–339.

Schmitz-Peiffer, C., Reeves, M.L. and Denton, R.M. (1992). Characterization of the cyclic nucleotide phosphodiesterase isoenzymes present in rat epididymal fat cells. Cell. Signal. 4, 37–49.

Sculptoreanu, A., Totman, E., Takahashi, M., Scheuer, T. and Catterall, W.A. (1993). Voltage-dependent potentiation of the activity of cardiac L-type calcium channel α_1 subunits due to phosphorylation by cAMP-dependent protein kinase. Proc. Natl Acad. Sci. USA 90, 10135–10139.

Sekhar, K.R., Hatchett, R.J., Shabb, J.B., Wolfe, L., Francis, S.H., Wells, J.N., Jastorff, B., Butt, E., Chakinala, M.M. and Corbin, J.D. (1992). Relaxation of pig coronary arteries by new and potent cGMP analogs that selectively activate type-1α compared with type-1β, cGMP-dependent protein kinase. Mol. Pharmacol. 42, 103–108.

Shah, P.K., Amin, D.K., Hulse, S., Shellock, F. and Swan, H.J.C. (1985). Inotropic therapy of refractory congestive heart failure with oral fenoximone (MDL-17,043): poor long term results despite early clinical and hemodynamic improvement. Circulation 71, 326–331.

Shahid, M. and Nicholson, C.D. (1990). Comparison of cyclic nucleotide phosphodiesterase isoenzymes in rat and rabbit ventricular myocardium: positive inotropic and phosphodiesterase inhibitory effects of Org 30029, milrinone and rolipram. Naunyn-Schmiedebergs Arch. Pharmacol. 342, 698–705.

Shahid, M., van Amersterdam, R.G.M., de Boer, J., ten Berge, R.G., Nicholson, C.D. and Zaagsma, J. (1991). The presence of five cyclic nucleotide phosphodiesterases in bovine tracheal smooth muscle and the functional effects of selective inhibitors. Br. J. Pharmacol. 104, 471–477.

Sham, J.S.K., Jones, L.R. and Morad, M. (1991). Phospholamban mediates the β-adrenergic-enhanced Ca^{2+} uptake in mammalian ventricular myocytes. Am. J. Physiol. 261, H1344–H1349.

Shibata, H. and Kono, T. (1990a). Stimulation of the insulin-sensitive cAMP phosphodiesterase by an ATP-dependent soluble factor from insulin-treated rat adipocytes. Biochem. Biophys. Res. Commun. 167, 614–620.

Shibata, H. and Kono, T. (1990b). Cell-free stimulation of the insulin-sensitive phosphodiesterase by the joint actions of ATP and the soluble fraction from insulin-treated rat liver. Biochem. Biophys. Res. Commun. 170, 533–540.

Silver, P.J., Hamel, L.T., Perrone, M.H., Bentley, R.G., Bushover, C.R. and Evans, D.B. (1988a). Differential pharmacologic sensitivity of cyclic nucleotide phosphodiesterase isozymes isolated from cardiac muscle, arterial and airway smooth muscle. Eur. J. Pharmacol. 150, 85–94.

Silver, P.J., Lepore, R.E., O'Connor, B., Lemp, B.M., Hamel, L.T., Bentley, R.G. and Harris, A.L. (1988b). Inhibition of the low K_m cyclic AMP phosphodiesterase and activation of the cyclic AMP system in vascular smooth muscle by milrinone. J. Pharmacol. Exp. Ther. 247, 34–42.

Silver, P.J., Allan, P., Etzler, J.H., Hamel, L.T., Bentley, R.G and Pagani, E.D. (1990). Cellular distribution and pharmacological sensitivity of low K_m cyclic nucleotide phosphodiesterase isozymes in human cardiac muscle from normal and cardiomyopathic subjects. Second Messengers Phosphoproteins 13, 13–25.

Silver, P.J., Gordon, R.J., Bucholz, R.A., Dundore, R.L., Ferguson, E.W., Harris, A.L. and Pagani, E.D. (1992). Comparative studies on cyclic nucleotide phosphodiesterases and inhibitors in experimental models of hypertension, congestive heart failure, and allergic asthma. Adv. Second Messenger Phosphoprotein Res. 25, 341–351.

Simpson, A.W., Reeves, M.L. and Rink, T.J. (1988). Effects of SKF 94120, an inhibitor of cyclic nucleotide phosphodiesterase type III, on human platelets. Biochem. Pharmacol. 37, 2315–2320.

Sircar, I., Duell, B.L., Bobowski, G., Bristol, J.A. and Evans, D.B. (1985). Cardiotonic agents. 2. Synthesis and structure–activity relationships of 4,5-dihydro-6-[4-(H-imidazol-1-yl)phenyl-]-3(2H)-pyridazenones: a new class of positive inotropic agents. J. Med. Chem. 28, 1405–1413.

Sircar, I., Weishaar, R.E., Kobylarz, D., Moos, W.H. and Bristol, J.A. (1987). Cardiotonic agents. 7. Inhibition of separated forms of cyclic nucleotide phosphodiesterase from guinea pig cardiac muscle by 4,5-dihydro-6-[4-(1H-imidazole-1-yl)phenyl]-3(2H)-pyridazinones and related compounds. Structure–activity relationships and correlations with *in vivo* positive inotropic activity. J. Med. Chem. 30, 1955–1962.

Smith, C.J. and Manganiello, V.C. (1988). The role of hormone-sensitive low K_m cAMP phosphodiesterase (PDE) in regulation of cAMP-dependent protein kinase and lipolysis in rat adipocytes. Mol. Pharmacol. 35, 381–386.

Smith, C.J., Vasta, V., Degerman, E., Belfrage, P. and Manganiello, V.C. (1991). Hormone-sensitive cyclic GMP-inhibited cyclic AMP phosphodiesterase in rat adipocytes. J. Biol. Chem. 266, 13385–13390.

Smith, C.J., Krall, J., Manganiello, V.C. and Movsesian, M.A. (1993). Cytosolic and sarcoplasmic reticulum-associated low K_m, cGMP-inhibited cAMP phosphodiesterase in mammalian myocardium. Biochem. Biophys. Res. Commun. 190, 516–521.

Souness, J.E., Hassall, G.A. and Parrott, D.P. (1992). Inhibition of pig aortic smooth muscle cell cDNA synthesis by selective type III and type IV cyclic AMP phosphodiesterase inhibitors. Biochem. Pharmacol. 44, 857–866.

Sutherland, E.W. and Rall, T.W. (1958). Fractionation and characterization of a cyclic adenine ribonucleotide formed by tissue particles. J. Biol. Chem. 232, 1077–1091.

Taira, M., Hockman, S.C., Calvo, J.C., Taira, M., Belfrage, P. and Manganiello, V.C. (1993). Molecular cloning of the rat adipocyte hormone-sensitive cyclic GMP-inhibited cyclic nucleotide phosphodiesterase. J. Biol. Chem. 268, 18573–18579.

Tang, K.M., Jang, E.K. and Haslam, R.J. (1994). Photoaffinity labelling of cGMP-inhibited cyclic nucleotide phosphodiesterase (PDE III) in human and rat platelets and rat tissues: effects of phosphodiesterase inhibitors. Eur. J. Pharmacol. 268, 105–114.

Tani, T., Sakurai, K., Kimura, Y., Ishikawa, T. and Hidaka, H. (1992). Pharmacological manipulation of tissue cyclic AMP by inhibitors: effects of phosphodiesterase inhibitors on functions of platelets and endothelial cells. Adv. Second Messenger Phosphoprotein Res. 25, 215–229.

Taniguchi, T., Fujiwara, M. and Ohsumi, K. (1977). Possible involvement of cyclic adenosine 3′,5′-monophosphate in the genesis of catecholamine-induced tachycardia in isolated rabbit sinoatrial node. J. Pharmacol. Exp. Ther. 201, 678–688.

Tenor, H., Hatzelmann, A., Kupferschmidt, R., Stanciu, L., Djukanović, R., Schudt, C., Wendel, A., Church, M.K. and Shute, J.K. (1995a). Cyclic nucleotide phosphodiesterase isoenzyme activities in human alveolar macrophages. Clin. Exp. Allergy 25, 625–633.

Tenor, H., Staniciu, L., Schudt, C., Hatzelmann, A., Wendel, A., Djukanović, R., Church, M.K. and Shute, J.K. (1995b). Cyclic nucleotide phosphodiesterases from purified human $CD4^+$ and $CD8^+$ T lymphocytes. Clin. Exp. Allergy 25, 616–624.

Thompson, W.J. (1991). Cyclic nucleotide phosphodiesterases: pharmacology, biochemistry, and function. Pharmacol. Ther. 51, 13–33.

Thompson, W.J. and Appleman, M.M. (1971a). Multiple cyclic nucleotide phosphodiesterase activities from rat brain. Biochemistry 10, 311–316.

Thompson, W.J. and Appleman, M.M. (1971b). Characterization of cyclic nucleotide phosphodiesterase of rat tissues. J. Biol. Chem. 246, 3145–3150.

Tomkinson, A., Karlsson, J.-A. and Raeburn, D. (1993). Comparison of the effects of selective inhibitors of phosphodiesterase type III and IV in airway smooth muscle with differing β adrenoreceptor subtypes. Br. J. Pharmacol. 108, 57–61.

Torphy, T.J. and Cieslinski, L.B. (1990). Characterization and selective inhibition of cyclic nucleotide phosphodiesterase isoenzymes in canine tracheal smooth muscle. Mol. Pharmacol. 37, 206–214.

Torphy, T.J. and Undem, B.J. (1991). Phosphodiesterase inhibitors: new opportunities for the treatment of asthma. Thorax 46, 512–523.

Torphy, T.J., Zhou, H.-L., Burman, M. and Huang, L.F. (1991). Role of cyclic phosphodiesterase isozymes in intact canine trachealis. Mol. Pharmacol. 39, 376–384.

Torphy, T.J., Undem, B.J., Cieslinski, L.B., Luttman, M.A., Reeves, M.L. and Hay, D.W.P. (1993). Identification, characterization, and functional role of phosphodiesterase isozymes in human airway smooth muscle. J. Pharmacol. Exp. Ther. 265, 1213–1223.

Trovati, M., Massucco, P., Mattiello, L., Mularoni, E., Cavalot, F. and Anfossi, G. (1994). Insulin increases guanosine 3′,5′-cyclic monophosphate in human platelets: a mechanism involved in the insulin anti-aggregating effect. Diabetes 43, 1015–1019.

Tsien, R.S. (1977). Cyclic AMP and contractile activity in heart. Adv. Cyclic Nucleotide Res. 8, 363–419.

Uretsky, B.F., Generalovich, T., Reddy, P.S., Spangenbery, R.B. and Follansbee, W.P. (1983). The acute hemodynamic effects of a new agent, MDL 17,043, in the treatment of congestive heart failure. Circulation 67, 823–828.

Uretsky, B.F., Jessup, M., Konstam, M.A., Dec, G.W., Leier, C.V., Benotti, J., Murali, S., Herrmann, H.C. and Sandberg, J.A. (1990). Multicenter trial of oral enoximone in patients with moderate to moderately severe congestive heart failure: lack of benefit compared with placebo. Circulation 82, 774–780.

Venuti, M.C., Jones, G.H., Alvarez, R. and Bruno, J.J. (1987). Inhibitors of cyclic AMP phosphodiesterase. 2. Structural variations of *N*-cyclohexyl-*N*-methyl-4-[(1,2,3,5-tetrahydro-2-oxoimidazo[2,1-b]quinazolin-7-yl)-oxy]butyramide (RS-82856). J. Med. Chem. 30, 303–318.

Venuti, M.C., Stephenson, R.A., Alvarez, R., Bruno, J.J. and Strosberg, A.M. (1988). Inhibitors of cyclic AMP phosphodiesterase. 3. Synthesis and biological evaluation of pyrido and imidazolyl analogues of 1,2,3,5-tetrahydro-2-oxoimidazo[2,1-b]guanizoline. J. Med. Chem. 31, 2326–2145.

Weber, H.W. and Appleman, M.M. (1982). Insulin-dependent low K_m cyclic AMP phosphodiesterase from rat adipose tissue. J. Biol. Chem. 257, 5339–5341.

Weishaar, R.E., and Bristol, J.A. (1989). Selective inhibitors of phosphodiesterase. In "Comprehensive Medicinal Chemistry" (eds. P. Sammes and J. Taylor), pp. 1–32. Pergamon, New York.

Weishaar, R.E., Carn, M.H. and Bristol, J.A. (1985a). A new generation of phosphodiesterase inhibitors: multiple molecular forms and the potential for drug selectivity. J. Med. Chem. 28, 537–545.

Weishaar, R.E., Quade, M., Schenden, J.A., Boyd, D.K. and Evans, D.B. (1985b). Studies aimed at elucidating the mechanism of CI-914, a new cardiotonic agent. Eur. J. Pharmacol. 119, 205–215.

Weishaar, R.E., Burrows, S.D., Kobylarz, D.C., Quade, M.M. and Evans, D.B. (1986). Multiple molecular forms of cyclic nucleotide phosphodiesterase in cardiac and smooth muscle and platelets. Biochem. Pharmacol. 35, 787–800.

Weishaar, R.E., Kobylarz-Singer, D.C., Steffen, R.P. and Kaplan, H.R. (1987). Subclasses of cyclic AMP-specific phosphodiesterase in left ventricular muscle and their involvement in regulating myocardial contractility. Circ. Res. 61, 539–547.

Weishaar, R.E., Kobylarz-Singer, D., Keiser, J.A., Wright, C.D., Cornicelli, J. and Panek, R. (1992). Cyclic nucleotide phosphodiesterases in the circulatory system: biochemical, pharmacological, and functional characteristics. Adv. Second Messenger Phosphoprotein Res. 25, 249–271.

Yamamoto, T., Lieberman, F., Osborne, J.C., Jr, Manganiello, V.C., Vaughan, M. and Hidaka, H. (1984). Selective inhibition of two soluble adenosine cyclic 3′,5′-phosphodiesterases partially purified from calf liver. Biochemistry 23, 670–675.

Zinman, B. and Hollenberg, C.H. (1974). Effect of insulin and lipolytic agents on fat adenosine 3′,5′-monophosphate phosphodiesterase. J. Biol. Chem. 240, 2182–2187.

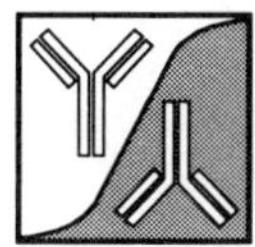

7. Interaction of PDE4 Inhibitors with Enzymes and Cell Functions

Gordon Dent ***and*** **Mark A. Giembycz**

1.	The PDE4 Isoenzyme Family	111
	1.1 Enzyme Characteristics	111
	1.2 Enzyme Distribution	112
	1.3 Selective Inhibitors	112
2.	Pharmacology of PDE4 Inhibitors	115
	2.1 *In Vitro*	115
	2.2 *In Vivo*	117
3.	Adverse Effects of PDE4 Inhibitors	119
4.	PDE4 Alterations in Allergic Diseases	119
5.	Summary and Future Directions	120
6.	References	121

1. The PDE4 Isoenzyme Family

Low K_m, cAMP-specific PDEs fall into two isoenzyme families, PDE4 and the recently identified PDE7 (see Chapter 1). With the exception of PDE7, the isoenzymes forming the PDE4 family are the most poorly understood of the phosphodiesterases. Considering that a large body of pre-clinical pharmacological data exists for selective inhibitors of PDE4 which may have significant immunomodulatory and anti-inflammatory properties (Palfreyman, 1995; see also Chapters 2, 10 and 11), the relative paucity of biochemical data is somewhat ironic. PDE4s occur in trace amounts in cells and it was only the development of selective inhibitors of this isoenzyme family that allowed these enzymes to be studied in detail (see Chapter 2).

More recently, molecular biological techniques have been applied to the study of this isoenzyme family and there has been a great expansion in the number of distinct PDE4s recognized. Four mammalian cDNA homologues of the *Drosophila melanogaster* "dunce" cAMP PDE (described by Chen *et al.*, 1986) have been identified and cloned, establishing a molecular basis for the observed heterogeneity of gene products within this PDE family (Colicelli *et al.*, 1989; Davis *et al.*, 1989; Swinnen *et al.*, 1989a,b). The molecular cloning of PDE4 isoenzymes has revealed the existence of mRNA transcripts of different sizes for each of the four variants, whose expression differs between tissues (Colicelli *et al.*, 1989; Davis *et al.*, 1989; Swinnen *et al.*, 1989a,b; Conti *et al.*, 1992). This heterogeneity is attributable both to alternative mRNA splicing and to the presence in PDE4 genes of multiple promoter regions and the consequent multiplicity of start codons for gene transcription (Monaco *et al.*, 1994).

Although early studies of PDE4 gene heterogeneity were performed in rat tissues, evidence has also been provided recently for the existence of at least four human genes encoding PDE4 isoenzymes (Livi *et al.*, 1990; Bolger *et al.*, 1993; McLaughlin *et al.*, 1993; Obernolte *et al.*, 1993; Sullivan *et al.*, 1994; Baecker *et al.*, 1995; Engels *et al.*, 1995). In common with their rat counterparts, the mRNA transcripts of these genes exhibit restricted localization between tissues (Livi *et al.*, 1990; McLaughlin *et al.*, 1993).

1.1 Enzyme Characteristics

PDE4 was first identified as a cAMP-specific PDE activity in canine kidney that was unaffected by cGMP but inhibited by the alkoxybenzyl-substituted imidazoline, Ro 20-1724 (Thompson *et al.*, 1979; Epstein *et al.*, 1982). PDE4s have subsequently been purified and partially characterized from several sources,

Phosphodiesterase Inhibitors
ISBN 0-12-210720-9

including human monocytes and leucocytes (Torphy *et al.*, 1993; Truong and Muller, 1994). All PDE4 isoenzymes characterized to date are acid proteins (pI = 4–6) that preferentially or exclusively hydrolyse cAMP (K_m = 1–20 μM) (Conti and Swinnen, 1990; Bolger *et al.*, 1993). The quaternary structure – and even the monomeric size – of PDE4 isoenzymes remains unclear (see Chapter 1).

As described in detail in other chapters of this volume, PDE4 activity is subject to short-term regulation by phosphorylation catalysed by protein kinases (Conti *et al.*, 1995; see also Chapters 1, 2 and 11). Long-term regulation of PDE4 is also observed in many cells following prolonged elevation of intracellular cAMP levels (see Chapter 2). In addition, stimulation of immune cells can lead to an up-regulation of PDE4 activity. For example, rat T lymphocytes and mouse peritoneal macrophages display increased PDE4 activity within one hour of stimulation with concanavalin A (Con A) or bacterial lipopolysaccharide (LPS), respectively (Valette *et al.*, 1990; Okonogi *et al.*, 1991). A similar up-regulation is observed in monocytes obtained from subjects with atopic dermatitis (see section 4; see also Chapter 2).

1.2 ENZYME DISTRIBUTION

With the possible exception of platelets (see Chapter 2), high affinity cAMP-specific PDE4 is present in all cell types implicated in allergic and inflammatory diseases. Some degree of differential expression of the four PDE4 subtypes, as detected by reverse transcription/polymerase chain reaction (RT-PCR) is observed between cell types, although this is not marked (Table 7.1).

1.3 SELECTIVE INHIBITORS

As described elsewhere (see Chapters 2, 10, 11, 12 and 13), selective inhibitors of PDE4 have been developed by several pharmaceutical companies with the aim of producing agents effective in the treatment of inflammatory diseases. A list of selective PDE4 inhibitors currently in use in the laboratory or under clinical development is given in Table 7.2.

PDE4 inhibitors fall into three broad groups: xanthine derivatives, rolipram analogues and quinazolinediones.

Table 7.1 PDE isoenzyme profiles and PDE4 subtype expression in immune cells

Cell type	*Source*	*PDE isoenzyme(s) present*	*Comments*	*Reference*	*PDE4 subtype expression*[a]			
					4A	*4B*	*4C*	*4D*
B Lymphocyte		Not known						
T Lymphocyte	Human peripheral blood	4		Epstein and Hachisu (1984)	++	++	?	++
		3,4	PDE4 soluble and particulate	Robicsek *et al.* (1989,1991)				
	Human peripheral blood CD4⁺	2, 3, 4, 5, 7	PDE4 predominantly soluble	Tenor *et al.* (1995d)	++	++	–	++
	Human peripheral blood CD8⁺	2, 3, 4, 5, 7		Giembycz *et al.* (1996)	++	++	–	++
	Human T-cell clone (HUT 78)	4, 7		Ichimura and Kase (1993)				
	Rat	2, 3, 4		Valette *et al.* (1990)				
		2, 3, 4, 5		Marcoz *et al.* (1993)				
Neutrophil	Human peripheral blood	1, 4	Evidence for PDE1 not subsequently corroborated	Engerson *et al.* (1986) Grady and Thomas (1986)	±	++	–	±
		4	Particulate, non-linear kinetics	Wright *et al.* (1990)				
		4	Enzyme apparently soluble	Nielson *et al.* (1990)				
				Schudt *et al.* (1991a)				
Eosinophil	Human peripheral blood	4	Particulate, non-linear kinetics	Dent *et al.* (1994)	++	++	–	++
		4	Enzyme ~70% soluble	Hatzelmann *et al.* (1995)				
	Guinea-pig peritoneal	4	Particulate, non-linear kinetics	Souness *et al.* (1991)	–	–	–	++
				Dent *et al.* (1991)				
Monocyte	Human peripheral blood	4		Thompson *et al.* (1976)	++	++	–	++
				White *et al.* (1990)				
		1, (3), 4		Seldon *et al.* (1995)				
				Verghese *et al.* (1995)				
Macrophage	Human alveolar	1, 3, 4, 5	PDE4 predominantly soluble	Tenor *et al.* (1995a)				
	Guinea pig peritoneal	1, 4	PDE4 particulate	Turner *et al.* (1993)	–	±	–	++
	Mouse peritoneal	2, 3, 4		Okonogi *et al.* (1991)				
Basophil	Human peripheral blood	3, 4, 5		Peachell *et al.* (1992)				
Mast cell	Mouse bone marrow	1, 4		Torphy and Undem (1991)				
	Rat peritoneal	2, 3, 4		Bergstrand *et al.* (1978)				
Platelet	Human peripheral blood	1, 2, 3, 5		Hagiwara *et al.* (1984)				
				MacPhee *et al.* (1986)				

[a] Subtype expression assessed by RT-PCR and summarized from Engels *et al.* (1994) ±, weak mRNA expression; –, no mRNA expression.

Table 7.2 PDE4 inhibitors under development

Company	*Drug*	*Isoenzyme selectivity*	*Indication*	*Development stage*
Sandoz	Benafentrine (AH 21,132)	3/4	Asthma	Discontinued
Byk Gulden	Tolafentrine	3/4	Asthma	Discontinued
Byk Gulden	Zardaverine	3/4	Asthma	Discontinued
Organon	Org 20241	3/4	Asthma	Phase I
Troponwerke	Nitraquazone	4	Inflammation	Discontinued
Roche (Syntex)	RS 5344	4	Asthma	Pre-clinical
SmithKline Beecham	BRL 1063	4	Asthma	Phase I
SmithKline Beecham	SB 207,499	4	Asthma	Phase I
Sandoz	SDZ MKS 492	4	Asthma	Phase I
Celltech/Merck	CDP 840	4	Asthma	Discontinued
Pfizer	CP 80,633	4	Asthma, Atopic dermatitis	Phase II
Rhône-Poulenc Rorer	RP 73401	4	Asthma	Phase II
Wyeth-Ayerst	WAY-PDA-641	4	Asthma	Phase II
Almirall	LAS 31025	4	Asthma	Phase III
Eli Lilly	Tibenelast	4	Asthma	Phase III

1.3.1 Xanthine Derivatives

Although many xanthines possess PDE inhibitory activity (see Chapter 3), the development of this class of compounds has been retarded by their lack of potency and the confounding factor of their adenosine antagonistic properties. Denbufylline (Fig. 7.1) was the first xanthine derivative to be identified as a selective inhibitor of PDE4 with negligible adenosine antagonism (Nicholson *et al.*, 1989). Although a series of further derivatives displaying PDE4 selectivity has been synthesized (see Palfreyman, 1995), no structure–activity relationship studies were reported. More recently, however, a range of 1,3,7-substituted xanthines has been studied, revealing that, whereas alkyl substitution at N3 leads to increased non-selective PDE inhibitory potency, alkyl substitution at N1 and oxypropyl substitution at N7 lead to increased selectivity for PDE4 (Miyamoto *et al.*, 1994); this PDE4-selectivity correlates with increased selectivity for relaxation of guinea-pig tracheal smooth muscle, compared to augmentation of the rate of right atrial contraction (Sakai *et al.*, 1992; Miyamoto *et al.*, 1994). Studies of the influence of structural changes on PDE4 inhibitory and adenosine antagonistic activity are continuing, with the aim of developing effective PDE4 inhibitors that do not exert side-effects due to inhibition of other isoenzyme families or blockade of adenosine receptors (Buckle *et al.*, 1994).

1.3.2 Rolipram Analogues

Rolipram is the most extensively studied selective PDE4 inhibitor and is the drug with which novel compounds are compared (see Chapters 11 and 12). Rolipram itself (Fig. 7.1) is a 4-substituted pyrrolidinone and a range of analogues of this molecule has been studied to evaluate the structural requirements for PDE4 inhibitor potency. Briefly, a large alkoxy substituent at the 3′ position, a small alkoxy (ideally methoxy) group at 4′ and a carbonyl group at position 2 of the pyrrolidinone ring are required for PDE inhibitor activity. While non-specific PDE inhibitor potency depends on the presence of the pyrrolidinone ring, selectivity for PDE4 can be increased by enlarging the alkoxy substituent at 3′ while retaining a small alkoxy group at the 4′ position (C. Schudt, personal communication). Substitution at position 1 or 3 of the pyrrolidinone ring can lead to changes in activity. Replacement of pyrrolidinone with imidazolinone, pyrazolidinone, spirolactam, oximocarbamate or oxamide leads to increased activity against PDE4 from a variety of sources, with the molecules containing the open oximocarbamate or oxamide structures being particularly potent (Palfreyman, 1995). Recently, a rolipram derivative in which the pyrrolidinone ring is replaced by 4-(2′-phenylethyl)pyridine (CDP 840) has been described as a potent, selective and orally active PDE4 inhibitor (Hughes *et al.*, 1995).

Although removal of the carbonyl carbon group in rolipram's pyrrolidinone ring leads to loss of PDE4 inhibitory activity, the nitrogen atom is not required: if the pyrrolidinone ring is replaced with a cycloalkane-based ring, such as cyclopentanone, activity is retained. The cyanoacid derivative, SB 207499 (Fig. 7.1), is a highly potent and selective inhibitor of PDE4 (see Chapter 13), and cyclopropane and cyclic sulphoxide derivatives also inhibit PDE4 at nanomolar concentrations (Palfreyman, 1995).

Replacement of the pyrrolidinone ring of rolipram with benzamide groups also produces a series of selective PDE4 inhibitors in which substitution of the *N*-phenyl ring at position 2 – and, additionally, at position 6 – leads to increased inhibitory potency.

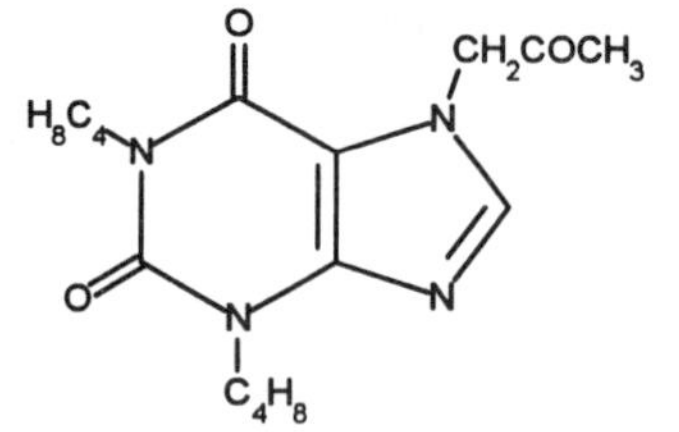

Denbufylline (1,3-di-*n*-butyl-7-[2'-oxopropyl] xanthine)

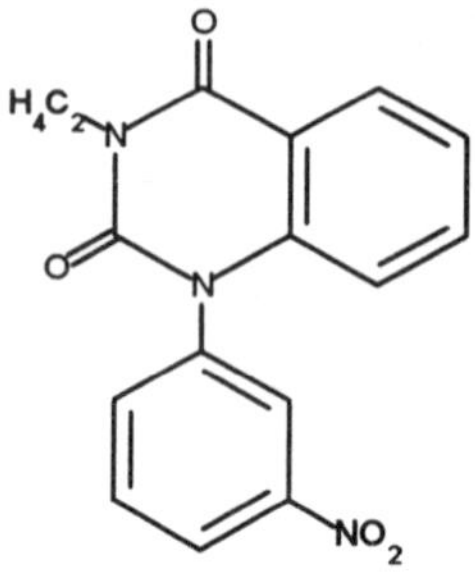

Nitraquazone (3-[3'-nitrophenyl] *N*-ethylquinazoline-2,6-dione)

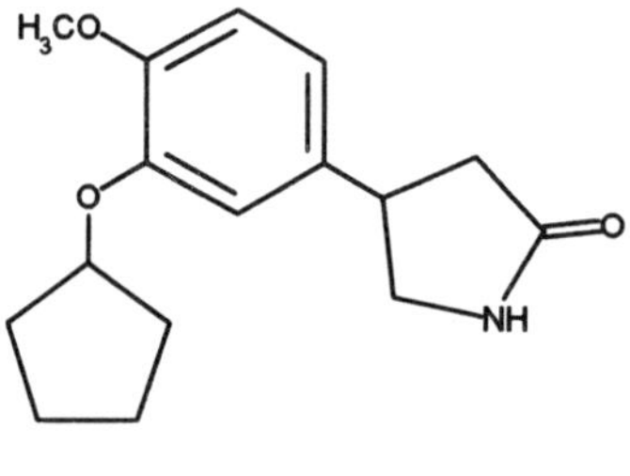

Rolipram (4-[3'-cyclopentyloxy-4'-methoxyphenyl]-2-pyrrolidinone)

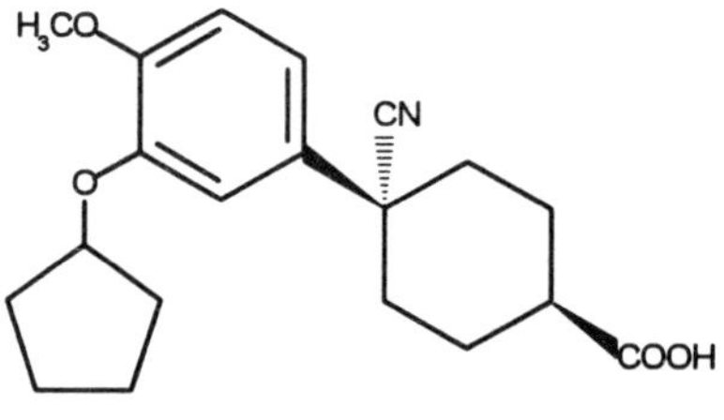

SB 207499 (*c*-4-cyano-4-[3'-cyclopentyloxy-4'-methoxyphenyl]-*r*-1-cyclohexanecarboxylic acid)

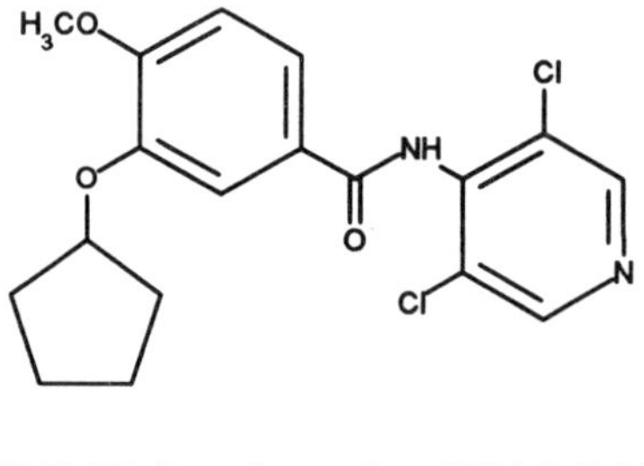

RP 73401 (3-cyclopentyloxy-*N*-[3',5'-dichloro-4'-pyridyl]-4-methoxybenzamide)

Figure 7.1 Structures of some representative PDE4 inhibitors: a xanthine derivative, denbufylline; a quinazolinedione, nitraquazone; a pyrrolidinone derivative, rolipram, and two of its analogues, SB 207499 (see Chapter 13) and RP 73401 (see Chapter 12).

Replacement of the *N*-phenyl ring by heterocyclic rings alters the activity; maximal potency is achieved with pyridyl analogues such as RP 73401 (Fig. 7.1) (Ashton *et al.*, 1994; Palfreyman, 1995; see also Chapter 12).

1.3.3 Quinazolinediones

The quinazolinediones are structurally distinct from rolipram but may be classified along with denbufylline, for the purposes of structure–activity relationship, as "fused ring" compounds (see Chapter 13). Compounds based on nitraquazone (Fig. 7.1) have been studied and replacement of the *N*-ethyl group with more lipophilic groups was shown to increase PDE4 inhibitory potency. Loss of the dione structure does not necessarily lead to loss of inhibitor activity and some activity can be retained even with loss of the fused ring structure (Palfreyman, 1995).

2. *Pharmacology of PDE4 Inhibitors*

2.1 IN VITRO

Extensive study has been made of the actions of both non-selective PDE inhibitors and selective inhibitors of specific isoenzyme families upon cells of the immune system (see Chapters 2 and 3). In addition, several dual-selective inhibitors – mainly inhibitors of PDE3 and PDE4 – have been studied (see Chapter 10). Most of these studies have been undertaken using cells obtained from laboratory animals or from normal human volunteers. Given the evidence indicating a defective regulation of cAMP metabolism in immune cells from atopic subjects (see section 4), the relevance of these findings to cell function in disease states is not clear. Furthermore, the results of experiments conducted using isolated, purified cells of a single type cannot necessarily be extrapolated to the same cells in a physiological environment, where mixed cell populations occur and the role of cytokines and mediators derived from other cells may be of great importance. The data summarized in the following sections must, therefore, be regarded as preliminary but they appear to indicate therapeutic potential for selective inhibitors of the PDE4 family.

2.1.1 Lymphocytes

2.1.1.1 B lymphocytes

The effects of PDE inhibitors on B-cell function have not been studied in depth. The role of cAMP in the regulation of immunoglobulin production is unclear and cAMP can increase or decrease production *in vitro*, depending on the time of drug addition (Kammer, 1988). Cyclic AMP inhibits proliferation of B-cells (Kammer, 1988) but can promote antibody class-switching, leading to increased production of IgG_1 and IgE (Lycke *et al.*, 1990; Phipps *et al.*, 1990; Roper *et al.*, 1990; Lycke, 1993; Paul-Eugène *et al.*, 1993). The PDE4 inhibitor, Ro 20-1724, however, has been reported to inhibit spontaneous IgE release from mononuclear cells obtained from atopic donors; this action is clearly indirect, since it is not observed in purified B cells (Cooper *et al.*, 1985), and may be due to the suppression of interleukin-4 (IL-4) release from T cells or monocytes (Chan *et al.*, 1993b). IL-2 may also contribute to the production of IgE in mixed lymphocyte populations and the inhibition of IL-2 release may account for the suppression of IgE synthesis in mononuclear cell preparations (Phipps *et al.*, 1990).

2.1.1.2 T Lymphocytes

Inhibitors of PDE4 exhibit broadly similar actions to non-selective methylxanthine PDE inhibitors on lymphocyte function (see Chapter 3). PDE4 inhibitors suppress phytohaemagglutinin (PHA)-induced blastogenesis of mixed human T lymphocytes; the inhibition is only partial (50–70%) but a greater effect is elicited when the drugs are administered conjointly with inhibitors of PDE3 (Robicsek *et al.*, 1991), reflecting the presence in these cells of a significant PDE3 activity in addition to the PDE4 (see Chapter 2). Similar results are obtained with purified $CD4^+$ or $CD8^+$ human T cells and in a human T_H2 clone established from aeroallergen-specific T cells (Giembycz *et al.*, 1994; Crocker *et al.*, 1994). Anti-CD3-induced secretion of IL-4 by the allergen-specific T_H2-cells is also suppressed by a PDE4 inhibitor, WAY-PDA-641 (Crocker *et al.*, 1994), and it has been suggested that reduced production of cytokines – particularly IL-2 – may underlie the anti-proliferative actions of cAMP PDE inhibitors (Averill *et al.*, 1988; Thanhauser *et al.*, 1993; Giembycz *et al.*, 1994). In fact, the selective PDE4 inhibitor, rolipram, inhibits potently the generation of IFNγ and IL-2 from $CD4^+$ and $CD8^+$ T cells stimulated with PHA under conditions where the PDE3-selective drug, SK&F 95654, is ineffective (Giembycz *et al.*, 1994). The inhibition of Con A-induced proliferation of mouse splenocytes, however, occurs at lower concentrations of rolipram than those required to suppress IL-2 gene transcription (Lewis *et al.*, 1993), calling into question the involvement of reduced IL-2 production in the anti-suppressive action of the PDE4 inhibitor. Furthermore, rolipram does not affect steady-state levels of IL-2 mRNA in Jurkat T cells (Lewis *et al.*, 1993). If reduced IL-2 generation is not the mechanism through which PDE inhibitors exert their anti-proliferative effects, mitogenic signal transduction pathways may be worthy of consideration (van Tits *et al.*, 1991; Anastassiou *et al.*, 1992). Agents that elevate intracellular cAMP levels inhibit anti-CD3-induced tyrosine phosphorylation of a 100 kD protein implicated in T-cell activation as well as decreasing IL-2 biosynthesis and IL-2 receptor expression (Anastassiou *et al.*, 1992). The decrease in IL-2 synthesis resulting from cAMP elevation may be related to an effect on IL-2 gene transcription and a decrease in the $t_{1/2}$ for IL-2 mRNA degradation (Anastassiou *et al.*, 1992).

Some evidence exists for a selective action of PDE inhibitors against T_H1 versus T_H2 $CD4^+$ T-cell function. The non-selective inhibitor, pentoxifylline (see Chapter 3), is more effective in inhibiting the release of a T_H1-derived cytokine (IL-2) than of a cytokine derived from T_H2 cells (Rott *et al.*, 1993). This finding is consistent with demonstrations that cAMP is a more effective inhibitor of the release of IL-2 and IFNγ release than of IL-4 and IL-5 (Munoz *et al.*, 1990; Novak and Rothenberg, 1990; Betz and Fox, 1991; van der Poow-Kraan *et al.*, 1992; Lee *et al.*, 1993; Hilkens *et al.*, 1995; Tsuruta *et al.*, 1995) while, under certain conditions, cAMP actually activates transcription of the IL-4 and IL-5 genes (Lee *et al.*, 1993; Watanabe *et al.*, 1994; Hilkens *et al.*, 1995; Tsurata *et al.*, 1995). On the other hand, appearance of IL-4 and IL-5 in the bronchoalveolar lavage (BAL) fluid and spleen cells of sensitized BALB/c mice following allergen challenge is inhibited by rolipram pretreatment (Longchampt *et al.*, 1995; Foissier *et al.*, 1995). Rolipram – but not a PDE3 inhibitor, siguazodan, or a PDE5 inhibitor, zaprinast – blocks tetanus toxoid-induced (T_H1-driven) proliferation of peripheral blood mononuclear cells more effectively than that induced by ragweed pollen (T_H2-driven) (Essayan *et al.*, 1994). RT-PCR reveals attenuation of allergen-stimulated IL-5 and IFNγ gene transcription by rolipram while IL-4 gene transcription is unaffected (Essayan *et al.*, 1994, 1995). The ability of rolipram to suppress T_H2 cell proliferation may depend on the selective expression of PDE4 isoenzymes, since the ragweed-sensitive T_H2 cell line expresses mRNA for both the PDE4A and PDE4B variants whereas the T_H1-like Jurkat cell line expresses low levels of PDE4A mRNA but no message for PDE4B (Livi *et al.*, 1990; McLaughlin *et al.*, 1993; Essayan *et al.*, 1994).

2.1.2 Neutrophils

Multiple neutrophil functions are sensitive to suppression by agents that elevate intracellular cAMP, including non-selective PDE inhibitors (see Chapter 3). Activation of the human neutrophil respiratory burst is sensitive to selective PDE4 inhibitors such as rolipram, Ro 20-1724 and tibenelast, as well as mixed PDE3/4 inhibitors such as zardaverine, but is unaffected by the selective PDE3 inhibitors, amrinone and cilostamide, or a PDE5 inhibitor, zaprinast (Ho *et al.*, 1990; Nielson *et al.*, 1990; Wright *et al.*, 1990; Schudt *et al.*, 1991a,c). Suppression of respiratory burst activity can be observed in neutrophils primed with tumour necrosis factor (TNF-α), as well as basally active cells (Sullivan *et al.*, 1995).

Selective inhibitors of PDE4, but not of PDE3, also reduce *N*-formylmethionyl-L-leucinyl-L-phenylalanine (FMLP)-stimulated adhesion of human neutrophils to human umbilical vein endothelial cells (HUVEC), apparently as a result of down-regulation of the β_2 integrin CD11b/CD18 (Mac-1) on the neutrophils (Fig. 7.2) (Derian *et al.*, 1995). These findings reflect the exclusive presence of PDE4 in neutrophils (see Chapter 2). Interestingly, the suppression of neutrophil adhesion to HUVEC was entirely dependent on endogenous adenosine: in the presence of adenosine deaminase both rolipram and Ro 20-1724 failed to inhibit adhesion, indicating that an adenylate cyclase (AC) activator is required to uncover the functional effects of PDE4 inhibitors (Derian *et al.*, 1995).

2.1.3 Eosinophils

Both selective PDE4 inhibitors and mixed inhibitors of PDE3 and PDE4 (see Chapter 10) cause suppression of a range of eosinophil functions *in vitro*. Superoxide anion and hydrogen peroxide (H_2O_2) generation by human and guinea-pig eosinophils activated with soluble or particulate stimuli are reduced after treatment of the cells with drugs including rolipram, denbufylline, WAY-PDA-641, RP 73401, zardaverine and Org 20241 (Dent *et al.*, 1991, 1994; Souness *et al.*, 1991, 1995; Maruo *et al.*, 1994; Barnette *et al.*, 1995; Nicholson *et al.*, 1995), although another group observed no inhibition of complement fragment C5a-induced oxygen radical generation by rolipram, RP 73401 or zardaverine unless the cells were concurrently exposed to the β_2-adrenoceptor agonst, salbutamol (Hatzelmann *et al.*, 1995).

The generation of thromboxane by guinea-pig peritoneal eosinophils in response to leukotriene B_4 (LTB_4) is also inhibited by rolipram and the mixed PDE3/4 inhibitor, Org 20241 (Souness *et al.*, 1994; Nicholson *et al.*, 1995). Pretreatment of human peripheral blood eosinophils with the PDE4 inhibitors, rolipram and RP 73401, or the mixed PDE3/4 inhibitors, tolafentrine and zardaverine, leads to suppression of FMLP-induced generation of LTC_4, although tolafentrine and RP 73401 exhibit lower potency for inhibition of LTC_4 generation than of PDE activity and this relationship is reversed in the case of rolipram and zardaverine (Tenor *et al.*, 1995c).

Degranulation of guinea-pig eosinophils stimulated by LTB_4 is inhibited by RP 73401 and rolipram (Souness *et al.*, 1995); these drugs, as well as zardaverine and tolafentrine, have been shown to inhibit C5a-induced degranulation – like C5a-induced oxygen radical production (see above) – of human eosinophils only in the presence of salbutamol (Hatzelmann *et al.*, 1995). Selective PDE4 inhibitors also suppress chemotaxis of guinea-pig and human eosinophils (Cohan *et al.*, 1992; Tanimoto *et al.*, 1994) but have not been shown to promote eosinophil apoptosis (Hallsworth *et al.*, 1996), as has been described for theophylline (see Chapter 3).

2.1.4 Monocytes and Macrophages

Although human peripheral blood monocytes express mainly PDE4 (Thompson *et al.*, 1976; Elliott and Leonard, 1989), they also contain minor PDE1 and PDE3 activities (Table 7.1). Human alveolar macrophages, in contrast, contain predominantly PDE1, with significant PDE3 and PDE4 activities and a smaller PDE5 activity (Tenor *et al.*, 1995a).

Rolipram is a weak inhibitor of FMLP-stimulated superoxide generation in monocytes and of opsonized zymosan-stimulated H_2O_2 generation in alveolar macrophages (Elliott and Leonard, 1989; Dent *et al.*, 1993) but a more effective inhibitor of LPS-induced TNF-α release from both cell types (Semmler *et al.*, 1993; Seldon *et al.*, 1995; Schudt *et al.*, 1992). In alveolar macrophages, suppression of TNF-α production is much more pronounced when PDE4 and PDE3 are inhibited simultaneously (Schudt *et al.*, 1993).

It may be of great importance that, while LPS-induced TNF-α production by a macrophage cell line, RAW 264.7, is suppressed by rolipram, the production of nitric oxide (NO) is increased under the same conditions, possibly as a result of cyclic AMP-dependent protein

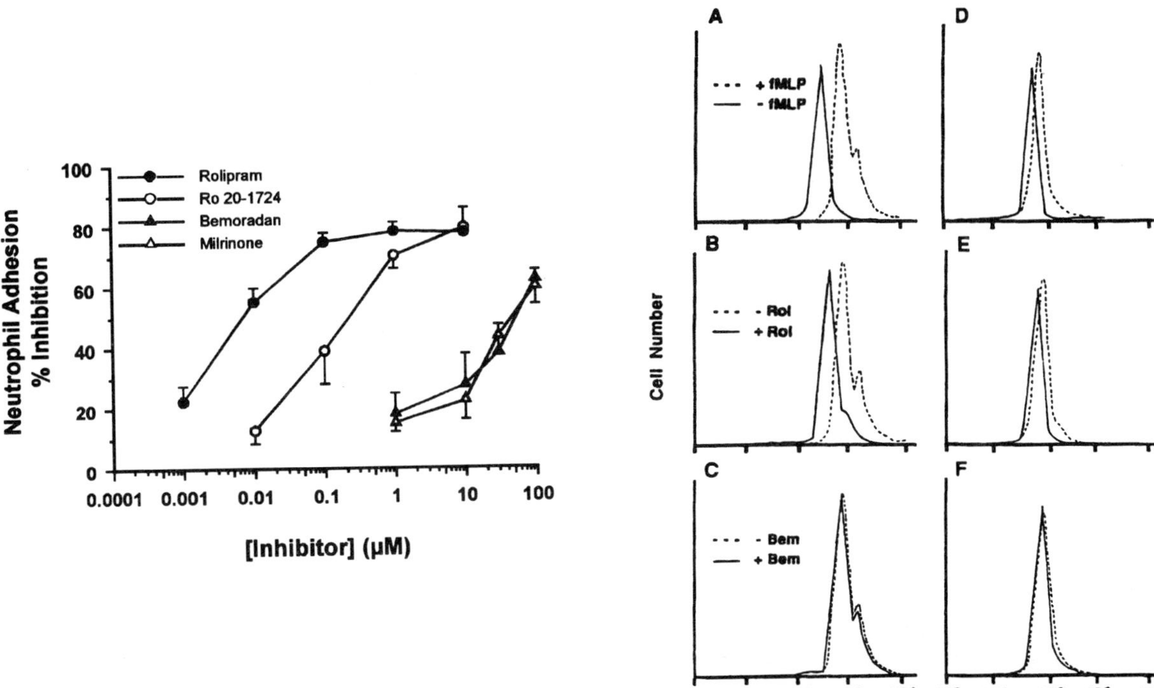

Figure 7.2 Effects of selective PDE4 and PDE3 inhibitors on adherence of human neutrophils to HUVEC. Left panel: inhibition of FMLP-stimulated adherence by PDE4 inhibitors, rolipram and Ro 20-1724, and PDE3 inhibitors, bemoradan and milrinone, which are effective only at much higher concentrations. Right panel: fluorescence-activated cell sorter analysis of the effects of a PDE4 inhibitor, rolipram (Rol), and a PDE3 inhibitor, bemoradan (Bem), on FMLP-stimulated expression of the β_2-integrins, Mac-1 (CD11b/CD18) and LFA-1 (CD11a/CD18). FMLP induces an increase in expression (rightward shift) of Mac-1 and a smaller increase in LFA-1 expression; both effects are reversed by rolipram but not by bemoradan. Figures reproduced, with permission, from Derian *et al.* (1995) © 1995, The American Association of Immunologists.

kinase (PKA)-mediated phosphorylation of inducible NO synthase (Greten *et al.*, 1995). Thus, some bactericidal and vasodilator actions of macrophages may be maintained or improved by treatment with drugs that simultaneously reduce other pro-inflammatory functions of the cells. Since NO has been postulated to amplify and perpetuate T_H2 cell-mediated inflammation, however (Barnes and Liew, 1995), an increase in NO production may itself represent a pro-inflammatory mechanism.

2.1.5 Mast Cells and Basophils

Human peripheral blood basophils contain predominantly PDE4, with some additional PDE3 and PDE5 activity. Selective inhibitors of PDE4 suppress histamine secretion and LTC_4 generation from basophils stimulated with anti-IgE or allergen, as well as platelet-activating factor (PAF)-induced histamine release. The effects of rolipram on anti-IgE-induced mediator release are enhanced in the presence of PDE3 inhibitors, even though the latter drugs are ineffective when given alone (Peachell *et al.*, 1992; Columbo *et al.*, 1993).

The PDE isoenzyme complement of mast cells remains unclear. Although rolipram inhibits antigen-induced LTC_4 release from mouse bone marrow-derived mast cells, it is ineffective against antigen-induced histamine secretion from rat peritoneal mast cells (Frossard *et al.*, 1981; Torphy and Undem, 1991), despite an earlier enzymological study suggesting the presence of PDE4 in the rat peritoneal cells (Bergstrand *et al.*, 1978). Anti-IgE-induced release of histamine from human lung mast cells is also barely affected by rolipram, even at concentrations that cause significant increases in intracellular cAMP concentration (Anderson and Peachell, 1994), suggesting some spatial or functional separation between PDE4-dependent cAMP and mast cell degranulation.

2.2 IN VIVO

The observation of suppressive effects of PDE4 inhibitors on inflammatory cell function has led to studies of the actions of these drugs in inflammatory reactions *in vivo*. Although no demonstration has yet been made of any influence of PDE4 inhibitors on IgE production *in vivo*, several IgE-mediated responses are affected. Similar responses can be evoked by exogenous inflammatory mediators but the response to allergen is invariably greater than those to individual mediators. The ability of PDE4 inhibitors to suppress allergen-induced reactions is equal to – or even greater than – their ability to inhibit mediator responses. Rolipram, for example, effectively blocks bronchoconstriction induced in guinea pigs by allergen but is poorly effective against LTD_4-induced bronchospasm; in contrast, a selective inhibitor of PDE3, CI 930, blocks both responses equally (Howell *et al.*, 1993), suggesting that the PDE3 inhibitor acts by antagonizing smooth muscle contraction whereas the PDE4 inhibitor blocks IgE-dependent mediator release from immune cells in the airways.

Two further important actions of anti-inflammatory drugs, which have been studied in more detail, are the inhibition of inflammatory cells to the sites of allergic reactions and the reduction of leakage of plasma from the microcirculation. PDE4 inhibitors have been demonstrated clearly to exert both of these actions.

2.2.1 Inflammatory Cell Infiltration

A mixed inhibitor of PDE3 and PDE4, zardaverine, suppresses allergen-induced infiltration of eosinophils, macrophages and neutrophils into the BAL fluid of sensitized guinea pigs (Schudt *et al.*, 1991b) and similar results have been obtained after chronic dosing with the PDE3/4 inhibitor, benafentrine, against PAF- and allergen-induced pulmonary eosinophil recruitment (Sanjar *et al.*, 1989, 1990a,b). More recently, selective inhibitors of PDE4 have been studied; intragastric administration of rolipram or RS 25344 to conscious guinea pigs has been demonstrated to attenuate allergen-induced influx of eosinophils to the BAL and airway tissues (Underwood *et al.*, 1993; see also Chapter 11), whereas inhalation of a micronized dry powder formulation of rolipram prevents the recruitment of leucocytes to the BAL fluid of allergen-challenged anaesthetized guinea pigs (Raeburn *et al.*, 1993). Although acute intraperitoneal dosing of sensitized guinea pigs with lower doses of zardaverine or the selective PDE4 inhibitor, Ro 20-1724, 1 hour before challenge does not suppress inflammatory cell recruitment to the BAL fluid, chronic low dosing for 7 days prior to challenge causes a reduction to baseline levels of BAL eosinophil numbers and an even greater reduction of mononuclear cells (Banner and Page, 1995).

Treatment of sensitized cynomolgus monkeys with rolipram does not affect the immediate bronchoconstriction following acute antigen provocation but causes significant inhibition of the pulmonary eosinophilia and neutrophilia and of the increase in BAL levels of several cytokines measured 4 hours after antigen exposure (Fig. 7.3). Airways hyperresponsiveness and BAL eosinophil, neutrophil, lymphocyte and monocyte/macrophage numbers following repeated exposure to antigen are also reduced (Turner *et al.*, 1994), supporting the hypothesis that PDE4 inhibitors may be beneficial in the treatment of bronchial asthma by virtue of their anti-inflammatory action, rather than any direct bronchodilator property (Howell *et al.*, 1993).

Sensitivity to PDE4 inhibitors is also exhibited in other *in vivo* models of inflammation, including allergen-induced pulmonary eosinophilia in brown Norway rats (Elwood *et al.*, 1995), histamine- or LTB_4/D_4-induced eosinophilia and arachidonic acid-induced leucocyte accumulation in the guinea-pig conjunctiva (Newsholme and Schwartz, 1993; Griswold *et al.*, 1993) and uric acid/LTB_4-induced neutrophilia in the guinea-pig peritoneal cavity (Griswold *et al.*, 1993). In the skin of guinea pigs, accumulation of ^{111}In-labelled eosinophils in response to intradermal injections of zymosan-activated plasma (ZAP), PAF or histamine is blocked by systemic administration of rolipram – but not of selective PDE3 (SK&F 94120) or PDE5 inhibitors (zaprinast) – whereas neutrophil recruitment is unaffected (Texeira *et al.*, 1994), possibly reflecting the low sensitivity to rolipram of guinea-pig neutrophils when compared to human neutrophils *in vitro* (Boucheron *et al.*, 1991).

The pulmonary eosinophilia resulting from exposure of guinea pigs to allergen is significantly greater than that elicited by inflammatory mediators, such as PAF, histamine and LTB_4, and cannot be abolished by a combination of specific mediator antagonists (Aoki *et al.*, 1988). This probably reflects the contribution of additional pro-inflammatory molecules, particularly cytokines and chemokines, to the process of eosinophil recruitment, since IL-3, IL-5, TNF-α, granulocyte/macrophage colony-stimulating factor (GM-CSF), the "regulated upon activation, normal T-cell expressed and secreted" chemokine (RANTES) and macrophage inflammatory protein 1α (MIP-1α) have all been demonstrated to induce pulmonary accumulation and activation of eosinophils (Dahinden *et al.*, 1993; Holtzman *et al.*, 1994; Baggiolini *et al.*, 1995). Pretreatment of guinea pigs with benafentrine suppresses the recruitment of eosinophils to the airways in response to human recombinant IL-3 or GM-CSF or to mouse TNF-α (Kings *et al.*, 1991), whereas zardaverine inhibits the TNF-mediated recruitment of neutrophils to the airways of rats in response to LPS (Kips *et al.*, 1993). Benafentrine and zardaverine are both inhibitors of PDE3 and PDE4 (see Chapter 10). To date, similar experiments using monoselective inhibitors of PDE4 have not been reported.

2.2.2 Actions in the Microvasculature

Allergens and inflammatory mediators can induce increases in the permeability of postcapillary venules to plasma proteins, leading to local oedema. PDE4 inhibitors,

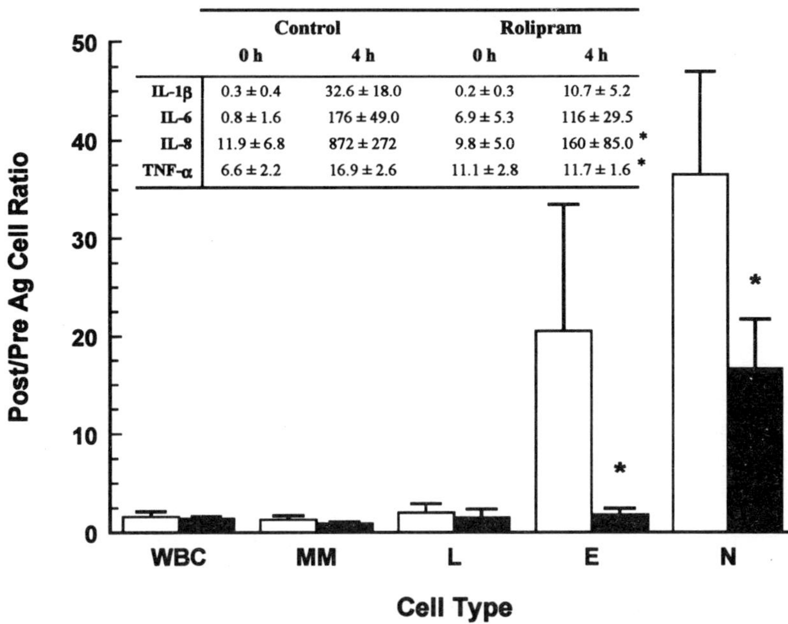

	Control 0 h	Control 4 h	Rolipram 0 h	Rolipram 4 h
IL-1β	0.3 ± 0.4	32.6 ± 18.0	0.2 ± 0.3	10.7 ± 5.2
IL-6	0.8 ± 1.6	176 ± 49.0	6.9 ± 5.3	116 ± 29.5
IL-8	11.9 ± 6.8	872 ± 272	9.8 ± 5.0	160 ± 85.0 *
TNF-α	6.6 ± 2.2	16.9 ± 2.6	11.1 ± 2.8	11.7 ± 1.6 *

Figure 7.3 BAL leucocyte increases (expressed as ratio of post- to pre-antigen numbers) in control (open bars) and rolipram (10 mg/kg)-treated monkeys (solid bars) following exposure to an antigen aerosol. Abbreviations: WBC, white blood cells; MM, monocytes/macrophages; L, lymphocytes; E, eosinophils; N, neutrophils. Reproduced, with permission, from Turner *et al.* (1994). Inset: Mean ± SEM cytokine levels (pg/ml) in the BAL fluid of control and rolipram-treated monkeys before and 4 h after antigen exposure. Data taken from Turner *et al.* (1994). * $P < 0.05$ compared with corresponding control.

such as rolipram and RP 73401, suppress the PAF-induced exudation of serum albumin into the large and small airways and the BAL of anaesthetized guinea pigs whereas rolipram is also effective against allergen-induced microvascular leakage in sensitized animals (Raeburn and Karlsson, 1991; Raeburn *et al.*, 1991, 1994; Ortiz *et al.*, 1992). Rolipram does not inhibit oedema formation in guinea-pig skin in response to PAF, ZAP or histamine, however (Texeira *et al.*, 1994); there may be some dependence of this action of PDE4 inhibitors on the local environment or the blood vessel type, leading to site differences within the same species. In other species, PDE4 and mixed PDE3/4 inhibitors attenuate bradykinin-induced microvascular leakage in the hamster cheek pouch (Svensjo *et al.*, 1992) whereas arachidonic acid-induced ear oedema is inhibited in the rat by rolipram (Raeburn *et al.*, 1993) and in the mouse by denbufylline (Crummey *et al.*, 1987) and RS 25344 (see Chapter 11).

3. *Adverse Effects of PDE4 Inhibitors*

The development of isoenzyme-selective PDE inhibitors was prompted, in part, by the desire to obtain drugs with the therapeutic actions of non-selective drugs such as the methylxanthines but devoid of their side-effects (see Chapter 3). The selective drugs do, however, exhibit adverse actions that usually represent a subset of theophylline's side-effects. The most recognized adverse action of PDE4 inhibitors is their propensity to cause nausea and vomiting, possibly via stimulation of neurones in the area postrema of the brain (Zeller *et al.*, 1984; Carpenter *et al.*, 1988; Brunée *et al.*, 1992). The drugs also act, like theophylline, to enhance gastric acid secretion from the parietal cells of the stomach (Black *et al.*, 1988) and exocrine secretion of the pancreas (Iwatsuki *et al.*, 1991). Rolipram causes a transient fall in plasma osmolality (Sturgess and Searle, 1990) which may relate to the predominance of PDE4 in the kidney.

4. *PDE4 Alterations in Allergic Diseases*

As discussed in Chapter 2, cAMP PDE activity is elevated in mononuclear leucocytes obtained from patients with atopic diseases, including dermatitis, urticaria pigmentosa, allergic rhinitis and asthma (Grewe *et al.*, 1982; Holden, 1990; Chan and Hanifin, 1993; Townley, 1993). A similar, though smaller, influence of atopy is observed in T lymphocytes and neutrophils (Chan and Hanifin, 1993; Goldberg *et al.*, 1994).

Two distinct cAMP PDE isoenzymes exist in monocytes from patients with atopic dermatitis whereas only a single enzyme is expressed in monocytes from non-atopic subjects (Chan *et al.*, 1993c). Both isoenzymes found in the atopic subjects' monocytes are inhibited by Ro 20-1724 and rolipram, indicating that they are both members of the PDE4 family, but the proposed unique atopic monocyte isoenzyme exhibits distinct kinetic characteristics and is stimulated by Ca^{2+} and calmodulin (Holden *et al.*, 1989; Chan and Hanifin, 1993; Chan *et al.*, 1993c). This enzyme is also more potently inhibited by rolipram and several other PDE4 inhibitors – and, notably, by theophylline – than is the enzyme found in both normal and atopic monocytes (Chan and Hanifin, 1993).

The cause of the presence in atopic monocytes of this additional PDE4 is uncertain. Although a primary gene defect may lead to an hereditarily elevated PDE4 activity in the monocytes of the offspring of atopic parents (Heskel *et al.*, 1984), atopic dermatitis in young children is not invariably associated with elevated PDE activity (Coulson *et al.*, 1989). Furthermore, whereas the raised PDE activity persists in the monocytes of patients whose disease is in complete remission following topical steroid therapy (Holden and Yuen, 1989), suggesting that the elevation does not result from inflammation, more prolonged steroid treatment does result in a restoration of normal PDE levels (Holden *et al.*, 1989). It appears, therefore, that the increase in monocyte PDE4 might be a transient consequence of atopy or the associated inflammation.

Substances, such as inflammatory mediators and cytokines, that are released during allergic reactions are known to influence cAMP PDE activity in lymphoid and myeloid cells. Histamine, for example, increases PDE activity in normal monocytes to levels seen in atopic donors' cells, (Holden *et al.*, 1987), possibly via an elevation of cAMP which is known to lead to increased expression of specific PDE4 isoenzymes (see Chapter 11). Exposure of monocytes to IFNγ for 1 hour also leads to an increase in the cells' PDE activity and this effect is enhanced in the presence of IL-4 (Li *et al.*, 1992, 1993). Thus, local mediator and cytokine release at sites of allergic reactions might lead to an increase in monocyte PDE4 and a consequent increase in inflammatory cell activity, since increased PDE activity is likely to lead to the decreased responsiveness of atopic monocytes to agents, such as prostaglandin (PG)E_1, PGE_2 and β-adrenoceptor agonists, that stimulate cAMP synthesis (Safko *et al.*, 1981; Chan *et al.*, 1982; Grewe *et al.*, 1982). A similar down-regulation of AC-mediated cell responses can be induced both by inflammatory cytokines (Beckner and Farrar, 1986; van Oosterhout *et al.*, 1992; Zicari *et al.*, 1995) and by prolonged exposure to cAMP-elevating drugs (see Chapter 11).

Elevated PDE4 activity in monocytes has been suggested to result in lowered intracellular cAMP and consequently enhanced production of PGE_2, leading to suppression of IFNγ production by T_H1 lymphocytes and thereby removing a brake from T_H2 cytokine production (Chan *et al.*, 1993a). The resulting increase in IL-4 production by T_H2-cells would lead to B lymphocyte antibody class switching to increase the production of IgE (Chan *et al.*, 1993a,b). High levels of PGE_2 are released from atopic monocytes and suppression of this function by a cyclooxygenase inhibitor, indomethacin, leads to a very large increase in IFNγ from mixed mononuclear cells (monocytes and lymphocytes) *in vitro* that is not observed in cells from non-atopic donors (Chan *et al.*, 1993a). A highly significant negative correlation exists between PGE_2 and IFNγ levels in supernatants of atopic mononuclear cells in culture, whereas anti-CD3-stimulated IL-4 production correlates positively with the elevated monocyte PDE activity whose inhibition by Ro 20-1724 leads to suppression of IL-4 release (Chan *et al.*, 1993a).

Since the anomalous behaviour of monocytes from atopic subjects and of those treated with cAMP-elevating drugs are thought to result from an overexpression of PDE4, it is logical to assume that it could be corrected by PDE4 inhibitors. In fact, as well as suppressing the elevated IL-4 release from atopic mononuclear cells, selective inhibitors of this isoenzyme family both suppress the elevated histamine and IgE production in mixed leucocytes from atopic donors (Butler *et al.*, 1983; Cooper *et al.*, 1985) and reduce substantially the deficit in intracellular cAMP synthesis in monocyte-like U937 cells in response to PGE_2 (by reducing the accelerated cAMP breakdown) following prolonged exposure to the β-adrenoceptor agonist, salbutamol (Torphy *et al.*, 1995). A similar up-regulation of PDE4 following treatment with salbutamol or other cAMP elevators (rolipram or forskolin) is observed in a human keratinocyte cell line, where the resulting subsensitivity to β-adrenoceptor agonists is also reversed by selective PDE4 inhibitors (Tenor *et al.*, 1995b).

5. *Summary and Future Directions*

PDE4 inhibitors exhibit a range of actions upon cells of the immune system, both *in vitro* and *in vivo*, that suggest a possible therapeutic use for such drugs in inflammatory allergic diseases such as bronchial asthma and atopic dermatitis. Several major pharmaceutical companies have developed drugs of this class and many such compounds are currently undergoing clinical or advanced preclinical trials for use in these conditions (Table 7.2); information should soon be available to indicate whether these drugs are likely to fulfil their early promise and provide a new therapeutic approach to allergic disease.

The adverse actions of selective PDE4 inhibitors, as summarized in section 3, may present a substantial obstacle. The identification of subtypes of PDE4 (see

Chapters 1 and 11) and the recognition of the ability of inhibitors preferentially to suppress the activity of particular isoenzymes or specific subtypes thereof (Verghese *et al.*, 1995; see also Chapters 11 and 12), might present an opportunity for the development of more precisely targeted inhibitors that can act on immune cells without affecting the cells – such as area postrema neurones, renal tubules and gastric parietal cells – that mediate the drugs' side-effects.

The elevated PDE4 activity in atopic monocytes remains unexplained, although mechanisms for increasing expression of specific PDE4 subtypes in these cells have been clarified (Verghese *et al.*, 1995). The influence of this elevated activity on IgE production *in vitro*, if extrapolated to the *in vivo* situation, could have a major impact on allergic reactions; the, apparently unique, isoenzyme involved could present an important target for new anti-inflammatory drugs.

While the results of clinical trials with potent and selective PDE4 inhibitors are awaited, a new generation of drugs is emerging that inhibits differentially the subtypes of PDE4 and/or discriminates between states of the enzymes that occur at different stages of cell activation (see Chapters 11, 12 and 13). The potential of these drugs in the therapy of allergic disorders can, at present, only be imagined.

6. *References*

Anastassiou, E.D., Paliogianni, F., Balow, J.P., Yamada, H. and Boumpas, D.T. (1992). Prostaglandin E_2 and other cyclic AMP-elevating agents modulate IL-2 and IL-2R gene expression at multiple levels. J. Immunol. 148, 2845–2852.

Anderson, N. and Peachell, P.T. (1994). Effect of isoenzyme-selective inhibitors of phosphodiesterase (PDE) on human lung mast cells (HLMC). FASEB J. 8, A239. [Abstract]

Aoki, S., Boubekeur, K., Kristersson, A., Morley, J. and Sanjar, S. (1988). Is allergic airway hyperactivity of the guinea pig dependent on eosinophil accumulation in the lung? Br. J. Pharmacol. 94, 3658. [Abstract]

Ashton, M.J., Cook, D.C., Fenton, G, Karlsson, J.-A., Palfreyman, M.N., Raeburn, D., Ratcliffe, A.J., Souness, J.E., Thurairaitnam, S. and Vicker, N. (1994). Selective type IV phosphodiesterase inhibitors as anti-asthmatic agents: the synthesis and biological activities of 3-cyclopentyloxy-4-methoxybenzamides and analogues. J. Med. Chem. 37, 1696–1703.

Averill, L.E., Stein, R.L. and Kammer, G.M. (1988). Control of human T-lymphocyte interleukin-2 production by a cyclic AMP-dependent pathway. Cell. Immunol. 115, 88–99.

Baecker, P.A., Obernolte, R., Bach, C., Yee, C. and Shelton, E.R. (1995). Isolation of a cDNA encoding a human rolipram-sensitive cyclic AMP phosphodiesterase (PDE IVD). Gene 138, 253–256.

Baggiolini, M., Loetscher, P. and Moser, B. (1995). Interleukin-8 and the chemokine family. Int. J. Immunopharmacol. 17, 103–108.

Banner, K.H. and Page, C.P. (1995). Acute versus chronic administration of phosphodiesterase inhibitors on allergen-induced pulmonary cell influx in sensitized guinea-pigs. Br. J. Pharmacol. 114, 93–98.

Barnes, P.J. and Liew, F.Y. (1995). Nitric oxide and asthmatic inflammation. Immunol. Today 16, 128–130.

Barnette, M.S., Manning, C.D., Cieslinski, L.B., Burman, M., Christensen, S.B. and Torphy, T.J. (1995). The ability of phosphodiesterase IV inhibitors to suppress superoxide production in guinea-pig eosinophils is correlated with inhibition of phosphodiesterase IV catalytic activity. J. Pharmacol. Exp. Ther. 273, 674–679.

Beckner, S.K. and Farrar, W.L. (1986). Interleukin 2 modulation of adenylate cyclase: potential role of protein kinase C. J. Biol. Chem. 261, 3043–3047.

Bergstrand, H., Lundqvist, B. and Schurman, A. (1978). Rat mast cell high affinity cyclic nucleotide phosphodiesterases: separation and inhibitory effects of two anti-allergic agents. Mol. Pharmacol. 14, 848–855.

Betz, M. and Fox, B.S. (1991). Prostaglandin E_2 inhibits production of Th1 lymphokines but not of Th2 lymphokines. J. Immunol. 146, 108–113.

Black, E.W., Strada, S.J. and Thompson, W.J. (1988). Relationships between secretagogue-induced cAMP accumulation and acid secretion in elutriated rat gastric parietal cells. J. Pharmacol. Methods 20, 57–78.

Bolger, G., Michaeli, T., Martins, T., St. John, T., Steiner, B., Rodgers, L., Riggs, M., Wigler, M. and Ferguson, K. (1993). A family of human phosphodiesterases homologous to the dunce learning and memory gene products of *Drosophila melanogaster* are potential targets for antidepressant drugs. Mol. Cell. Biol. 13, 6558–6571.

Boucheron, J.A., Verghese, M.W., Irsula, O. and Stacy, L. (1991). Species differences in neutrophil superoxide modulation by cyclic nucleotide phosphodiesterase inhibitors. FASEB J. 5, A510. [Abstract]

Brunée, T., Engelstätter, R., Steinijans, V.W. and Kunkel, G. (1992). Bronchodilatory effect of inhaled zardaverine, a phosphodiesterase III and IV inhibitor, in patients with asthma. Eur. Respir. J. 5, 982–985.

Buckle, D.R., Arch, J.R.S., Connolly, B.J., Fenwick, A.E., Foster, K.A., Murray, K.J., Readshaw, S.A., Smallridge, M. and Smith, D.G. (1994). Inhibition of cyclic nucleotide phosphodiesterase by derivatives of 1,3-bis(cyclopropylmethyl)xanthine. J. Med. Chem. 37, 476–485.

Butler, J.M., Chan, S.C., Stevens, S. and Hanifin, J.M. (1983). Increased leukocyte histamine release with elevated cyclic AMP-phosphodiesterase activity in atopic dermatitis. J. Allergy Clin. Immunol. 71, 490–497.

Carpenter, D.O., Briggs, D.B., Knox, A.P. and Strominger, N. (1988). Excitation of area postrema neurons by transmitters, peptides, and cyclic nucleotides. J. Neurophysiol. 59, 358–369.

Chan, S.C. and Hanifin, J.M. (1993). Differential inhibitor effects on cyclic adenosine monophosphate-phosphodiesterase isoforms in atopic and normal leukocytes. J. Lab. Clin. Med. 121, 44–51.

Chan, S.C., Grewe, S., Stevens, S.R. and Hanifin, J.M. (1982). Functional desensitization due to stimulation of cyclic AMP phosphodiesterase in human mononuclear leukocytes. J. Cyclic Nucleotide Res. 8, 211–224.

Chan, S.C., Kim, J.-W., Henderson, W.R., Jr and Hanifin, J.M. (1993a). Altered prostaglandin E_2 regulation of cytokine production in atopic dermatitis. J. Immunol. 151, 3345–3352.

Chan, S.C., Li, S.-H. and Hanifin, J.M. (1993b). Increased interleukin-4 production by atopic mononuclear leukocytes correlates with increased cyclic adenosine monophosphate-phosphodiesterase activity and is reversible by phosphodiesterase inhibition. J. Invest. Dermatol. 100, 681–684.

Chan, S.C., Reifsnyder, D., Beavo, J.A. and Hanifin, J.M. (1993c). Immunochemical characterization of the distinct monocyte cyclic AMP-phosphodiesterase from patients with atopic dermatitis. J. Allergy Clin. Immunol. 91, 1179–1188.

Chen, C.N., Denome, S. and Davis, R.L. (1986). Molecular analysis of cDNA clones and the corresponding genomic coding sequences of the *Drosophila* dunce gene$^+$, the structural gene for cAMP phosphodiesterase. Proc. Natl Acad. Sci. USA 83, 9313–9317.

Cohan, V.L., Johnson, K.L., Breslow, R., Cheng, J.B. and Showell, H.J. (1992). PDE IV is the predominant PDE isozyme regulating chemotactic factor-mediated guinea-pig eosinophil functions in vitro. J. Allergy Clin. Immunol. 89, 663. [Abstract]

Colicelli, J., Birchmeier, C., Michaeli, T. O'Neill, K., Riggs, M. and Wigler, M. (1989). Isolation and characterization of a mammalian gene encoding a high affinity cAMP phosphodiesterase. Proc. Natl. Acad. Sci. USA 86, 3599–3603.

Columbo, M., Horowitz, E.M., McKenzie-White, J., Kagey-Sobotka, A. and Lichtenstein, L.M. (1993). Pharmacologic control of histamine release from human basophils induced by platelet-activating factor. Int. Arch. Allergy Immunol. 102, 383–390.

Conti, M. and Swinnen, J.V. (1990). Structure and function of the rolipram-sensitive, low K_m cyclic AMP phosphodiesterases: a family of highly related enzymes. In "Cyclic Nucleotide Phosphodiesterases: Structure, Regulation and Drug Action" (eds. M.D. Houslay and J. Beavo) pp. 243–266. Wiley, Chichester.

Conti, M., Swinnen, J.V., Tsikalas, K.E. and Jin, S.-L.C. (1992). Structure and regulation of the rat high affinity cyclic AMP phosphodiesterase. Adv. Cyclic Nucleotide Protein Phosphorylation Res. 25, 87–99.

Conti, M., Némoz, G., Sette, C. and Vicini, E. (1995). Recent progress in understanding the hormonal regulation of phosphodiesterases. Endocr. Rev. 16, 370–389.

Cooper, K.D., Kang, K., Chan, S.C. and Hanifin, J.M. (1985). Phosphodiesterase inhibition by Ro 20-1724 reduces hyper-IgE synthesis by atopic dermatitis cells in vitro. J. Invest. Dermatol. 84, 477–482.

Coulson, I.H., Duncan, S.N. and Holden, C.A. (1989). Peripheral blood mononuclear leukocyte cyclic adenosine monophosphate specific phosphodiesterase activity in atopic dermatitis. Br. J. Dermatol. 120, 607–612.

Crocker, I.C., Townley, R.G. and Khan, M.M. (1994). Phosphodiesterase inhibitors modulate cytokine secretion and proliferation. J. Allergy Clin. Immunol. 93, 286. [Abstract]

Crummey, A., Harper, G.P., Boyle, E.A. and Mangan, F.R. (1987). Inhibition of arachidonic acid-induced ear oedema as a model for assessing topical anti-inflammatory compounds. Agents Actions 20, 69–76.

Dahinden, C.A., Brunner, T., Krieger, M., Bischoff, S.C. and De Weck, A.L. (1993). Cytokines in allergic inflammation. Agents Actions Suppl. 43, 189–196.

Davis, R.L., Tadayasu, H., Eberwine, M. and Myers, J. (1989). Cloning and characterization of mammalian homologues of the *Drosophila* dunce$^+$ gene. Proc. Natl Acad. Sci. USA 86, 3604–3608.

Dent, G., Giembycz, M.A., Rabe, K.F. and Barnes, P.J. (1991). Inhibition of eosinophil cyclic nucleotide PDE activity and opsonised zymosan-induced respiratory burst by type IV-selective PDE inhibitors. Br. J. Pharmacol. 103, 1339–1346.

Dent, G., Giembycz, M.A., Rabe, K.F., Barnes, P.J. and Magnussen, H. (1993). Effects of selective phosphodiesterase inhibitors on human alveolar macrophage cyclic nucleotide hydrolysis and respiratory burst. Br. J. Pharmacol. 111, 74P. [Abstract]

Dent, G., Giembycz, M.A., Evans, P.M., Rabe, K.F. and Barnes, P.J. (1994). Suppression of human eosinophil respiratory burst and cyclic AMP hydrolysis by inhibitors of type IV phosphodiesterase: interaction with the beta adrenoceptor agonist albuterol. J. Pharmacol. Exp. Ther. 271, 1167–1174.

Derian, C.K., Santulli, R.J., Rao, P.E., Soloman, H.F. and Barrett, J.A. (1995). Inhibition of chemotactic peptide-induced neutrophil adhesion to vascular endothelium by cAMP modulators. J. Immunol. 154, 308–317.

Elliott, K.R.F. and Leonard, E.J. (1989). Interactions of formylmethionyl-leucyl-phenylalanine, adenosine, and phosphodiesterase inhibitors in human monocytes: effects on superoxide release, inositol phosphates and cAMP. FEBS Lett. 254, 94–98.

Elwood, W., Sun, J., Barnes, P.J., Giembycz, M.A. and Chung, K.F. (1995). Inhibition of allergen-induced lung eosinophilia by type-IV and combined type III- and IV-selective phosphodiesterase inhibitors in brown-Norway rats. Inflamm. Res. 44, 83–86.

Engels, P., Fichtel, K. and Lubbert, H. (1994). Expression and regulation of human and rat phosphodiesterase type IV isogenes. FEBS Lett. 350, 291–295.

Engels, P., Sullivan, M., Muller, T. and Lubbert, H. (1995). Molecular cloning and functional expression in yeast of a human cAMP-specific phosphodiesterase subtype (PDE IV-C). FEBS Lett. 358, 305–310.

Engerson, T., Legendre, J.L. and Jones, H.P. (1986). Calmodulin-dependency of human neutrophil phosphodiesterase. Inflammation 10, 31–35.

Epstein, P.M. and Hachisu, R. (1984). Cyclic nucleotide phosphodiesterase in normal and leukemic human lymphocytes and lymphoblasts. Adv. Cyclic Nucleotide Protein Phosphorylation Res. 16, 303–324.

Epstein, P.M., Strada, S.J., Sarada, K. and Thompson, W.J. (1982). Catalytic and kinetic properties of purified high affinity cyclic AMP phosphodiesterase from dog kidney. Arch. Biochem. Biophys. 218, 119–133.

Essayan, D.M., Huang, S.-K., Undem, B.J., Kagey-Sobotka, A. and Lichtenstein, L.M. (1994). Modulation of antigen- and mitogen-induced proliferative responses of peripheral blood mononuclear cells by non-selective and isozyme-selective cyclic nucleotide phosphodiesterase inhibitors. J. Immunol. 153, 3408–3416.

Essayan, D.M., Huang, S.-K., Kagey-Sobotka, A. and Lichtenstein, L.M. (1995). Effects of nonselective and isozyme selective cyclic nucleotide phosphodiesterase inhibitors on antigen-induced cytokine gene expression in peripheral blood mononuclear cells. Am. J. Respir. Cell Mol. Biol. 13, 692–702.

Foissier, M., Longchampt, F., Cogé, E. and Canet, E. (1995). *In vitro* effect of cAMP-modulating agents on antigen-induced interleukin-4 (IL-4) and IL-5 mRNA expression and protein production by spleen cells from BALB/c mice. Am. J. Respir. Crit. Care Med. 151, A827. [Abstract]

Frossard, N., Landry, Y., Pauli, G. and Ruckstuhl, M. (1981). Effects of cyclic AMP- and cyclic GMP-phosphodiesterase inhibitors on immunological release of histamine and on lung contraction. Br. J. Pharmacol. 73, 933–938.

Giembycz, M.A., Corrigan, C.J., Kay, A.B. and Barnes, P.J. (1994). Inhibition of CD4 and CD8 T-lymphocytes (T-LC) proliferation and cytokine secretion by isoenzyme-selective phosphodiesterase (PDE) inhibitors: correlation with intracellular cyclic AMP (cAMP) concentrations. J. Allergy Clin. Immunol. 93, 167. [Abstract]

Giembycz, M.A., Corrigan, C.J., Seybold, J., Newton, R. and Barnes, P.J. (1996). Identification of cyclic AMP phosphodiesterases 3, 4, and 7 in human $CD4^+$ and $CD8^+$ T-lymphocytes: role in regulating proliferation and the biosynthesis of interleukin-2. J. Immunol. [In press]

Goldberg, B.J., Lad, P.M. and Ghekiere, L. (1994). Phosphodiesterase activity and superoxide production in normal and asthmatic subjects neutrophils. J. Allergy Clin. Immunol. 93, 166. [Abstract]

Grady, P.G. and Thomas, L.L. (1986). Characterization of cyclic-nucleotide phosphodiesterase activities in resting and *N*-formyl-methionyl-leucyl-phenylalanine-stimulated human neutrophils. Biochim. Biophys. Acta 855, 282–293.

Greten, T.F., Eigler, A., Sinha, B., Moeller, J. and Endres, S. (1995). The specific type IV phosphodiesterase inhibitor rolipram differentially regulates the proinflammatory mediators TNF-α and nitric oxide. Int. J. Immunopharmacol. 17, 605–610.

Grewe, S., Chan, S.C. and Hanifin, J.M. (1982). Elevated leukocyte cyclic AMP-phosphodiesterase in atopic disease. J. Allergy Clin. Immunol. 70, 452–457.

Griswold, D.E., Webb, E.F., Breton, J., White, J.R., Marshall, P.J. and Torphy, T.J. (1993). Effect of selective phosphodiesterase type IV inhibitor, rolipram, on fluid and cellular phases of inflammatory responses. Inflammation 17, 333–344.

Hagiwara, M., Endo, T., Kanayama, T. and Hidaka, H. (1984). Effect of 1-(3-chloroanilo)-4-phenylphthalazine (MY 5445), a specific inhibitor of cyclic GMP phosphodiesterase on human platelet aggregation. J. Pharmacol. Exp. Ther. 228, 467–471.

Hallsworth, M.P., Giembycz, M.A., Barnes, P.J. and Lee, T.H. (1996). Cyclic AMP-elevating agents prolong or inhibit eosinophil survival depending on prior exposure to GM-CSF. Br. J. Pharmacol. 117, 79–86.

Hatzelmann, A., Tenor, H. and Schudt, C. (1995). Differential effects of non-selective and selective phosphodiesterase inhibitors on human eosinophil functions. Br. J. Pharmacol. 114, 821–831.

Heskel, N.S., Chan, S.C., Thiel, M.L., Stevens, S.R., Casperson, L.S. and Hanifin, J.M. (1984). Elevated umbilical cord blood leukocyte cyclic adenosine monophosphate-phosphodiesterase activity in children with atopic parents. J. Am. Acad. Dermatol. 11, 422–426.

Hilkens, C.M., Vermeulen, H., van Neerven, R.J., Snijdewint, F.G., Wierenga, E.A. and Kapsenberg, M.L. (1995). Differential modulation of T helper type 1 (Th1) and T helper type 2 (Th2) cytokine secretion by prostaglandin E_2 critically depends on interleukin-2. Eur. J. Immunol. 25, 59–63.

Ho, P.P.K., Wang, L.Y., Towner, R.D., Hayes, S.J., Pollock, D., Bowling, N., Wyss, V. and Ranetta, J.A. (1990). Cardiovascular effect and stimulus-dependent inhibition of superoxide generation from human neutrophils by tibenelast 5,6-diethoxybenzo(*b*)thiophene 2-carboxylic acid, sodium salt (LY 186655). Biochem. Pharmacol. 40, 2085–2092.

Holden, C.A. (1990). Atopic dermatitis: a defect of intracellular secondary messenger systems? Clin. Exp. Allergy 20, 131–136.

Holden, C.A. and Yuen, C.-T. (1989). Responses of mononuclear leukocyte cyclic adenosine monophosphate-phosphodiesterase activity to treatment with topical fluorinated steroid ointment in atopic dermatitis. J. Am. Acad. Dermatol. 21, 69–74.

Holden, C.A., Chan, S.C., Norris, S. and Hanifin, J.M. (1987). Histamine-induced elevation of cyclic AMP phosphodiesterase activity in human monocytes. Agents Actions 22, 36–42.

Holden, C.A., Yuen, C.-T. and Coulson, I.H. (1989). The effect of in vitro exposure to histamine on mononuclear leukocyte phosphodiesterase activity in atopic dermatitis. Clin. Exp. Dermatol. 14, 186–190.

Holtzman, M.J., Shannon, V.R., Masi, A.M. and Look, D.C. (1994). Mechanisms of airway inflammation and implications for treatment of asthma and bronchitis. In "Drugs and the Lung" (eds. C.P. Page and W.J. Metzger), pp. 549–583. Raven, New York.

Howell, R.E., Sickles, B.D. and Woeppel, S.L. (1993). Pulmonary antiallergic and bronchodilator effects of isozyme-selective phosphodiesterase inhibitors in guinea-pigs. J. Pharmacol. Exp. Ther. 264, 609–615.

Hughes, B., Holbrook, M., Allen, R., Owens, R., Perry, M., Warrellow, G., Howat, D., Bloxham, D. and Higgs, G. (1995). Suppression of antigen-induced airway responses and bronchial hyperresponsiveness by CDP840, a novel stereo-selective inhibitor of phosphodiesterase type IV. Br. J. Pharmacol. 116, 6P. [Abstract]

Ichimura, M. and Kase, H. (1993). A new cyclic nucleotide phosphodiesterase isozyme expressed in the T-lymphocyte cell lines. Biochem. Biophys. Res. Commun. 193, 985–990.

Iwatsuki, K., Horiuci, A., Ren, L.M. and Chiba, S. (1991). Effects of the cyclic nucleotide phosphodiesterase inhibitors, rolipram, 3-isobutyl-1-methylxanthine, amrinone and zaprinast, on pancreatic exocrine secretion in dogs. Eur. J. Pharmacol. 209, 63–68.

Kammer, G.M. (1988). The adenylate cyclase-cyclic AMP-protein kinase A pathway and regulation of the immune response. Immunology Today 9, 222–229.

Kings, M.A., Chapman, I., Kristersson, A., Sanjar, S. and Morley, J. (1991). Human recombinant lymphokines and cytokines induce pulmonary eosinophilia in the guinea-pig which is inhibited by ketotifen and AH 21-132. Int. Arch. Allergy Appl. Immunol. 91, 354–361.

Kips, J.C., Joos, G.F., Peleman, R.A. and Pauwels, R.A. (1993). The effect of zardaverine, an inhibiter of phosphodiesterase isoenzymes III and IV, on endotoxin-induced airway changes in rats. Clin. Exp. Allergy 23, 518–523.

Lee, H.-J., Koyano-Nakagawa, N., Naito, Y., Nishada, J., Arai, K.-I. and Yokota, T. (1993). Cyclic AMP activates the IL-5 promoter synergistically with phorbol ester through the signaling pathway involving protein kinase A in mouse thymoma line EL-4. J. Immunol. 151, 6135–6142.

Lewis, G.M., Caccese, R.G., Heaslip, R.L. and Bansbach, C.C. (1993). Effects of rolipram and CI-930 on IL-2 mRNA transcription in human Jurkat cells. Agents Actions 39, C89–C92.

Li, S.-H., Chan, S.C., Toshitani, A., Leung, D.Y.M. and Hanifin, J.M. (1992). Synergistic effects of interleukin 4 and interferon-gamma on monocyte phosphodiesterase activity. J. Invest. Dermatol. 99, 65–70.

Li, S.-H., Chan, S.C., Kramer, S.M. and Hanifin, J.M. (1993). Modulation of leukocyte cyclic AMP phosphodiesterase activity by recombinant interferon-γ: evidence for a differential effect on atopic monocytes. J. Interferon. Res. 13, 197–202.

Livi, G.P., Kmetz, P., McHale, M.M., Cieslinksi, L.B., Sathe, G.M., Taylor, D.P., Davis, R.L., Torphy, T.J. and Balcarek, J.M. (1990). Cloning and expression of a human low-K_m rolipram-sensitive cyclic AMP phosphodiesterase. Mol. Cell. Biol. 10, 2678–2686.

Longchampt, M., Pennel, L. and Canet, E. (1995). *In vivo* effect of rolipram on antigen-induced interleukin-4 (IL-4) and interleukin-5 (IL-5) production in bronchoalveolar lavage of sensitized BALB/c mice. Am. J. Respir. Crit. Care Med. 151, A827. [Abstract]

Lycke, N.Y. (1993). Cholera toxin promotes B cell isotype switching by two different mechanisms: cAMP induction augments germ-line Ig H-chain RNA transcripts whereas membrane ganglioside GM1-receptor binding enhances later events in differentiation. J. Immunol. 150, 4810–4821.

Lycke, N.Y., Severinson, E. and Strober, W. (1990). Cholera toxin acts synergistically with IL-4 to promote IgG_1 switch differentiation. J. Immunol. 145, 3316–3324.

MacPhee, C.H., Harrison, S.A. and Beavo, J. (1986). Immunological identification of the major platelet low K_m cAMP phosphodiesterase: probable target for antithrombotic agents. Proc. Natl Acad. Sci. USA 83, 6660–6663.

Marcoz, P., Prigent, A.F., Lagarde, M. and Némoz, G. (1993). Modulation of rat thymocyte proliferative response through the inhibition of different cyclic nucleotide phosphodiesterase isoforms by means of selective inhibitors and cGMP-elevating agents. Mol. Pharmacol. 44, 1027–1035.

Maruo, H., Tanimoto, Y., Bewtra, A.K. and Townley, R.G. (1994). Effect of phosphodiesterase IV inhibitor (WAY-PDA-641) on PAF-induced superoxide generation from human eosinophils. Am. J. Respir. Crit. Care Med. 149, A114. [Abstract]

McLaughlin, M.M., Cieslinski, L.B., Burman, M., Torphy, T.J. and Livi, G.P. (1993). A low-K_m, rolipram-sensitive, cAMP phosphodiesterase from human brain: cloning and expression of cDNA, biochemical characterization of recombinant protein, and tissue distribution of mRNA. J. Biol. Chem. 268, 6470–6476.

Miyamoto, K.-i., Kurita, M., Ohmae, S., Sakai, R., Sanae, F. and Takagi, K. (1994). Selective tracheal relaxation and phosphodiesterase-IV inhibition by xanthine derivatives. Eur. J. Pharmacol. 267, 317–322.

Monaco, L., Vicini, E. and Conti, M. (1994). Structure of two rat genes coding for closely related rolipram-sensitive cAMP phosphodiesterases: multiple mRNA variants originate from alternative splicing and multiple start sites. J. Biol. Chem. 269, 347–357.

Munoz, E., Zubiaga, A.M., Merrow, M., Sauter, N.P. and Huber, B.T. (1990). Cholera toxin discriminates between T helper 1 and 2 cell receptor-mediated activation: role of cyclic AMP in T cell proliferation. J. Exp. Med. 172, 95–103.

Newsholme, S.J. and Schwartz, L. (1993). cAMP-specific phosphodiesterase inhibitor, rolipram, reduces eosinophil infiltration evoked by leukotrienes or by histamine in guinea-pig conjunctiva. Inflammation 17, 25–31.

Nicholson, C.D., Jackman, S.A. and Wilke, R. (1989). The ability of denbufylline to inhibit cyclic nucleotide phosphodiesterase and its affinity for adenosine receptors and the adenosine re-uptake site. Br. J. Pharmacol. 97, 889–897.

Nicholson, C.D., Shahid, M., Bruin, J., Barron, E., Spiers, I., de Boer, J., van Amsterdam, R.G.M., Zaagsma, J., Kelly, J.J., Dent, G., Giembycz, M.A. and Barnes, P.J. (1995). Characterization of Org 20241, a combined phosphodiesterase IV/III cyclic nucleotide phosphodiesterase inhibitor for asthma. J. Pharmacol. Exp. Ther. 274, 678–687.

Nielson, C.P., Vestal, R.E., Sturm, R.J. and Heaslip, R.J. (1990). Effects of selective phosphodiesterase inhibitors on the polymorphonuclear leukocyte respiratory burst. J. Allergy Clin. Immunol. 86, 801–808.

Novak, J.P. and Rothenberg, E.V. (1990). Cyclic AMP inhibits induction of interleukin 2 but not of interleukin 4 in T cells. Proc. Natl Acad. Sci. USA 87, 9353–9357.

Obernolte, R., Bhakta, S., Alvarez, R., Bach, C., Zuppan, P., Mulkins, M., Jarnagin, K. and Shelton, E.R. (1993). The cDNA of a human lymphocyte cyclic-AMP phosphodiesterase (PDE IV) reveals a multigene family. Gene 129, 239–247.

Okonogi, K., Gettys, T.W., Uhing, R.J., Tarry, C., Adams, P.O. and Prpic, V. (1991). Inhibition of prostaglandin E_2-stimulated cAMP accumulation by lipopolysaccharide in murine peritoneal macrophages. J. Biol. Chem. 266, 10305–10312.

Ortiz, J.L., Cortijo, J., Valles, J.M., Bou, J. and Morcillo, E.J. (1992). Rolipram inhibits PAF-induced airway microvascular leakage in guinea-pig: a comparison with milrinone and theophylline. Fundam. Clin. Pharmacol. 6, 247–249.

Palfreyman, M.N. (1995). Phosphodiesterase type IV inhibitors as antiinflammatory agents. Drugs Fut. 20, 793–804.

Paul-Eugène, N., Kolb, J.P., Calenda, A., Gordon, J., Kikutani, H., Kishimoto, T., Mencia-Huerta, J.M., Braquet, P. and Dugas, B. (1993). Functional interaction between β_2-adrenoceptor agonists and interleukin-4 in the regulation of CD23 expression and release and IgE production in human. Mol. Immunol. 30, 157–164.

Peachell, P.T., Undem, B.J., Schleimer, R.P., MacGlashan, D.W., Jr, Lichtenstein, L.M., Cieslinski, L.B. and Torphy, T.J. (1992). Preliminary identification and role of phosphodiesterase isozymes in human basophils. J. Immunol. 148, 2503–2510.

Phipps, R.P., Roper, R.L. and Stein, S.H. (1990). Regulation of B-cell tolerance and triggering by macrophages and lymphoid dendritic cells. Immunol. Rev. 117, 135–158.

Raeburn, D. and Karlsson, J.-A. (1991). Effects of isoenzyme-selective inhibitors of cyclic nucleotide phosphodiesterase on microvascular leak in guinea-pig airways. J. Pharmacol. Exp. Ther. 267, 1147–1152.

Raeburn, D., Woodman, V., Buckley, G. and Karlsson, J.-A. (1991). Inhibition of PAF-induced microvascular leakage in the guinea-pig *in vivo*: the effects of rolipram and theophylline. Eur. Respir. J. 4 (Suppl. 14), 590s. [Abstract]

Raeburn, D., Souness, J.E., Tomkinson, A. and Karlsson, J.-A. (1993). Isozyme-selective cyclic nucleotide phosphodiesterase inhibitors: biochemistry, pharmacology and therapeutic potential in asthma. Prog. Drugs Res. 40, 9–31.

Raeburn, D., Underwood, S.L., Lewis, S.A., Woodman, V.R., Battram, C.H., Tomkinson, A., Sharma, S., Jordan, R., Souness, J.E., Webber, S.E. and Karlsson, J.-A. (1994). Anti-inflammatory and bronchodilator properties of RP 73401, a novel and selective phosphodiesterase IV inhibitor. Br. J. Pharmacol. 113, 1423–1431.

Robicsek, S.A., Krzanowski, J.J., Szentivanyi, A. and Polson, J.B. (1989). High pressure liquid chromatography of cyclic AMP phosphodiesterase from purified human T-lymphocytes. Biochem. Biophys. Res. Commun. 163, 554–560.

Robicsek, S.A., Blanchard, D.K., Djeu, J.Y., Krzanowski, J.J., Szentivanyi, A. and Polson, J.B. (1991). Multiple high-affinity cyclic AMP phosphodiesterases in human T-lymphocytes. Biochem. Pharmacol. 42, 869–877.

Roper, R.L., Conrad, D.H., Brown, D.M., Warner, C.L. and Phipps, R.P. (1990). Prostaglandin E_2 promotes IL-4-induced IgE and IgG_1 synthesis. J. Immunol. 145, 2644–2651.

Rott, O., Cash, E. and Fleischer, B. (1993). Phosphodiesterase inhibitor pentoxifylline, a selective suppressor of T-helper type 1- but not type 2-associated lymphokine production, prevents induction of experimental autoimmune encephalomyelitis in Lewis rats. Eur. J. Immunol. 23, 1745–1751.

Safko, M.J., Chan, S.C., Cooper, K.D. and Hanifin, J.M. (1981). Heterologous desensitization of leukocytes: a possible mechanism of beta-adrenoceptor blockade in atopic dermatitis. J. Allergy Clin. Immunol. 68, 218–225.

Sakai, R., Konno, K., Yamamoto, Y., Sanae, F., Takagi, K., Hasegawa, T., Iwasaki, N., Kakiuchi, M., Kato, H. and Miyamoto, K.-i. (1992). Effects of alkyl substitutions of xanthine skeleton on bronchodilation. J. Med. Chem. 35, 4039–4044.

Sanjar, S., Aoki, S., Boubekeur, K., Burrows, L., Colditz, I., Chapman, I. and Morley, J. (1989). Inhibition of PAF-induced eosinophil accumulation in pulmonary airways of guinea-pigs by anti-asthma drugs. Jpn. J. Pharmacol. 51, 167–172.

Sanjar, S., Aoki, S., Boubekeur, K., Chapman, I.D., Smith, D., Kings, M.A. and Morley, J. (1990a). Eosinophil accumulation in pulmonary airways of guinea-pigs induced by exposure to an aerosol of platelet-activating factor: effect of anti-asthma drugs. Br. J. Pharmacol. 99, 267–272.

Sanjar, S., Aoki, S., Kristersson, A., Smith, D. and Morley, J. (1990b). Antigen challenge induces pulmonary airway eosinophil accumulation and airway hyperreactivity in sensitized guinea-pigs: the effect of anti-asthma drugs. Br. J. Pharmacol. 99, 679–686.

Schudt, C., Winder, B., Forderkuntz, S., Hatzelmann, A. and Ullrich, V. (1991a). Influence of selective phosphodiesterase inhibitors on human neutrophil functions and levels of cAMP and Ca_i. Naunyn-Schmiedebergs Arch. Pharmacol. 344, 682–690.

Schudt, C., Winder, B., Litze, M., Kilian, U. and Beume, R. (1991b). Zardaverine: a cyclic AMP specific PDEIII/IV inhibitor. Agents Actions 34, 161–177.

Schudt, C., Winder, B., Muller, B. and Ukena, D. (1991c). Zardaverine as a selective inhibitor of phosphodiesterase isoenzymes. Biochem. Pharmacol. 42, 153–162.

Schudt, C., Tenor, H., Loos, U., Mallmann, P., Szamel, M. and Resch, K. (1993). Effect of selective phosphodiesterase (PDE) inhibitors on activation of human macrophages and lymphocytes. Eur. Respir. J. 6 (Suppl. 17), 367s. [Abstract]

Seldon, P.M., Barnes, P.J., Meja, K. and Giembycz, M.A. (1995). Suppression of lipopolysaccharide-induced tumor necrosis factor-α generation from human peripheral blood monocytes by inhibitors of phosphodiesterase 4: interaction with stimulants of adenylyl cyclase. Mol. Pharmacol. 48, 747–757.

Semmler, J., Wachtel, H. and Endres, S. (1993). The specific type IV phosphodiesterase inhibitor rolipram suppresses tumor necrosis factor-α production by human mononuclear cells. Int. J. Immunopharmacol. 15, 409–413.

Souness, J.E., Carter, C.M., Diocee, B.K., Hassall, G.A., Wood, L.J. and Turner, N.C. (1991). Characterization of guinea-pig eosinophil phosphodiesterase activity: assessment of its involvement in regulating superoxide generation. Biochem. Pharmacol. 42, 937–945.

Souness, J.E., Villamil, M.E., Scott, L.C., Tomkinson, A., Giembycz, M.A. and Raeburn, D. (1994). Possible role of cyclic AMP phosphodiesterases in the actions of ibudilast on eosinophil thromboxane generation and airways smooth muscle tone. Br. J. Pharmacol. 111, 1081–1088.

Souness, J.E., Maslen, C., Webber, S., Foster, M., Raeburn, D., Palfreyman, M.N., Ashton, M.J. and Karlsson, J.-A. (1995). Suppression of eosinophil function by RP 73401, a potent and selective inhibitor of cyclic AMP-specific phosphodiesterase: comparison with rolipram. Br. J. Pharmacol. 115, 39–46.

Sturgess, I. and Searle, G.F. (1990). The acute toxic effect of the phosphodiesterase inhibitor rolipram on plasma osmolality. Br. J. Clin. Pharmacol. 29, 369–370.

Sullivan, G.W., Carper, H.T. and Mandell, G.L. (1995). The specific type IV phosphodiesterase inhibitor rolipram combined with adenosine reduces tumor necrosis factor-α-primed neutrophil oxidative activity. Int. J. Immunopharmacol. 17, 793–803.

Sullivan, M., Egerton, M., Shakur, Y., Marquardsen, A. and Houslay, M. (1994). Molecular cloning and expression, in both COS-1 cells and *S. cerevisiae*, of a human cytosolic type-IVA, cyclic AMP specific phosphodiesterase (hPDE-IVA-h6.1). Cell. Signal. 6, 793–812.

Svensjo, E., Andersson, K.E., Bouskela, E., Cyrino, F.Z.G.A. and Lindgren, S. (1992). Effects of two vasodilatory phosphodiesterase inhibitors on bradykinin induced permeability increase in the hamster. Int. J. Microcirc. 11, A179. [Abstract]

Swinnen, J.V., Joseph, D.R. and Conti, M. (1989a). Molecular cloning of rat homologues of the *Drosophila melanogaster* dunce cAMP phosphodiesterase: evidence for a family of genes. Proc. Natl Acad. Sci. USA 86, 5325–5329.

Swinnen, J.V., Joseph, D.R. and Conti, M. (1989b). The mRNA encoding a high affinity cAMP phosphodiesterase is regulated by hormones and cAMP. Proc. Natl Acad. Sci. USA 86, 8197–8201.

Tanimoto, Y., Maruo, H., Hopp, R.J., Bewtra, A.K. and Townley, R.G. (1994). Effects of phosphodiesterase IV inhibitor (WAY-PDA-641) on human eosinophil and neutrophil migration in vitro. Am. J. Respir. Crit. Care Med. 149, A113. [Abstract]

Tenor, H., Hatzelmann, A., Kupferschmidt, R., Stanciu, L., Djukanović, R., Schudt, C., Wendel, A., Church, M.K. and Shute, J.K. (1995a). Cyclic nucleotide phosphodiesterase isoenzyme activities in human alveolar macrophages. Clin. Exp. Allergy 25, 625–633.

Tenor, H., Hatzelmann, A., Wendel, A. and Schudt, C. (1995b). Identification of phosphodiesterase IV activity and its cyclic adenosine monophosphate-dependent up-regulation in a human keratinocyte cell line (HaCaT). J. Invest. Dermatol. 105, 70–74.

Tenor, H., Shute, J.K., Church, M.K., Hatzelmann, A. and Schudt, C. (1995c). Inhibition of human peripheral blood eosinophil LTC_4 production by PDE inhibitors. Eur. Respir. J. 8 (Suppl. 19), 9s. [Abstract]

Tenor, H., Staniciu, L., Schudt, C., Hatzelmann, A., Wendel, A., Djukanović, R., Church, M.K. and Shute, J.K. (1995d). Cyclic nucleotide phosphodiesterases from purified human $CD4^+$ and $CD8^+$ T lymphocytes. Clin. Exp. Allergy 25, 616–624.

Texeira, M.M., Rossi, A.G., Williams, T.J. and Hellewell, P.G. (1994). Effects of phosphodiesterase isoenzyme inhibitors on cutaneous inflammation in the guinea-pig. Br. J. Pharmacol. 112, 332–340.

Thanhauser, A., Reiling, N., Bohle, A., Toellner, K.-M., Duchrow, M., Scheel, D., Schluter, C., Ernst, M., Flad, H.-D. and Umler, A.J. (1993). Pentoxifylline: a potent inhibitor of IL-2 and IFNγ biosynthesis and BCG-induced cytotoxicity. Immunology 80, 151–156.

Thompson, W.J., Ross, C.P., Pledger, W.J., Strada, S.J., Banner, R.L. and Hersh, E.M. (1976). Cyclic adenosine 3′:5′-monophosphate phosphodiesterase: distinct forms in human lymphocytes and monocytes. J. Biol. Chem. 251, 4922–4929.

Thompson, W.J., Epstein, P.M. and Strada, S.J. (1979). Purification and characterization of a high affinity cyclic adenosine monophosphate phosphodiesterase from dog kidney. Biochemistry 18, 5228–5237.

Torphy, T.J. and Undem, B.J. (1991). Phosphodiesterase inhibitors: new opportunities for the treatment of asthma. Thorax 46, 512–523.

Torphy, T.J., De Wolf, W.E., Green, D.W. and Livi, G.P. (1993). Biochemical characteristics and cellular regulation of phosphodiesterase IV. Agents Actions Suppl. 43, 51–71.

Torphy, T.J., Zhou, H. -L., Foley, J.J., Sarau, H.M., Manning, C.D. and Barnette, M.S. (1995). Salbutamol up-regulates PDE4 activity and induces a heterologous desensitization of U937 cells to prostaglandin E_2: implications for the therapeutic use of β-adrenoceptor agonists. J. Biol. Chem. 270, 23598–23604.

Townley, R.G. (1993). Elevated cyclic AMP-phosphodiesterase in atopic disease: cause or effect? J. Lab. Clin. Med. 121, 15–17.

Truong, V.H. and Muller, T. (1994). Isolation, biochemical characterization and N-terminal sequence of rolipram-sensitive cAMP phosphodiesterase from human mononuclear leukocytes. FEBS Lett. 353, 113–118.

Tsuruta, L., Lee, H.-J., Masuda, E.S., Koyano-Nakagawa, N., Arai, N., Arai, K. and Yokota, T. (1995). Cyclic AMP inhibits expression of the IL-2 gene through the nuclear factor of activated T-cells (NF-AT) site, and transfection of NF-AT cDNAs abrogates the sensitivity of EL-4 cells to cyclic AMP. J. Immunol. 154, 5255–5264.

Turner, C.R., Andresen, C.J., Smith, W.B. and Watson, J.W. (1994). Effects of rolipram on responses to acute and chronic antigen exposure in monkeys. Am. Respir. Crit. Care Med. 149, 1153–1159.

Turner, N.C., Wood, L.J., Burns, F.M., Gueremy, T. and Souness, J.E. (1993). The effect of cyclic AMP and cyclic GMP phosphodiesterase inhibitors on the superoxide burst of guinea-pig peritoneal macrophages. Br. J. Pharmacol. 108, 876–883.

Underwood, D.C., Osborn, R.R., Novak, L.B., Matthews, J.K., Newsholme, S.J., Undem, B.J., Hand, J.M. and Torphy, T.J. (1993). Inhibition of antigen-induced bronchoconstriction and eosinophil infiltration in the guinea-pig by the cyclic AMP-specific phosphodiesterase inhibitor, rolipram. J. Pharmacol. Exp. Ther. 266, 306–313.

Valette, L., Prigent, A.F., Némoz, G., Anker, G., Macovschi, O. and Lagarde, M. (1990). Concanavalin A stimulates the rolipram-sensitive isoforms of cyclic nucleotide phosphodiesterase in rat thymic lymphocytes. Biochem. Biophys. Res. Commun. 169, 864–872.

van der Poow-Kraan, T., van Kooten, C., Rensink, I. and Aarden, L. (1992). Interleukin (IL)-4 production by human T cells: differential regulation of IL-4 vs IL-2 production. Eur. J. Immnunol. 22, 1237–1241.

van Oosterhout, A.J.M., Stam, W.B., Vanderschueren, R.G.R.A. and Nijkamp, F.P. (1992). Effects of cytokines on β-adrenoceptor function of human peripheral blood mononuclear cells and guinea pig trachea. J. Allergy Clin. Immunol. 90, 340–348.

van Tits, L.J.H., Michel, M.C., Motulsky, H.J., Maisel, A.S. and Brodde, O.-E. (1991). Cyclic AMP counteracts mitogen-induced inositol phosphate generation and increases in intracellular Ca^{2+} concentrations in human lymphocytes. Br. J. Pharmacol. 103, 1288–1294.

Verghese, M.W., McConnell, R.T., Lenhard, J.M., Hamacher, L. and Jin, S.-L.C. (1995). Regulation of distinct cyclic AMP-specific phosphodiesterase (phosphodiesterase type 4) isozymes in human monocytic cells. Mol. Pharmacol. 47, 1164–1171.

Watanabe, S., Yssel, H., Harada, Y. and Arai, K. (1994). Effects of prostaglandin E_2 on Th0-type T cell clones: modulation of functions of nuclear proteins involved in cytokine production. Int. Immunol. 6, 523–532.

White, J.R., Torphy, T.J., Christensen, S.B., Lee, J.A. and Mong, S. (1990). Purification and characterization of the rolipram sensitive, low-K_m phosphodiesterase from human monocytes. FASEB J. 4, A1987. [Abstract]

Wright, C.D., Kuipers, P.J., Kobylarz-Singer, D., Devall L.J., Klinkefus, B.A. and Weishaar, R.E. (1990). Differential inhibition of neutrophil functions: role of cyclic AMP-specific, cyclic GMP-insensitive phosphodiesterase. Biochem. Pharmacol. 40, 699–707.

Zeller, E., Stief, H.J., Pflug, B. and Sastre-y-Hernandez, M. (1984). Results of a phase II study of the antidepressant effect of rolipram. Pharmacopsychiatry 17, 188–190.

Zicari, A., Lipari, M., Di Renzo, L., Salerno, A., Losardo, A. and Pontieri, G.M. (1995). Stimulation of macrophages with IFNγ or TNFα shuts off the suppressive effect played by PGE_2. Int. J. Immunopharmacol. 17, 779–786.

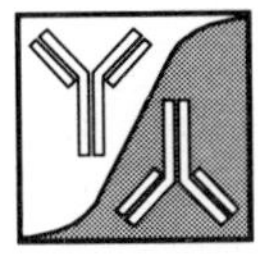

8. Inhibition of Phosphodiesterase Isoenzymes and Cell Function by Selective PDE5 Inhibitors

Paul J. Silver

1. Introduction 127
2. Scientific Rationale for PDE5 Inhibitors 127
3. PDE5 Inhibition and Vasorelaxation 128
4. Potential Therapeutic Applications of PDE5 Inhibitors 130
5. Additional Indications for PDE5 Inhibitors 131
6. Newer PDE5 Inhibitors 131
7. Combination Inhibitors 132
8. Summary 132
9. Acknowledgement 132
10. References 132

1. Introduction

In the past decade, numerous selective inhibitors of cardiovascular cAMP PDE3 have been developed as therapeutic agents. PDE3 inhibitors increase cAMP in target tissues, produce positive inotropic/lusitropic/vasodilator effects *in vivo* and are used clinically in the acute management of heart failure (Weishaar *et al.*, 1987; Silver, 1989; Goldstein and Fleming, 1993; see also Chapter 6). PDE4 inhibitors, which also serve to increase cAMP, are now in development for the treatment of asthma and other inflammatory conditions (see reviews by Christensen and Torphy, 1994; Nicholson and Shahid, 1994; see also Chapter 7). However, relatively little has been reported on the design, synthesis and testing of inhibitors of PDE5. Because PDE5 inhibitors should potentiate the effects of nitric oxide (NO) or atrial natriuretic factor (ANF) and increase cGMP in target tissues, they should offer a different profile of pharmacological activity compared to PDE3 or PDE4 inhibitors which increase cAMP in target tissues.

The purpose of this chapter is to review the potential therapeutic applications and scientific rationale for PDE5 inhibitors, discuss the preclinical enzymology, cellular biology and *in vivo* pharmacology that support these potential applications, and preview some of the PDE5 inhibitors which have been described in the literature. In addition, a brief discussion of combined PDE1/PDE5, PDE4/PDE5 and PDE3/PDE5 inhibitors is included.

2. Scientific Rationale for PDE5 Inhibitors

As with cAMP and PDE3 or PDE4 inhibitors, the therapeutic rationale for PDE5 inhibitors derives solely from the ability to potentiate or perpetuate the activity of increased cGMP levels brought about by the activation of guanylate cyclase (GC). Simplistically, PDE5 must be in the target cell or tissues and cGMP must play an important role in regulating cellular function in that tissue. Since cGMP has long been recognized as having smooth muscle relaxant activity as well as anti-

Phosphodiesterase Inhibitors
ISBN 0-12-210720-9

aggregatory activity in platelets, a logical place for therapeutic intervention involves these two cell types. Moreover, PDE5 (cGMP-PDE) was originally identified in lung, aorta and platelets (Coquil *et al.*, 1980; Francis *et al.*, 1980).

The basis for activity of PDE5 inhibition is shown schematically in Fig. 8.1. Activation of GC in a target cell can occur via local regulation by NO. Typically, NO is produced by NO synthase, is likely coupled with an endogenous protein containing sulfhydryl groups (Stamler *et al.*, 1992) and increases cGMP via activation of soluble GC. Similarly, exogenously applied NO donors, such as the nitrovasodilators, work at the target cell via initial conversion to NO and subsequent similar activation of soluble GC. Another endogenous regulator of cGMP is the hormone ANF, which activates particulate GC in target cells.

After either soluble or particulate GC is activated, intracellular cGMP levels increase. Cyclic GMP then produces its cellular effect, mainly via activation of cGMP-dependent protein kinase (PKG) and subsequent enhanced phosphorylation of key phosphoprotein substrates. Increased cGMP also activates cGMP-PDE, which hydrolyses cGMP to the inactive 5′-GMP (Fig. 8.1).

Thus, a selective PDE5 inhibitor should potentiate or perpetuate the effects of NO, nitrovasodilators or ANF in target cells. Some examples of target cells and the effects of increased cGMP are: vascular smooth muscle, in which vasodilation is effected and cell proliferation repressed by cGMP; airway smooth muscle, in which bronchodilation can be effected by cGMP; platelets, whose aggregation is inhibited by cGMP; renal collecting duct cells, where inhibition of sodium resorption is mediated by cGMP.

3. *PDE5 Inhibition and Vasorelaxation*

Two types of PDE which hydrolyse cGMP are present in smooth muscle (Lugnier *et al.*, 1986; Silver and Harris, 1988). Both have K_m values for cGMP in the 0.2–1 μM range. One isoenzyme is a member of the PDE1 family, which is characterized by an increased *V*max in the presence of activation by Ca^{2+} and calmodulin (CaM). Vinpocetine, a cerebral vasodilator, has been reported to be a selective inhibitor of PDE1, but this is questionable given the other known mechanisms of this compound (Souness *et al.*, 1989).

PDE5 is the second cGMP-PDE found in most smooth muscles. This isoenzyme is related to the cGMP-PDE found in rod and cone photoreceptors (now classified as PDE6) and is not regulated by CaM (Beavo, 1988). Zaprinast, also known as M&B 22,948, is a selective inhibitor of PDE5 that has been extensively used in the literature as a reference agent (Fig. 8.2). Potency for inhibition of PDE5 is in the 0.3–1 μM range (IC_{50} value) but selectivity vs. PDE1 (IC_{50} value

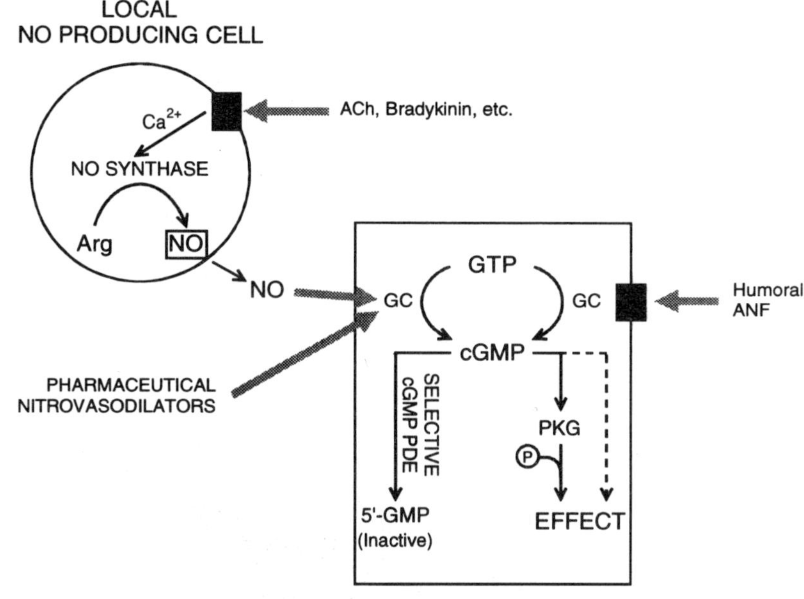

Figure 8.1 Schematic representation for the activity of cGMP-PDE inhibitors. Cyclic GMP levels are increased in the target cell via activation of soluble guanylate cyclase (GC) by local NO producing cells or exogenous NO donors such as nitrovasodilators. Particulate GC is activated via ANF binding to a receptor. Increased intracellular levels of cGMP produce their effect via activation of cGMP-dependent protein kinase (PKG). Intracellular cGMP is also degraded by cGMP-specific PDE. Inhibition of this PDE potentiates or prolongs the activity of these GC activators.

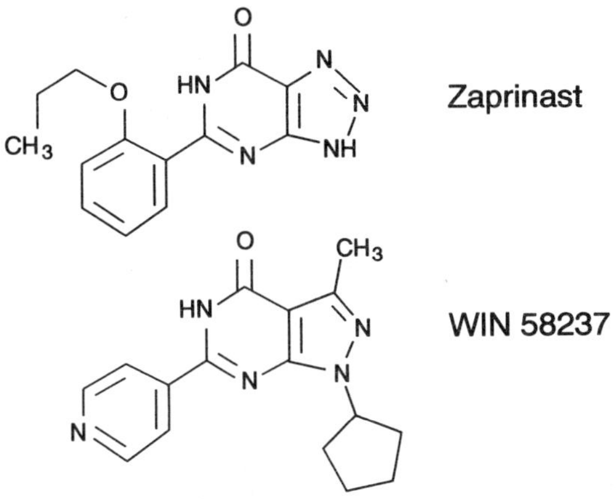

Zaprinast

WIN 58237

Figure 8.2 Chemical structures of zaprinast (M&B 22,948) and WIN 58237.

10–30 μM) is not great. A more recent set of PDE5 inhibitors has been described (Silver *et al.*, 1994). A prototypical compound from this chemical series is WIN 58237 (Fig. 8.2), which is a competitive inhibitor for PDE5 (K_i = 170 nM) but with a similar lack of selectivity over PDE1 (IC_{50} = 1.5 μM). A fuller discussion of more potent analogues of WIN 58237, as well as other PDE5 inhibitors such as E-4021, follows later in this chapter.

Both zaprinast and WIN 58237 have been used as tools to elucidate the role of cGMP-PDE in vasorelaxation and possibly to predict therapeutic applications. To demonstrate cellular activity in vascular tissue, both PDE5 inhibitors were examined for their ability to potentiate the activity of either a nitrovasodilator (sodium nitroprusside, SNP) or ANF in vascular smooth muscle rings (Fig. 8.3) (Martin *et al.*, 1986; Harris *et al.*, 1989). In these experiments, the endothelium is typically removed to eliminate the influence of NO. Both zaprinast and WIN 58237 potentiate the activity of SNP and ANF at concentrations which inhibit PDE5. WIN 58237 is more potent than zaprinast in intact cells, probably as a result of greater cellular penetration.

A second type of experiment to demonstrate PDE5 activity is to show that the putative PDE5 inhibitor potentiates the ability of a GC activator to increase cGMP levels. In this case, a direct measurement of cGMP is made. In a similar vascular ring preparation, both zaprinast and SNP are able to increase cGMP levels (approximately 3-fold over basal levels). However, the combination of zaprinast and SNP causes an 18–20-fold increase in cGMP.

PDE5 inhibitors also produce direct vasodilatory effects in the presence of an intact endothelium (Martin *et al.*, 1986; Harris *et al.*, 1989; McMahon *et al.*, 1989;

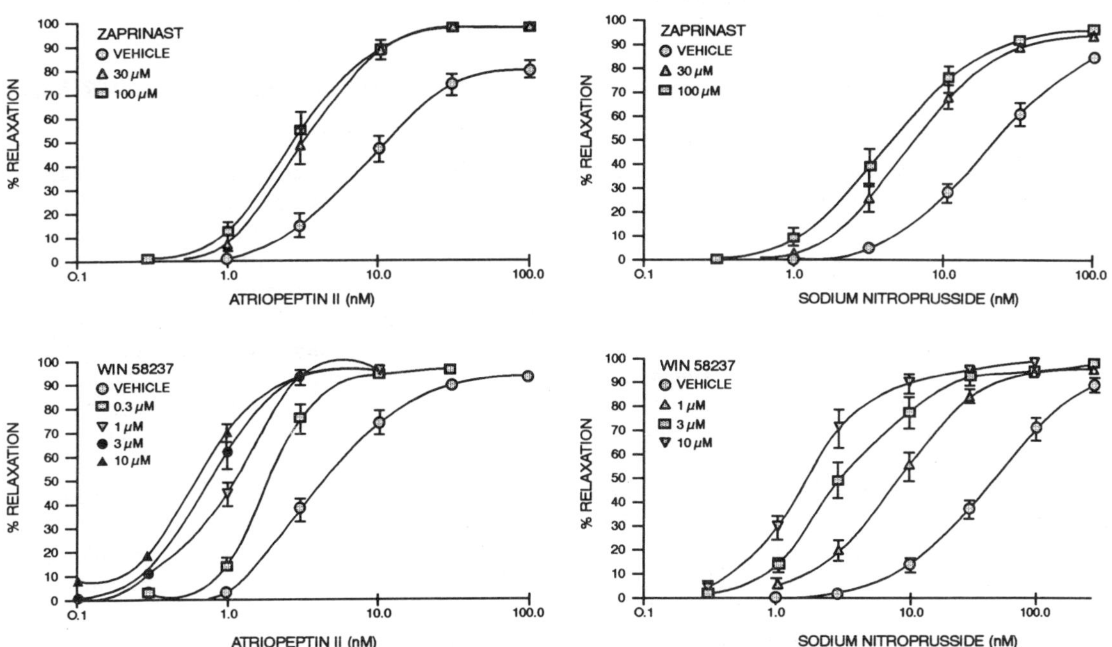

Figure 8.3 A typical experimental paradigm showing activity of PDE5 inhibitors in an intact tissue model. In these experiments rat aortic rings, which were denuded of endothelium to remove NO, are contracted with phenylephrine to induce tone and then relaxed by adding increasing concentrations of either ANF (atriopeptin II) or sodium nitroprusside. Experiments are conducted in the presence of the PDE5 inhibitors zaprinast (upper panels) or WIN 58237 (lower panels). Since there is no NO present, there is no relaxation by the PDE5 inhibitors alone. Potentiation – indicated by a leftward shift in the concentration–response curve – is evident for both sodium nitroprusside and ANF by both PDE5 inhibitors.

Silver *et al.*, 1994). As predicted by their mechanism of action, NO is required to stimulate basal GC activity for a PDE5 inhibitor to work. Also, consistent with an intracellular mechanism, efficacy vs. a wide variety of vasoconstrictors is evident. *In vivo*, zaprinast, WIN 58237 and other PDE5 inhibitors lower blood pressure, decrease peripheral vascular resistance and promote natriuresis in rats and dogs (McMahon *et al.*, 1989; Wilkins *et al.*, 1990; Trapani *et al.*, 1991; Pagani *et al.*, 1992; Dundore *et al.*, 1993).

To further elucidate the hypotensive effect of zaprinast, Dundore *et al.* (1991) determined whether $N\omega$-nitro-L-arginine (NNA), an inhibitor of NO synthase, alters the hypotensive response to zaprinast. Zaprinast or vehicle was given to conscious spontaneously hypertensive rats (SHR) in cumulative i.v. doses after pretreatment with NNA or saline. Mean arterial pressure was measured 5 min after each dose of zaprinast. At 5 min after the last dose of zaprinast, the rats were anaesthetized with pentobarbital and a segment of the abdominal aorta was rapidly freeze-clamped *in situ* and removed for the determination of cGMP levels. NNA significantly decreased basal aortic cGMP levels by 54% and increased mean arterial pressure by 37 mmHg. Zaprinast increased aortic cGMP by 187% and decreased mean arterial pressure by 49 mmHg. More importantly, NNA reduced the accumulation of cGMP in aortic tissue and attenuated the depressor response produced by zaprinast. These data are consistent with the hypothesis that NNA inhibits the tonic release of NO and that the depressor effects of zaprinast are due to potentiation of the vasodilator effects of NO *in vivo*. Moreover, since the changes in mean arterial pressure produced by NNA and zaprinast were significantly correlated with cGMP levels in aortic tissue, the concentration of cGMP in vascular smooth muscle may be one of several factors that determine blood pressure in SHR. Similar findings were also evident with the PDE5 inhibitor WIN 58237 (Silver *et al.*, 1994).

Although the measurement of cGMP in aortic tissue of rats can detect changes in cGMP levels at a site of action after PDE5 administration, it is a terminal procedure which produces a single sample from each animal and requires an extensive extraction procedure to isolate cGMP. Moreover, it is not a procedure which can be applied in a clinical setting. Accordingly, Dundore *et al.* (1993) examined the relationships between plasma cGMP levels, aortic cGMP levels and mean arterial blood pressure in conscious SHR. Overall, there was a significant correlation ($r = 0.76$) between plasma and aortic cGMP content with multiple doses of zaprinast. *In vitro*, it has been shown that cGMP is extruded from platelets and vascular tissue into the extracellular medium (Schini *et al.*, 1989; Wu *et al.*, 1993). Assuming that this extrusion also occurs *in vivo*, cGMP in plasma may arise from several cellular or tissue sources. Since the levels of cGMP in plasma were correlated to levels of cGMP at a pharmacodynamic site of action (vascular smooth muscle) of zaprinast, the measurement of plasma cGMP may be used to confirm the mechanism of action of PDE5 inhibitors. Similarly, WIN 58237 has been shown to increase plasma cGMP levels (Silver *et al.*, 1994).

An inhibitor of PDE5, such as zaprinast, would also be expected to potentiate the renal effects of ANF, since ANF produces natriuresis by stimulating particulate GC coupled to the ANF receptor (Huang *et al.*, 1986). In porcine LLC-PK1 cells, PDE5 was identified and ANF-mediated increases in cGMP in this cell line were more than tripled in the presence of zaprinast (Pagani *et al.*, 1992). *In vivo*, zaprinast produces natriuresis in rats and potentiates the natriuresis in response to ANF (McMahon *et al.*, 1989; Wilkins *et al.*, 1990). Moreover, the natriuresis produced by zaprinast is dose- and time-related to urinary cGMP excretion. Thus, urinary cGMP may be indicative of a renal response to zaprinast and this finding is similar to the aforementioned link of plasma cGMP levels to *in vivo* vasorelaxation by zaprinast (Dundore *et al.*, 1993). These data indicate that the levels of cGMP may provide biochemical indices related to the hypotensive and natriuretic actions of PDE5 inhibitors *in vivo*. Moreover, plasma or urinary cGMP levels may be easy and useful biochemical markers for assessing biodistribution or efficacy of PDE5 inhibitors in clinical trials.

4. Potential Therapeutic Applications of PDE5 Inhibitors

From these preclinical *in vitro* and *in vivo* data, it is clear that PDE5 inhibitors can potentiate the effects of NO (vasorelaxation, anti-vasoconstriction) and ANF (natriuresis). This combination of vasodilatory and natriuretic activity would be useful for the therapy of hypertension or congestive heart failure. For hypertension, however, the modes of current therapy, including angiotensin-converting enzyme (ACE) inhibitors, β-adrenoceptor blockers and Ca^{2+} channel antagonists, are adequate. Moreover, in heart failure the pharmacopoeia of agents currently used makes it difficult to assess the impact of mechanistically unique agents. Since high circulating ANF levels and ANF receptor down-regulation are serious sequelae in heart failure, a PDE5 inhibitor which potentiates the intracellular effect of ANF and thus reduces the amount of circulating ANF needed might offer promise for reducing down-regulation of ANF receptors.

Other potential acute vascular indications for PDE5 inhibitors include ischaemic cardiovascular or cerebrovascular disease. The added effect of inhibition of vascular smooth muscle cell proliferation/migration by increased cGMP offers potential advantages for post-angioplastic restenosis or subarachnoid haemorrhage.

An emerging major cardiovascular indication for PDE5 inhibitors involves reversal of vascular tolerance induced by nitrovasodilators. A limitation to the therapeutic use of nitroglycerine or other nitrovasodilators is the development of vasodilatory tolerance which occurs with repeated use of these agents (Thadani *et al.*, 1982; Flaherty, 1989). Although the exact mechanism for the development of tolerance is not entirely understood, most evidence indicates that oxidation of sulfhydryl groups of key proteins inactivates an essential enzymatic pathway responsible for the conversion of nitrovasodilators to NO (Needleman and Johnson, 1973; Axelsson and Ahlner, 1987). Other possible mechanisms include desensitized or diminished GC activity (Waldman *et al.*, 1986; Axelsson and Ahlner, 1987; Romanin and Kukovetz, 1989). However, the nitrate–cGMP-vasodilator pathway remains functional, although blunted, in tolerant vascular smooth muscle. Recently, several investigators have reported that nitrate tolerance can be reversed both *in vitro* and *in vivo* in several animal models by the PDE5 inhibitors zaprinast and WIN 58237 (Silver *et al.*, 1991, 1994; Merkel *et al.*, 1992; Pagani *et al.*, 1993). These studies showed that PDE5 inhibitors do not prevent the development of tolerance but do reinstate vascular responsiveness to nitroglycerine via restoration of intracellular cGMP levels. Thus, in the face of diminished GC activity in tolerance, prevention of hydrolysis of cGMP by PDE5 inhibitors serves to maintain enhanced cGMP levels and vasorelaxation. Since tolerance to the vasodilatory and anti-platelet effects of nitroglycerine is a significant acute and chronic clinical problem (Thadani *et al.*, 1982; Flaherty, 1989; Stamler and Loscalzo, 1992), it will be interesting in future studies to see if PDE5 inhibitors reverse tolerance in patients.

5. *Additional Indications for PDE5 Inhibitors*

In pulmonary medicine, Zapol and colleagues (Rossaint *et al.*, 1993; Zapol *et al.*, 1994) have pioneered the use of inhaled nitric oxide in several acute clinical situations of ventilation/perfusion deficit. NO increases cGMP levels in both airway and vascular smooth muscles so that both bronchodilation and pulmonary vasodilation occur. PDE5 inhibitors have the potential to perpetuate the duration of action of inhaled NO, since PDE5 is present and pharmacologically responsive in airway smooth muscle (Christensen and Torphy, 1994). A combination of inhaled NO and a PDE5 inhibitor might be useful for treatment of chronic obstructive pulmonary disease.

Another area in which NO plays a key role is penile erection. Rafjer *et al.* (1992) have shown that the non-adrenergic, non-cholinergic (NANC) transmitter is NO and that vascular endothelial NO in the corpus cavernosum elicits penile erection via vasodilation. Impotence may be related to decreased release of NO. Both NO donors and PDE5 inhibitors can reverse impotence by cGMP-related vasodilation in preclinical models. In addition, some pharmaceutical companies are purportedly having some success in treating impotence with PDE5 inhibitors in early clinical trials.

6. *Newer PDE5 Inhibitors*

In addition to zaprinast, other early inhibitors of cGMP PDE have been described. FK453 (Sakate *et al.*, 1992) and 4-(2-n-butyl-5-chloro-1-[2-chlorobenzyl]) imidazoylmethyl acetate (Booth *et al.*, 1990) inhibit cGMP-PDE with potencies in the 5–10 μM range. As with other PDE5 inhibitors, these agents possess vasorelaxant as well as depressor activity *in vivo* in various pharmacological models.

More recently, three new chemical series of inhibitors have been described which possess nanomolar potency for inhibition of PDE5. The first series, described in a patent by Pfizer (Brown and Terrett, 1993) consists of pyrazolo[3,4-d]-pyrimidin-7-ones. Structurally, these agents are somewhat related to zaprinast and possess potency in the 4–100 nM range. Secondary activity in inhibiting rabbit platelet aggregation is also evident with some of these compounds.

A second chemical series is described in a patent by Sterling–Winthrop (Bacon *et al.*, 1994). These agents are analogues of WIN 58237 but possess enhanced potency (IC_{50} values as low as 2 nM) and selectivity, particularly for PDE5 relative to PDE1 or PDE4, when compared with WIN 58237. One of the more interesting compounds of this series is 1-cyclopentyl-3-ethyl-6-(3-ethyoxy-4-pyrridyl)-pyrazolo[3,4-d]pyrimidin-4-one. This compound is a competitive inhibitor of PDE5 with an IC_{50} value of 1.6 nM. Vasorelaxation *in vitro* is endothelium-dependent, with an EC_{50} value of 60 nM. This compound also possesses *in vivo* activity in SHR following intravenous administration, decreasing mean arterial blood pressure by 24% at 3 mg/kg. As with other compounds of this series, it also reverses nitroglycerine tolerance in SHR and increases the concentration of cGMP in plasma.

A third new inhibitor is 1-[6-chloro-4-(3,4-methylenedioxybenzyl)-aminoquinolin-2-yl] piperidine-4-carboxylate sesquihydrate (also known by the code number E-4021; Saeki *et al.*, 1995). E-4021 is also a competitive inhibitor of PDE5 with an IC_{50} value of 4 nM. This compound is highly selective for PDE5, with inhibition of other PDE isoenzymes occurring in the 9–45 μM range. This compound possesses the same spectrum of activity as previously described for zaprinast and WIN 58237. It potentiates the vasorelaxant activity of either a nitrovasodilator (nitroglycerine) or ANF and potentiates cGMP accumulation by nitrovasodilators in isolated smooth muscle. Vasorelaxation by E-4021 is endothelium-dependent and can be blocked with an inhibitor of NO synthase. Interestingly, the potency of E-4021 as a vasorelaxant (EC_{50} = 110 nM) relative to its PDE5 inhibitory potency (IC_{50} = 4 nM) parallels the relative potencies for the aforementioned Sterling compound (EC_{50} = 60 nM; IC_{50} = 1.6 nM). These data further suggest that complete inhibition of PDE5 is necessary to achieve vasorelaxation in intact vascular smooth muscle. Moreover, since both agents are highly selective for PDE5 relative to CaM-sensitive PDE1, these data controvert the hypothesis that inhibition of both PDE1 and PDE5 in vascular smooth muscle (Ahn *et al.*, 1992) is necessary to obtain vasorelaxation.

E-4021 also possesses *in vivo* activity. Intravenous administration to conscious pigs mimicked the haemodynamic changes seen with isosorbide dinitrate in this model (Saeki *et al.*, 1995). A preferential effect on large coronary artery dilation, common to cGMP-increasing agents, was observed. In addition, a substantial reduction in mean pulmonary arterial pressure occurred with E-4021. These data are similar to observations with zaprinast in lambs (Braner *et al.*, 1993) or WIN 58237 in dogs (P.J. Silver, unpublished observations) and suggest that PDE5 inhibitors may have potential utility as selective pulmonary vasodilators.

7. Combination Inhibitors

PDE5 inhibition has been combined with inhibition of other PDE isozymes to yield classes of PDE inhibitors with distinct pharmacological activity. Combined PDE1/PDE5 inhibitors have been reported by Ahn *et al.* (1992) at Schering. These agents are effective vasorelaxants and inhibit both cGMP-degrading PDEs in smooth muscle.

Combined PDE4/5 inhibition is actually observed with WIN 58237 (Silver *et al.*, 1994). This class of inhibitor may be more useful for a disease such as asthma, since inhibition of PDE4 and PDE5 in airway smooth muscle will inhibit bronchoconstriction whereas inhibition of PDE4 in inflammatory cells will inhibit activation and release of cytokines (Christensen and Torphy, 1994; Nicholson and Shahid, 1994).

Combined PDE3/PDE5 inhibition has been reported for a series of substituted 1,6-naphthyridin-2-ones (Bacon *et al.*, 1995). These agents increase both cAMP and cGMP in myocardium and vascular smooth muscle. Relative to selective PDE3 inhibitors, these combined inhibitors are primarily vasodilators which produce modest increases in positive inotropy and little increase in heart rate.

8. Summary

In summary, several new, potent and selective inhibitors of PDE5 have been synthesized in recent years and shown to be effective in several different preclinical animal models. Some of these agents are now advancing into clinical trials. Given the roles of NO and ANF in human physiology, it seems likely that potentiators of NO or ANF, such as these PDE5 inhibitors, will have a role as therapeutic agents in certain diseases. However, clinical hypothesis testing remains to be conducted and reported before this new class of selective PDE isozyme inhibitor can gain acceptance as a new therapeutic modality.

9. Acknowledgement

The author acknowledges the assistance of Ms Jessica Silver in compiling this manuscript.

10. References

Ahn, H.S., Crim, W., Pitts, B. and Sybertz, E.J. (1992). Calcium-calmodulin-stimulated and cyclic GMP-specific phosphodiesterases: tissue distribution, drug sensitivity, and regulation of cyclic GMP levels. Adv. Second Messenger Phosphoprotein Res. 25, 271–288.

Axelsson, K.L. and Ahlner, J. (1987). Nitrate tolerance from a biochemical point of view. Drugs 33, 63–74.

Bacon, E.R., Singh, B. and Lesher, G.Y. (1994). 6-Heterocyclyl pyrazolo[3,4-D] pyrimidin-4-ones and compositions and method of use thereof. US Patent 5, 294, 612.

Bacon, E.R, Singh, B., Lesher, G.Y., Gruett, M., Pluncket, K., Lee, K.C., Ezrin, A M., Pagani, E.D., Buchholz, R.A., Fort, D.J., Connell, M.J., Hamel, L.T., Bentley, R.G., Canniff, P.C., Hamel, D.W. and Silver, P.J. (1995). 5-Alkylamino-substituted-1,6-naphthyridin-2-ones: novel vasotropic agents with low K_m cyclic adenosine monophosphate (cAMP) and cyclic guanosine monophosphate (cGMP) phosphodiesterase inhibitory activities. J. Med. Chem. 38. [In press]

Beavo, J.A. (1988). Multiple isozymes of cyclic nucleotide phosphodiesterases. Adv. Second Messenger Phosphoprotein Res. 22, 1–38.

Booth, R.F.G., Lunt, D.O., Lad, N., Buckham, S.P., Oswald, S., Clough, D.P., Floyd, C.D. and Dickens, J. (1990). A structurally novel inhibitor of cGMP phosphodiesterase with vasodilator activity. Biochem. Pharmacol. 40, 2315–2321.

Braner, D.A., Fineman, J.R., Chang, R. and Soifer, S.J. (1993). M&B 22948, a cGMP phosphodiesterase inhibitor, is a pulmonary vasodilator in lambs. Am. J. Physiol. 264, H252–H258.

Brown, D. and Terrett, N.K. (1993). Pyrazolopyrimidinone antianginal agents. Patent number WO 9306104.

Christensen, S.B. and Torphy, T.J. (1994). Isozyme selective phosphodiesterase inhibitors as antiasthmatic agents. Ann. Rep. Med. Chem. 29, 185–194.

Coquil, J.F., Franks, D.J., Wells, J.N., Dupuis, M. and Hamet, P. (1980). Characteristics of a new binding protein distinct from the kinase for guanosine 3′,5′-monophosphate in rat platelets. Biochim. Biophys. Acta 631, 148–165.

Dundore, R.L., Pratt, P.F., O'Conner, B., Buchholz, R.A. and Pagani, E.D. (1991). Nω-nitro-L-arginine attenuates the accumulation of aortic cGMP and the hypotension produced by zaprinast. Eur. J. Pharmacol. 200, 83–87.

Dundore, R.L., Clas, D.M., Wheeler, L.T., Habeeb, P.G., Bode, D.C., Buchholz, R.A., Silver, P.J. and Pagani, E.D. (1993). Zaprinast increases the cyclic GMP levels in plasma and aortic tissue of rats. Eur. J. Pharmacol. 249, 293–297.

Flaherty, J.T. (1989). Nitrate tolerance, a review of the evidence. Drugs 37, 523–535.

Francis, S.H., Lincoln, T.M. and Corbin, J.D. (1980). Characterization of a novel cGMP binding phosphodiesterase from rat lung. J. Biol. Chem. 255, 620–626.

Goldstein, R.A. and Fleming, R.M. (1993). Clinical aspects of phosphodiesterase inhibitors. In "Heart Failure: Basic Science And Clinical Aspects" (eds. J.K. Gwathmey, G.M. Briggs and P.D. Allen), pp. 387–398. Marcel Dekker, New York.

Harris, AL., Lemp, B.M., Bentley, R.G., Perrone, M.H., Hamel, L.T. and Silver, P.J. (1989). Phosphodiesterase isozyme inhibition and the potentiation by zaprinast of endothelium-derived relaxing factor and guanylate cyclase stimulatory agents in vascular smooth muscle. J. Pharmacol. Exp. Ther. 249, 394–400.

Huang, C.-L., Ives, H.E. and Cogan, M.G. (1986). In vivo evidence that cGMP is the second messenger for atrial natriuretic factor. Proc. Natl Acad. Sci. USA 83, 8015–8024.

Lugnier, C., Schoffter, P., LeBec, A., Strouthou, E. and Stoclet, J.C. (1986). Selective inhibition of cyclic nucleotide phosphodiesterases of human, bovine, and rat aorta. Biochem. Pharmacol. 35, 1743–1751.

Martin, W., Furchgott, R.F., Villani, G.M. and Jothianandan, D. (1986). Phosphodiesterase inhibitors induce endothelium-dependent relaxation of rat and rabbit aorta by spontaneously released endothelium-derived relaxing factor. J. Pharmacol. Exp. Ther. 237, 539–547.

McMahon, E.G., Palomo, M.A., Mehta, P. and Olins, G.M. (1989). Depressor and natriuretic effects of M&B 22,948, a guanosine cyclic 3′,5′-monophosphate-selective phosphodiesterase inhibitor. J. Pharmacol. Exp. Ther. 251, 1000–1005.

Merkel, L.A., Rivera, L.M., Perrone, M.H. and Lappe, R.W. (1992). *In vitro* and *in vivo* interactions of nitrovasodilators and zaprinast, a cGMP-selective phosphodiesterase inhibitor. Eur. J. Pharmacol. 216, 29–36.

Needleman, P.J. and Johnson, E.M. (1973). Mechanism of tolerance development to organic nitrates. J. Pharmacol. Exp. Ther. 184, 709–715.

Nicholson, C.D. and Shahid, M. (1994). Inhibitors of cyclic nucleotide phosphodiesterase isoenzymes – their potential utility in the therapy of asthma. Pulmonary Pharmacol. 7, 1–17.

Pagani, E.D., Buchholz, R.A. and Silver, P.J. (1992). Cardiovascular cyclic nucleotide phosphodiesterases and their role in regulating cardiovascular function. In "Cellular and Molecular Alterations in the Failing Human Heart" (eds. G. Hasenfuss, C. Holubarsch, H. Just and N. Alpert), pp. 73–86. Steinkopff, Darmstadt.

Pagani, E.D., Van Aller, G.S., O'Connor, B. and Silver, P.J. (1993). Reversal of nitroglycerin tolerance *in vitro* by the cGMP-phosphodiesterase inhibitor zaprinast. Eur. J. Pharmacol. 243, 141–147.

Rafjer, J., Aronson, W.J., Bush, P.A., Dorey, F.J. and Ignarro, L.J. (1992). Nitric oxide as a mediator of relaxation of the corpus cavernosum in response to nonadrenergic, noncholinergic neurotransmission. N. Engl. J. Med. 326, 90–94.

Romanin, C. and Kukovetz, W.R. (1989). Tolerance to nitroglycerin is caused by reduced guanylate cyclase activation. J. Mol. Cell. Cardiol. 21, 1–9.

Rossaint, R., Falke, K.J., Lopez, F., Slama, K., Pison, U. and Zapol, W.M. (1993). Inhaled nitric oxide for the adult respiratory distress syndrome. N. Engl. J. Med. 328, 399–405.

Saeki, T., Adachi, H., Takase, Y., Yoshitake, S., Souda, S. and Saito, I. (1995). A selective type V phosphodiesterase inhibitor, E4021, dilates porcine large coronary artery. J. Pharmacol. Exp. Ther. 272, 825–831.

Sakate, N., Zhou, Q., Sato, N., Matsuo, M., Sawada, T. and Shibata, S. (1992). Characteristics of vasoinhibitory action of FK453 (a pyrazolo-pyridine derivative), a new antihypertensive agent with diuretic action in isolated rabbit aorta. Pharmacology 44, 206–214.

Schini, V., Schoeffter, P. and Miller, R.C. (1989). Effect of endothelium on basal and on stimulated accumulation and efflux of cyclic GMP in rat isolated aorta. Br. J. Pharmacol. 97, 853–858.

Silver, P.J. (1989). Biochemical aspects of inhibition of cardiovascular low K_m cyclic adenosine monophosphate phosphodiesterase. Am. J. Cardiol. 63, 2A-8A.

Silver, P.J. and Harris, A.L. (1988). Phosphodiesterase isozyme inhibition and vascular smooth muscle. In "Resistance Arteries" (ed. W. Halpern), pp. 284–291. Perinatology Press, New York.

Silver, P.J., Pagani, E.D., de Garavilla, L., Van Aller, G.S., Volberg, M.L., Pratt, P.F. and Buchholz, R.A. (1991). Reversal of nitroglycerin tolerance by the cGMP phosphodiesterase inhibitor zaprinast. Eur. J. Pharmacol. 199, 141–142.

Silver, P.J., Dundore, R.L., Bode, D.C., de Garavilla, L., Buchholz, R.A., Van Aller, G.S., Hamel, L.T., Bacon, E.R., Singh, B., Lesher, G.Y., Hlasta, D. and Pagani, E.D. (1994). Cyclic GMP potentiation by WIN 58237, a novel cyclic nucleotide phosphodiesterase inhibitor. J. Pharmacol. Exp. Ther. 271, 1143–1149.

Souness, J.E., Brazdil, R., Diocee, B.K. and Jordan, R. (1989). Role of selective cyclic GMP phosphodiesterase inhibition in the myorelaxant actions of M&B 22,948, MY 5545, vinpocetine and 1-methyl-3-isobutyl-8-(methylamino)xanthine. Br. J. Pharmacol. 98, 725–734.

Stamler, J.S. and Loscalzo, J. (1992). The antiplatelet effects of organic nitrates and related nitroso compounds in vitro and in vivo and their relevance to cardiovascular disorders. J. Am. Coll. Cardiol. 18, 1529–1537.

Stamler, J.S., Simon, D.L., Osborne, J.A., Mullins, M.E., Jaraki, O., Michel, T., Singel, D.J. and Loscalzo, J. (1992). *S*-Nitrosylation of proteins with nitric oxide: synthesis and characterization of biologically active compounds. Proc. Natl Acad. Sci. USA 89, 444–448.

Thadani, U., Fung, H.-L., Darke, A.C. and Parker, J.L. (1982). Oral isosorbide dinitrate in angina pectoris: comparison of duration of action and dose–response relation during acute and sustained therapy. Am. J. Cardiol. 49, 411–419.

Trapani, A.J., Smits, G.J., McGraw, D.E., McMahon, E.G. and Blaine, E.H. (1991). Hemodynamic basis for the depressor activity of zaprinast, a selective cyclic GMP phosphodiesterase inhibitor. J. Pharmacol. Exp. Ther. 258, 269–277.

Waldman, S.A., Rapoport, R.M., Ginsburg, R. and Murad, F. (1986). Desensitization to nitroglycerin in vascular smooth muscle from rat and human. Biochem. Pharmacol. 35, 3525–3531.

Weishaar, R.E., Kobylarz-Singer, D.C., Steffen, R.P. and Kaplan, H.R. (1987). Subclasses of cyclic AMP-specific phosphodiesterase in left ventricular muscle and their involvement in regulating myocardial contractility. Circ. Res. 61, 539–547.

Wilkins, M.R, Settle, S.L. and Needleman, P. (1990). Augmentation of the natriuretic activity of exogenous and endogenous atriopeptin by inhibition of guanosine 3′,5′-cyclic monophosphate degradation. J. Clin. Invest. 85, 1274–1279.

Wu, X.-B., Brune, B., Von Appen, F. and Ullrich, V. (1993). Efflux of cyclic GMP from activated human platelets. Mol. Pharmacol. 43, 564–570.

Zapol, W.M., Rimar, S., Gillis, N., Marletta, M. and Bosken, C.H. (1994). Nitric oxide and the lung. Am. J. Respir. Crit. Care Med. 149, 1375–1380.

9. Design and Synthesis of Xanthines and Cyclic GMP Analogues as Potent Inhibitors of PDE5

Konjeti R. Sekhar, Pascal Grondin, Sharron H. Francis *and* Jackie D. Corbin

1. Introduction 135
2. Strategy for the Design of PDE5 Inhibitors 136
3. Synthesis of IBMX and cGMP Analogues as PDE Inhibitors 136
4. Selectivity of the IBMX Analogues as PDE Inhibitors 138
   4.1 IBMX Analogues As PDE5 Inhibitors 138
   4.2 Effects of Hydrophobic Substitutions on IBMX 141
   4.3 IBMX Analogues as Inhibitors of Other PDEs 141
5. cGMP Analogues as PDE Inhibitors 142
6. Smooth Muscle Relaxation by IBMX and cGMP Analogues 143
7. Conclusions 144
8. Acknowledgements 145
9. References 145

1. Introduction

Cyclic nucleotides serve as the second messengers for a variety of cellular responses to biological stimuli, including smooth muscle relaxation (Hardman, 1984; Murad, 1986). The signals are often recognized by specific cell-surface proteins which regulate the activities of enzymes that synthesize or degrade second messenger molecules. Changes in the intracellular second messenger levels result in alterations in the activities of specific target proteins within the cell and thereby evoke a physiological response. The intracellular concentrations of cyclic nucleotides are determined by the balance between the activities of adenylate or guanylate cyclases (AC and GC, respectively) and cyclic nucleotide phosphodiesterases (PDE). The hydrolysis of adenosine 3′:5′-cyclic monophosphate (cAMP) or guanosine 3′:5′-cyclic monophosphate (cGMP) is catalysed by a family of structurally related enzymes (Beavo, 1988) that are relatively tissue specific. The PDEs differ in their relative substrate affinities for cAMP and cGMP. Based on the regulatory features and substrate specificities, the PDEs are divided into seven families: PDE1, Ca^{2+}/calmodulin-dependent PDEs (CaM-PDE); PDE2, cGMP-stimulated PDEs (cGS-PDE); PDE3, cGMP-inhibited PDEs (cGI-PDE); PDE4, cAMP-specific, rolipram-sensitive PDEs; PDE5, cGMP-specific PDEs; PDE6, cGMP-specific photoreceptor PDE; PDE7, cAMP-specific, rolipram-insensitive PDEs (Beavo *et al.*, 1994; see also Chapters 1 and 2). Because PDE constitutes a diverse superfamily of enzymes, the rate of degradation of cAMP and cGMP can be controlled by a wide variety of intracellular and extracellular signals.

The major intracellular receptor for cAMP is cAMP-dependent protein kinase (PKA). Cyclic AMP can exert physiological effects via activation of proteins other than PKA. It has been shown that the relaxant effects of cAMP-elevating agents on pig coronary arteries is

Phosphodiesterase Inhibitors
ISBN 0-12-210720-9

mediated by activation of cGMP-dependent protein kinase (PKG), an enzyme which is related to, but distinct from, PKA (Jiang *et al.*, 1992). A second example of PKA-independent cAMP effects occurs in olfactory sensory neurones, which depolarize and fire an action potential in response to chemical odourants. In this case, cAMP elevation leads to the opening of a cyclic nucleotide-gated non-specific cation channel and the increased conductance through this channel leads to depolarization (Firestein *et al.*, 1991; Zufall *et al.*, 1991). As compared to cAMP, less progress has been made in the understanding of cGMP-mediated signal transduction. This is because of several complexities of the cGMP action. One intracellular target for cGMP is PKG, which, like PKA, phosphorylates and therefore modulates the activities of intracellular proteins. In retinal rods and cones cGMP acts as an internal ligand that directly binds and activates a plasma membrane cation channel (Kaupp, 1991). In addition to kinases and ion channels, PDEs may also serve as important mediators of cGMP action. Several PDEs contain allosteric cGMP-binding sites which are distinct from sites of cyclic nucleotide hydrolysis (Martins *et al.*, 1982; Gillespie and Beavo, 1988, 1989; Francis *et al.*, 1990). The functional role of the allosteric sites is not completely understood in all cases but cGMP binding to the allosteric sites of PDE2 has been shown to stimulate hydrolysis of cAMP at the catalytic site, thereby providing a means of communication between cGMP and cAMP signal transduction systems (Martins *et al.*, 1982; see also Chapter 5). Since this enzyme also actively hydrolyses cGMP, it could serve as a negative feedback system when cellular cGMP levels are elevated.

Effects of both PDE inhibition and PKG activation by xanthine compounds and cGMP analogues for intact cell effects have not been addressed previously. The methylxanthines and other PDE inhibitors are believed to act competitively with the cyclic nucleotide substrates. As such, these compounds are cyclic nucleotide analogues even though they do not possess the ribose-phosphate moieties that are required for activation of cyclic nucleotide-dependent protein kinases (Miller, 1981). The latter enzymes are also important pharmacological targets for smooth muscle relaxation (Hardman, 1984; Francis *et al.*, 1988; Jiang *et al.*, 1992). Comprehensive studies of the effects of newly synthesized cGMP analogues on PKG activation, PDE inhibition and smooth muscle relaxation have been performed (Beebe *et al.*, 1985, 1988; Francis *et al.*, 1988; Thomas *et al.*, 1992). Some new insights were uncovered concerning the modifications of the cGMP molecule that are efficacious for these effects. It seemed logical that similar modifications might prove efficacious for the methylxanthine family of compounds. In the present study, a variety of cyclic nucleotide analogues and new xanthine analogues with modifications at C-8, N-1 and N-3 were studied for their potencies to inhibit PDE5 and to relax vascular smooth muscle. Cellular factors that might influence their potencies are discussed.

2. *Strategy for the Design of PDE5 Inhibitors*

The present study concentrated mainly on cGMP-specific PDEs and it utilized the logical approach of designing compounds that should mimic cGMP. From previous work 3-isobutyl-1-methylxanthine (IBMX) is known to be a potent and non-selective inhibitor of PDE. Despite efforts by many scientists, modification of IBMX has produced inhibitors that are only slightly more potent than IBMX itself. This may be due to the lack of necessary substitution that will mimic the cyclic phosphate group of cGMP. The new compounds reported here have been designed so that the substitution at the C-8 position of IBMX mimics the cyclic phosphate group of cGMP. The various modifications of the xanthine nucleus are shown in Fig. 9.1.

3. *Synthesis of IBMX and cGMP Analogues as PDE Inhibitors*

IBMX analogues were synthesized either by literature methods or by slightly modified methods. 8-Substituted phenylthio-IBMX analogues (**3**) were synthesized by reacting 8-bromo-IBMX (**1**) (Kramer *et al.*, 1977) with substituted thiophenols (**2**) overnight in methanol in the presence of sodium acetate and trace amounts of water (Fig. 9.2). The reaction between 8-bromo-IBMX and 4-hydroxythiophenol occurred in the absence of water. This method is not useful for the preparation of 8-substituted phenylamino-IBMX analogues; these were prepared by fusing 8-bromo-IBMX and substituted anilines (**4**) at 160°C for 3 h

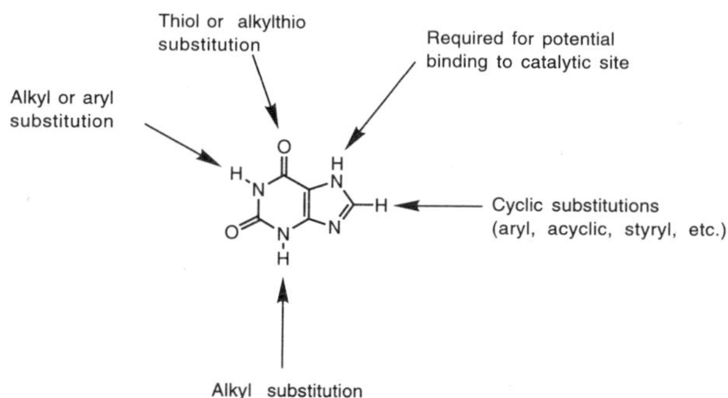

Figure 9.1 Some modifications of the xanthine nucleus used for PDE inhibition.

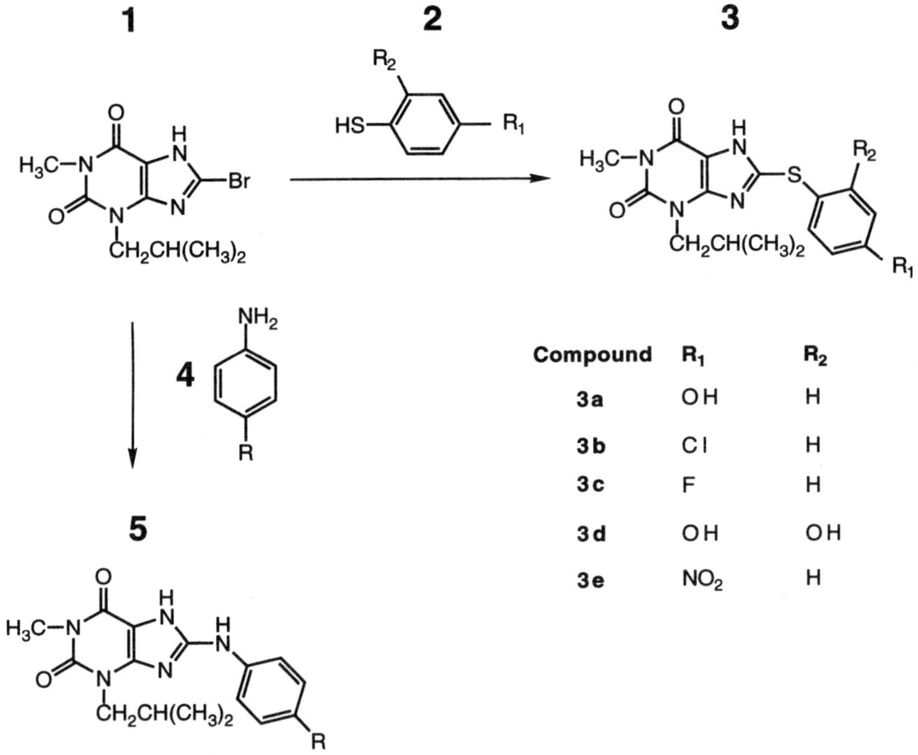

Compound	R_1	R_2
3a	OH	H
3b	Cl	H
3c	F	H
3d	OH	OH
3e	NO_2	H

Compound	R
5a	Cl
5b	F

Figure 9.2 Synthesis of 8-phenylthio-IBMX and 8-phenylamino-IBMX analogues.

(Fig. 9.2). The 8-substituted benzyl-IBMX analogues (**8**), 8-substituted cycloalkylmethyl-IBMX (**10**) and 8-substituted styryl-IBMX analogues (**12**) were obtained from the reaction between 1-isobutyl-3-methyl-5,6-diaminouracil (**6**) with substituted phenylacetic acids (**7**) (Fig. 9.3), alicyclic carboxylic acids (**9**) and substituted cinnamic acids (**11**) (Fig. 9.4), respectively. The two reagents were stirred overnight in the presence of condensing agent 1-(3-dimethylaminopropyl)-3-ethylcarbodiimide hydrochloride. Sodium hydroxide (2M) was added and the reaction mixture heated at 90°C for 2 h to get the desired products (Katsushima *et al.*, 1990). A similar method was used to prepare theophylline analogues. The compounds were purified either by recrystallization or by column chromatography using silica gel. Pure *trans* isomer of 8-(4-methoxystyryl)-IBMX was synthesized by reacting commercial *trans* 4-methoxycinnamic acid with 1-isobutyl-3-methyl-5,6-diaminouracil (**6**) in the dark. This *trans* isomer was dissolved in methanol (concentration of analogue ≈50 μM) and exposed to light for 1 week. The solvent was then removed and the solute dissolved in a minimum quantity of methanol. The *cis* isomer was separated from the *trans* isomer by HPLC (Whatman Partisil M20 10/25 ODS-3, mobile phase CH_3CN : water (7 : 3), flow rate 2.5 ml/min, detection at 246 nm) in the dark. The *cis* and *trans* isomers of 8-(2-methoxystyryl)-IBMX were prepared from commercially available *cis* and *trans* isomers of 4-methoxycinnamic acids. The structures of the compounds were confirmed by nuclear magnetic resonance, mass spectrometry and elemental analysis.

Synthesis of cGMP analogues was achieved by following the literature methods (Beaman and Robins, 1962; Miller *et al.*, 1973, 1981; Muneyama *et al.*, 1974; Sekhar *et al.*, 1992). 8-Substituted phenylthio-cGMP analogues (**13**) were synthesized as described above for the preparation of IBMX analogues (water was added in all cases). β-Phenyl-1,N^2-etheno (1,N^2-PET)-cGMP analogues (**14**) were synthesized by reacting cGMP or 8-bromo-cGMP with substituted phenacyl bromide in dimethyl sulfoxide (DMSO) in the presence of 1,8-diazabicyclo[5.4.0]undec-7-ene. Cyclic GMP analogues were purified on a G-25 superfine column using 0.05 M ammonium bicarbonate as eluting buffer (Corbin *et al.*, 1985).

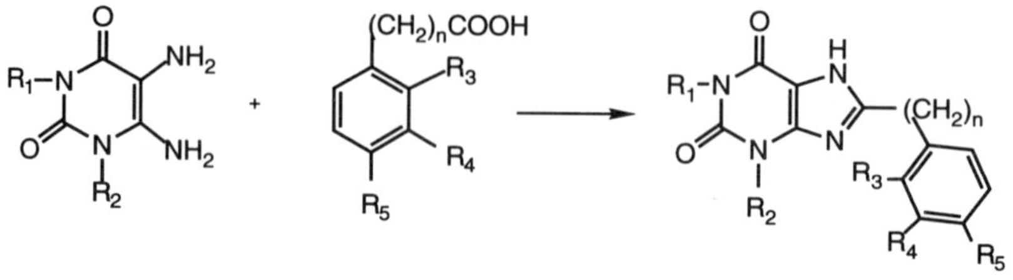

Compound	R_1	R_2	R_3	R_4	R_5	n
8a	CH_3	$CH_2CH(CH_3)_2$	H	H	Cl	1
8b	CH_3	$CH_2CH(CH_3)_2$	H	H	F	1
8c	CH_3	$CH_2CH(CH_3)_2$	H	H	OCH_3	1
8d	CH_3	$CH_2CH(CH_3)_2$	H	H	OH	1
8e	CH_3	$CH_2CH(CH_3)_2$	Cl	H	H	1
8f	CH_3	$CH_2CH(CH_3)_2$	OH	H	H	1
8g	CH_3	$CH_2CH(CH_3)_2$	OCH_3	H	H	1
8h	CH_3	$CH_2CH(CH_3)_2$	H	H	OH	2
8i	CH_3	CH_3	OCH_3	H	H	1
8j	CH_3	CH_3	H	H	Cl	1
8k	$CH_3CH_2CH_2$	$CH_3CH_2CH_2$	H	H	Cl	1

Figure 9.3 Synthesis of 8-benzyl-IBMX analogues.

4. *Selectivity of the IBMX Analogues as PDE Inhibitors*

Several newly synthesized IBMX analogues were tested for potency and specificity to inhibit the hydrolysis of cyclic nucleotides by PDEs.

4.1 IBMX ANALOGUES AS PDE5 INHIBITORS

IBMX was selected as the main structure for modification because of its well-established potency as an inhibitor of PDE. The effects of various substitutions of IBMX on potency of inhibition of cGMP-PDE (PDE5) activity are summarized in Table 9.1. Many of the new IBMX analogues were 20–1000-fold more potent than IBMX ($IC_{50} = 10\ \mu M$) for inhibition of PDE5 activity and 10–30-fold more potent than zaprinast ($IC_{50} = 0.25\ \mu M$) and dipyridamole ($IC_{50} = 1.0\ \mu M$), which are the classical inhibitors of PDE5.

Electron-donating groups such as OH^-, Cl^- and F^- on the phenyl ring enhanced the inhibitory potency of phenylthio-IBMX analogues. Within this series, 8-(4-hydroxyphenylthio)-IBMX (**3a**) was the most potent and most selective inhibitor of PDE5, with an IC_{50} of 0.10 μM. In this series each of the 4-halo analogues, 8-(4-chlorophenylthio)-IBMX (**3b**) and 8-(4-fluorophenylthio)-IBMX (**3c**) analogues followed with IC_{50} values of 0.25 μM. In contrast, an electron-withdrawing nitro group on the phenyl ring dramatically decreased the inhibitory potency of the IBMX analogue (**3e**). Although most of these compounds exhibited their highest potency in inhibiting PDE5, they were also excellent inhibitors of PDE1.

Experiments were designed to better define the explanation for these patterns of inhibition. Electron-donating groups on the phenyl ring may alter the electron density on the xanthine nucleus so that it binds more tightly to the catalytic site of PDE5. Alternatively, the substituents on the phenyl ring could directly interact with components of the catalytic site. Therefore, the electron flow from the phenyl ring to the xanthine nucleus was obstructed by substituting a methylene group for the sulphur. Four 4-substituted benzyl-IBMX analogues (**8a–d**) were synthesized, with electron-donating substituents (chloro-, fluoro-, hydroxy- and methoxy-) on the phenyl ring. All of the resultant

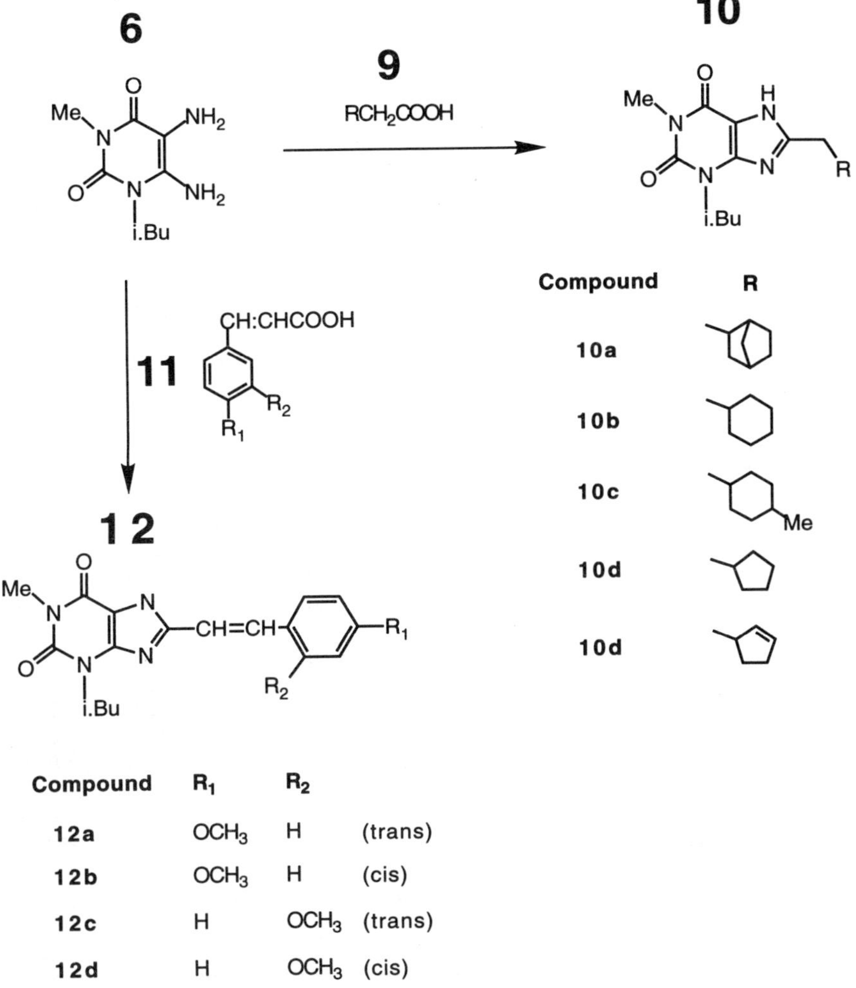

Figure 9.4 Synthesis of 8-styryl-IBMX and 8-alicyclylmethyl-IBMX analogues.

analogues were more potent than the corresponding phenylthio-IBMX analogues. The most potent compounds in the series of benzyl-IBMX analogues were 8-(4-chlorobenzyl)-IBMX (**8a**), with an IC_{50} of 0.01 μM, and 8-(4-fluorobenzyl)-IBMX (**8b**), with an IC_{50} of 0.024 μM, which were 25-fold and 10-fold more potent than the corresponding thio analogues (**3a, 3c**). These were followed in potency by 8-(4-methoxybenzyl)-IBMX (**8c**) and 8-(4-hydroxybenzyl)-IBMX (**8d**), with IC_{50} values of 0.046 μM and 0.054 μM, respectively. These results suggested that electronegative substituents on the phenyl ring may interact directly with elements within or near the catalytic site. Furthermore, all of these analogues were more potent than the best existing selective inhibitor of PDE5, zaprinast (IC_{50} = 0.25 μM).

Comparison of modelled structures of these IBMX analogues with that of cGMP suggested that the 4-substitution of the phenyl ring might simulate either the 2′-OH or the phosphate group of cGMP. If so, substitutions at the 2-position of the phenyl ring should produce less potent inhibitors than the corresponding 4-substituted analogues. Three 2-substituted benzyl-IBMX analogues, which had chloro-, hydroxy- or methoxy- at the 2-position, were synthesized and tested for inhibition of cGMP-PDE catalytic activity. These compounds were in fact 4–30-fold less potent than the corresponding 4-substituted analogues. The most potent compound was 8-(2-methoxybenzyl)-IBMX (**8g**), with an IC_{50} of 0.17 μM, followed by 8-(2-chlorobenzyl)-IBMX (**8e**) and 8-(2-hydroxybenzyl)-IBMX (**8f**), with IC_{50} values of 0.30 μM and 0.47 μM, respectively. These data emphasize the importance of the electron-donating substituent at the 4-position, which may simulate a portion of the cGMP molecule.

In the analogues described above (**3a–e, 8a–g**) the bridging atoms between the xanthine nucleus and the phenyl ring were either sulphur or carbon. The bridging atom was changed to nitrogen to observe the changes in the potencies of the analogues for inhibition of PDEs.

Table 9.1 Potency of IBMX analogues for inhibition of phosphodiesterases and relaxation of pig coronary artery

Compound number	Analogue	Inhibition (IC_{50}, μM)					Artery relaxation
		PDE1	PDE2	PDE3	PDE4	PDE5	(EC_{50}, μM)
3a	8-(4-Hydroxyphenylthio)-IBMX	0.4	6.0	6.6	>10.0	0.10	12.5
3b	8-(4-Chlorophenylthio)-IBMX	0.5	>10.0	>10.0	>10.0	0.25	6.5
3c	8-(4-Fluorophenylthio)-IBMX	0.1	>10.0	10.0	>10.0	0.25	8.2
3d	8-(2,4-Dihydroxyphenylthio)-IBMX	0.6	2.0	3.0	10.0	0.35	12.7
3e	8-(4-Nitrophenylthio)-IBMX	–	–	–	–	>10.00	55.5
5a	8-(4-Chlorophenylamino)-IBMX	10.0	>10.0	>10.0	>10.0	0.13	*
5b	8-(4-Fluorophenylamino)-IBMX	–	–	–	–	0.17	*
8a	8-(4-Chlorobenzyl)-IBMX	0.06	10.0	0.3	>10.0	0.010	23.9
8b	8-(4-Fluorobenzyl)-IBMX	0.07	5.0	0.8	>10.0	0.024	48.4
8c	8-(4-Methoxybenzyl)-IBMX	–	–	–	–	0.046	33.9
8d	8-(4-Hydroxybenzyl)-IBMX	0.1	0.8	2.0	10.0	0.054	31.1
8e	8-(2-Chlorobenzyl)-IBMX	0.7	>10.0	2.0	5.0	0.30	53.3
8f	8-(2-Hydroxybenzyl)-IBMX	–	–	–	–	0.47	61.0
8g	8-(2-Methoxybenzyl)-IBMX	1.0	3.0	1.0	>10.0	0.17	28.5
8h	8-(2-[4-Hydroxyphenyl]ethyl)-IBMX	0.7	0.4	10.0	>10.0	1.20	*
8i	8-(2-Methoxybenzyl)-theophylline	>10.0	>10.0	10.0	>10.0	8.00	52.3
8j	8-(4-Chlorobenzyl)-theophylline	8.0	10.0	5.0	>10.0	3.50	34.4
8k	1,3-Dipropyl-8-(4-chlorobenzyl)-xanthine	>10.0	10.0	>10.0	>10.0	0.60	*
10a	8-(Norbornylmethyl)-IBMX	0.03	>10.0	10.0	>10.0	0.0015	34.5
10b	8-(Cyclohexylmethyl)-IBMX	0.3	>10.0	>10.0	>10.0	0.04	30.7
10c	8-([4-Methylcyclohexyl]methyl)-IBMX	0.3	10.0	8.0	>10.0	0.15	43.8
10d	8-(Cyclopentylmethyl)-IBMX	0.2	>10.0	5.0	>10.0	0.08	33.7
10e	8-([2-Cyclopentene]methyl)-IBMX	0.2	>10.0	5.0	>10.0	0.30	33.8
12a	*trans* 8-(4-Methoxystyryl)-IBMX	–	–	–	–	0.38	–
12b	*cis* 8-(4-Methoxystyryl)-IBMX	–	–	–	–	0.016	–
12c	*trans* 8-(2-Methoxystyryl)-IBMX	–	–	–	–	>10.0	–
12d	*cis* 8-(2-Methoxystyryl)-IBMX	–	–	–	–	0.20	–
	IBMX	7.0	>10.0	<10.0	10.0	10.0	12.5
	Zaprinast	–	–	–	–	0.25	18.0
	Dipyridamole	–	–	–	–	1.00	–

* Compound precipitates in the test buffer even at low concentrations.
–, not determined.

Two analogues, 8-(4-chlorophenylamino)-IBMX (**5a**) and 8-(4-fluorophenylamino)-IBMX (**5b**), were synthesized with a nitrogen link between IBMX and the phenyl ring. These two compounds were somewhat more potent than the corresponding sulphur analogues (**3b, 3c**) but they were far less potent when compared to the corresponding benzyl analogues (**8a, 8b**). Thus, a carbon link between xanthine C-8 and the phenyl substituent enhanced the potencies of IBMX analogues as inhibitors of PDE5, as compared to either a nitrogen or sulphur link. In addition, with increasing hydrophobicity of the group on the phenyl ring at the C-8 position the inhibitory potency of the analogue increased accordingly. It is also possible that these alterations in the IBMX compound could enhance aromatic–aromatic interactions of the phenyl ring with structures in the catalytic site and thereby contribute to the increased potency of the IBMX analogues.

It is obvious from the above results that substitution at the C-8 position plays an important role in PDE inhibition. In order to address the stereospecific requirements of the substitution at the C-8 position, *cis* and *trans* isomers of 8-(4-methoxystyryl)-IBMX and 8-(2-methoxystyryl)-IBMX were synthesized and tested for their potency as inhibitors of PDE5. The *cis* isomers (**12b, 12d**) were much more potent inhibitors than were the *trans* isomers (**12a, 12c**) for PDE5. In the *trans* isomers the substituted phenyl ring extends from the IBMX ring in a different position from that of the cyclic phosphate group of cGMP (Fig. 9.5). This clearly suggests that the substitution at position C-8 should have a definite orientation, perhaps resembling cGMP more closely, in order to inhibit PDE5 potently.

The contribution of the 1-methyl and the 3-isobutyl groups of IBMX to the inhibitory potencies of these analogues was examined. Replacement of the isobutyl group on 8-(4-chlorobenzyl)-IBMX (**8a**) with a methyl group to produce 8-(4-chlorobenzyl)-theophylline (**8j**) reduced the inhibitory potency 350-fold

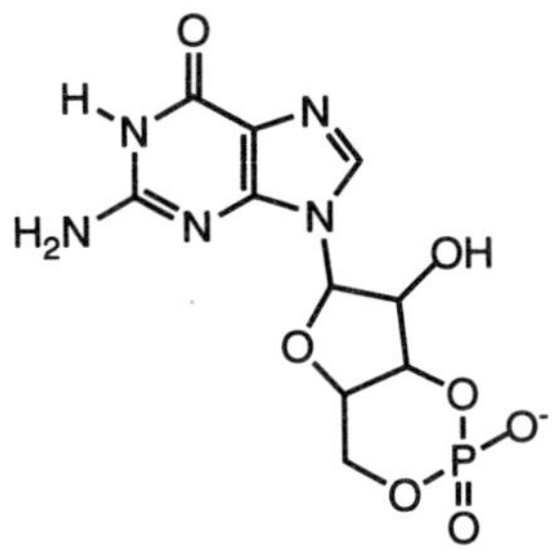

cGMP

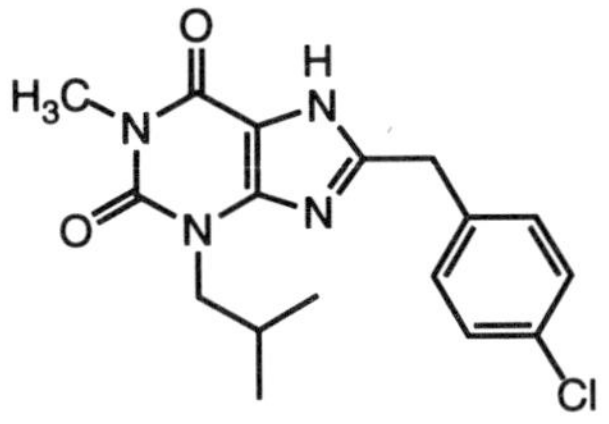

8a

8-(4-Chlorobenzyl)-IBMX

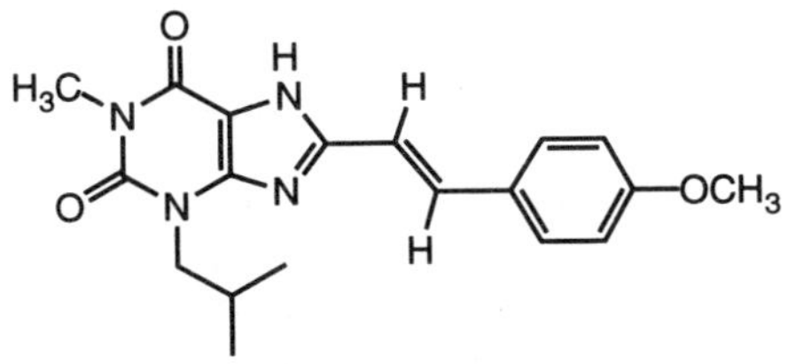

12a

Trans 8-(4-Methoxystyryl)-IBMX

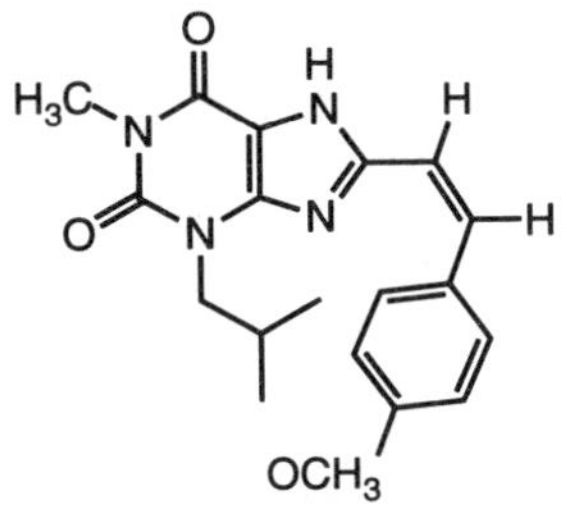

12b

Cis 8-(4-Methoxystyryl)-IBMX

Figure 9.5 Comparison of the structures of cGMP with 8-(4-chlorobenzyl)-IBMX, and *trans* and *cis* isomers of 8-(4-methoxystyryl)-IBMX.

(IC_{50} = 3.5 μM). Both substituents of IBMX, i.e. the methyl and isobutyl groups on 8-(4-chlorobenzyl)-IBMX (**8a**), were changed to propyl groups to yield 1,3-dipropyl-8-(4-chlorobenzyl)-xanthine (**8k**). This analogue (IC_{50} = 0.6 μM) was also less potent than the corresponding IBMX analogue (**8a**). When the 1-methyl group on compound **8a** was changed to a propyl or isoamyl group the resulting compounds, 1-propyl-3-isobutyl-8-(4-chlorobenzyl)-xanthine and 1-isoamyl-3-isobutyl-8-(4-chlorobenzyl)-xanthine (with IC_{50} values of 2.0 μM and 0.65 μM, respectively), were less potent than **8a**. From these results, it was concluded that the isobutyl and methyl groups contribute significantly to the inhibition of PDE5.

4.2 EFFECTS OF HYDROPHOBIC SUBSTITUTIONS ON IBMX

Several alicyclic groups were introduced at the C-8 position of IBMX to examine the effects of hydrophobic substitutions on PDE inhibition. The results (Table 9.1, **10a–e**) indicated that these compounds were very potent inhibitors of PDEs when compared to the 8-substituted benzyl-IBMX analogues. 8-(Norbarnyl-methyl)-IBMX (**10a**) was the most potent inhibitor of PDE5 with an IC_{50} of 1.5 nM. These compounds were also potent inhibitors of PDE1.

4.3 IBMX ANALOGUES AS INHIBITORS OF OTHER PDES

Some of the xanthine analogues synthesized in the present investigation were also tested for inhibition of several other PDEs (PDE1–PDE4) (Table 9.1). It can be seen that some of the analogues that inhibited PDE5 were in most cases also highly potent inhibitors of PDE1. Previous studies of these two PDEs have also suggested strong similarities in their catalytic sites (Charbonneau, 1990). Among all the compounds tested, 8-(4-chlorophenylamino)-IBMX (**5a**) was the most selective inhibitor of PDE5 (IC_{50} = 0.13 μM, versus 10 μM for PDE1 and >10 μM for PDE2–4, respectively). Two analogues, 8-(2-[4-hydroxyphenyl]ethyl)-IBMX (**8h**) and 8-(4-hydroxybenzyl)-IBMX (**8d**), were relatively potent inhibitors of PDE2, with IC_{50} values of 0.4 μM and 0.8 μM, respectively, whereas 8-(4-chlorobenzyl)-IBMX (**8a**) and 8-(4-fluorobenzyl)-IBMX (**8b**) were potent PDE3 inhibitors (IC_{50} values of 0.3 μM and 0.8 μM, respectively). All compounds were relatively poor inhibitors of PDE4 ($IC_{50} \geqslant 10$ μM).

5. cGMP Analogues as PDE Inhibitors

Our previous work demonstrated a strong correlation between the potencies with which cGMP analogues activate PKG and relax pig coronary arteries (Sekhar *et al.*, 1992). However, in addition to the effects of cGMP analogues in activating PKG, these compounds might also be predicted to inhibit PDE catalytic activity, thereby elevating cGMP levels in this tissue and further enhancing the activation of PKG. Since the catalytic site of PDE5 is highly specific for cGMP, selective modification of cGMP has the potential to produce potent and specific inhibitors of this enzyme. The potencies of various cGMP analogues in inhibiting PDE5 are shown in Table 9.2.

The analogues in the 1,N^2-PET-cGMP series (Fig. 9.7) were significantly more potent than 8-phenylthio-cGMP analogues (Fig. 9.6). The most potent inhibitors of cGMP-PDE were 8-(4-hydroxyphenylthio)-1,N^2-PET-cGMP (**14a**) and β-(2-naphthyl)-1,N^2-etheno (1,N^2-NET)-cGMP (**14b**), with IC_{50} values of 0.012 μM and 0.7 μM, respectively, whereas 8-(4-hydroxyphenylthio)-cGMP (**13a**) was the best inhibitor in the 8-phenylthio-cGMP series (Fig. 9.6), with an IC_{50} of 15 μM. There was no apparent correlation between the potency with which a compound inhibited PDE5 and the potency with which it elicited relaxation (Table 9.1).

Despite the strong selectivity of PDE5 for cGMP as opposed to cAMP (~100-fold), 8-(4-chlorophenylthio)-cAMP (**15a**) (Fig. 9.8) inhibited the catalytic activity of this enzyme with an IC_{50} of 4.4 μM. Similar observations were made by Connolly *et al.* (1992). This value was much lower than that for the corresponding cGMP analogue, 8-(4-chlorophenylthio)-cGMP (**13c**, $IC_{50} = 91$ μM). The corresponding inosine 3′:5′-cyclic monophosphate (cIMP) derivative, 8-(4-chlorophenylthio)-cIMP (**15b**) (Fig. 9.5) was also a very poor inhibitor ($IC_{50} = 135$ μM) of PDE5. These data suggested that the C-6-amino group on 8-(4-chlorophenylthio)-cAMP forms an important contact with PDE5.

Four of the cGMP analogues, including 8-(4-hydroxyphenylthio)-cGMP (**13a**), 8-(2,4-dihydroxyphenylthio)-cGMP (**13b**), 8-(2-aminophenylthio)-cGMP (**13d**) and 8-bromo-1,N^2-PET-cGMP (**14e**), were also tested for their potencies to inhibit other PDEs. Two of these analogues (**13a, 13b**) were potent inhibitors of PDE1, with IC_{50} values of 0.4 μM and 0.8 μM, respectively. Compound **13d** was a potent inhibitor of PDE4, with an IC_{50} of 2 μM, whereas 8-bromo-1,N^2-PET-cGMP (**14e**) was a potent inhibitor of both PDE2 and PDE5, with IC_{50} values of 6 μM and 4 μM, respectively. These results suggested that cGMP analogues could modulate several enzymes including PKG (type Iα and type Iβ) and several PDEs in intact tissue.

It was possible that the apparently lower inhibitory potency of cyclic nucleotide analogues compared to IBMX analogues could be an artefact due to hydrolysis of these

Table 9.2 Potency of cGMP analogues for inhibition of PDE5 and relaxation of pig coronary artery

Compound number	*Analogue*	*cGMP PDE inhibition (IC_{50}, μM)*	*Artery relaxation*[a] *(EC_{50}, μM)*
	8-Phenylthio-cGMP analogues		
13a	8-(4-Hydroxyphenylthio)-cGMP	15.0	1.1
13b	8-(2,4-Dihydroxyphenylthio)-cGMP	48.0	2.0
13c	8-(4-Chlorophenylthio)-cGMP	91.0	17.8
13d	8-(2-Aminophenylthio)-cGMP	125.0	4.7
13e	8-(4-Aminophenylthio)-cGMP	140.0	11.8
	1,N^2-PET-cGMP analogues		
14a	8-(4-Hydroxyphenylthio)-1,N^2-PET-cGMP	0.012	0.6
14b	1,N^2-,β-NET-cGMP	0.7	20.0
14c	1,N^2-(4-Methoxy-PET)-cGMP	1.6	11.4
14d	1,N^2-PET-cGMP	5.0	11.0
14e	8-Bromo-1,N^2-PET-cGMP	6.5	0.7
14f	8-Iodo-1,N^2-PET-cGMP	7.0	0.4
14g	8-Bromo-1,N^2-β-PMET-cGMP	Inactive	11.4
14h	8-Bromo-1,N^2-β-NET-cGMP	Inactive	50.0
	Miscellaneous analogues		
15a	8-(4-Chlorophenylthio)-cAMP	4.4	–
15b	8-(4-Chlorophenylthio)-cIMP	135.0	–

[a] Data taken from Sekhar *et al.* (1992).

–, Not determined.

Abbreviations: cIMP, inosine 3′,5′-cyclic monophosphate; 1,N^2-NET, β-(2-naphthyl)-1,N^2-etheno; 1,N^2-PET, β-phenyl-1,N^2-etheno; 1,N^2-PMET, β-phenyl-α-methyl-1,N^2-etheno.

13

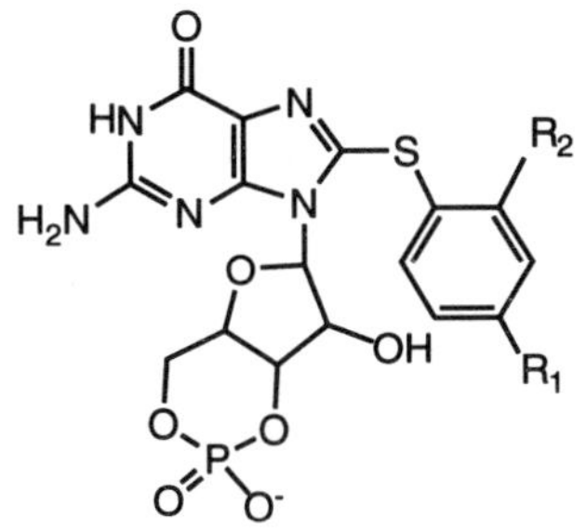

Compound

	13a	13b	13c	13d	13e
R_1	OH	OH	Cl	H	NH_2
R_2	H	OH	H	NH_2	H

Figure 9.6 8-Phenylthio-cGMP analogues.

14

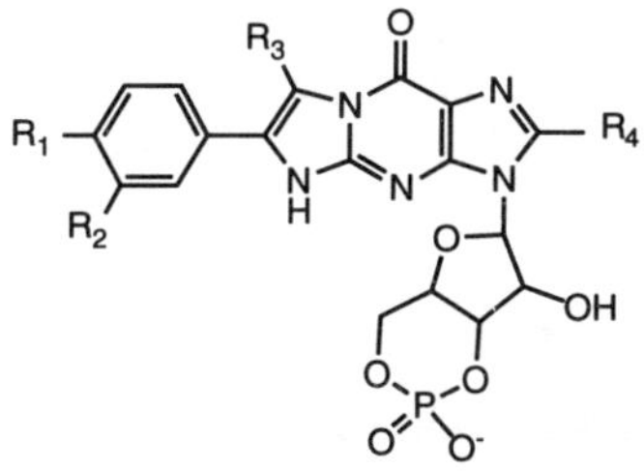

Compound

	14a	14b	14c	14d	14e	14f	14g	14h
R_1	H	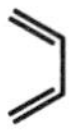	CH_3O	H	H	H	H	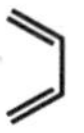
R_2	H		H	H	H	H	H	
R_3	H	H	H	H	H	H	CH_3	H
R_4	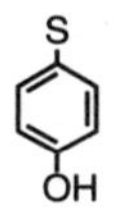 S / OH	H	H	H	Br	I	Br	Br

Figure 9.7 1,N²-PET-cGMP analogues.

15

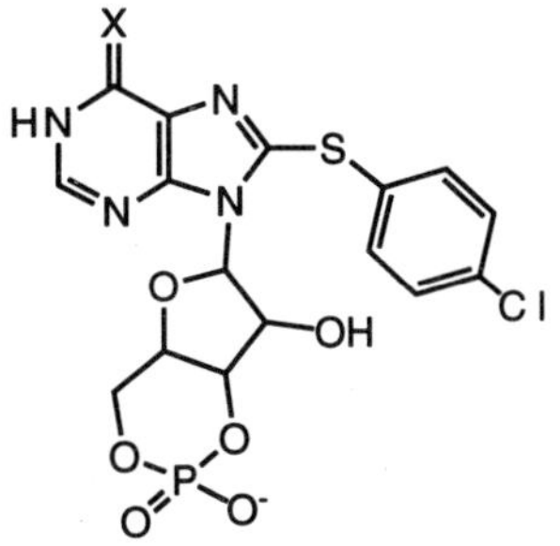

Compound	X
15a	NH
15b	O

Figure 9.8 8-(4-Chlorophenylthio)-cAMP/cIMP.

analogues by PDE5 during the test reactions. Therefore, the extent of hydrolysis of various cGMP analogues by PDE5 was determined (Table 9.3). The 8-substituted phenylthio-cGMP analogues were highly resistant to hydrolysis by high concentrations of PDE5 even after 2 h incubation at 30°C, whereas cGMP was completely (100%) hydrolysed within 1 min. The cGMP analogues most commonly used in experimentation, N^2-monobutyryl-cGMP and N^2,2′-*O*-dibutyryl-cGMP were also completely hydrolysed in 2 h. The analogues of 1,N^2-PET-cGMP were susceptible to hydrolysis, since typical PDE assays use a 10 min incubation time and much lower concentrations of PDE, the extent of hydrolysis of cGMP analogues would be negligible. These analogues were poor inhibitors of PDE5 as well, suggesting that their affinity for the catalytic site is poor compared to the best of the IBMX analogues (Table 9.2).

6. *Smooth Muscle Relaxation by IBMX and cGMP Analogues*

Previous studies have demonstrated the effects of PDE inhibitors to relax pig coronary artery smooth muscle and to potentiate the effects of sodium nitroprusside and

Table 9.3 Hydrolysis of cGMP analogues by PDE5

Compound number	*Analogue*	*% Hydrolysis in 2.5 h*
13a	8-(4-Hydroxyphenylthio)-cGMP	0
13b	8-(2,4-Dihydroxyphenylthio)-cGMP	20
13c	8-(4-Chlorophenylthio)-cGMP	85
13d	8-(2-Aminophenylthio)-cGMP	45
13e	8-(4-Aminophenylthio)-cGMP	5
14b	1,N^2-β-NET-cGMP	70
14c	1,N^2-(4-Methoxy-PET)-cGMP	70
14d	1,N^2-PET-cGMP	73
14e	8-Bromo-1,N^2-PET-cGMP	45
14f	8-Iodo-1,N^2-PET-cGMP	25
	cGMP[a]	100
	N^2-Monobutyryl-cGMP	100
	N^2,2′-*O*-Dibutyryl-cGMP	100

[a] cGMP was hydrolysed completely within 1 min.

isoprenaline on this tissue (Lorenz and Wells, 1983; Kramer and Wells, 1979). Since the xanthine analogues synthesized in the present investigation inhibited the catalytic activity of PDE5, the possibility that these analogues would also relax vascular smooth muscle by raising cGMP levels was investigated. The results are presented in Table 9.2.

The most potent relaxant of the IBMX series was 8-(4-chlorophenylthio)-IBMX (**3b**), with an EC_{50} value of 6.5 μM. This was followed by 8-(4-fluorophenylthio)-IBMX (**3c**), 8-(4-hydroxyphenylthio)-IBMX (**3a**) and 8-(2,4-dihydroxyphenylthio)-IBMX (**3d**), with EC_{50} values of 8.2 μM, 12.5 μM and 12.7 μM, respectively. Zaprinast had an EC_{50} of 18.0 μM. Some of the benzyl analogues were highly potent inhibitors of PDE5 and might be predicted to be good relaxants, but several of these compounds (**5a, 5b, 8h, 8k**) formed visible precipitates even at low concentrations when added to the buffers used for studies of smooth muscle relaxation. Although these analogues caused relaxation of the coronary arteries, the concentration that elicited the relaxation could not be determined.

When designing compounds to be used for smooth muscle relaxation, it seems logical to consider the theoretical limit of their potencies. This limit would depend in part on the ability of the analogue to penetrate the cell, on the identity of the cellular target and on the affinity of that target for a particular compound. Another important limiting factor – often ignored in such inhibitor studies but which will be scrutinized below – is the cellular concentration of the target protein. Earlier studies determined that the relaxant potencies of cGMP analogues correlated well with their potencies for activating PKG, particularly the type Iα isoform (Sekhar *et al.*, 1992). There was a poor correlation between inhibitory potency (IC_{50}) towards PDE5 and relaxant potency (EC_{50}) of either xanthine analogues or cGMP analogues. Thus, it appears from the analysis that cGMP analogues cause relaxation mainly by direct activation of PKG. Although some cGMP analogues could also interact with PDEs to competitively inhibit catalysis and elevate endogenous cGMP, the kinase is the more likely target of these analogues according to the following considerations. The intracellular concentrations of PKG and PDE5 are each ~0.15 μM (0.6 μM cGMP binding sites for each dimeric enzyme) in smooth muscle cells (Francis *et al.*, 1988; Corbin *et al.*, 1993) whereas cGMP in the basal state is about 0.1 μM in concentration (Francis *et al.*, 1988). Assuming that the intracellular affinity of analogues for either PKG or PDE5 is rather high, there should be little free analogue at low concentrations. Then it would be expected that the high-affinity analogues should produce a significant (~10%) activation of PKG when the intracellular concentration of analogue is about 10% of the cGMP binding site concentration, which would be 0.06 μM analogue. Assuming that the intracellular analogue equilibrates with the extracellular analogue concentration, this value of ~0.06 μM should be the theoretical limit for potency of any cGMP analogue that acts strictly by PKG activation. It is of interest that the potency of the most efficacious cGMP analogue is about 0.4 μM in pig coronary arteries (Sekhar *et al.*, 1992), which should allow for some further improvement in cGMP analogue potency. If cGMP-PDE is also a target for analogues, and if it is assumed that the mechanism is competitive inhibition, then relatively higher concentrations of extracellular analogue might be required to generate a signal by cGMP elevation. This is because the PDE5 holoenzyme is present at 0.15 μM (0.3 μM catalytic sites) and it would be expected that the analogue should have to achieve nearly saturating levels, i.e. 0.3 μM, in order to compete with intracellular cGMP sufficiently to cause cGMP to accumulate. By this reasoning, ~5 times higher concentrations of analogue would be required for cellular effects if PDE5 rather than PKG is the target for analogues. Therefore, the potential for developing increasingly potent drugs for smooth muscle relaxation is constrained significantly by the concentration of the intracellular receptor for those compounds. Even so, it seems prudent to design a new generation of cGMP analogues that will inhibit PDE as well as activate PKG. Such a family of analogues might be predicted to exhibit highly potent relaxation properties due to the direct activation of PKG as well as to the elevation of cGMP through inhibition of PDE.

7. *Conclusions*

This study has produced several new and potent inhibitors for PDE5 and has provided insight into structural features of IBMX analogues that contribute to enhanced PDE inhibitory potency. Although IBMX is a non-selective inhibitor of PDEs, simple modifications of this molecule produce inhibitors that are quite selective. The amino acid sequence alignment of several PDEs has been reported (Charbonneau, 1990; McAllister-Lucas *et al.*, 1993; see also Chapter 1). The sequence identity is 28–40% between the catalytic site of PDE5 and the catalytic sites of other PDEs. The IBMX analogue studies suggest that the structural features of the catalytic sites of PDE1 and PDE5 are more similar than that of PDE5 and other PDEs, even though the sequence homology between these two catalytic sites is not noticeably greater. In conclusion, IBMX analogues are more potent than cGMP analogues for inhibition of PDEs, whereas cGMP analogues are more potent smooth muscle relaxants.

IBMX is a potent inhibitor of PDEs in many tissues including coronary arteries (Wells *et al.*, 1975) and produces smooth muscle relaxation that is associated with increases in the levels of both cAMP and cGMP in these tissues (Schultz *et al.*, 1973; Sutherland *et al.*,

1973). Several 7-substituted IBMX analogues have been synthesized in order to produce an inhibitor selective for one of the smooth muscle PDEs (Garst *et al.*, 1976). 7-Benzyl-IBMX was identified as a relatively specific inhibitor of peak I PDE, which hydrolyses both cGMP and cAMP. Furthermore, several IBMX analogues have been reported with modifications at positions 1, 3, 7 and 8 that were significantly more potent than IBMX as inhibitors of peak I PDE (Kramer *et al.*, 1977). The 8-substituted-IBMX analogues were poor inhibitors of peak II. The IC_{50} values of several alkyl derivatives of xanthine for a "low K_m" cAMP-PDE have been shown to correlate well with EC_{50} values for tracheal smooth muscle relaxation (Takagi *et al.*, 1992). Similar observations were made with 3-alkylxanthines (Sakai *et al.*, 1992). In most of these cases, the IBMX analogues were modified at 1-, 3-, 7- and 8-positions with various small groups. Although the cyclic phosphate group of cyclic nucleotide is important for the interaction of cGMP or cAMP with the catalytic site, in most cases the compounds described above lack a substituent that mimics the cyclic phosphate group in the molecule. However, the IBMX analogues synthesized in the present investigation have a substitution at the 4-position of the phenyl ring and this group appears to mimic the cyclic phosphate group of cGMP (Fig. 9.5). In addition, the relative potencies of the xanthine analogues synthesized in the present study emphasize the importance of the orientation of the substitution at the C-8 position, the nature of the atom linking this substitution to the IBMX moeity, and substitutions at the N-1 and N-3 positions.

8. *Acknowledgements*

We are grateful to Dr Jack Wells for the generous gifts of 1-isobutyl-3-methyl-5-nitroso-6-aminouracil, 8-trifluoromethyl-IBMX and zaprinast, and to Dr Janet Colbran for PDE5. We also thank Mr Brian Nobes and Professor Ian Blair for mass spectral analysis. We also wish to thank ICOS Corporation for generous supply of human recombinant PDE2 and PDE4. We thank Dr K. Grimes and Dr Der Ming Chu for assistance in the smooth muscle preparations. This work was supported by a Glaxo Cardiovascular Discovery grant.

9. *References*

Beaman, A.G. and Robins, R.K. (1962). Potential purine antagonists. XXXIII. Synthesis of chloropurines. J. Appl. Chem. 12, 432–437.

Beavo, J.A. (1988). Multiple isozymes of cyclic nucleotide phosphodiesterase. Adv. Second Messenger Phosphoprotein Res. 22, 1–38.

Beavo, J.A., Conti, M. and Heaslip, R.J. (1994). Multiple cyclic nucleotide phosphodiesterases. Mol. Pharmacol. 46, 399–405.

Beebe, S.J., Redmon, J.B., Blackmore, P.F. and Corbin, J.D. (1985). Discriminative insulin antagonism of stimulatory effects of various cAMP analogs on adipocyte lipolysis and hepatocyte glycogenolysis. J. Biol. Chem. 260, 15781–15788.

Beebe, S.J., Beasley-Leach, A. and Corbin, J.D. (1988). cAMP analogs used to study low-K_m, hormone-sensitive phosphodiesterase. Methods Enzymol. 159, 531–540.

Charbonneau, H. (1990). Structure–function relationships among cyclic nucleotide phosphodiesterases. Mol. Pharmacol. Cell Regulation 2, 267–296.

Connolly, B.J., Willits, P.B., Warrington, B.H. and Murray, K.J. (1992). 8-(4-Chlorophenyl)thio-cyclic AMP is a potent inhibitor of the cyclic GMP-specific phosphodiesterase (PDE V_A). Biochem. Pharmacol. 44, 2303–2306.

Corbin, J.D., Beebe, S.J. and Blackmore, P.F. (1985). cAMP-dependent protein kinase activation lowers hepatocyte cAMP. J. Biol. Chem. 260, 8731–8735.

Corbin, J.D., Woodall, C.C., Colbran, J.L., McAllister, L.M., Sekhar, K.R. and Francis, S.H. (1993). Identifying protein kinases in crude extracts that phosphorylate cyclic GMP-binding cyclic GMP-specific phosphodiesterase. Agents Actions Suppl. 43, 27–33.

Firestein, S., Zufall, F. and Shepherd, G.M. (1991). Single odor-sensitive channels in olfactory neurons are also gated by cyclic nucleotides. J. Neurosci. 11, 3565–3572.

Francis, S.H., Noblett, B.D., Todd, B.W., Wells, J.N. and Corbin, J.D. (1988). Relaxation of vascular and tracheal smooth muscle by cyclic nucleotide analogs that preferentially activate purified cGMP-dependent protein kinase. Mol. Pharmacol. 34, 506–517.

Francis, S.H., Thomas, M.K. and Corbin, J.D. (1990). Cyclic GMP-binding cyclic GMP-specific phosphodiesterase from lung. In "Cyclic Nucleotide Phosphodiesterases: Structure, Regulation and Drug Action" (eds. M.D. Houslay and J. Beavo), pp. 117–140. Wiley, Chichester.

Garst, J.E., Kramer, G.L., Wu, Y.J. and Wells, J.N. (1976). Inhibition of separated forms of phosphodiesterases from pig coronary arteries by uracils and by 7-substituted derivatives of 1-methyl-3-isobutylxanthine. J. Med. Chem. 19, 499–503.

Gillespie, P.G. and Beavo, J.A. (1988). Characterization of a bovine cone photoreceptor phosphodiesterase purified by cyclic GMP-sepharose chromatography. J.Biol. Chem. 263, 8133–8141.

Gillespie, P.G. and Beavo, J.A. (1989). cGMP is tightly bound to bovine retinal rod phosphodiesterase. Proc. Natl Acad. Sci. USA 86, 4311–4315.

Hardman, J.G. (1984). Cyclic nucleotides and regulation of vascular smooth muscle. J. Cardiovasc. Pharmacol. 6, 5639–5645.

Jiang, H., Colbran, J.L., Francis, S.H. and Corbin, J.D. (1992). Direct evidence for cross activation of cGMP-dependent protein kinase by cAMP in pig coronary arteries. J. Biol. Chem. 267, 1015–1019.

Katsushima, T., Nieves, L. and Wells, J.N. (1990). Structure–activity relationships of 8-cycloalkyl-1,3-dipropylxanthines as antagonists of adenosine receptors. J. Med. Chem. 33, 1906–1910.

Kaupp, U.B. (1991). The cyclic nucleotide-gated channels of vertebrate photoreceptors and olfactory epithelium. Trends Neurosci. 14, 150–157.

Kramer, G.L. and Wells, J.N. (1979). Effects of phosphodiesterase inhibitors on cyclic nucleotide levels and relaxation of pig coronary arteries. Mol. Pharmacol. 16, 813–822.

Kramer, G.L., Garst, J.E., Mitchel, S.S. and Wells, J.N. (1977). Selective inhibition of cyclic nucleotide phosphodiesterases by analogs of 1-methyl-3-isobutylxanthine. Biochemistry 16, 3316–3321.

Lorenz, K.L. and Wells, J.N. (1983). Potentiation of the effects of sodium nitroprusside and of isoproterenol by selective phosphodiesterase inhibitors. Mol. Pharmacol. 23, 424–430.

Martins, T.J., Mumby, M.C. and Beavo, J.A. (1982). Purification and characterization of cyclic GMP-stimulated cyclic nucleotide phosphodiesterase from bovine tissues. J. Biol. Chem. 257, 1973–1979.

McAllister-Lucas, L.M., Sonnenburg, W.K., Kadlecek, A., Seger, D., Trong, H.L., Colbran, J.L., Thomas, M.K., Walsh, K.A., Francis, S.H., Corbin, J.D. and Beavo, J.A. (1993). The structure of a bovine lung cGMP-binding, cGMP-specific phosphodiesterase deduced from cDNA clone. J. Biol. Chem. 268, 22863–22873.

Miller, J.P. (1981). Cyclic AMP derivatives as tools for mapping cyclic AMP binding sites of cyclic AMP-dependent protein kinases I and II. Adv. Cyclic Nucleotide Res. 14, 335–344.

Miller, J.P., Boswell, K.H., Muneyama, K., Simon, L.N., Robins, R.K. and Shuman, D.A. (1973). Synthesis and biological studies of various 8-substituted derivatives of guanosine 3′,5′-cyclic phosphate, inosine 3′,5′-cyclic phosphate, and xanthosine 3′,5′-cyclic phosphate. Biochemistry 12, 5310–5319.

Miller, J.P., Uno, H., Christensen, L.F., Robins, R.K. and Meyer, R.B., Jr (1981). Effect of modification of the 1-, 2-, and 6-positions of 9-β-D-ribofuranosylpurine cyclic 3′,5′-phosphate on the cyclic nucleotide specificity of adenosine cyclic 3′,5′-phosphate- and guanosine cyclic 3′,5′-phosphate-dependent protein kinases. Biochem. Pharmacol. 30, 509–515.

Muneyama, K., Shuman, D.A., Boswell, K.H., Robins, R.K., Simon, L.N. and Miller, J.P. (1974). Synthesis and biological activity of 8-haloadenosine 3′,5′-cyclic phosphates. J. Carbohydrates Nucleosides Nucleotides 1, 55–60.

Murad, F. (1986). Cyclic guanosine monophosphate as a mediator of vasodilation. J. Clin. Invest. 78, 1–5.

Sakai, R., Konno, K., Yamamoto, Y., Sanae, F., Takagi, K., Hasegawa, T., Iwasaki, N., Kakiuchi, M., Kato, H. and Miyamoto, K. (1992). Effects of alkyl substitutions of xanthine skeleton on bronchodilation. J. Med. Chem. 35, 4039–4044.

Schultz, G., Hardman, J.G., Schultz, K., Davis, J.W. and Sutherland, E.W. (1973). A new enzymatic assay for guanosine 3′:5′-cyclic monophosphate and its application to the ductus deferens of the rat. Proc. Natl Acad. Sci. USA 70, 1721–1725.

Sekhar, K.R., Hatchett, R.J., Shabb, J.B., Wolfe, L., Francis, S.H., Wells, J.N., Jastorff, B., Butt, E., Chakinala, M.M. and Corbin, J.D. (1992). Relaxation of pig coronary arteries by new and potent cGMP analogs that selectively activate type Iα, compared with type Iβ, cGMP-dependent protein kinase. Mol. Pharmacol. 42, 103–108.

Sutherland, C.A., Schultz, G., Hardman, J.G. and Sutherland, E.W. (1973). Effects of vasoactive agents on cyclic nucleotide levels in pig coronary arteries. Fed. Proc. 32, 773.

Takagi, K., Ogawa, K., Tanaka, H., Satake, T., Watanabe, Y., Chijiwa, T. and Hidaka, H. (1992). Relaxant effects of various xanthine derivatives: relationship to cyclic nucleotide phosphodiesterase inhibition. Adv. Second Messenger Phosphoprotein Res. 25, 353–362.

Thomas, M.K., Francis, S.H., Beebe, S.J., Gettys, T.W. and Corbin, J.D. (1992). Partial mapping of cyclic nucleotide sites and studies of regulatory mechanisms of phosphodiesterases using cyclic nucleotide analogs. Adv. Second Messenger Phosphoprotein Res. 25, 45–53.

Wells, J.N., Wu, Y.J., Baird, C.E. and Hardman, J.G. (1975). Phosphodiesterases from porcine coronary arteries: inhibition of separated forms by xanthines, papaverine, and cyclic nucleotides. Mol. Pharmacol. 11, 775–783.

Zufall, F., Firestein, S. and Shepherd, G.M. (1991). Analysis of single cyclic nucleotide-gated channels in olfactory receptor cells. J. Neurosci. 11, 3573–3580.

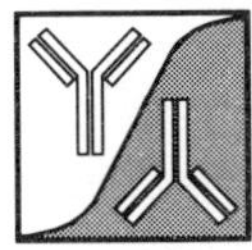

10. Enzymatic and Functional Aspects of Dual-selective PDE3/4 Inhibitors

Armin Hatzelmann, Renate Engelstätter, John Morley *and* Lazzarro Mazzoni

1. Introduction 147
2. Dual Inhibitors of PDE3/4 for Asthma Therapy 148
   - 2.1 Airway Smooth Muscle Relaxation 148
   - 2.2 Modulation of Inflammatory Cell Functions 148
   - 2.3 Side-Effects 148
3. Preclinical Pharmacology of PDE 3/4 Inhibitors 149
   - 3.1 AH 21–132 (Benafentrine) 149
   - 3.2 Zardaverine 151
   - 3.3 Others (Tolafentrine, Org 20241, Org 30029, EMD 54622) 151
4. Clinical Experience with PDE 3/4 Inhibitors 152
   - 4.1 AH 21–132 152
   - 4.2 Zardaverine 153
   - 4.3 Other Selective PDE Inhibitors (Tibenelast, Enoximone/Isomazole/SDZ-MKS 492, Zaprinast) 157
5. Conclusions 157
6. References 158

1. *Introduction*

Present anti-asthma therapy is based on two principles: relief of symptoms by bronchodilator drugs such as β_2-adrenoceptor agonists or anticholinergics, and suppression of the chronic inflammatory process by glucocorticoids (Barnes, 1989; McFadden, 1989). There is growing evidence that theophylline, the most widely prescribed anti-asthma drug worldwide, combines both bronchodilator and anti-inflammatory activities (Barnes and Pauwels, 1994). The mode of action of theophylline remains unclear but inhibition of phosphodiesterase (PDE) activity is likely to contribute to therapeutic activities and to the side-effects that restrict the use of this drug (see Chapter 3). Because non-selective inhibition of PDE enzymes may cause undesired side-effects, the concept arose that isoenzyme-selective PDE inhibitors combined with higher potency should lead to effective anti-asthmatic drugs with a higher therapeutic index (Morley, 1991; Nicholson and Shahid, 1994; Torphy *et al.*, 1994).

Over the last decade, inhibitors of type 4 PDE isoenzymes predominate in the patent literature. This reflects the emphasis that has recently been given to the role of inflammatory cells in asthma pathogenesis and to the recognition that the type 4 PDE predominates in haematogenous inflammatory cells (Giembycz, 1992). Consequently, there is an expectation that PDE4 inhibitors may exhibit anti-inflammatory and immunomodulatory activities. Since bronchial smooth muscle preparations also contain PDE4 there is an expectation that selective PDE4 inhibitors may have bronchodilator as well as anti-inflammatory potential. In this circumstance, it is reasonable to question whether dual inhibitors of type 3 and 4 PDE isoenzymes have any additional potential in asthma therapy. This chapter outlines three lines of evidence which indicate that combined inhibition of PDE3 and PDE4 isoenzymes is advantageous by comparison with selective inhibition of PDE4. First, there is synergy of PDE3 and PDE4 inhibition in airway smooth muscle relaxation; secondly, inhibition of type 3 PDE isoenzymes may contribute to

Phosphodiesterase Inhibitors
ISBN 0–12–210720–9

the regulation of inflammatory cell activity; thirdly, synergic pharmacological effects of inhibitors of type 3/4 PDE isoenzymes may lower the therapeutic dose and thereby reduce the frequency and intensity of side-effects.

2. *Dual Inhibitors of PDE3/4 for Asthma Therapy*

2.1 AIRWAY SMOOTH MUSCLE RELAXATION

The relaxation of airway smooth muscle by PDE inhibitors has been studied intensively in a range of species both *in vitro* and *in vivo*. It is beyond the scope of this chapter to discuss this issue thoroughly and the reader is referred to recent reviews (Torphy and Undem, 1991; Torphy *et al.*, 1994; Nicholson and Shahid, 1994). However, some aspects are important to mention in the context of a discussion of dual PDE3/4 inhibitors.

Synergism between type 3 and 4 PDE inhibitors as relaxants of guinea-pig airway smooth muscle was first described by Harris *et al.* (1989) who reported such interaction both *in vitro* (trachea contracted by carbachol, histamine or LTD_4) and *in vivo* (histamine-induced bronchoconstriction). Two other groups have reported that rolipram, a selective inhibitor of PDE4, is a more potent inhibitor of anaphylactic bronchoconstriction in the guinea pig than of comparable responses to histamine and leukotriene D_4 (LTD_4) (Howell *et al.*, 1993; Underwood *et al.*, 1993). On the other hand the bronchodilator efficacy of CI-930, a selective inhibitor of PDE3, appears not to be influenced by the spasmogenic stimulus (Howell *et al.*, 1993). These studies might be interpreted as evidence that inhibition of allergic bronchospasm by PDE4 may involve impaired release or generation of spasmogens and/or selective suppression of processes responsible for hyper-responsiveness (Morley, 1994). The latter explanation might be favoured since bronchoconstriction induced by histamine or LTD_4 is inhibited by a combination of inhibitors of type 3 and 4 PDE synergically (Turner *et al.*, 1994; Underwood *et al.*, 1994). Comparable data are not yet available for man; however, study of the isolated human bronchus supports this proposal. Thus, in precontracted (histamine, methacholine/carbachol, LTD_4) tissue, selective inhibition of PDE4 is equipotent to (de Boer *et al.*, 1992) or weaker than (Torphy *et al.*, 1993) selective inhibition of PDE3 isoenzymes in relaxing isolated human bronchus, whereas combined inhibition of type 3 and 4 PDE isoenzymes is synergic. In one study, it has been noted that intrinsic tone of the isolated human bronchus was not affected by selective inhibition of PDE4 isoenzymes, but was relaxed by selective inhibition of PDE3 isoenzymes, an effect that was even more pronounced when dual-selective inhibitors of PDE3 and PDE4 were employed (Rabe *et al.*, 1993). Other studies have reported that selective inhibitors of PDE4 were able to relax human bronchus under resting tone (Qian *et al.*, 1993; Cortijo *et al.*, 1993). The reasons for this discrepancy are unclear (Dent *et al.*, 1994b).

On the basis of the available evidence from studies *in vitro* it can be proposed that, as in the guinea pig, inhibition of PDE3 may be of importance in the relaxation of bronchial smooth muscle in man and may effect synergism in combination with inhibition of PDE4 isoenzymes.

2.2 MODULATION OF INFLAMMATORY CELL FUNCTIONS

The question of which PDE isoenzymes are involved in the regulation of cellular activities of inflammatory cells has been discussed thoroughly elsewhere (see Chapter 2). PDE isoenzyme profiles in human cells reveal the existence of cell types in which PDE4 isoenzymes predominate (neutrophils, eosinophils, monocytes and epithelial cells) and others in which both PDE3 and PDE4 isoenzymes are present (T lymphocytes, macrophages and endothelial cells); in addition to these isoenzymes, endothelial cells contain type 2 and macrophages contain type 1 PDE isoenzymes (Schudt *et al.*, 1995). Therefore, although type 4 is the predominant PDE isoenzyme in human inflammatory cells, the presence of PDE3 isoenzymes may be of relevance to the regulation of leucocyte responses. In this context, there are publications which have suggested that inhibition of PDE3 is additive to, or synergic with, inhibition of PDE4 isoenzymes in suppression of inflammatory cell responses. Such interaction has been suggested for the inhibition of tumour necrosis factor-α (TNF-α) synthesis and superoxide anion (O_2^-) formation in alveolar macrophages (Loos *et al.*, 1994), as well as for the inhibition of proliferation of peripheral blood lymphocytes (Robicsek *et al.*, 1991; Schudt *et al.*, 1993a; Banner and Page, 1994; Giembycz *et al.*, 1994).

It is therefore likely that inhibition of PDE3 isoenzymes contributes to anti-inflammatory effects when mononuclear cell activation predominates but that selective inhibition of PDE4 may suffice to inhibit the activity of granulocytic effector cells (eosinophils and neutrophils).

2.3 SIDE-EFFECTS

The number of clinical studies of selective PDE inhibitors is insufficient to allow a detailed description of the side-effect profile for the various classes of PDE inhibitors (Christensen and Torphy, 1994). It is, however, possible to indicate areas of concern. Thus, inhibitors of PDE3 isoenzymes may increase cardiac contractility and cause vasodilation or even arrhythmias by influencing the regulation of cAMP in myocardium and vascular smooth muscle. On the other hand, selective PDE4 inhibitors, such as rolipram, cause nausea, vomiting and changes in plasma osmolality which can be presumed to reflect inhibition of type 4 PDE isoenzymes in the brain, gastrointestinal tract and/or kidney. It remains to be established whether these side-effects can be overcome by choosing compounds that are selective for particular tissues or subclasses of PDE4. However, it should not be presumed that the risk of adverse side-effects associated with inhibition of PDE4 isoenzymes will be increased by inhibition of PDE3, since it can reasonably be expected that required doses of dual inhibitors of PDE3 and PDE4 isoenzymes will be proportionately lower owing to synergic interaction at the therapeutic target cells (see sections 2.1 and 2.2).

3. *Preclinical Pharmacology of PDE 3/4 Inhibitors*

Of the available dual PDE3/4 inhibitors, six have been studied extensively and are discussed in the following sections. The chemical structures of these compounds are shown in Fig. 10.1.

3.1 AH 21–132 (BENAFENTRINE)

AH 21–132 (*cis*-6-[*p*-acetamidophenyl]-1,2,3,4,4*a*,10*b*-hexahydro-8,9-dimethoxy-2-methylbenzo-[*c*][1,6]-naphthyridine; benafentrine) inhibits PDE3 from guinea-pig platelets with an IC_{50} of 1.74 μM PDE4 from guinea-pig neutrophils with an IC_{50} of 1.76 μM (data on file). The inhibitory effect on type 4 PDE isoenzymes was confirmed using bovine trachea, for which an IC_{50} of 3.6 μM was estimated, with relatively minor inhibition of type 2 and 5 PDE isoenzymes from this tissue (IC_{50} estimates of 53 μM and >300 μM, respectively) (Giembycz and Barnes, 1991). Both spasmolytic and anti-inflammatory actions have been demonstrated for AH 21–132.

3.1.1 Relaxation of Airway Smooth Muscle

AH 21–132 relaxes isolated guinea-pig trachea contracted by endogenous spasmogens or following addition of histamine or carbachol with IC_{50} values of 1.6, 3.2 and 6.9 μM, respectively (Bewley and Chapman, 1988). The lower potency of AH 21–132 in relaxing airway smooth muscle precontracted with carbachol *in vitro* may reflect the progressive loss of potency and efficacy that is observed as the concentration of muscarinic receptor agonist is increased, a phenomenon that is also evidenced by sympathomimetics (Offermeier and van den Brink, 1974; Torphy *et al.*, 1983). For tracheal smooth muscle under spontaneous tone, the relaxant action is enhanced slightly but significantly (IC_{50} = 0.8 μM) when the airway epithelium has been removed, a property not shared with sympathomimetics, from which AH 21–132 may

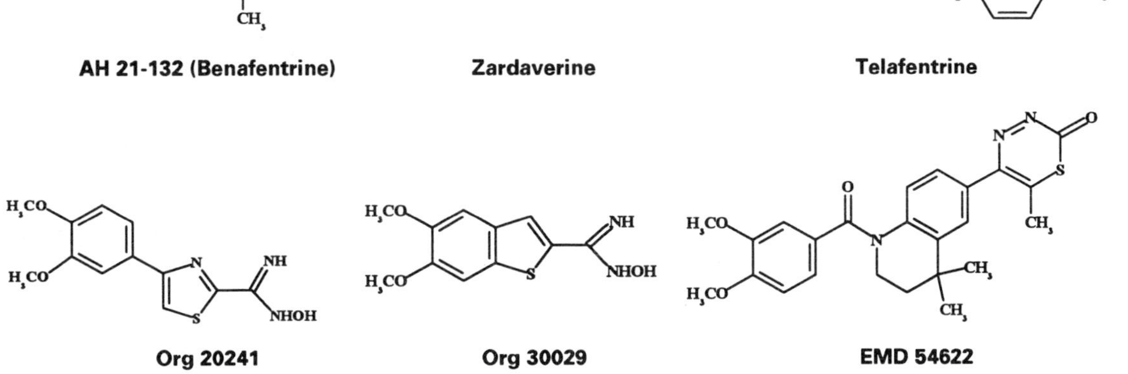

AH 21-132 (Benafentrine) — Zardaverine — Telafentrine

Org 20241 — Org 30029 — EMD 54622

Figure 10.1 Chemical structures of PDE3/4 inhibitors.

be distinguished by having spasmolytic effects that are not diminished by propranolol (Bewley and Chapman, 1988; Small *et al.*, 1989). Isolated human bronchus is also relaxed by AH 21–132, when contracted by endogenous spasmogens or following addition of histamine or carbachol, with IC_{50} values of 4.7, 4.0 and 8.0 μM, respectively (data on file). As in guinea-pig trachea, the relaxant action of AH 21–132 against spontaneous tone is enhanced slightly but significantly (IC_{50} = 1.3 μM) after removal of the airway epithelium.

Although there are several lines of evidence suggesting that the spasmolytic actions of AH 21–132 are determined by inhibition of PDE, there remains some uncertainty. The capacity of the two enantiomers of AH 21–132 to relax guinea-pig trachea corresponds to their ability to inhibit cAMP hydrolysis; however, the concentrations required to cause significant accumulation of cAMP comfortably exceed the concentration needed to achieve maximal relaxation (Small *et al.*, 1991). Similarly, although AH 21–132 augments the capacity of isoprenaline to relax airway smooth muscle, this effect is also exhibited by agents (e.g. cromokalim) that are considered to effect relaxation through other mechanisms (Small *et al.*, 1991). The possibility that direct spasmolysis by AH 21–132 is not exclusively determined by PDE isoenzyme inhibition has yet to be excluded. Actions that may contribute include activation of cAMP-dependent protein kinase (Giembycz and Barnes, 1991) or formation of inositol phosphates (Giembycz *et al.*, 1990), although present evidence indicates that such effects are only detected when concentrations of AH 21–132 considerably exceed those needed to cause relaxation of airway smooth muscle.

In vivo observations provide some indication as to whether spasmolytic actions *in vitro* might be translated into actions on the intact airways that anticipate therapeutic efficacy. In the anaesthetized, ventilated guinea pig, sustained airway obstruction (due to infusion of bombesin) is reduced by AH 21–132 (ID_{50} = 0.08 mg/kg). This ability to reduce established tone is complemented by a capacity to prevent obstruction due to intravenous injection of platelet activating factor (PAF), acetylcholine, histamine, serotonin, substance P or LTC_4 when administered intravenously as a bolus (0.1–1.0 mg/kg). These findings were extended to include demonstration of bronchodilator efficacy following intraduodenal application and following inhalation. In this latter situation, it could be shown that inhalation of AH 21–132 (0.001–0.1 mg/kg) diminishes allergic reactions and at 1.0 mg/kg prevents lethal anaphylaxis in the passively sensitized guinea pig.

Existing bronchodilators suppress hyperresponsiveness as a means of effecting bronchodilation in obstructed airways (Morley, 1994). Hence, it is interesting to ascertain whether, in addition to effecting relaxation of normally responsive airways, AH 21–132 may also effect reduction of hyperresponsiveness. AH 21–132 was a most effective inhibitor of airway obstruction due to histamine in animals rendered hyperresponsive by exposure to PAF (ID_{50} = 0.15 mg/kg), *rac*-isoprenaline (ID_{50} = 0.32 mg/kg) and immune complexes (ID_{50} = 0.32 mg/kg). No direct comparison was made, but it seems likely that AH 21–132 is less effective in normally responsive animals. Unusually, there is some direct evidence of a selective effect of AH 21–132 upon airway responsiveness. It is known that PAF can induce hyperresponsiveness in the guinea pig through the action of a labile product of platelet activation (Sanjar *et al.*, 1989). The presence of AH 21–132 in the platelet preparation caused a dose-related reduction of hyperresponsiveness whereas the corresponding final doses *in vivo* (1 and 10 μg/kg) did not influence airway responsiveness. Since the predominant PDE(s) in platelets is/are type 3, it may be presumed that inhibition of these enzymes determines the suppression of this form of hyperresponsiveness.

3.1.2 Anti-inflammatory Effects

Anti-inflammatory actions are implied by the capacity of AH 21–132 to inhibit PDE isoenzymes of type 3 and 4 in platelets and eosinophils, respectively. The role of platelets in inflammatory responses is poorly defined, although accumulation of platelets precedes accumulation of mononuclear cells in atherosclerotic plaques whereas, in the guinea pig, eosinophil accumulation in response to PAF is impaired when intrathoracic accumulation of platelets is prevented (I.D. Chapman and J. Morley, unpublished observations). In accordance with expectations, the accumulation of eosinophils in the airway lumen of animals exposed to PAF was diminished in animals exposed to AH 21–132 by subcutaneous infusion (1.0 mg/kg/day for five days prior to exposure to PAF and for one day thereafter). Similarly, in sensitized animals, the eosinophil accumulation in the airway lumen that follows inhalation of antigen was diminished in animals exposed to AH 21–132 by subcutaneous infusion (1.0 mg/kg/day for six days and for one day following exposure to antigen). Inhibition extended also to spontaneous eosinophilia in the airway lumen and the peritoneal cavity, which could be reduced substantially by daily oral administration of AH 21–132 (1.5 mg/kg). Some indication of a possible mechanism for selective suppression of eosinophil migration into the airways is provided by studies which used human recombinant (rh) cytokines to induce eosinophilia. rhIL-3 and rhGM-CSF caused a modest selective eosinophilia that could be suppressed by daily oral administration of AH 21–132 (1.5 mg/kg for five days before exposure to cytokine and for one day thereafter). It seems likely that inhibition of PDE4 contributes to these effects of AH 21–132, although it must be pointed out that actions on PDE3 in platelets and effects upon vascular elements cannot be excluded.

3.2 ZARDAVERINE

Zardaverine (6-[4-difluoromethoxy-3-methoxyphenyl]-3[2H]pyridazinone) inhibits PDE3 from human platelets and PDE4 from canine trachea and human polymorphonuclear leucocytes (PMN) with respective IC_{50} values of 0.6, 0.8 and 0.2 μM; inhibition of PDE1, 2 and 5 isoenzymes is only marginal, with concentrations up to 100 μM being necessary to achieve significant inhibition (Schudt *et al.*, 1991c). Both the bronchodilator and the anti-inflammatory potential of this inhibitor of PDE3/4 isoenzymes have been demonstrated in various species *in vitro* and *in vivo*.

3.2.1 Relaxation of Airway Smooth Muscle

Zardaverine relaxes isolated guinea-pig trachea, whether contracted by endogenous spasmogens or by a variety of agonists (histamine, carbachol, ovalbumin, prostaglandin $F_{2\alpha}$ ($PGF_{2\alpha}$), LTC_4 or the thromboxane analogue U46619) (Kilian *et al.*, 1989). Similarly, isolated trachea from guinea pigs sensitized to ovalbumin relax on addition of zardaverine irrespective of the contractile agent (histamine, LTD_4 or ovalbumin) (Underwood *et al.*, 1994). Although individually, rolipram (PDE4-selective) and siguazodan (PDE3-selective) are ineffective as relaxants of airway smooth muscle contracted by histamine or LTD_4, in combination they are effective, suggesting a synergic interaction. Comparable results have been reported for isolated human bronchi under endogenous tone, which are relaxed by zardaverine or a combination of rolipram and SKF 94120 (PDE3-selective) but not by rolipram alone (Rabe *et al.*, 1993). On the other hand, cilostamide and CI-930 (PDE3-selective), rolipram and denbufylline (PDE4-selective) and zardaverine were found to be equieffective in relaxing tracheal strips (guinea pig, ferret) or bronchial rings (guinea pig, human) (Knowles *et al.*, 1994). Oral zardaverine (3–30 μmol/kg) is a bronchodilator in the rat (Hoymann *et al.*, 1994). In the guinea pig, intravenous zardaverine (1–60 μmol/kg) inhibits histamine-induced bronchospasm in anaesthetized animals and, after oral application (1–100 μmol/kg), protects against dyspnoea induced by acetylcholine or ovalbumin (Kilian *et al.*, 1989). Similarly, zardaverine (5 mg/kg, intragastrically) markedly inhibits bronchoconstriction induced by histamine, LTD_4 or ovalbumin in conscious guinea pigs (Underwood *et al.*, 1994).

3.2.2 Anti-inflammatory Effects

The anti-inflammatory potential of zardaverine can be inferred from a range of *in vitro* studies. Zardaverine inhibits O_2^- generation in opsonized zymosan (OZ)- and C5a-stimulated human eosinophils (Dent *et al.*, 1994a; Hatzelmann *et al.*, 1995), as well as in *N*-formyl-methionyl-L-leucinyl-L-phenylalanine (FMLP)- or OZ-stimulated human PMN (Schudt *et al.*, 1991a,b) and PAF-stimulated guinea pig or bacterial lipopolysaccharide (LPS)-stimulated human alveolar macrophages (Schmidt *et al.*, 1992; Loos *et al.*, 1994). Leukotriene synthesis in LPS-stimulated murine peritoneal macrophages is inhibited (Schade and Schudt, 1993) as well as in FMLP-stimulated human PMN (Schudt *et al.*, 1991b). Synthesis of TNF-α by LPS-stimulated murine peritoneal – as well as human alveolar – macrophages is inhibited (Fischer *et al.*, 1993; Schade and Schudt, 1993; Loos *et al.*, 1994), as is permeability in hydrogen peroxide (H_2O_2)-treated porcine pulmonary endothelial cells (Suttorp *et al.*, 1993), IgE-mediated histamine release from human peripheral leucocytes (Kleine-Tebbe *et al.*, 1992), C5a-stimulated secretion of cationic proteins (ECP and EDN) from human eosinophils in the presence of salbutamol (Hatzelmann *et al.*, 1995) and phytohaemagglutinin-stimulated proliferation of human peripheral blood mononuclear cells (Banner and Page, 1994).

In vitro evidence of inhibition of inflammatory cell functions by zardaverine cannot be adduced necessarily as evidence of anti-inflammatory activity. There is, however, some evidence of inhibition from *in vivo* studies. In the guinea pig, acute (Schudt *et al.*, 1991a; Underwood *et al.*, 1994) or chronic (Banner and Page, 1995) administration of zardaverine reduces the influx of eosinophils into airways during an allergic reaction. Oral zardaverine ($ED_{50} \approx 30$ mg/kg) protects against LPS-induced liver injury in mice (Fischer *et al.*, 1993). Similarly, zardaverine (30 μmol/kg i.p.) suppresses airway hyperresponsiveness resulting from exposure of rats to LPS (Kips *et al.*, 1993).

3.3 OTHERS (TOLAFENTRINE, Org 20241, Org 30029, EMD 54622)

3.3.1 Tolafentrine

Tolafentrine ((-)*cis*-8,9-dimethoxy-2-methyl-6-[4-*p*-toluenesulfonamidophenyl]-1,2,3,4,4*a*,10*b*-hexahydro-benzo-[c][1,6]-naphthyridine; B9004–070) is structurally related to AH 21–132 and has a higher potency for inhibition of PDE activities *in vitro*. Its IC_{50} values for inhibition of PDE isoenzymes 1–5 are 18, 0.8, 0.09, 0.06 and 2 μM, respectively (Schudt *et al.*, 1993b). In accordance with this profile of PDE inhibition, tolafentrine relaxes spontaneous tone in guinea-pig trachea and human bronchus (Schudt *et al.*, 1993b) as well as histamine-precontracted guinea-pig trachea (Beume *et al.*, 1993) at sub-micromolar concentrations *in vitro*. In the anaesthetized guinea pig *in vivo*, tolafentrine inhibits histamine-induced bronchoconstriction most effectively after local administration to the lung and is less effective after oral administration, possibly due to low bioavailability via the oral route in this species (Beume *et al*, 1993). The potential of tolafentrine as an anti-inflammatory agent may be indicated by *in vitro* studies which have revealed inhibition of LPS-stimulated TNF-α synthesis in human alveolar macrophages, as well as inhibition of BMA031-stimulated human peripheral blood lymphocyte proliferation (Schudt *et al.*, 1993a) and of C5a-stimulated human eosinophil functions (O_2^- formation, granule protein secretion) when used in combination with the β_2-adrenoceptor agonist salbutamol (Hatzelmann *et al.*, 1995).

3.3.2 Org 20241

Org 20241 (*N*-hydroxy-4-[3,4-dimethoxyphenyl]-thiazole-2-carboximidamide) has been reported to inhibit PDE3 isoenzymes from rabbit heart and human platelets with IC_{50} values of 25 and 40 μM, respectively, and PDE4 isoenzymes from rabbit heart, bovine trachea, and guinea-pig peritoneal eosinophils and macrophages with IC_{50} values of 6, 0.8, 2 and 3 μM, respectively. At concentrations up to 250 μM, Org 20241 does not affect PDE1, 2 or 5 activities from bovine trachea and rabbit heart (Nicholson *et al.*, 1995). *In vitro* Org 20241 relaxes guinea-pig trachea contracted by histamine or methacholine and inhibits LTB_4-induced H_2O_2 and thromboxane B_2 production by guinea-pig eosinophils (Nicholson *et al.*, 1995). *In vivo*, Org 20241 at a non-bronchodilator dose (5 mg/kg, i.p.) shows anti-inflammatory potential by reducing the number of eosinophils and macrophages in bronchoalveolar lavage, as well as bronchial hyperreactivity, in a guinea-pig model of airway inflammation (Santing *et al.*, 1995). Similarly, Org 20241 (30 μmol/kg, i.p.) abolishes allergen-induced airway eosinophilia and neutrophilia in Brown Norway rats (Elwood *et al.*, 1995).

3.3.3 Org 30029

In rat and rabbit cardiac ventricle, Org 30029 (*N*-hydroxy-5,6-dimethoxybenzo-[b]- thiophene-2-carboximidamide) has been shown to inhibit type 3 and 4 PDE isoenzymes selectively, with IC_{50} values of 25 (rat) and 31 μM (rabbit), and 16 and 15 μM, respectively; PDE1 and PDE2 isoenzymes are unaffected by concentrations up to 250 μM (Shahid and Nicholson, 1990). Likewise, in human atrial myocardium, Org 30029 selectively inhibits PDE3 and PDE4 isoenzymes with IC_{50} values of 67 and 17 μM, respectively (Shahid *et al.*, 1991a). A similar potency of Org 30029 was reported for inhibition of type 4 PDE isoenzymes in porcine aorta (34 μM) and bovine trachea (8 μM) (Cottney *et al.*, 1990), as well as in guinea-pig eosinophils (9 μM) (Dent *et al.*, 1991). In agreement with these studies, Org 30029 inhibits PDE3A and PDE3B isoenzymes (see Chapters 1 and 6 for classification) in human platelets with IC_{50} values of 16 and 17 μM, whereas PDE2 is unaffected at concentrations up to 250 μM. However, in addition to inhibiting PDE3 isoenzymes, Org 30029 also inhibits PDE5 with an IC_{50} of 40 μM (Shahid *et al.*, 1990).

In addition to inhibition of PDE isoenzymes, Org 30029 also has calcium-sensitizing properties and increases the sensitivity of cardiac contractile proteins to calcium (Cottney *et al.*, 1990). By exhibiting such a combination of activities, Org 30029 may be considered as a member of a class of new cardiotonic agents including pimobendan, sulmazole, isomazole, UK-61,260, MCI-154 and EMD53998: compounds which are postulated to be of advantage in the treatment of congestive heart failure compared to type 3-selective PDE inhibitors (Remme, 1993; Endoh, 1995). Consequently, orally administered Org 30029 has been selected for clinical trials for this indication (Cottney *et al.*, 1990). Nevertheless, some preclinical studies have also been conducted to assess the role of Org 30029 in lung pharmacology.

In various species Org 30029 has been shown to relax airway smooth muscle preparations precontracted with either histamine or methacholine *in vitro*; these include guinea-pig tracheal and lung strips (Cottney *et al.*, 1990), bovine trachea (Shahid *et al.*, 1991b) and human bronchi (de Boer *et al.*, 1992). *In vivo*, intravenous Org 30029 inhibits histamine- and methacholine-induced bronchoconstriction in a modified Konzett–Roessler preparation in anaesthetized guinea pigs (Cottney *et al.*, 1990).

3.3.4 EMD 54622

This compound (5-[1-(3,4-dimethoxybenzoyl)-4,4-dimethyl-1,2,3,4-tetrahydrochinolin-6-yl]-6-methyl-3,6-dihydro-1,3,4-thiadiazin-2-one) has been reported to inhibit PDE3 and PDE4 isoenzymes of guinea-pig heart ventricle with IC_{50} values of 1 and 7 μM, respectively (Klockow and Jonas, 1989).

4. *Clinical Experience with PDE 3/4 Inhibitors*

4.1 AH 21–132

An important factor in selecting AH 21–132 for development as an anti-asthma drug was the anticipation of safety, based upon earlier experiences when this compound (as benafentrine) was being developed for treatment of congestive heart failure. Thus, there were no untoward effects when single doses of up to 800 mg were administered to normal subjects or when a daily dose of 600 mg was administered to a group of six patients for three days and a daily dose of 200 mg was administered to a group of 15 patients for 14 days.

Normal subjects who have inhaled methacholine at regular intervals experience obstruction of the airways which can be reversed by sympathomimetics (Foster *et al.*, 1991). In such individuals, oral administration of

AH 21–132 (up to 90 mg) did not produce a bronchodilator effect. However, intravenous administration (20 and 40 mg) produced transient bronchodilation whereas inhalation produced a dose-related bronchodilation with an ID_{50} of 9.2 mg (Foster *et al.*, 1992). In a group of 14 asthma patients, oral administration of AH 21–132 produced significant bronchodilation after a single dose of 400 mg but not after a single dose of 100 mg.

In allergic asthmatics, regular administration of AH 21–132 (3 mg t.i.d., 10 mg t.i.d. or 30 mg t.i.d. for seven days prior to inhalation of antigen) did not inhibit acute allergic bronchospasm but was associated with suppression of the late response in 0/3, 1/2 and 2/3 respectively as compared with 0/8 in patients receiving placebo (data on file). The implication that anti-inflammatory effects might be manifest at doses that were not overtly bronchodilator received support from a further study in which the effect of withdrawal of steroid therapy in steroid-dependent asthma patients could be maintained in 2/3 patients receiving AH 21–132 (30 mg t.i.d. over the four-week period of withdrawal) as compared with 1/5 patients treated with placebo for the same period.

4.2 ZARDAVERINE

Approximately 250 patients were included in clinical studies with inhaled zardaverine. Single doses up to 10 mg and repeated doses up to 24 mg/day administered by a metered-dose inhaler were tested. When the clinical development was discontinued because the dosing frequency and the risk–benefit ratio were not considered to be ideal, not all of the studies had been completed in order to allow reliable conclusions. However, some of the studies which gave certain insights in the clinical profile of zardaverine are summarized in Table 10.1.

4.2.1 Asthma

4.2.1.1 Bronchospasmolytic Effects

In the first study, 12 asthmatic patients with a forced expiratory volume in one second (FEV_1) of 50–85% predicted and a ΔFEV_1 of 15–36% after inhalation of 200 μg salbutamol were included; 11 of the 12 patients were available for efficacy analysis (Brunneé *et al.*, 1992). Four doses of 1.5 mg zardaverine were inhaled at 15 min intervals and compared with placebo. The maximum improvement of lung function was seen after the third puff (i.e. 4.5 mg zardaverine) and amounted to 0.23 kPa/s for SG_{aw} (specific airways conductance) and 0.31 for FEV_1. Increases of at least 40% in SG_{aw} and of at least 15% in FEV_1 were seen in eight and seven patients, respectively. The improvements in SG_{aw} and FEV_1 were statistically significant (at least $P < 0.05$) during the first hour of inhalation but not during the entire recording period of almost 5 hours, although some individuals showed a sustained improvement. In a second study, 14 asthmatics inhaled a single dose of 4.5 mg zardaverine (data on file). The FEV_1 time-averaged 15 to 240 min after drug inhalation improved significantly compared with placebo ($P < 0.05$).

Both studies suggested – and this was confirmed later on in further trials – that only a subpopulation of asthmatics respond to zardaverine, although no criteria were obvious to further characterize the responders. Overall, improvements in FEV_1 of at least 10% and 15% were seen in approximately 55% and 30%, respectively, of the patients studied. Due to this finding,

Table 10.1 Summary of clinical studies performed with inhaled zardaverine

Reference	*Design*	*Reference drug(s)*	*No. of patients*	*Duration of treatment*	*Zardaverine dosage*
Asthma					
Brunée *et al.*, (1992)	db, r, co	placebo	12	single dose	6 mg
Data on file	db, r, co	placebo	14	single dose	4.5 mg
Wempe *et al.* (1992)	db, r, co	placebo, 300 μg salbutamol	10	single dose	1.5, 3, 6 mg
Jordan *et al.* (1993)	db, r, co	placebo	13	single dose	2, 4 mg
Data on file	db, r, co	placebo	3	single dose	6 mg
Data on file	open		30	single dose	6, 8, 10 mg
Data on file	db, r, co	placebo	7	1 week	6 mg q.i.d.
Koper *et al.* (1993)	db, r, pg	placebo	21	4 weeks	4 mg q.i.d.
Data on file	db, r, pg	placebo, 100 μg BDP q.i.d.	33	4 weeks	4 mg q.i.d.
Chronic obstructive pulmonary disease					
Data on file	db, r, co	placebo	11	single dose	4.5 mg
Ukena *et al.* (1992)	db, r, co	placebo, 300 μg salbutamol	38	single dose	1.5, 3, 6 mg

Abbreviations: co, crossover; db, double-bind; pg, parallel group; r, randomized.

it was decided to start investigations of the dose–effect relationship in zardaverine responders only. For this purpose a responder was defined as showing an improvement in FEV_1 of more than 10% during the first 30 min after inhalation.

In the first study (Wempe *et al.*, 1992), 10 asthmatic patients (FEV_1 = 52–102% predicted) were selected based on an improvement in FEV_1 of more than 15% after inhalation of 200 μg salbutamol (range 16–78%) and more than 10% after inhalation of 6 mg zardaverine (range 11–33%); FEV_1 was measured during 5 hours after inhalation. A short-lasting improvement in FEV_1 on zardaverine treatment was seen, which reached its maximum at 10–20 min after inhalation and amounted to a mean of 0.41 l after 6 mg versus baseline (Fig. 10.2). The change in FEV_1 was significant ($P < 0.05$) for 1.5 and 6 mg zardaverine 20 min after inhalation, but not for 3 mg zardaverine. The FEV_1 time averages during the first hour and the 5 hours after inhalation did not differ significantly from placebo. The responses to 6 mg zardaverine during the pre-test and during the double-blind recording session were well comparable. No dose–effect relationship could be established for zardaverine. The response to 300 μg salbutamol was statistically superior to both placebo and zardaverine throughout the recording time ($P < 0.001$).

In another trial, 13 asthmatic patients (FEV_1 = 45–81% predicted) who showed a mean improvement in FEV_1 of 32% on baseline to 400 μg fenoterol and 20% to 6 mg zardaverine during the baseline period of the study inhaled 2 and 4 mg zardaverine and placebo in a randomized order (Jordan *et al.*, 1993). The FEV_1 increases at both zardaverine doses were statistically significant compared to placebo throughout the 4 hour recording interval (at least $P < 0.05$) and amounted maximally to 0.46 and 0.49 l for 2 and 4 mg zardaverine, respectively (Fig. 10.3). Again, however, no significant difference between the two zardaverine doses was detected. Furthermore, no dose–response relationship was apparent in 30 patients who inhaled single doses of 6, 8 and 10 mg zardaverine (data on file).

As it is well-known for β_2-adrenoceptor agonists that the extent of response is dependent on the time of administration (reviewed in Smolenski and D'Alonzo, 1993), a study was set up to investigate the bronchospasmolytic effect of 6 mg zardaverine at 6 a.m. and 6 p.m. (data on file). Patients selected for showing a $\Delta FEV_1 > 10\%$ after inhalation of 6 mg zardaverine at noon showed a rapid and pronounced bronchospasmolysis when the same dose was given at 6 a.m. (ΔFEV_1 = 0.55 l 15 min post-inhalation). However, when the drug was given at 6 p.m., no FEV_1 change at all was observed compared to placebo. Although the limited number of patients (n = 3) does not allow any valid conclusions, one might speculate that the time point of administration is a critical factor for the efficacy of PDE inhibitors. As PDE activity has been shown to follow a circadian pattern in rats (Lemmer *et al.*, 1987) and circadian rhythms also play an important role in asthma (Smolenski and D'Alonzo, 1993), it might be worthwhile to study this issue further in humans.

Taken together, these data suggest that inhaled zardaverine causes a modest bronchodilation in asthmatic patients. However, the extent as well as the duration of effects are quite variable.

4.2.1.2 *Anti-inflammatory Effects*

The effect of zardaverine on early (EAR) and late asthmatic reactions (LAR) after challenge with house

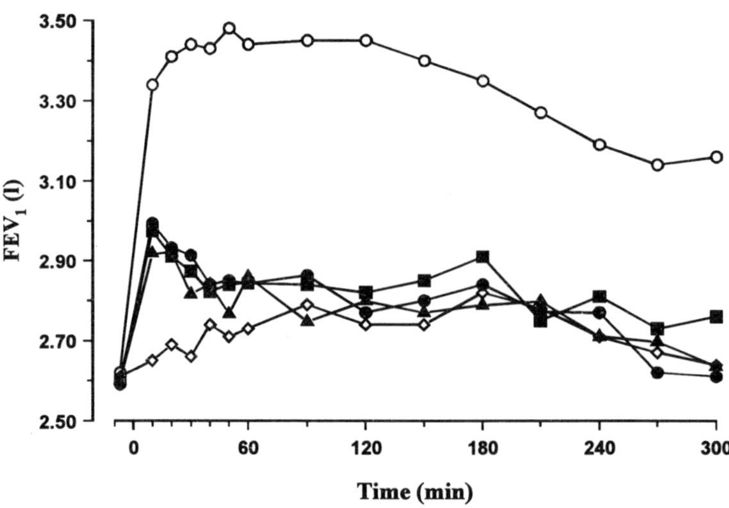

Figure 10.2 Effect of placebo (◇), zardaverine 1.5 mg (■), 3 mg (▲) and 6 mg (●) and salbutamol 300 µg (○) on mean FEV_1 ($n = 10$).

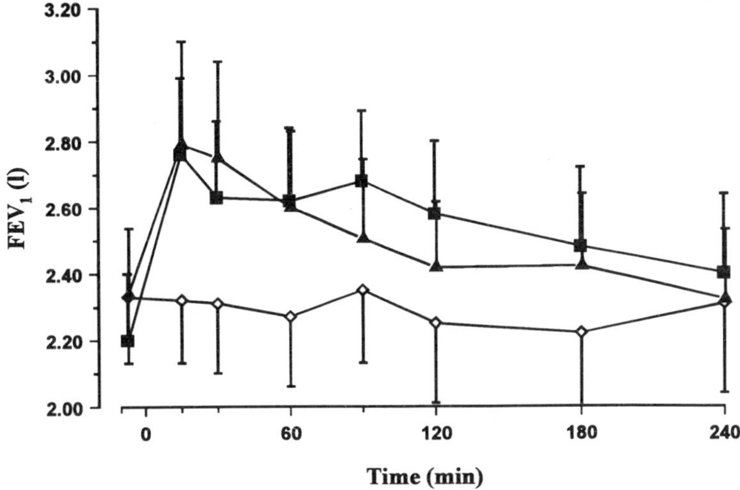

Figure 10.3 Effects of placebo (◇) or a single dose of 2 mg (■) or 4 mg zardaverine (▲) on mean (±SEM) FEV_1 ($n = 13$). Data taken from Jordan *et al.* (1993).

dust mite was addressed in seven patients who inhaled 6 mg zardaverine q.i.d. and placebo for one week each in a randomized order. The last dose of the trial medication was inhaled on day 8, 30 min prior to challenge. During the provocation test, the allergen dose was individually increased up to the dose that had increased (sR_{aw}) specific airways resistance by at least 100% during the baseline period. Five patients completed the study; two withdrew because of adverse effects of zardaverine (see section 4.2.3). Individual data as well as the mean value curve are shown in Fig. 10.4. Overall, the EAR was reduced by zardaverine by about 50%, whereas the LAR was only slightly affected. Individual data, however, suggest that zardaverine decreased LAR in three patients.

In order to study the long-term effects on FEV_1, diurnal peak expiratory flow rate variation, symptoms, use of β_2-adrenoceptor agonists and hyperreactivity to histamine, 21 asthmatic patients (FEV_1 = 51–103% predicted) were treated for four weeks with either 4 mg zardaverine q.i.d. or placebo (Koper *et al.*, 1993). The provocative concentration of histamine reducing FEV_1 by 20% ($PC_{20}FEV_1$) was determined at the beginning and the end of the treatment period, the latter 8 hours after the last use of the trial medication. Compared to baseline, $PC_{20}FEV_1$ values were virtually unchanged (zardaverine: before 0.12 mg/ml (range 0.04–0.44 mg/ml), after 0.11 mg/ml (0.04–0.3 mg/ml); placebo: before 0.13 mg/ml (0.04–0.42 mg/ml), after 0.17 mg/ml (0.04–0.77 mg/ml)). There were also no differences between drug and placebo for the other parameters tested.

In addition, the effect of a 4-week treatment with 4 mg zardaverine q.i.d. on hyperreactivity to carbachol was compared with placebo and 100 μg beclomethasone dipropionate q.i.d. (data on file). Although, owing to the small number of patients, no significant effect was found for any of the groups, zardaverine might have improved hyperreactivity in some of the patients (Fig. 10.5).

4.2.2 Chronic Obstructive Pulmonary Disease (COPD)

The acute effect of zardaverine on pulmonary function in patients with COPD was evaluated in two studies. In the first (data on file), 11 patients (FEV_1 = 46–79% predicted, $\Delta FEV_1 \geq 15\%$ after inhalation of 200 μg salbutamol) were included, who inhaled a single dose of 4.5 mg zardaverine and placebo in a randomized order. No significant improvement compared with placebo was found for FEV_1 R_{aw}, MEF_{25} or MEF_{75} (maximal expiratory flow at 25% and 75% of forced vital capacity) . This result was confirmed in a second trial (Ukena *et al.*, 1992).

4.2.3 Safety

A total of 202 patients inhaled either a single dose or up to three doses of zardaverine (1.5–10 mg). Most frequently reported adverse events were local (unpleasant taste, paraesthesia of tongue, itchy throat; 14 cases). Otherwise nausea (5 cases), headache (4 cases), dizziness (4 cases) and vomiting (2 cases) occurred. Thus, in general, single doses of zardaverine were well tolerated. This was probably due to the use of a spacer device which was used in all studies except the first (Brunnée *et al.*, 1992). With repeated administration, however, the spacer did not prevent the more frequent occurrence of nausea and

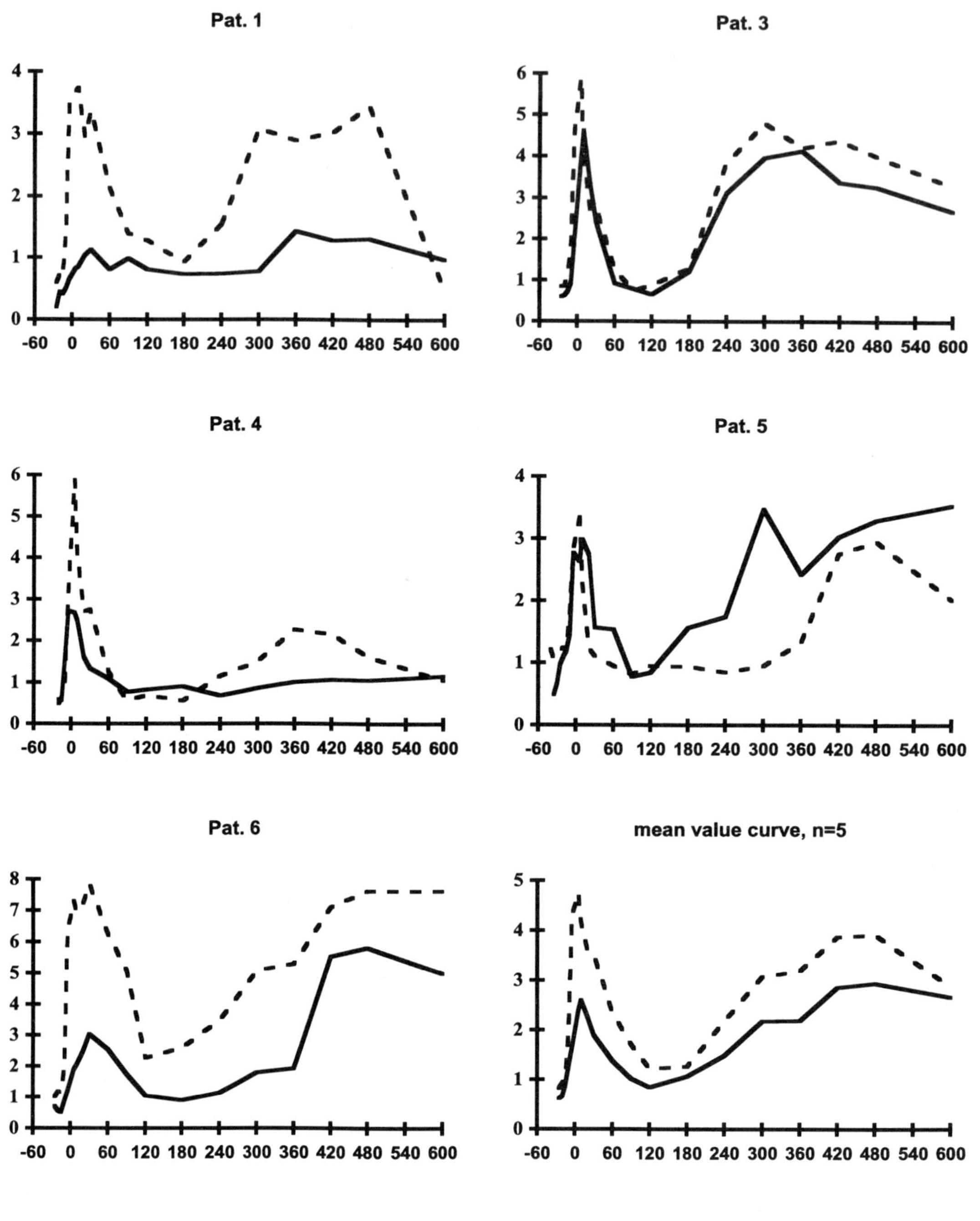

Figure 10.4 Effects of zardaverine on early and late asthmatic reactions after challenge with house dust mite. Solid lines: zardaverine; broken lines: placebo.

vomiting. For example, in the allergen challenge study four out of seven subjects inhaling 6 mg zardaverine q.i.d. complained of nausea and two withdrew from the trial because of this adverse effect. In the 4 week studies nausea also occurred with unacceptably high frequency (eight out of 17 patients inhaling 4 mg zardaverine q.i.d.) (Koper *et al.*, 1993). Hence, 4 or 6 mg q.i.d. was associated with an excessively high incidence of adverse events and was shown to be unsuitable for long-term treatment. It can only be speculated as to whether these events also affected compliance and thus efficacy data.

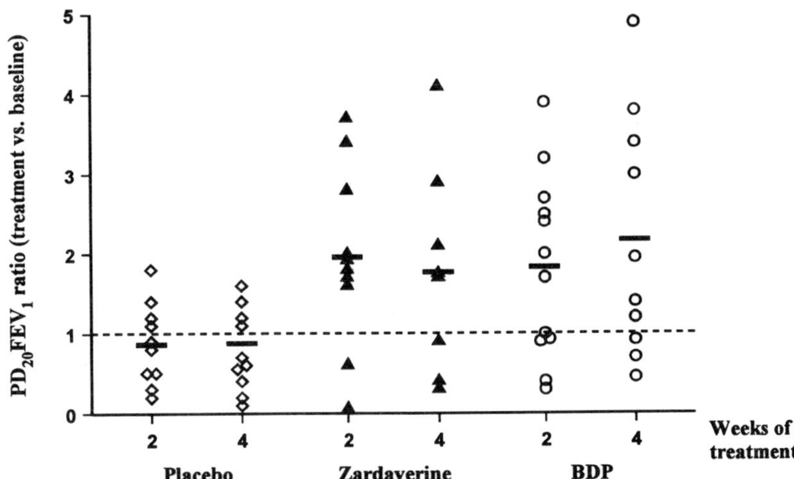

Figure 10.5 Ratios of the $PD_{20}FEV_1$ values for carbachol measured after 2 and 4 weeks of treatment versus baseline for placebo (◇), 4 mg zardaverine (▲) and 100 µg beclomethasone dipropionate (BDP, ○) q.i.d. A ratio of 1 (broken line) indicates that no change has occurred. Points are data for individual subjects; horizontal lines represent geometric mean.

4.3 OTHER SELECTIVE PDE INHIBITORS (TIBENELAST, ENOXIMONE/ISOMAZOLE/ SDZ-MKS 492, ZAPRINAST)

As reviewed by Torphy *et al.* (1994), the published clinical data on the effect of monoselective PDE inhibitors in airway diseases are very limited.

There is only one reported study on the partial prevention of spontaneous bronchospasm in asthmatic patients by a selective PDE4 inhibitor (tibenelast, LY186655) (Israel *et al.*, 1988). Among PDE3-selective inhibitors, enoximone (MDL17,043) displays bronchodilating and pulmonary vasodilating properties in COPD patients (Leeman *et al.*, 1987). The latter effect on the pulmonary vascular system was confirmed by another study with patients undergoing cardiac surgery; in this study milrinone significantly reduced pulmonary vascular resistance concomitant with an increase in cardiac index (Harris *et al.*, 1992). The bronchodilatory potential of PDE3 inhibitors is also evident from recent clinical studies using isomazole in asthmatic patients (Koper *et al.*, 1994) as well as SDZ-MKS 492 in allergen-challenged asthmatic subjects (Bardin *et al.*, 1994). Two clinical trials have been reported for the PDE5(1)-selective inhibitor zaprinast (M&B 22,948; see Chapter 8). Whereas in the first study (Rudd *et al.*, 1983) zaprinast reduced exercise-induced bronchospasm in adult asthmatics, no effects on exercise-induced bronchoconstriction were found in a subsequent evaluation in asthmatic children (Reiser *et al.*, 1986).

5. *Conclusions*

Despite having prominent side-effects, theophylline has long been recognized as an effective therapy for asthma whose benefits arise largely – if not exclusively – from inhibition of phosphodiesterases. Hence, the recognition that PDEs comprise a series of distinct isoenzymes raised the possibility that superior anti-asthma therapies might be developed on the basis of a more selective inhibition of PDE isoenzymes than that evidenced by theophylline. To achieve this strategy, it is necessary to identify those PDE isoenzymes for which inhibition is a necessary part of anti-asthma activity.

Type 4 PDE isoenzymes are represented both in inflammatory cells (e.g. eosinophil, mast cell), whose activation contributes to inflammatory features of asthma, and in airway smooth muscle. Type 3 PDE isoenzymes are represented in platelets and in airway smooth muscle. Thus, inhibition of PDE4 isoenzymes is associated with both anti-inflammatory and spasmolytic activities whereas inhibition of PDE3 is associated with spasmolysis and suppression of hyperresponsiveness of the airways. The reductionistic approach of selective inhibition of PDE4 isoenzymes stems from a belief that inflammatory events necessarily determine hyperresponsiveness (British Thoracic Society, 1990). If, on the other hand, inflammatory events are viewed as being associated with – as opposed to being determinants of – hyperresponsiveness (Chapman *et al.*, 1993), then inhibition of both PDE3 and PDE4 isoenzymes will be preferred. At present, there is insufficient clinical or laboratory evidence to establish that inhibition of type 4 PDE isoenzymes will be as effective as inhibition of both enzyme categories. In electing to focus upon inhibition of PDE4 rather than PDE3, it has been opined that inhibition of PDE3 must necessarily be associated with cardiovascular side-effects. As suggested by AH 21–132, such presumption may be invalid. Until there is a clarification of the respective roles of type 3 and type 4 isoenzymes in side-effect profiles of PDE inhibitors, it would seem prudent to retain inhibition of both enzymes as a basis of asthma therapy, especially as it seems likely that there may be synergic interactions between the two categories of inhibitor when present together.

6. References

Banner, K.H. and Page, C.P. (1994). Effect of type III and IV isozyme selective phosphodiesterase (PDE) inhibitors on proliferation of human peripheral blood mononuclear cells. Br. J. Pharmacol. 113, 89P. [Abstract]

Banner, K.H. and Page, C.P. (1995). Acute versus chronic administration of phosphodiesterase inhibitors on allergen-induced pulmonary cell influx in sensitized guinea-pigs. Br. J. Pharmacol. 114, 93–98.

Bardin, P.G., Dorward, M.A., Franke, B. and Holgate, S.T. (1994). Inhibition of the early and late asthmatic responses to inhaled allergen by MKS 492, a selective phosphodiesterase inhibitor. S. Afr. Med. J. 84, 769.

Barnes, P.J. (1989). A new approach to the treatment of asthma. N. Engl. J. Med. 321, 1517–1527.

Barnes, P.J. and Pauwels, R.A. (1994). Theophylline in the management of asthma: time for reappraisal? Eur. Respir. J. 7, 579–591.

Beume, R., Kilian, U., Brand, U., Hafner, D., Eltze, M. and Flockerzi, D. (1993). The bronchospasmolytic effect of the PDE III/IV inhibitors B9004–070 and zardaverine – dependency on the route of administration in guinea pigs. Am. Rev. Respir. Dis. 147, A184. [Abstract]

Bewley, J.S. and Chapman, I.D. (1988). AH 21–132 a novel relaxant of airway smooth muscle. Br. J. Pharmacol. 93, 52P. [Abstract]

British Thoracic Society, Research Unit of the Royal College of Physicians, King's Fund Centre and National Asthma Campaign (1990). Guidelines for management of asthma in adults. I. Chronic persistent asthma. Br. Med. J. 301, 651–653.

Brunneé, T., Engelstätter, R., Steinijans, V.W. and Kunkel, G. (1992). Bronchodilatory effect of inhaled zardaverine, a phosphodiesterase III and IV inhibitor, in patients with asthma. Eur. Respir. J. 5, 982–985.

Chapman, I.D., Foster, A. and Morley, J. (1993). The relationship between inflammation and hyperreactivity of the airways in asthma. Clin. Exp. Immunol. 23, 168–171.

Christensen, S.B. and Torphy, T.J. (1994). Isozyme-selective phosphodiesterase inhibitors as antiasthmatic agents. Ann. Rep. Med. Chem. 29, 185–194.

Cortijo, J., Bou, J., Beleta, J., Cardelús, I., Llenas, J., Morcillo, E. and Gristwood, R.W. (1993). Investigation into the role of phosphodiesterase IV in bronchorelaxation, including studies with human bronchus. Br. J. Pharmacol. 108, 562–568.

Cottney, J., Logan, R., Marshall, R., Nicholson, D., Shahid, M. and Walker, G. (1990). Org30029: a new cardiotonic agent possessing both phosphodiesterase inhibitory and calcium-sensitising properties. Cardiovasc. Drug Rev. 8, 179–202.

de Boer, J., Philpott, A.J., van Amsterdam, R.G.M., Shahid, M., Zaagsma, J. and Nicholson, C.D. (1992). Human bronchial cyclic nucleotide phosphodiesterase isoenzymes: biochemical and pharmacological analysis using selective inhibitors. Br. J. Pharmacol. 106, 1028–1034.

Dent, G., Giembycz, M.A., Rabe, K.F. and Barnes, P.J. (1991). Inhibition of eosinophil cyclic nucleotide PDE activity and opsonized zymosan-stimulated respiratory burst by "type IV" – selective PDE inhibitors. Br. J. Pharmacol. 103, 1339–1346.

Dent, G., Giembycz, M.A., Evans, P.M., Rabe, K.F. and Barnes, P.J. (1994a). Suppression of human eosinophil respiratory burst and cyclic AMP hydrolysis by inhibitors of type IV phosphodiesterase: interaction with the *beta* adrenoceptor agonist albuterol. J. Pharmacol. Exp. Ther. 271, 1167–1174.

Dent, G., Magnussen, H. and Rabe, K.F. (1994b). Cyclic nucleotide phosphodiesterases in the human lung. Lung 172, 129–146.

Elwood, W., Sun, J., Barnes, P.J., Giembycz, M.A. and Chung, K.F. (1995). Inhibition of allergen-induced lung eosinophilia by type-III and combined type III- and IV-selective phosphodiesterase inhibitors in Brown-Norway rats. Inflamm. Res. 44, 83–86.

Endoh, M. (1995). The effects of various drugs on the myocardial inotropic response. Gen. Pharmacol. 26, 1–31.

Fischer, W., Schudt, C. and Wendel, A. (1993). Protection by phosphodiesterase inhibitors against endotoxin-induced liver injury in galactosamine-sensitized mice. Biochem. Pharmacol. 45, 2399–2404.

Foster, R.W., Atanga, G.K., Carpenter, J.R., Evans, D.E., Rakshi, K. and Small, R.C. (1991). A method for bioassay of potency and effectiveness of inhaled bronchodilators in normal subjects. Br. J. Clin. Pharmacol. 31, 445–455.

Foster, R.W., Rakshi, K., Carpenter, J.R. and Small, R.C. (1992). Trials of the bronchodilator activity of the isoenzyme-selective phosphodiesterase inhibitor AH 21–132 in healthy volunteers during a methacholine challenge test. Br. J. Clin. Pharmacol. 34, 527–534.

Giembycz, M.A. (1992). Could isoenzyme-selective phosphodiesterase inhibitors render bronchodilator therapy redundant in the treatment of bronchial asthma? Biochem. Pharmacol. 43, 2041–2051.

Giembycz, M.A. and Barnes. P.J. (1991). Selective inhibition of a high affinity type IV cyclic AMP phosphodiesterase in bovine trachealis by AH 21–132. Biochem. Pharmacol. 42, 663–677.

Giembycz, M.A., Chilvers, E.R. and Barnes, P.J. (1990). Biochemical and pharmacological interactions between the papaverine derivative AH 21–132 and methacholine in bovine tracheal smooth muscle. Eur. J. Pharmacol. 183, 2157–2158.

Giembycz, M.A., Corrigan, C.J., Kay, A.B. and Barnes, P.J. (1994). Inhibition of CD4 and CD8 T lymphocyte (T-LC) proliferation and cytokine secretion by isoenzyme-selective phosphodiesterase (PDE) inhibitors: correlation with intracellular cyclic AMP (cAMP) concentrations. J. Allergy Clin. Immunol. 93, 167. [Abstract]

Harris, A.L., Connell, M.J., Ferguson, E.W., Wallace, A.M., Gordon, R.J., Pagani, E.D. and Silver, P.J. (1989). Role of low K_M cyclic AMP phosphodiesterase inhibition in tracheal relaxation and bronchodilation in the guinea pig. J. Pharmacol. Exp. Ther. 251, 199–206.

Harris, M.N.E., Dabom, A.K. and O'Dwyer, J.P. (1992). Milrinone and the pulmonary vascular system. Eur. J. Anaesthesiol. 5 (Suppl.), 27–30.

Hatzelmann, A., Tenor, H. and Schudt, C. (1995). Differential effects of non-selective and selective phosphodiesterase inhibitors on human eosinophil functions. Br. J. Pharmacol. 114, 821–831.

Howell, R.E., Sickels, B.D. and Woeppel, S.L. (1993). Pulmonary antiallergic and bronchodilator effects of isozyme-selective phosphodiesterase inhibitors in guinea pigs. J. Pharmacol. Exp. Ther. 264, 609–615.

Hoymann, H.G., Heinrich, U., Beume, R. and Kilian, U. (1994). Comparative investigation of the effects of zardaverine and theophylline on pulmonary functions in rats. Exp. Lung Res. 20, 235–250.

Israel, E., Mathur, P.N., Tashkin, D. and Drazen, J.M. (1988). LY186655 prevents bronchospasm in asthma of moderate severity. Chest 94, 71S. [Abstract]

Jordan, K., Fischer, J., Engelstätter, R. and Steinijans, V. (1993). Einfluβ eines inhalierbaren selektiven Phosphodiesterase-Hemmers (PDE III/IV-Hemmer Zardaverin) auf die Lungenfunktion von Patienten mit Asthma bronchiale. Atemw. -Lungenkrkh. 19, 358–359.

Kilian, U., Beume, R., Eltze, M. and Schudt, C. (1989). Is phosphodiesterase inhibition a relevant bronchospasmolytic principle? Agents Actions Suppl. 28, 331–348.

Kips, J.C., Joos, G.F., Peleman, R.A. and Pauwels, R.A. (1993). The effect of zardaverine, an inhibitor of phosphodiesterase isoenzymes III and IV, on endotoxin-induced airway changes in rats. Clin. Exp. Allergy 23, 518–523.

Kleine-Tebbe, J., Wicht, L., Gagne, H., Friese, A., Schunack, W., Schudt, C. and Kunkel, G. (1992). Inhibition of IgE-mediated histamine release from human peripheral leukocytes by selective phosphodiesterase inhibitors. Agents Actions 36, 200–206.

Klockow, M. and Jonas, R. (1989). Particulate cAMP-specific phosphodiesterase (P-PDE) in cardiac ventricle of guinea pig. Naunyn-Schmiedebergs Arch. Pharmacol. 339, R53. [Abstract]

Knowles, I.D., Ball, D.I., Nials, A.T. and Coleman, R.A. (1994). A comparison of the relaxant activity of a range of phosphodiesterase (PDE) inhibitors on human bronchus, ferret trachea and guinea-pig trachea and bronchus. Br. J. Pharmacol. 113, 162P. [Abstract]

Koper, I., Braun, H., Ukena, D., Reiber, C., Engelstätter, R. and Sybrecht, G.W. (1993). Effect of zardaverine, a phosphodiesterase type III/IV inhibitor, on hyperreactivity, lung function and symptoms in asthmatic patients. Am. Rev. Respir. Dis. 147, A295. [Abstract]

Koper, I., Stelmer, B., Ukena, D., Krause, G. and Sybrecht, G.W. (1994). Isomazole, a phosphodiesterase III selective inhibitor and its bronchodilatory effects in patients with obstructive lung disease. Eur. Respir. J. 7 (Suppl. 18), 150s. [Abstract]

Leeman, M., Lejeune, P., Melot, C. and Naeije, R. (1987). Reduction in pulmonary hypertension and in airway resistances by enoximone (MDL17,043) in decompensated COPD. Chest 91, 662–666.

Lemmer, B., Lang, P.-H., Schmidt, S. and Baermeier, H. (1987). Evidence for circadian rhythmicity of the β-adrenoceptor–adenylcyclase–cAMP–phosphodiesterase system in the rat. J. Cardiovasc. Pharmacol. 10 (Suppl. 4), s138–s140.

Loos, U., Mallmann, P., Schudt, C. and Luederitz, B. (1994). Anti-inflammatory activity of newer phosphodiesterase (PDE) inhibitors on alveolar macrophages. Tuber. Lung Dis. 75 (Suppl. 1), 121–122.

McFadden, E.R. (1989). Therapy of acute asthma. J. Allergy Clin. Immunol. 84, 151–158.

Morley, J. (1991). New drug developments for asthma. In "Preventive Therapy in Asthma" (ed. J. Morley), pp. 253–270. Academic Press, London.

Morley, J. (1994). Potassium channel openers and suppression of airway hyperreactivity. Trends Pharmacol. Sci. 15, 463–468.

Nicholson, C.D. and Shahid, M. (1994). Inhibitors of cyclic nucleotide phosphodiesterase isoenzymes – their potential utility in the therapy of asthma. Pulmonary Pharmacol. 7, 1–17.

Nicholson, C.D., Shahid, M., Bruin, J., Barron, E., Spiers, I., de Boer, J., van Amsterdam, R.G.M., Zaagsma, J., Kelly, J.J., Dent, G., Giembycz, M.A. and Barnes, P.J. (1995). Characterization of ORG 20241, a combined phosphodiesterase IV/III cyclic nucleotide phosphodiesterase inhibitor for asthma. J. Pharmacol. Exp. Ther. 274, 678–687.

Offermeier, J. and van den Brink, F.G. (1974). The antagonism between cholinomimetic agonists and β-adrenoceptor stimulants: the differentiation between functional and metaffinoid antagonism. Eur. J. Pharmacol. 27, 206–213.

Qian, Y., Naline, E., Karlsson, J.-A., Raeburn, D. and Advenier, C. (1993). Effects of rolipram and siguazodan on the human isolated bronchus and their interaction with isoprenaline and sodium nitroprusside. Br. J. Pharmacol. 109, 774–778.

Rabe, K.F., Tenor, H., Dent, G., Schudt, C., Liebig, S. and Magnussen, H. (1993). Phosphodiesterase isozymes modulating inherent tone in human airways: identification and characterization. Am. J. Physiol. 264, L458–L464.

Reiser, J., Yeang, Y. and Warner, J.O. (1986). The effect of zaprinast (M&B22,948, an orally absorbed mast cell stabilizer) on exercise-induced asthma in children. Br. J. Dis. Chest 80, 157–163.

Remme, W.J. (1993). Inodilator therapy for heart failure: early, late, or not at all? Circulation 87 (Suppl. IV), 97–107.

Robicsek, S.A., Blanchard, D.K., Djeu, J.Y., Krzanowski, J.J., Szentivanyi, A. and Polson, J.B. (1991). Multiple high affinity cAMP-phosphodiesterases in human T-lymphocytes. Biochem. Pharmacol. 42, 869–877.

Rudd, R.M., Gellert, A.R., Studdy, P.R. and Geddes, D.M. (1983). Inhibition of exercise-induced asthma by an orally absorbed mast cell stabilizer (M&B22,948). Br. J. Dis. Chest 77, 78–86.

Sanjar, S., Smith, D., Kristersson, A. and Morley, J. (1989). Incubation of platelets with PAF produces a factor which causes airway hyperreactivity in guinea-pigs. Br. J. Pharmacol. 96, 75P. [Abstract]

Santing, R.E., Olymulder, C.G., van der Molen, K., Meurs, H. and Zaagsma, J. (1995). Phosphodiesterase inhibitors reduce bronchial hyperreactivity and airway inflammation in unrestrained guinea pigs. Eur. J. Pharmacol. 275, 75–82.

Schade, F.U. and Schudt, C. (1993). The specific type III and IV phosphodiesterase inhibitor zardaverine suppresses formation of tumor necrosis factor by macrophages. Eur. J. Pharmacol. 230, 9–14.

Schmidt, J., Lindstaedt, R. and Szelenyi, I. (1992). Characterization of platelet-activating factor induced superoxide anion generation by guinea-pig alveolar macrophages. J. Lipid Mediat. 5, 13–22.

Schudt, C., Winder, S., Eltze, M., Kilian, U. and Beume, R. (1991a). Zardaverine: a cyclic AMP specific PDE III/IV inhibitor. Agents Actions Suppl. 34, 379–402.

Schudt, C., Winder, S., Forderkunz, S., Hatzelmann, A. and Ullrich, V. (1991b). Influence of selective phosphodiesterase inhibitors on human neutrophil functions and levels of cAMP and Ca_i. Naunyn-Schmiedebergs Arch. Pharmacol. 344, 682–690.

Schudt, C., Winder, S., Mueller, B. and Ukena, D. (1991c). Zardaverine as a selective inhibitor of phosphodiesterase isozymes. Biochem. Pharmacol. 42, 153–162.

Schudt, C., Tenor, H., Loos, U., Mallmann, P., Szame, M. and Resch, K. (1993a). Effect of selective phosphodiesterase (PDE) inhibitors on activation of human macrophages and lymphocytes. Eur. Respir. J. 6, 367s. [Abstract]

Schudt, C., Tenor, H., Wendel, A., Eltze, M., Magnussen, H. and Rabe, K.F. (1993b). Influence of the PDE III/IV inhibitor B9004–070 on contraction and PDE activities in airway and vascular smooth muscle. Am. Rev. Respir. Dis. 147, A183. [Abstract]

Schudt, C., Tenor, H. and Hatzelmann, A. (1995). PDE isoenzymes as targets for anti-asthma drugs. Eur. Respir. J. 8, 1179–1183.

Shahid, M. and Nicholson, C.D. (1990). Comparison of cyclic nucleotide phosphodiesterase isoenzymes in rat and rabbit ventricular myocardium: positive inotropic and phosphodiesterase inhibitory effects of Org30029, milrinone and rolipram. Naunyn-Schmiedebergs Arch. Pharmacol. 342, 698–705.

Shahid, M., Holbrook, M., Coker, S.J. and Nicholson, C.D. (1990). Characterization of human platelet cyclic nucleotide phosphodiesterase (PDE) isoenzymes and their sensitivity to a variety of selective inhibitors. Br. J. Pharmacol. 100, 443P. [Abstract]

Shahid, M., Bruin, J.C., Walker, G.B., Cottney, J.E. and Nicholson, C.D. (1991a). Positive inotropic, Ca^{2+} sensitizing and cyclic nucleotide phosphodiesterase (PDE) isoenzyme inhibitory effects of Org30029 in human atrial myocardium. Br. J. Pharmacol. 104, 17P. [Abstract]

Shahid, M., van Amsterdam, R.G.M., de Boer, J., ten Berge, R.E. and Nicholson, C.D. (1991b). The presence of five cyclic nucleotide phosphodiesterase isoenzyme activities in bovine tracheal smooth muscle and the functional effects of selective inhibitors. Br. J. Pharmacol. 104, 471–477.

Small, R.C., Boyle, J.P., Duty, S., Elliott, K.R.F., Foster, R.W. and Watt, A.J. (1989). Analysis of the relaxant effects of AH 21–132 in guinea-pig isolated trachealis. Br. J. Pharmacol. 97, 1165–1173.

Small, R.C., Berry, J.L., Boyle, J.P., Chapman, I.D., Elliott, K.R.F., Foster, R.W. and Watt, A.J. (1991). Biochemical and electrical aspects of the tracheal relaxant action of AH 21–132. Eur. J. Pharmacol. 192, 417–426.

Smolenski, M.H. and D'Alonzo, G.E. (1993). Medical chronobiology: concepts and applications. Am. Rev. Respir. Dis. 147, 2–19.

Suttorp, N., Weber, U., Welsch, T. and Schudt, C. (1993). Role of phosphodiesterases in the regulation of endothelial permeability in vitro. J. Clin. Invest. 91, 1421–1428.

Torphy, T.J. and Undem, B.J. (1991). Phosphodiesterase inhibitors: new opportunities for the treatment of asthma. Thorax 46, 512–523.

Torphy, T.J., Rinard, G.A., Rietow, M.G. and Mayer, S.E. (1983). Functional antagonism in canine tracheal smooth muscle: inhibition by methacholine of the mechanical and biochemical responses to isoproterenol. J. Pharmacol. Exp. Ther. 227, 694–699.

Torphy, T.J., Undem, B.J., Cieslinski, L.B., Luttmann, M.A., Reeves, M.L. and Hay, D.W.P. (1993). Identification, characterization and functional role of phosphodiesterase isozymes in human airway smooth muscle. J. Pharmacol. Exp. Ther. 265, 1213–1223.

Torphy, T.J., Murray, K.J. and Arch, J.R.S. (1994). Selective phosphodiesterase isozyme inhibitors. In "Drugs and the Lung" (eds. C.P. Page and W.J. Metzger), pp. 397–447. Raven, New York.

Turner, N.C., Dolan, J.S., Grimsditch, D., Lamb, J., Worby, A., Murray, K.J., Coates, W.J. and Warrington, B.H. (1994). Pulmonary effects of type V cyclic GMP specific phosphodiesterase inhibition in the anaesthetized guinea-pig. Br. J. Pharmacol. 111, 1198–1204.

Ukena, D., Renz, K., Reiber, C., Engelstätter, R. and Sybrecht, G.W. (1992). Bronchodilator effect of a phosphodiesterase type III/IV selective inhibitor, zardaverine, in patients with chronic obstructive airway disease (COPD). Am. Rev. Respir. Dis. 145, A757. [Abstract]

Underwood, D.C., Osbom, R.-R., Novak, L.B., Matthews, J.K., Newsholme, S.J., Undem, B.J., Hand, J.M. and Torphy, T.J. (1993). Inhibition of antigen-induced bronchoconstriction and eosinophil infiltration in the guinea pig by the cyclic AMP-specific phosphodiesterase inhibitor, rolipram. J. Pharmacol. Exp. Ther. 266, 306–313.

Underwood, D.C., Kotzer, C.J., Bochnowicz, S., Osbom, R.R., Luttmann, M.A., Hay, D.W. and Torphy, T.J. (1994). Comparison of phosphodiesterase III, IV and dual III/IV inhibitors on bronchospasm and pulmonary eosinophil influx in guinea pigs. J. Pharmacol. Exp. Ther. 270, 250–259.

Wempe, J.B., Postma, D.S., Duipmans, J.C. and Koeter, G.H. (1992). Bronchodilating effect of zardaverine, a selective phosphodiesterase III/IV inhibitor. Eur. Respir. J. 5, 213s. [Abstract]

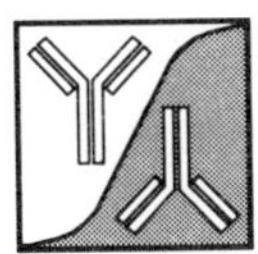

11. An Isoform-selective Inhibitor of Cyclic AMP-Specific Phosphodiesterase (PDE4) with Anti-inflammatory Properties

Robert Alvarez, Donald V. Daniels, Earl R. Shelton, Preston A. Baecker, T. Annie T. Fong, Bruce Devens, Robert Wilhelm, Richard M. Eglen *and* Marco Conti

1.	Introduction	161
2.	Materials and Methods	162
2.1	Cyclic Nucleotide PDE Assays	162
2.2	Accumulation of Cyclic AMP in Intact 43D Cells	163
2.3	Preparation of Recombinant Human PDE4 Isoforms	163
2.4	Phosphorylation of Human PDE4D3	164
2.5	Recombinant Human PDE7	164
2.6	Inflammatory Assays	164
2.7	Bronchoconstriction and Cell Infiltration in Guinea Pigs	165
2.8	Statistical Analysis	165
2.9	Chemicals	165
3.	Results	165
3.1	Biochemical Studies	165
3.2	Recombinant PDE4 Isoforms	166
3.3	Biological Responses	168
4.	Discussion	169
5.	Acknowledgements	170
6.	References	170

1. Introduction

Numerous studies have demonstrated that an increase in the intracellular concentration of cyclic AMP in human leucocytes inhibits the pro-inflammatory functions of these cells (Giembycz, 1992; Giembycz and Dent, 1992; Nicholson and Shahid, 1994). For example, cAMP inhibits the release of histamine from human basophils and mast cells (Peachell *et al.*, 1988), the chemotaxis of human eosinophils (Kaneko *et al.*, 1995), the proliferation of lymphocytes in response to antigen challenge (Essayan *et al.*, 1994) and the generation of superoxide anion and release of lysosomal enzymes in neutrophils (Simchowitz *et al.*, 1980).

An increase in the intracellular concentration of cAMP can be accomplished *in vitro* by incubating leucocytes with an appropriate inhibitor of cyclic nucleotide phosphodiesterase (PDE). To date, seven different gene families of PDE have been identified (Beavo *et al.*, 1994). These enzymes have different sensitivities to inhibitors, perform different biochemical roles and are not uniformly distributed in all cell types. In addition, they have different affinities for cAMP and cyclic GMP as substrates. Under appropriate conditions, catalytic activity can be increased by Ca^{2+} ions (calmodulin-dependent PDE1), nucleotides (cGMP-stimulated PDE2), phosphorylation (PDE3 and 4) and interaction with a guanine nucleotide-binding protein, transducin (PDE6) (see Chapter 1).

During the past 20 years, several studies have helped to clarify the distribution of the different cAMP PDEs. To facilitate these studies, an early objective was to obtain selective inhibitors of each major family. This goal has been achieved, in part, with the synthesis of potent compounds that are selective for PDE3, such as trequinsin, anagrelide and lixazinone (RS-82856).

Phosphodiesterase Inhibitors
ISBN 0-12-210720-9

Similarly, rolipram and Ro 20-1724 are selective for the inhibition of PDE4.

Theophylline, a non-selective PDE inhibitor, is useful in the treatment of asthma but has a narrow therapeutic range. The anti-inflammatory properties of theophylline appear to be due to the ability of this compound to inhibit PDE activity in leucocytes (Kuehl *et al.*, 1987; Bonta, 1984; Torphy *et al.*, 1992a). Unfortunately, doses of theophylline above the recommended range produce unwanted central nervous system and cardiovascular effects (Rall, 1990).

The predominant role of PDE4 in the hydrolysis of cAMP in human leucocytes (see Chapter 7) suggests that selective inhibitors of this enzyme may retain the anti-inflammatory properties of theophylline but produce fewer adverse effects. The available data indicate that selective PDE4 inhibitors do not elicit cardiovascular side-effects when administered under usual assay conditions (Giembycz and Dent, 1992). Several unpublished reports, however, indicate that structurally distinct PDE4 selective inhibitors produce emesis in dogs and monkeys.

Four isoforms of PDE4, designated A, B, C and D (Beavo *et al.*, 1994), have been cloned and expressed (Colicelli *et al.*, 1989; Davis *et al.*, 1989; Swinnen *et al.*, 1989; Bolger *et al.*, 1993; Obernolte *et al.*, 1993; Engels *et al.*, 1995) and an increase in intracellular cAMP induces the synthesis of the A, B and D isoforms (Engels *et al.*, 1994; Sette *et al.*, 1994a). These observations suggest a role for these enzymes in the long-term regulation of cAMP metabolism.

A variant of human PDE4D is subject to rapid activation following phosphorylation by cAMP-dependent protein kinase (PKA) (Alvarez *et al.*, 1995). A 132 amino acid domain, present in PDE4D3, is phosphorylated by the catalytic subunit of PKA (Sette *et al.*, 1994b). This domain contains two consensus sequences for PKA phosphorylation and is present in both the rat and human homologues. The rapid activation of PDE4D may serve to modulate the elevation in cAMP following the receptor-mediated stimulation of adenylate cyclase by hormones and neurotransmitters.

The studies described in this report were designed to identify novel and selective inhibitors of PDE4, to search for compounds with isoform selectivity and to determine whether selected compounds have potential anti-inflammatory properties.

2. *Materials and Methods*

2.1 Cyclic Nucleotide PDE Assays

PDE experiments were performed with enzyme from different sources. Unless stated otherwise, the incubation medium contained 1 μM of either [^{3}H]cAMP or [^{3}H]cGMP (0.2 μCi) in a total volume of 0.2 ml. Following addition of the enzyme, the contents were mixed and incubated for 10 min at 30°C. The reactions were terminated by immersing the tubes in a boiling-water bath for 60 s. After the tubes were cooled in an ice-water bath, 20 μl (20 μg) of 5′-nucleotidase from snake venom (*Crotalus atrox*) was added to each tube. The contents were mixed and incubated for 30 min at 30°C. The nucleotidase reaction was terminated by immersing the tubes in a boiling-water bath for 60 s. Radiolabelled adenosine was isolated from alumina columns according to a previously described method (Alvarez and Daniels, 1992). Assays were performed in triplicate. Hydrolysis of cAMP or cGMP ranged from 10 to 20%. The reactions were performed with five concentrations of inhibitor at log order intervals and were repeated at least three times.

2.1.1 Bovine Heart PDE (PDE1)

Bovine heart PDE (activator-deficient) and calmodulin (CaM) were obtained from Sigma Chemical Co. CaM (10 units) was added to the bovine heart stock solution that contained 0.05 U/mg in 2 ml of 10 mM Tris–HCl, pH 7.7. The reaction medium contained 10 mM Tris–HCl (pH 7.7), 0.1 mM $MgSO_4$, 10 μM $CaCl_2$, 1 μM [^{3}H]cGMP (0.2 μCi).

2.1.2 Mouse Splenocyte PDE (PDE2)

Spleens were dissected from five or six mice and homogenized with 2–3 ml RPMI 1640/spleen using a Dounce glass homogenizer. A splenocyte pellet was obtained by centrifugation of the homogenate at 200 *g* for 10 min at 22°C. The final pellet was resuspended in 7 ml of RPMI 1640 after the cells were washed one more time. The cell suspension was carefully layered over 3 ml of lympholyte M which was placed in the bottom of a 15 ml conical centrifuge tube. The upper layer of lymphocytes was extracted after the sample tube was centrifuged at 1200 *g* for 20 min at 22°C. The splenocytes were washed twice with 10 ml of RPMI 1640 by centrifugation at 200 *g* for 10 min. The final pellet was resuspended in 10 ml ice-cold 45 mM Tris–HCl, pH 7.7. The hypotonically lysed cell suspension was centrifuged at 12 000 *g* for 10 min at 4°C. The supernatant fraction contained PDE2 and PDE3 activity and was used as the soluble PDE. PDE3 activity was completely inhibited by 1 μM lixazinone (RS-82856–130). The PDE incubation medium contained 45 mM Tris–HCl (pH 7.7), 0.1 mM $MgSO_4$, 1 μM lixazinone and 1 μM [^{3}H]cAMP (0.2 μCi).

2.1.3 Human Platelet PDE (PDE3)

Platelet high affinity cAMP PDE was obtained from human blood according to previously described procedures (Alvarez *et al.*, 1981). The PDE incubation medium contained 45 mM Tris–HCl (pH 7.7), 10 mM $MgSO_4$, 1 μM [^{3}H]cAMP (0.2 μCi) in a total volume of 0.2 ml.

2.1.4 Human Lymphocyte PDE (PDE4)

A human B cell line (43D), isolated by Dr Mary Mulkins (Syntex Discovery Research), was used as a source of PDE4. The cells were cultured at 37°C in 7% CO_2 in RPMI 1640 with L-glutamine and 10% Nu-Serum (Collaborative Research). Prior to the assay ~1.5×10^8 cells were centrifuged at 250 *g* for 10 min. The pellet was resuspended in 2–3 ml of 45 mM Tris–HCl, pH 7.4. The suspension was centrifuged at 48 000 *g* at 4°C for 10 min. The supernatant was diluted to 28 ml with Tris–HCl buffer and stored at −20°C. The PDE incubation medium contained 45 mM Tris–HCl (pH 7.4), 0.1 mM $MgSO_4$ and 1 μM [^{3}H]cAMP (0.2 μCi) in a total volume of 0.2 ml.

2.1.5 High-throughput PDE Assay

A Tomtec Quadra 96 pipetting device was used to perform dilutions using microplates containing compounds at a concentration of 1% dimethyl sulfoxide (DMSO). The PDE reaction was initiated by the sequential addition of enzyme (80 μl) to the concentrated incubation medium (10 μl) and test compound (10 μl). The final reaction medium contained 10 mM Tris–HCl (pH 7.7), 0.1 mg/ml bovine serum albumin (BSA), 10 mM $MgSO_4$, 1 μM [^{3}H]cAMP (0.2 μCi), 1% DMSO, enzyme and test compound in a total volume of 100 μl. Following addition of the enzyme, the contents were mixed and incubated for 30 min at room temperature (~22°C). The reaction was terminated by heating in a microwave oven (Samsung Classic II) for 2.5 min. After the plates were cooled in an ice-water bath, 10 μl (10 μg) of 5′-nucleotidase was added to each well. The contents were mixed and incubated for 30 min at 22°C. The nucleotidase reaction was terminated by pipetting an aliquot (50 μl) from each well to a disposable column plate (Millipore, MADVS6510) containing 85 mg acidic alumina. The microcolumns of alumina were prepared using a 96-well powder dispenser (Millipore, MACL 09600). Radiolabelled adenosine was eluted with 80 μl of 5 mM HCl under vacuum into picoplates (Packard). Microscint 20 scintillation cocktail (270 μl) was added to each well and the plates sealed with adhesive plate seals. Hydrolysis of cAMP ranged from 5 to 15%. RS-25344 (10 pM–1 μM) was included in each assay as a reference standard. The radioactivity in each sample was measured with a liquid scintillation counter (Packard Topcount). Each plate contained four blank wells (no enzyme), four control wells (no test compound) and 88 test wells (compounds at given concentration). Each compound was tested at six concentrations (10 pM–1 μM) in log order increments and in quadruplicate. Data for each compound were transferred electronically via an RS-232 connection to a mainframe computer. The IC_{50} values were determined and the data stored in a database program.

2.2 ACCUMULATION OF CYCLIC AMP IN INTACT 43D CELLS

A cell suspension (~3×10^8 cells) was centrifuged at 120 *g* for 10 min at room temperature. Pellets were resuspended in 10 ml RPMI 1640 containing 15% Nu-Serum and 0.1 μM [2,8-^{3}H]adenine (50 μCi, DuPont-NEN). The cells were preincubated for 45 min at 37°C to permit incorporation of [^{3}H]adenine into the intracellular ATP pool. The suspension was then centrifuged at 120 *g* for 5 min and resuspended in fresh RPMI 1640 containing 10% Nu-Serum. Washing was repeated twice to remove extracellular [^{3}H]adenine. The final pellet was resuspended in 10 ml RPMI 1640 containing 15% Nu-Serum. The final reaction volume was 200 μl, consisting of 170 μl cell suspension, 20 μl PGE_2 and 10 μl rolipram or solvent. The incubation period was initiated by the addition of the cell suspension. The reaction was terminated with the addition of 20 μl of 2.2 M HCl containing 0.005 μCi [^{32}P]cAMP (DuPont-NEN) followed by rapid mixing. The isolation of cAMP was performed according to a previously described method (Alvarez and Daniels, 1992).

2.3 PREPARATION OF RECOMBINANT HUMAN PDE4 ISOFORMS

An insect cell line (Sf9), derived from pupal ovarian tissue of the fall army worm, *Spodoptera frugiperda*, was infected on day zero with a recombinant baculovirus expressing PDE4D3. Actively growing Sf9 cells at mid-log stage and a density of 1.2×10^6 cells/ml were infected (virus/cell ratio = 0.5) and grown at 27°C with orbital shaking at 160 r.p.m. in Ex-Cell 400 medium (JRH Biosciences). At the time of infection, the cell cultures were supplemented with 1% (final concentration) heat-inactivated foetal calf serum, 50 μg/ml gentamicin and 4% feed stock (80 ml of feed stock is prepared by mixing 20 ml Yeastolate Ultrafiltrate 50 × [Gibco-BRL], 10 ml lipid concentrate, 100 × [Gibco-BRL], 40 ml 2.5% L-glutamine in Ex-Cell 400 and 10 ml 20% glucose). The recombinant virus stock was made from an isolated plaque from the co-transfection of Sf9 cells with the plasmid expression vector pHPD-43T (derived from the parental plasmid pSYN XIV VI$^+$ X3/3 (O'Reilly *et al.*, 1992) and BaculoGold virus DNA (a derivative of the *Autographa californica* nuclear polyhedrosis virus, PharMingen). On day 3 after infection a cellular lysate was prepared with all steps performed at 0–4°C. The cells were harvested by centrifugation (10 min at 1000 *g*) and suspended in lysis buffer equivalent to 20% of the original culture volume. The lysis buffer contained: 5% glycerol, 45 mM Tris–HCl (pH 7.7), 0.5 mg/l leupeptin, 0.7 mg/l pepstatin, 0.2 mM phenylmethylsulphonyl fluoride (PMSF), 0.1 mM sodium vanadate. The cells were homogenized with five strokes of a Dounce homogenizer using pestle B (Kontes). The lysate was clarified by centrifugation (10 min at 10 000 *g*) and stored frozen at −80°C. Aliquots (0.3 ml) were stored in ethylene glycol (30% final concentration). Before use in the PDE assay, the enzyme was diluted 1 : 400 into 10 mM Tris-HCl, pH 7.4, containing 100 μg/ml BSA to stabilize the enzyme preparation. For inhibitor screening assays, the remaining human PDE4 isoforms were similarly prepared. Baculovirus expression of PDE4A5 from plasmid pHPA-3T (PDE sequence of 837 amino acids starting AEDE (Bolger *et al.*, 1993)) and PDE4C from plasmid pHPC-150 (Obernolte *et al.*, 1993), which includes UCR-2 (Bolger *et al.*, 1993) from intron B (Monaco *et al.*, 1994) to the carboxyl terminus, were prepared as described for PDE4D3 (above). Human PDE4B was expressed in the mouse Leydig cell line MA-10 by calcium phosphate-mediated transient transfection. MA-10 cells were cultured at 37°C and 5% CO_2 in Waymouth's MB 752/1 medium with L-glutamine supplemented with 15% horse serum, 20 mM HEPES (pH 7.7) and 50 μg/ml gentamycin. Near confluent cultures were split 1 : 10 into 100 mm tissue culture plates with 12 ml medium. The following day the medium was changed 2 hours before transfection. Cells were transfected for 6 hours by the calcium phosphate procedure with 20 μg of expression plasmid per plate. The expression plasmid was the vector pCMV5 with either a human PDE4B2 (pCMV5-B106) or the full-length PDE4B1 (pCMV5-B72S) cDNA insert. After 6 hours with the calcium phosphate co-precipitate, the cells were glycerol-shocked for 3 min at room temperature with 2 ml of complete medium plus 10% glycerol. Following three washes with phosphate-buffered saline (PBS), 12 ml of complete medium was added to each plate. After 18 hours the cells were washed twice with ice-cold PBS and scraped from the dish into 0.5 ml PDE buffer. The collected cells were transferred in PDE buffer to a Dounce homogenizer and disrupted with 30 strokes of the homogenizer. The lysate was centrifuged at 15 000 *g* for 15 min at 4°C and the supernatant removed. Cell lysate was assayed immediately or stored at −20°C after the addition of ethylene glycol to 30% final concentration.

2.4 PHOSPHORYLATION OF HUMAN PDE4D3

PDE4D3 was preincubated in the presence or absence of the catalytic subunit of PKA (1500 U/ml) for 8 min at 30°C. The tubes contained 40 mM Tris–HCl (pH 7.4), 0.2 mM ATP, 20 mM $MgSO_4$ and 0.05% BSA in a total volume of 0.1 ml. The kinase preincubation step was terminated by a 1 : 100 dilution into 10 ml ice-cold 10 mM Tris–HCl, pH 7.4, containing 0.2 mM EDTA. The diluted enzyme was added to test tubes containing 1 μM [^{3}H]cAMP, 100 mM NaCl and $MgSO_4$, as indicated, to initiate the PDE assay. The reaction was performed for 10 min at 30°C.

2.5 RECOMBINANT HUMAN PDE7

The cAMP-specific, very low K_m PDE7 was expressed using recombinant baculovirus as described above for PDE4 isoforms. The coding region was obtained by using reverse transcriptase to prepare cDNA from skeletal muscle mRNA followed by PCR using two pairs of nested primers. The previously identified HCP1 open reading frame starting at the first ATG (Michaeli *et al.*, 1993) was used to prepare the plasmid expression vector pHP-VII from the parental plasmid pSYN XIV VI$^+$ X3/4. DNA sequencing confirmed the expected deduced amino acid sequence except for two changes: F90 to L and A320 to V.

2.6 INFLAMMATORY ASSAYS

2.6.1 Arachidonic Acid-induced Ear Oedema

This procedure is as described by Young *et al.* (1984). Female CD-1 mice were treated with an oral dosage form of RS-25344 in Dulbecco's phosphate-buffered saline (D-PBS) 2 hours prior to challenge of the animals with arachidonic acid (1.5 mg) to the right ears. Animals in control groups received only D-PBS prior to challenge. 1 hour post-challenge, the animals were killed, their right ears removed and plugs punched. Oedema was quantified by calculating the mean increase in ear plug weights from animals treated with different doses of RS-25344 or D-PBS alone from the ear plug weights of D-PBS treated normal animals not challenged with arachidonic acid.

2.6.2 Carrageenan-induced Paw Oedema

This procedure was modified from Winter *et al.* (1962). Female CD-1 mice were treated with an oral dosage form of RS-25344 in D-PBS 2 hours prior to challenge of the animals with 30 μl of a 2% solution of carrageenan in the sub-plantar region of the right hind paw. Animals in control groups received only D-PBS prior to challenge. 3 hours post-challenge, the animals were killed and their right and left hind paws amputated at the tarsocrural joint. Oedema was quantified by calculating the mean increase in the right versus left hind paw weights from animals treated with different doses of RS-25344 or D-PBS alone compared to that obtained from D-PBS treated normal animals not challenged with carrageenan.

2.6.3 Platelet Activating Factor (PAF)-induced Mortality

This procedure was modified from Myers *et al.* (1988). Male CF-1 mice were treated with an oral dosage form of RS-25344 in D-PBS 2 hours prior to challenge of the animals with 55 mg/kg of PAF intravenously. Animals in control groups received only D-PBS prior to challenge. All animals were observed for approximately 5 hours post-PAF administration and mortality was recorded. The percentage mortality of animals treated with different doses of RS-25344 or D-PBS alone were compared.

2.6.4 Phenylquinone (PQ)-induced Writhing

This procedure was modified from Collier *et al.* (1968). Briefly, female CD-1 mice were treated with an oral dosage form of RS-25344 in D-PBS 2 hours prior to challenge of the animals with 0.25 ml of 0.02% solution of PQ intraperitoneally. Animals in control groups received only D-PBS prior to challenge. The number of writhes given by each animal during the first 10 min following injection of PQ was recorded. The average number of writhes given by groups of animals treated with different doses of RS-25344 or D-PBS were compared.

2.7 Bronchoconstriction and Cell Infiltration in Guinea Pigs

This procedure was modified from Hutson *et al.* (1988). Male Dunkin–Hartley guinea pigs were sensitized by exposure to aerosolized ovalbumin (2%) on two occasions. On the day of the experiment, each animal was placed into a double chamber whole body plethysmograph (BUXCO Electronics) and challenged with an ovalbumin aerosol following pretreatment with an oral dosage form of RS-25344 in D-PBS. Animals in control groups received the same volume of D-PBS. To protect the guinea pigs from immediate anaphylaxis following antigen challenge, mepyramine (10 mg/kg) was administered to the animals intraperitoneally 30 min before challenge. Ventilatory parameters and specific airway resistance (sR_{aw}) were recorded and analysed. The increase in sR_{aw} during the post-antigen challenge period compared to the pre-antigen period was calculated for each animal. Approximately 24 hours after antigen challenge, cellular infiltration into the broncheoalveolar lumen of the guinea pigs was assessed by broncheoalveolar lavage (BAL). Guinea pigs were killed and tracheotomy performed. The trachea was cannulated and the lungs lavaged with D-PBS. The total number of non-red blood cells in the BAL was quantified using a Coulter counter. Visual differential leucocyte counts were undertaken on cytocentrifuge preparations of cells following staining. Standard morphological criteria were used to classify the cells. The total number of eosinophils present in the BAL was calculated by multiplying the percentage of eosinophils present in the differential count by the total number of cells recovered from the BAL. The bronchoconstriction and cellular infiltration in animals treated with RS-25344 or D-PBS alone were compared.

2.8 Statistical Analysis

ED_{50} data from the biological assays were fitted to a 4-parameter model:

$$\text{Response} = \min + \frac{\max - \min}{1 + \left(\frac{[\text{drug}]}{ED_{50}}\right)^N}$$

where [drug] is the concentration of the test compound and N is the slope factor. The model fitting was performed using SAS on an IBM-compatible PC.

2.9 Chemicals

Rolipram, TVX-2706, denbufylline, Ro 20-1724, RX-RA 69, trequinsin, RS-82856, RS-14203, RS-14491 and RS-25344 (8-aza-1-[3-nitrophenyl]-3-[4-pyridylmethyl]-2,4-[lH,3H]-quinazoline dione) were synthesized and provided by the Institute of Organic Chemistry (Syntex). 3-Isobutyl-1-methylxanthine (IBMX) was from Sigma Chemical Co. Stock solutions of these inhibitors were prepared in 100% DMSO. The final concentration of DMSO in the enzyme assays was 1% (v/v).

3. *Results*

3.1 Biochemical Studies

The biochemical screen for these studies was designed to identify potent and selective inhibitors of PDE3 and PDE4. It consisted of a PDE3 preparation isolated from the cytosolic fraction of pooled human platelets and a PDE4 preparation isolated from the cytosolic fraction of 43D cells (a human B cell line). When assayed at 1 μM cAMP, PDE3 and PDE4 are the predominant isoenzymes in human platelets and 43D cells, respectively. There was no detectable PDE4 activity in human platelets and no detectable PDE3 activity in the human B cell line.

Over 700 compounds were synthesized and screened versus the two isoenzyme preparations. The compounds were primarily structural analogues of nitraquazone (TVX-2706) and trequinsin. Table 11.1 presents data for several selective PDE3 inhibitors (trequinsin, RX-RA 69 and RS-82856) and selective PDE4 inhibitors (rolipram, Ro 20-1724 and denbufylline). Structures of the PDE4 selective inhibitors are presented in Fig. 11.1.

RS-25344, an analogue of nitraquazone, is a potent inhibitor of PDE4 (IC_{50} = 0.28 nM) with little effect on PDE3 (IC_{50} = 330 μM). Similarly, RS-25344 had little or no effect on PDE1, PDE2 or recombinant PDE7 (Table 11.2).

Table 11.1 Selective inhibitors of PDE3 and PDE4

Compound	*IC_{50} (μM)* Platelet PDE3	*IC_{50} (μM)* Lymphocyte PDE4	*Selectivity ratio (PDE3/ PDE4)*[a]
RS-14203	260	0.00023	1 130 000
RS-25344	330	0.00028	1 179 000
RS-14491	120	0.001	120 000
Denbufylline	670	0.040	16 650
Rolipram	100	0.045	2220
Ro 20-1724	160	0.060	2670
IBMX	4.0	31	0.13
Trequinsin	0.00041	1.0	0.0004
RX RA 69	0.00074	2.2	0.0003
RS-82856-190	0.0034	>1000	<0.0000034

[a] Selectivity ratio >1 indicates selectivity for PDE4; ratio <1 indicates selectivity for PDE3.

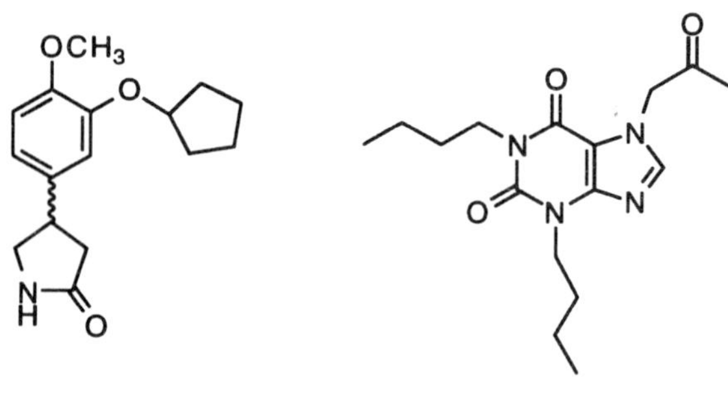

Rolipram

Denbufylline

TVX-2706

RS-25344

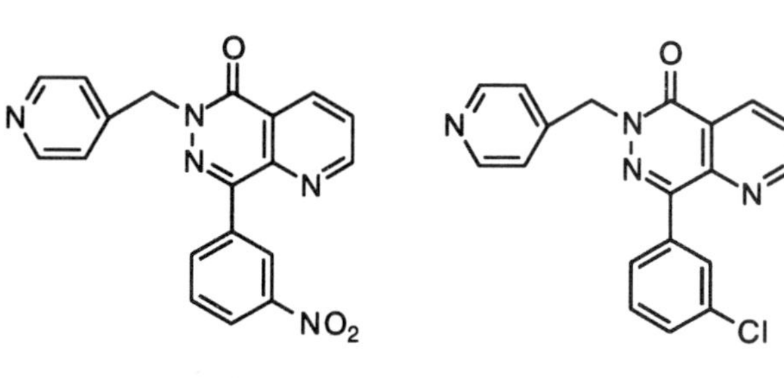

RS-14203

RS-14491

Figure 11.1 Structures of selective PDE4 inhibitors.

RS-25344 and two structurally related compounds (RS-14203 and RS-14491) inhibited PDE4 activity and increased intracellular cAMP in intact 43D cells in a concentration-dependent manner (Figs 11.2 and 11.3). The concentration required for half maximal inhibition of enzyme activity for each of these compounds was approximately 100 times lower than their EC_{50} values for the elevation of cAMP in intact cells. Presumably, this lower efficacy in cells reflects the poor ability of these compounds to cross the plasma membrane and inhibit the cytosolic PDE4.

3.2 RECOMBINANT PDE4 ISOFORMS

The discovery of multiple isoforms of PDE4 creates an opportunity to identify selective inhibitors of each isoform using recombinant enzymes and to examine their anti-inflammatory properties. A disadvantage of this approach is the large number of recombinant enzymes required to complete a comprehensive screening programme. Although only four genes encoding isoforms of PDE4 have been identified, there are nine known splice variants. In addition, multiple post-translational modifications of the enzymes are possible. In the initial biochemical screen the IC_{50} values for approximately 700 compounds were determined versus four genetic isoforms of PDE4. A high-throughput

Table 11.2 Inhibition of cyclic nucleotide phosphodiesterases by RS-25344

Isoenzyme family	*IC_{50} (μM)*
PDE1	>100
PDE2	160
PDE3	330
PDE4	0.00028
PDE7	Inactive

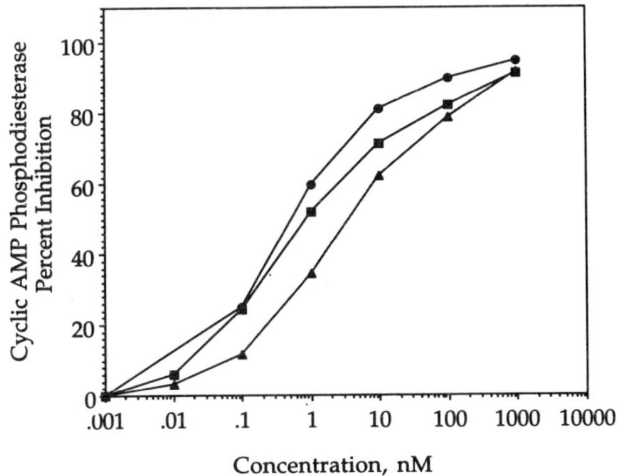

Figure 11.2 Concentration–response curves for the inhibition of PDE4 from a lymphoblastoid cell line (43D cells). The inhibitors were RS-25344 (■), RS-14203 (●) and RS-14491 (▲). The PDE assay was performed as described in section 2.

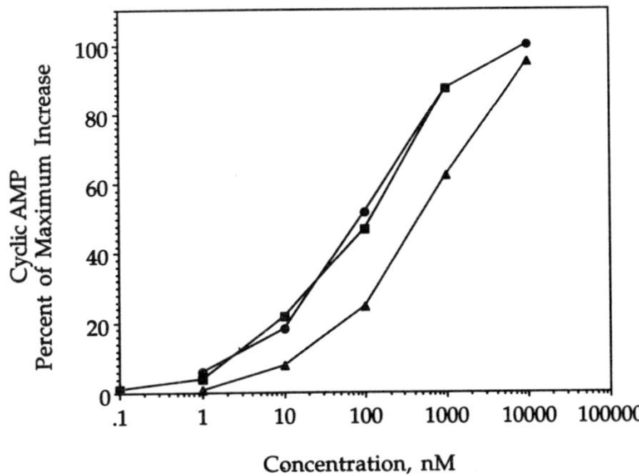

Figure 11.3 Concentration–response curves for the increase in intracellular cyclic AMP in intact 43D cells. The test compounds were RS-25344 (■), RS-14203 (●) and RS-14491 (▲). The cAMP assay was performed as described in section 2.

Table 11.3 Heterologous expression of human PDE4 isoforms

Source[a]	*Specific activity (pmol/ min/ µg)*	*n*	*Relative expression*	*Reference*
Recombinant				
E. coli lysate				
hPDE4B2	0.34		—	This work
rPDE4D1	0.36			Jin *et al.* (1992)
Yeast lysate				
hPDE4C	48.3 ± 10	3	166	Engels *et al.* (1995)
Baculovirus lysate				
Sf9 cells	0.29 ± 0.2	8	1	This work
hPDE4A5	956 ± 59	3	3300	This work
hPDE4B1	317 ± 39	3	1100	This work
hPDE4C150	2415 ± 506	1	8300	This work
hPDE4D3	41.6 ± 12	3	100	This work
Mammalian lysate				
MA-10 cells	0.086 ± 0.04	10	1	This work
hPDE4A5	1.98 ± 0.47	2	23	This work
hPDE4B1	1.11 ± 0.29	6	13	This work
hPDE4B2	0.43 ± 0.18	5	5	This work
hPDE4D3	0.43 ± 0.01	2	5	This work
rPDE4D1	0.716		—	Swinnen *et al.* (1989)
rPDE4D2	0.148		—	Sette *et al.* (1994b)
rPDE4D3	0.612		—	Sette *et al.* (1994b)
Natural Sources				
Tissues, cells	0.01–0.50			

[a] h, human gene product; r, rat gene product.

assay was designed to accommodate a large-scale screening programme. Microcolumns of aluminium oxide in a 96-well format were utilized to separate the substrate from the secondary product (adenosine). The pipetting steps were performed with a Quadra 96 Pipetting Station (see section 2). The assay used bar code technology to track each compound and the data were analysed by computer. With this system it was possible for one person to screen approximately 300 compounds in one week (six concentrations performed in quadruplicate).

The cloned A, C and D isoforms of human PDE4 used in these experiments were expressed at high levels in baculovirus infected Sf9 insect cells (Fig. 11.4, Table 11.3). PDE4B1 was expressed in MA-10 cells. This expression system was used because concentration–response curves for PDE4 inhibitors more closely matched the patterns obtained with endogenous enzyme from 43D cells. In all cases, PDE activity was specific for the hydrolysis of cAMP ($K_m \approx 3\ \mu M$) and catalytic activity was dependent upon Mg^{2+}.

RS-25344 was tested as an inhibitor of four cloned isoforms of PDE4 (A, B, C and D) and phosphorylated PDE4D. The results, presented in Table 11.4 reveal that RS-25344 selectively inhibits the phosphorylated form of PDE4D (Dp). The order of potency was Dp > D > A = C = B. A structurally related compound, RS-33793 showed a similar profile but was an even more selective inhibitor of the phosphorylated form of PDE4D than the control preparation (~330-fold).

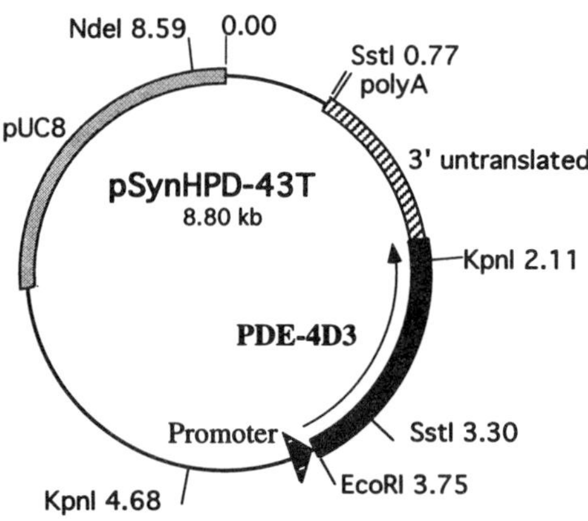

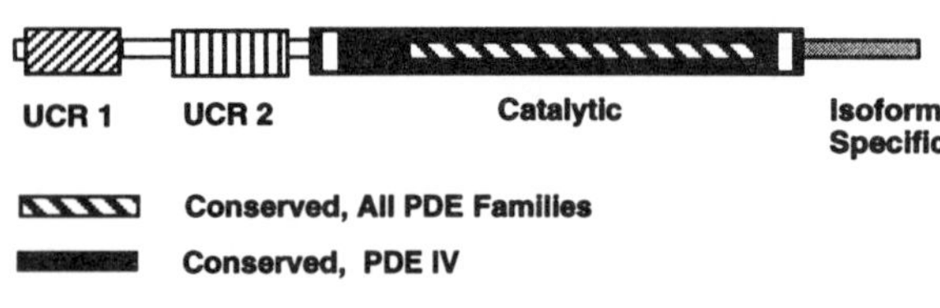

Figure 11.4 Circular map of the baculovirus expression plasmid pHPD-43T which encodes PDE4D3. A linear map illustrating the putative domain structure of the PDE4D3 protein is also shown. The sites of phosphorylation by PKA lie between the amino terminus and the first few amino acids in upstream conserved region 1 (UCR1).

Table 11.4 Selective inhibition of PDE4 isoforms by RS-25344 and RS-33793

	pIC_{50} [a]	
Isoenzyme	*RS-33793*	*RS-25344*
PDE4A	7.2	7.8
PDE4B	6.9	7.5
PDE4C	7.2	7.7
PDE4D	7.4	8.5
PDE4Dp	9.7	9.3

[a] $pIC_{50} = -(\log_{10}(\text{molar } IC_{50}))$.

3.3 BIOLOGICAL RESPONSES

To assess the potential activity of selective PDE4 inhibitors as therapeutic agents in inflammatory disease, RS-25344 was examined in several *in vivo* assays of inflammation. Oral administration of the compound inhibited the ear oedema produced by topical arachidonic acid to mice in a dose-dependent fashion with an ED_{50} of 0.02 mg/kg (Table 11.5). Similarly, RS-25344 inhibited carrageenan-induced paw oedema (ED_{50} = 1 mg/kg; Table 11.6), PQ-induced writhing (ED_{50} = 1.5 mg/kg) and PAF-induced mortality (ED_{50} = 1.0 mg/kg).

Theophylline has been shown to be useful in the treatment of asthma, an inflammatory disease of the airways (see Chapter 3). Several animal models of asthma have been developed that resemble the human disease. For example, sensitized guinea pigs exhibit an antigen-triggered bronchoconstriction and an influx of inflammatory cells (primarily eosinophils) to the lung. RS-25344 was tested for its ability to inhibit anaphylaxis and cellular influx using guinea pigs sensitized to ovalbumin. The results, presented in Table 11.7, indicate that RS-25344, administered orally 1 hour prior to antigen challenge, was effective in preventing anaphylaxis (ED_{50} = 0.5 mg/kg). This inhibition occurred in a dose-dependent fashion and antigen challenge could be delayed up to 24 hours after compound dosing without changing the dose–response curve (data not shown). Furthermore, RS-25344 inhibited the influx of inflammatory cells and eosinophils 24 hours after antigen challenge with ED_{50} values of 0.2 and 0.1 mg/kg, respectively. The potency of RS-25344 was greater than that of rolipram (ED_{50} values >10 mg/kg for all parameters) in this animal model.

Table 11.5 Increase in intracellular cAMP and inhibition of arachidonic acid-induced ear oedema by selective PDE4 inhibitors

Test compound	*Elevation of intracellular cAMP in human lymphocytes EC_{50} (μM)*	*Inhibition of arachidonic acid-induced ear oedema ED_{50} (mg/kg, p.o.)*
RS-14203	0.06	0.003
RS-25344	0.10	0.020
RS-14491	0.30	0.009

Table 11.6 Effects of rolipram and RS-25344 on three animal models of inflammation

	ED_{50} (mg/ kg, p.o.)		
Test compound	*PAF-induced mortality*	*PQ-induced writhing*	*Carageenan-induced paw oedema*
RS-25344	1.0	1.5	1.0
Rolipram	>100	>100	>100

Table 11.7 Ovalbumin-induced bronchoconstriction and cellular influx in sensitized guinea pigs

	ED_{50} (mg/kg, p.o.)		
		Cellular influx[b]	
Test compound[a]	*Anaphylaxis*	*Total cells*	*Eosinophils*
RS-25344	0.5	0.2	0.1
Rolipram	>10	>10	>10

[a] Test compound was administered 1 hour before antigen challenge.
[b] Measured 24 hours after antigen challenge.

4. Discussion

The results of this study reveal that RS-25344, a potent inhibitor of cAMP-specific phosphodiesterase (PDE4), has anti-inflammatory properties. The compound is selective for PDE4 with little or no inhibitory effect on PDE1, 2, 3 or 7. Within the PDE4 family, this compound exhibits isoenzyme selectivity with the following order of potency: Dp > D > A = C = B.

Previous studies demonstrated a rapid, cAMP-dependent activation of PDE4 in a promonocytic cell line (U937 cells). PDE4 activity was increased following exposure to either PGE_2 or histamine in these cells (Alvarez *et al.*, 1995). Thus, the activation of PDE4 by these agents accounts for the observed gradual decline in intracellular cAMP. Immunoprecipitation studies revealed that three of the four isoforms of PDE4 (A, B and D) are present in U937 cells (Alvarez *et al.*, 1995). The predominant isoenzyme in U937 cells is PDE4D, followed by PDE4B with PDE4A as a minor constituent. The activity of a variant of human recombinant PDE4D (PDE4D3) is stimulated by the catalytic subunit of PKA (Alvarez *et al.*, 1995). In contrast, neither human PDE4A nor PDE4B was significantly stimulated by the kinase. The phosphorylation of PDE4D3 changes the kinetic properties of the enzyme and increases its sensitivity to RS-25344 and RS-33793 (100-fold and 330-fold, respectively). Thus, these compounds selectively inhibit an activated (phosphorylated) form of PDE in human leucocytes.

The observation that PDE4 isoforms are not uniformly distributed in all cell types opens the possibility that an appropriate isoform-selective inhibitor may circumvent the emesis and nausea currently associated with PDE4 inhibitors.

Previous studies demonstrated that the synthesis of PDE4 was induced in U937 cells following prolonged (2–4 hours) exposure to prostaglandin E_2, salbutamol and 8-bromo-cAMP (Torphy *et al.*, 1992b). The increase in PDE4 was abolished by cycloheximide. In contrast, a brief (2–10 min) exposure to PGE_2 produced a rapid increase in PDE4 activity that was not blocked by 10 μM cycloheximide (Alvarez *et al.*, 1995). Thus, PGE_2 can increase PDE4 activity by two distinct mechanisms, a rapid activation and a delayed induction.

Clarification of the various roles of different PDE isoenzymes in normal cell physiology as well as inflammatory disease continues to be a challenge. As discussed above, various cell types differ with respect to the distribution of the PDE isoenzymes. Factors that regulate enzyme expression and post-translational modifications add another layer of complexity. Selective inhibitors should prove useful in understanding the role of this enzyme system in the regulation of cellular events.

RS-25344 has anti-inflammatory activity in several animal models. These assays were chosen to investigate the scope of the actions of the compound, since each of the assays differs with respect to the involvement of various mediators of inflammation. For example, the arachidonic acid ear oedema assay and the PAF-induced mortality assay are both mediated via a lipoxygenase pathway and are readily inhibited by lipoxygenase inhibitors whereas the carrageenan-induced paw oedema and the PQ-induced writhing assays are relatively insensitive to lipoxygenase inhibitors but highly sensitive to cyclooxygenase inhibitors. As predicted from the cellular data, PDE4 selective inhibitors are active in all of these assays, presumably reflecting the central role of cyclic AMP in the control of inflammatory processes.

The guinea-pig model provides additional information. This model is triggered by an immune, lymphocyte-based response to antigen. It has been shown that lymphocyte triggering can be inhibited *in vitro* by PDE4 inhibitors but it is not known whether RS-25344 is inhibiting the inflammatory response at the lymphocyte level, at the level of other pro-inflammatory cells, or both. In addition, using isolated tracheal rings, it has been shown that RS-25344 and other selective PDE4 inhibitors relax the muscle tone directly (data not shown). This may account for the dual effect on bronchoconstriction as well as inflammatory cell infiltration. In contrast, corticosteroids have no direct effect on bronchoconstriction, but effectively block the influx of cells.

The availability of isoform-selective PDE4 inhibitors should improve our understanding of the distribution and function of the target enzymes. Additional studies will be required to determine whether such compounds will have useful properties in the treatment of disease.

5. *Acknowledgements*

We gratefully acknowledge the expert assistance of Rena Obernolte, Diana Yang and Li-Feng Katy Chang.

6. *References*

Alvarez, R. and Daniels, D. (1992). A separation method for the assay of adenylyl cyclase, intracellular cyclic AMP, and cyclic AMP phosphodiesterase using tritium-labeled substrates. Anal. Biochem. 203, 76–82.

Alvarez, R., Taylor, A., Fazzari, J.J. and Jacobs, J. (1981). Regulation of cyclic AMP metabolism in human platelets, sequential activation of adenylate cyclase and cyclic AMP phosphodiesterase by prostaglandins. Mol. Pharmacol. 20, 302–309.

Alvarez, R., Sette, C., Yang, D., Eglen, R.M., Wilhelm, R., Shelton, E.R. and Conti, M. (1995). Activation and selective inhibition of a cyclic AMP-specific phosphodiesterase PDE-4D3. Mol. Pharmacol. 48, 616–622.

Beavo, J.A. (1988). Multiple isozymes of cyclic nucleotide phosphodiesterase. Adv. Second Messenger Phosphoprotein Res. 22, 1–38.

Beavo, J.A., Conti, M. and Heaslip, R.J. (1994). Multiple cyclic nucleotide phosphodiesterases. Mol. Pharmacol. 46, 399–405.

Bolger, G., Michaeli, T., Martins, T., St John, T., Steiner, B., Rodgers, L., Riggs, M., Wigler, M. and Ferguson, K. (1993). A family of human phosphodiesterases homologous to the dunce learning and memory gene product of *Drosophila melanogaster* are potential targets for antidepressant drugs. Mol. Cell. Biol. 13, 6558–6751.

Bonta, I.L. (1984). Macrophage regulation by arachidonic acid metabolites: can this physiological control process serve as a model for novel pharmacotherapy of immuno-inflammatory conditions? Agents Actions Suppl. 14, 5–19.

Colicelli, J., Birchmeier, C., Michaeli, T., O'Neill, K., Riggs, M. and Wigler, M. (1989). Isolation and characterization of a mammalian gene encoding a high-affinity cyclic AMP phosphodiesterase. Proc. Natl Acad. Sci. USA 86, 3599–3603.

Collier, H.O.J., Dinneen, L.C., Johnson, C.A. and Schneider, C. (1968). The abdominal constriction response and its suppression by analgesic drugs in the mouse. Br. J. Pharmacol. Chemother. 32, 295–310.

Davis, R.L., Takayasu, H., Eberwine, M. and Myres, J. (1989). Cloning and characterization of mammalian homologs of the *Drosophila* dunce[+] gene. Proc. Natl Acad. Sci. USA 86, 3604–3608.

Engels, P., Fichtel, K. and Lubbert, H. (1994). Expression and regulation of human and rat phosphodiesterase type IV isogenes. FEBS Lett. 350, 291–295.

Engels, P., Sullivan, M., Muller, T. and Lubbert, H. (1995). Molecular cloning and functional expression in yeast of a human cAMP-specific phosphodiesterase subtype (PDE IV-C). FEBS Lett. 358, 305–310.

Essayan, D.M., Huang, S.-K., Undem, B.J., Kagey-Sobotka, A. and Lichtenstein, L.M. (1994). Modulation of antigen- and mitogen-induced proliferative responses of peripheral blood mononuclear cells by nonselective and isozyme selective cyclic nucleotide phosphodiesterase inhibitors. J. Immunol. 153, 3408–3416.

Giembycz, M.A. (1992). Could isoenzyme-selective phosphodiesterase inhibitors render bronchodilator therapy redundant in the treatment of bronchial asthma? Biochem. Pharmacol. 43, 2041–2051.

Giembycz, M.A. and Dent, G. (1992). Prospects for selective cyclic nucleotide phosphodiesterase inhibitors in the treatment of bronchial asthma. Clin. Exp. Allergy 22, 337–344.

Hutson, P.A., Church, M.K., Clay, T.P., Miller, P. and Holgate, S.T. (1988). Early and late phase bronchoconstriction after allergen challenge of nonanesthetized guinea pigs. I. The association of disordered airway physiology to leukocyte infiltration. Am. Rev. Respir. Dis. 137, 548–557.

Jin, S.-L.C., Swinnen, J.V. and Conti, M. (1992). Characterization of the structure of a low Km, rolipram-sensitive cAMP phosphodiesterase. J. Biol. Chem. 267, 18929–18939.

Kaneko, T., Alvarez, R., Ueki, I.F. and Nadel, J.A. (1995). Elevated intracellular cyclic AMP inhibits chemotaxis in human eosinophils. Cell Signal. 7, 527–534.

Kuehl, F.A., Jr, Zanetti, M.E., Soderman, D.D., Miller, D.K. and Ham, E.A. (1987). Cyclic AMP-dependent regulation of lipid mediators in white cells, a unifying concept for explaining the efficacy of theophylline in asthma. Am. Rev. Respir. Dis. 136, 210–218.

Michaeli, T., Bloom, T.J., Martins, T., Loughney, K., Ferguson, K., Riggs, M., Rodgers, L., Beavo, J.A. and Wigler, M. (1993). Isolation and characterization of a previously undetected human cAMP phosphodiesterase by complementation of cAMP phosphodiesterase-deficient *Saccharomyces cerevisiae*. J. Biol. Chem. 268, 12925–12932.

Monaco, L., Vicini, E. and Conti, M. (1994). Structure of two rat genes coding for closely related rolipram-sensitive cyclic AMP phosphodiesterases, multiple mRNA variants originate from alternative splicing and multiple start sites. J. Biol. Chem. 269, 347–357.

Myers, A.K., Nakanishi, T. and Ramwell, P. (1988). Antagonism of PAF-induced death in mice. Prostaglandins. 35, 447–458.

Nicholson, C.D. and Shahid, M. (1994). Inhibitors of cyclic nucleotide phosphodiesterase isoenzymes – their potential utility in the therapy of asthma. Pulmonary Pharmacol. 7, 1–17.

Obernolte, R., Bhakta, S., Alvarez, R., Bach, C., Zuppan, P., Mulkins, M., Jarnagin, K. and Shelton, E.R. (1993). The cDNA of a human lymphocyte cyclic-AMP phosphodiesterase (PDE IV) reveals a multigene family. Gene 129, 239–247.

O'Reilly, D.R., Miller, L.K. and Luckow, V.A. (1992). Baculovirus Expression Vectors: a Laboratory Manual, pp. 66–67. Freeman, New York.

Peachell, P.T., MacGlashan, D.W., Jr, Lichtenstein, L.M. and Schleimer, R.P. (1988). Regulation of human basophil and lung mast cell function by cyclic adenosine monophosphate. J. Immunol. 140, 571–579.

Rall, T.W. (1990). Drugs used in the treatment of asthma: the methylxanthines, cromolyn sodium, and other agents. In "Goodman and Gilman's The Pharmacological Basis of Therapeutics", eighth edition (eds. A.G. Gilman, T.W. Rall, A.S. Nies and P. Taylor), pp. 618–637. Pergamon, New York.

Sette, C., Saveria, I. and Conti, M. (1994a). The short-term activation of a rolipram-sensitive, cyclic AMP-specific phosphodiesterase by thyroid-stimulating hormone in thyroid FRTL-5 cells is mediated by a cyclic AMP-dependent phosphorylation. J. Biol. Chem. 269, 1–8.

Sette, C., Vicini, E. and Conti, M. (1994b). The rat PDE3/IVd phosphodiesterase gene codes for multiple proteins differentially activated by cyclic AMP-dependent protein kinase. J. Biol. Chem. 269, 18271–18274.

Simchowitz, L., Fischbein, L.C., Spilberg, I. and Atkinson, J.P. (1980). Induction of a transient elevation in intracellular levels of adenosine-3′5′-cyclic monophosphate by chemotactic factors: an early event in human neutrophil activation. J. Immunol. 124, 1482–1491.

Swinnen, J.V., Joseph, D.R. and Conti, M. (1989). Molecular cloning of rat homologues of the *Drosophila melanogaster* dunce cyclic AMP phosphodiesterase, evidence for a family of genes. Proc. Natl Acad. Sci. USA 86, 5325–5329.

Torphy, T.J., Livi, G.P., Balcarek, J.M., White, J.R., Chilton, F.H. and Undem, B.J. (1992a). Therapeutic potential of isozyme-selective phosphodiesterase inhibitors in the treatment of asthma. Adv. Second Messenger Phosphoprotein Res. 25, 289–303.

Torphy, T.J., Zhou, H.-L. and Cieslinski, L. (1992b). Stimulation of beta adrenoceptors in a human monocyte cell line (U937) up-regulates cyclic AMP-specific phosphodiesterase activity. J. Pharmacol. Exp. Ther. 263, 1195–1205.

Winter, C.A., Risley, E.A. and Nuss, G.W. (1962). Carrageenan-induced edema in hindpaw of the rat as an assay for anti-inflammatory drugs. Proc. Soc. Exp. Biol. Med. 111, 544.

Young, J.M., Spires, D.A., Bedord, C.J., Wagner, B.M. Ballaron, S.J. and DeYoung, L.M. (1984). The mouse ear inflammatory response to topical arachidonic acid. J. Invest. Dermatol. 82, 367–371.

12. Characterization of Different States of PDE4 by Rolipram and RP 73401

John E. Souness

1. Introduction 173
2. Pharmacology of RP 73401 173
3. Interactions of RP 73401 and Rolipram with Eosinophil PDE4 175
   3.1 Poor Correlation Between PDE4 Inhibition and Suppression of Eosinophil Functions 175
   3.2 Effects of Solubilization and Vanadate/Glutathione Complex 176
   3.3 Possible Role of the High Affinity Rolipram-binding Site 178
4. Inhibition of PDE4 from other Cells and Tissues 178
   4.1 Enzyme Data 178
   4.2 Whole Cell Data 180
5. Possible Therapeutic Implications 181
6. Conclusions 182
7. References 182

1. Introduction

Inhibitors of cAMP-specific phosphodiesterase (PDE4) exhibit a range of anti-inflammatory/immunosuppressive activities which suggest potential in a wide range of inflammatory and autoimmune diseases (Torphy and Undem, 1991; Nicholson and Shahid, 1994; Genain *et al.*, 1995; Palfreyman and Souness, 1996; Sekut *et al.*, 1995; Sommer *et al.*, 1995; see also Chapters 7 and 11). Most interest to date has centred on the clinical evaluation of this class of compounds in asthmatic patients. This stems, predominantly, from the demonstration that rolipram and other PDE4 inhibitors exhibit anti-inflammatory activities in several acute animal models of asthma (Torphy and Undem, 1991; Nicholson and Shahid, 1994; Palfreyman and Souness, 1996). Furthermore, a number of reports document functional antagonism of the airways smooth muscle contractile effects of several agonists both *in vitro* and *in vivo*, suggesting that they may manifest bronchodilating effects in the clinic (Souness and Giembycz, 1994). The prospect of the dual benefits of anti-inflammatory and bronchodilator activities incorporated into a single molecule has prompted several major pharmaceutical companies to identify their own compounds, many of which are in various stages of development. These include: LAS-31025 (Almirall), WAY-PDA-641 (Wyeth Ayerst), SB 207449 (SmithKline Beecham), CDP 840 (Celltech/Mersk) and the Rhône-Poulenc Rorer compound, RP 73401.

This chapter briefly summarizes the published *in vitro* and *in vivo* pharmacology of RP 73401, which indicates that it may be useful in the treatment of asthma. Additionally, evidence suggesting that the interaction of RP 73401 with PDE4 differs from that of the archetypal inhibitor, rolipram (structures shown in Plate 12.1) is critically reviewed. Explanations for these differences are proposed and the possible impact on the therapeutic potential of RP 73401 is assessed.

2. Pharmacology of RP 73401

A series of very potent benzamide inhibitors was identified during the course of screening compounds against pig aortic PDE4 (Ashton *et al.*, 1994). Based on the apparent requirement of a 3,4-dialkoxy group in the phenyl ring of rolipram and its analogues for PDE4 inhibitory activity (Marivet *et al.*, 1989), initial searches were based on structures which incorporated these

Phosphodiesterase Inhibitors
ISBN 0-12-210720-9

features. A phenylbenzamide (compound A) was identified which exhibited similar potency to rolipram in the PDE4 assay (Table 12.1). 2,5-Dichloro substitution in the *N*-phenyl ring (compound B) increased PDE4 inhibitory potency by more than 100-fold and activity was increased a further 20-fold in the *N*-(3,5-dichloro-4-pyridyl) derivative (RP 73401). All compounds displayed great selectivity for PDE4 in comparison to other PDE isoenzymes (Ashton *et al.*, 1994) (see Table 12.1).

RP 73401 dampens functional responses in a variety of inflammatory cells: it potently inhibits leukotriene B_4 (LTB_4)-induced release of superoxide anion radical (O_2^-), major basic protein (MBP) and eosinophil cationic protein (ECP) from guinea-pig peritoneal eosinophils (Souness *et al.*, 1995). In anti-CD16 microbead-purified human peripheral blood eosinophils, RP 73401 potentiates the inhibitory effects of salbutamol on complement factor C5a-stimulated release of ECP and eosinophil-derived neurotoxin (EDN) as well as the chemiluminescence response (Hatzelmann *et al.*, 1995). In a mixed granulocyte population from human peripheral blood, inhibition of *N*-formyl-methionyl-L-leucyl-L-phenylalanine (FMLP)-induced generation of O_2^-, ECP and elastase by RP 73401 has been reported (Karlsson *et al.*, 1994). Recently, inhibitory effects of RP 73401 on lipopolysaccharide (LPS)-induced tumour necrosis factor (TNF-α) generation from human monocytes, interleukin-4 (IL-4) release from anti-CD3 stimulated murine splenocytes and

Table 12.1 Comparison of benzamide analogues and rolipram in inhibiting smooth muscle PDE4

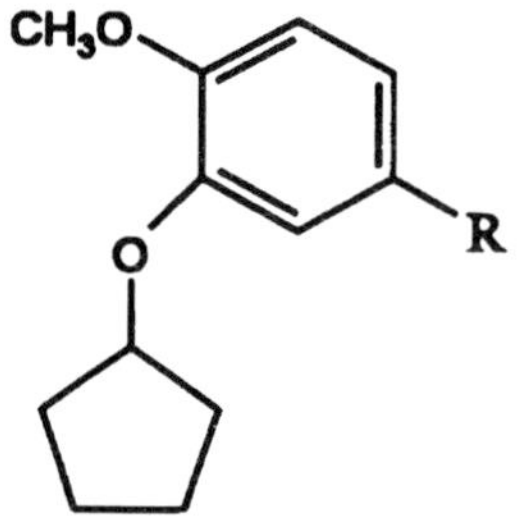

Compound	*R*	*PDE Inhibition – IC_{50} (µM)*				
		PDE1	*PDE2*	*PDE3*	*PDE4*	*PDE5*
Rolipram	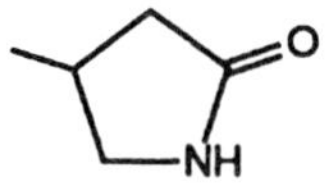 O, NH	>1000	>200	768	1.5	>1000
Compound A	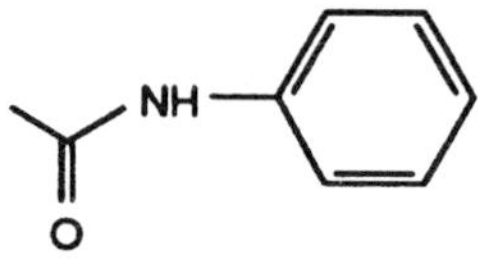 NH, O	>1000	ND	>1000	2.6	>1000
Compound B	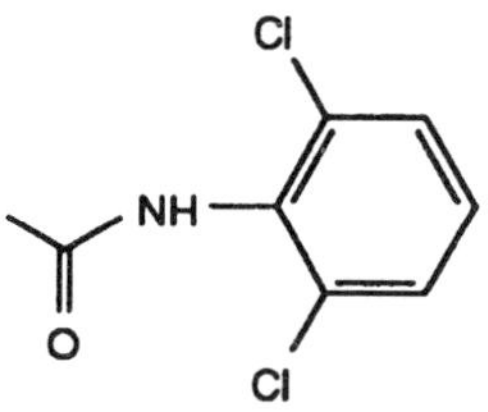 Cl, NH, O, Cl	>1000	ND	>1000	0.023	>1000
RP 73401	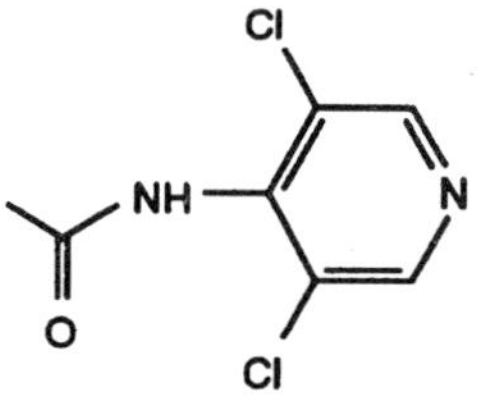 Cl, NH, N, O, Cl	49	73	267	0.001	19

Staphylococcus aureus enterotoxin-stimulated IL-2 generation from anti-CD3 differentiated mouse splenocytes have been demonstrated (Mirza *et al.*, 1994; Pollock *et al.*, 1995). RP 73401 also exhibits potent airways smooth muscle relaxant effects, inhibiting the contractile actions of histamine, LTD_4 and methacholine in guinea-pig trachealis (Raeburn *et al.*, 1994; Souness *et al.*, 1995) and relaxing guinea-pig trachealis under basal tone (Raeburn *et al.*, 1994).

The majority of the *in vivo* effects of RP 73401 have been investigated using dry-powder formulations on lactose administered directly into the tracheas (i.t.) of experimental animals. Administration of RP 73401 via this route inhibits antigen (ovalbumin)-induced bronchospasm in conscious sensitized guinea pigs and reduces inflammatory cell (total cells and eosinophils) influx into bronchoalveolar lavage (BAL) fluid (Raeburn *et al.*, 1994). Suppression of antigen-induced inflammatory cell influx is also observed in anaesthetized rats, although higher doses of RP 73401 are required to elicit similar effects (Raeburn *et al.*, 1994). Further evidence for *in vivo* anti-inflammatory activity is provided by its potent suppression of histamine-induced microvascular leakage into the tracheal tissue and airways (Raeburn *et al.*, 1994). In addition, RP 73401 very potently ($ID_{50} < 1$ mg/kg, i.v.) inhibits platelet activating factor (PAF)-induced bronchial hyperreactivity (BHR) to bombesin in anaesthetized guinea pigs (Raeburn *et al.*, 1994). Although much debate centres on the bronchodilating potential of PDE4 inhibitors (Souness and Giembycz, 1994), direct administration of RP 73401 into the lungs of guinea pigs effectively antagonizes the bronchoconstrictor effects of methacholine, histamine and LTD_4 (Raeburn *et al.*, 1994). These *in vitro* and *in vivo* actions of RP 73401 suggest potential in the treatment of asthma.

3. *Interactions of RP 73401 and Rolipram with Eosinophil PDE4*

3.1 POOR CORRELATION BETWEEN PDE4 INHIBITION AND SUPPRESSION OF EOSINOPHIL FUNCTIONS

The original screening data on pig aortic smooth muscle PDE4 demonstrated the identification of a compound at least 1000-fold more potent than rolipram (Ashton *et al.*, 1994; Souness *et al.*, 1995). In subsequent studies this potency difference failed to be translated into effects on several intact cells and tissues. For example, RP 73401 was only 3–4-fold more potent than rolipram in inhibiting the methacholine-induced contraction of guinea-pig trachealis and 5-fold more potent in suppressing LTB_4-elicited MBP release from guinea-pig eosinophils (Souness *et al.*, 1995). In human peripheral blood eosinophils, only 2–5-fold differences in the potencies of the two compounds are observed in potentiating the actions of salbutamol on C5a-induced release of O_2^-, ECP and EDN (Hatzelmann *et al.*, 1995). Such discrepancies might be interpreted as indicating that PDE4 inhibition and suppression of whole cell responses are not linked; however, this does not appear to be a satisfactory explanation since RP 73401 is also less than 4-fold more potent than rolipram in potentiating isoprenaline-induced cAMP accumulation (Souness *et al.*, 1995) and in increasing the cAMP-dependent protein kinase (PKA) activity ratio in intact guinea-pig eosinophils (see Fig. 12.1).

We and others have previously reported poor correlations between the potencies of compounds against PDE4 and in eliciting intact cell responses (Souness *et al.*, 1991; Dent *et al.*, 1991). For example, in intact eosinophils, cAMP accumulation and inhibition of O_2^- generation induced by a range of PDE inhibitors are not well correlated with inhibition of the eosinophil PDE4 (Souness *et al.*, 1991). Notably, although rolipram, denbufylline and trequinsin (HL-725) display similar PDE4 inhibitory potencies (Rupert and Weithmann, 1982), the latter is more than 100-fold less potent in inhibiting O_2^- release and increasing cAMP accumulation (Souness *et al.*, 1991). The simplest explanation – that the access of trequinsin to its site of action might be somehow impeded – was considered far from satisfactory in view of the very potent effects of this compound in other cell types (Ruppert and Weithmann, 1982).

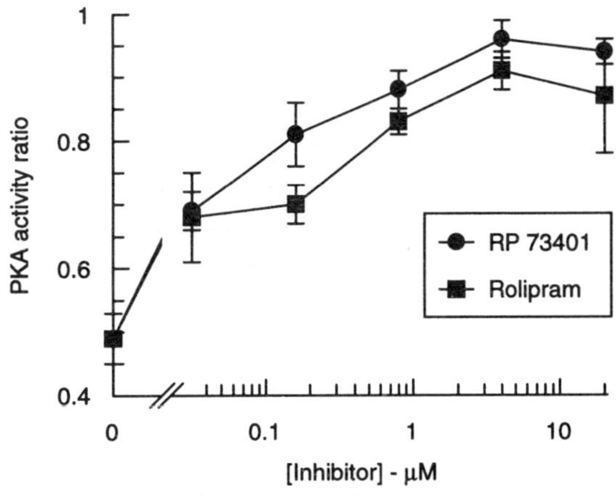

Figure 12.1 Stimulation of PKA by RP 73401 and rolipram in the presence of isoprenaline. Eosinophils were incubated with isoprenaline (10 μM) and the indicated concentrations of PDE inhibitors for 10 min. Measurement of PKA is described in Souness *et al.* (1991). The control PKA activity ratio was 0.28. Data are mean ± SEM from four experiments.

3.2 EFFECTS OF SOLUBILIZATION AND VANADATE/GLUTATHIONE COMPLEX

PDE4 is the predominant – and perhaps only – PDE isoenzyme in guinea-pig peritoneal eosinophils (Souness *et al.*, 1991). It is tightly membrane-bound and treatment of eosinophil membranes with Triton X-100 or high ionic strength buffers fails to dislodge any PDE4 activity. Although non-ionic detergents are ineffective, almost total solubilization of PDE4 can be achieved with the bile acid, deoxycholate, in the presence of NaCl or KCl (Souness *et al.*, 1992). Complex kinetics of cAMP hydrolysis are exhibited by the particulate PDE4 but these are largely lost upon solubilization (Souness *et al.*, 1992).

Vanadate/glutathione complex (V/GSH), which had previously been shown to activate PDE3 from adipocytes and hepatocytes (Souness *et al.*, 1985; Thompson *et al.*, 1991), activates the membrane-bound PDE4 by as much as 3-fold (Souness *et al.*, 1992). This treatment, like solubilization, also markedly alters the kinetic properties of the enzyme, increasing the *V*max for cAMP hydrolysis while slightly reducing the affinity for the substrate (Souness *et al.*, 1992). The ability of V/GSH to stimulate eosinophil PDE4 is markedly attenuated when examined against solubilized enzyme (Souness *et al.*, 1992). Moreover, the partially purified (anion-exchange chromatography), solubilized PDE4 is resistant to the effects of V/GSH (Souness *et al.*, 1992), suggesting, perhaps, that the effect is mediated indirectly through other membrane-associated components. It is not known how V/GSH stimulates cAMP PDEs. The active complex appears to consist minimally of a vanadyl ion and two oxidized electron donor compounds (Thompson *et al.*, 1991). Not all sulphydryl group-containing substances can substitute for GSH and it appears that a minimum structure of cysteamine (NH_2-CH_2-CH_2-SH) is required (Thompson *et al.*, 1991). Tungsten, niobium and tantalum – but not manganese, chromium or molybdenum – can replace vanadium to form hepatocyte PDE3-activating complexes (Thompson *et al.*, 1991). It is tempting to speculate that native PDE4 is under the influence of substances found within cells which complex with GSH or other sulphydryl donors in a manner similar to vanadate.

The information above suggests that the eosinophil PDE4 is capable of assuming different conformational states (Souness *et al.*, 1992; Giembycz and Souness, 1994). It is not known whether this is due to a conformational change in a monomeric form or whether

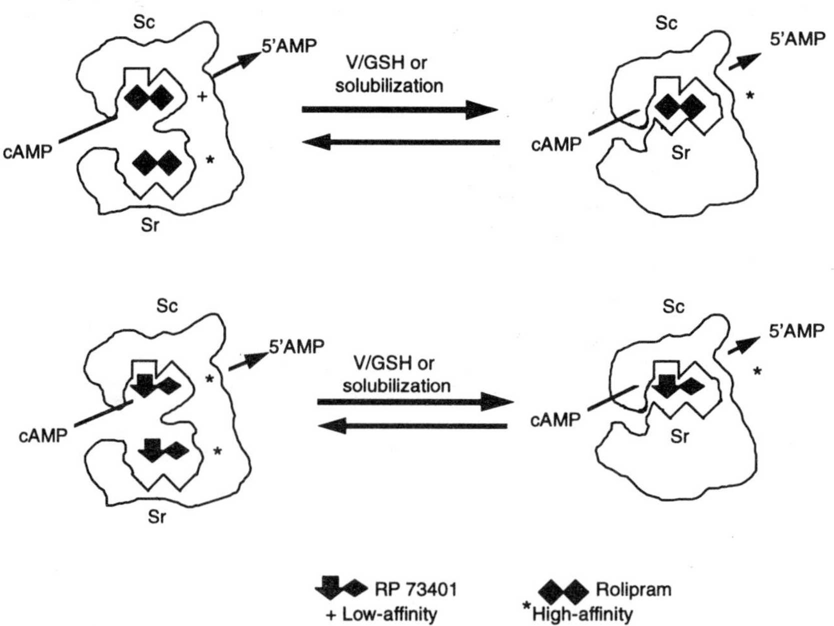

Figure 12.2 Two-site model to explain the different effects of solubilization and V/GSH on the potencies of RP 73401 and rolipram against eosinophil PDE4. Two sites, Sc and Sr, are proposed to be associated with PDE4. RP 73401 has similar affinities for the two sites, whereas rolipram is envisaged to exhibit a much higher affinity for Sr than Sc. In untreated membranes, Sc exerts a much greater influence on cAMP hydrolysis than Sr, so rolipram is only relatively weakly active. When PDE4 is exposed to V/GSH or solubilized, a conformational change occurs so that Sr exerts a much greater influence on catalytic activity and, as a result, the potency of rolipram increases. Since RP 73401 exhibits similar affinities for the two sites, the conformation of PDE4 does not influence its inhibitory potency.

multi-peptide interactions are involved. The quaternary structure of native PDE4 is uncertain and evidence indicating that the enzyme exists as a monomer (Thompson *et al.*, 1984; Némoz *et al.*, 1989), as well as more complex forms (Némoz *et al.*, 1989; Torphy, 1994; Torphy *et al.*, 1993), has been documented.

A potential explanation for the anomalies between the enzyme and whole cell data was suggested by the finding that the rank order of potency of PDE inhibitors is altered when the eosinophil PDE4 is solubilized or treated with V/GSH (Souness *et al.*, 1992). Notably, the potency of rolipram, denbufylline and Ro 20-1724 is increased by approximately 10-fold by solubilization and V/GSH treatment, whereas the IC_{50} values of other compounds (e.g. trequinsin, dipyridamole) are largely unaffected (Souness *et al.*, 1992). Furthermore, the correlation between enzyme and whole cell (cAMP) data was much improved by these treatments. It is noteworthy that, although V/GSH has no effect on the potencies of several standard inhibitors against PDE3, the inhibitory activity of cyclic GMP is increased (Thompson *et al.*, 1991).

To explain these results, different hypotheses can be advanced. If two PDE4 subtypes, for which rolipram-like compounds exhibit different affinities, exist in eosinophil membranes, the data might be rationalized if solubilization and V/GSH selectively activated the subtype for which rolipram, denbufylline and Ro 20-1724 have higher affinity. An equally plausible hypothesis proposes that PDE4 can exist in different conformational states against which rolipram-type compounds display different affinities. If this were the case, solubilization and V/GSH would convert PDE4 from a form(s) against which rolipram interacts with low affinity to one against which it interacts with high-affinity (Giembycz and Souness, 1994) (see Figs 12.2 and 12.3). A two-site model has been proposed (Souness *et al.*, 1992; Giembycz and Souness, 1994) in which the influence that one site, designated Sc, exerts on cAMP hydrolysis is unaffected by the conformation of PDE4, whereas that of the other, Sr, is dependent on the conformational state (Fig. 12.2). Thus, the potencies of compounds postulated to bind with high affinity to Sr and low affinity to Sc (rolipram, denbufylline, Ro 20-1724) are altered by treatments which change the conformation of the enzyme, whereas those with higher affinity for Sc or which do not discriminate between the two sites (trequinsin, dipyridamole) are unaffected by such treatments (Fig. 12.2). The requirement for two sites has been questioned (Torphy, 1994) and the altered potencies of rolipram-like compounds may simply reflect their different affinities for different

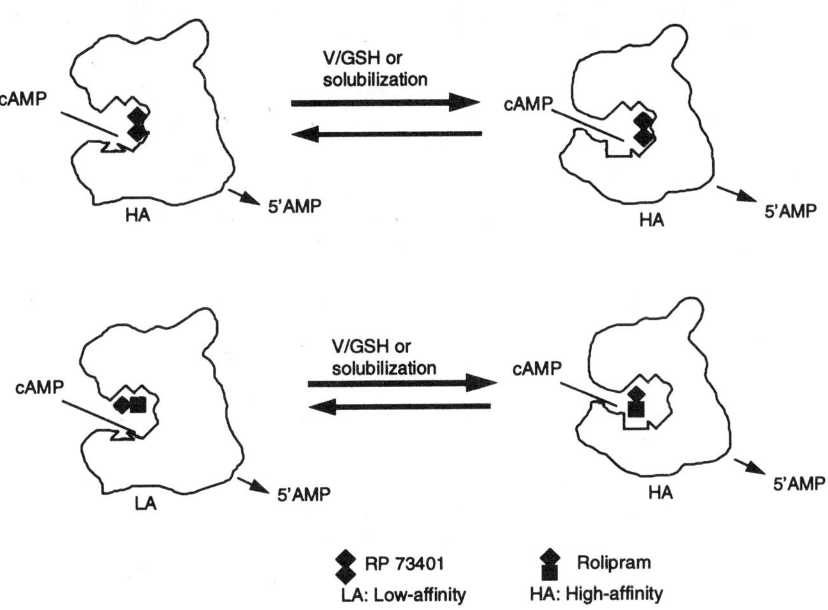

Figure 12.3 One-site model to explain the different effects of solubilization and V/GSH on the potencies of RP 73401 and rolipram against eosinophil PDE4. In a slight variation on the model proposed in Fig. 12.2, PDE4 is envisaged to exist in eosinophil membranes in two (or more?) conformational states. Both RP 73401 and rolipram act at the same site (catalytic site?), albeit through the formation of distinct chemical bonds. Rolipram has higher affinity for one conformational state than the other, whereas RP 73401 does not discriminate between the different forms. Solubilization and V/GSH, by changing the conformation of PDE4, alter the positions of amino acid groups with which rolipram interacts, thereby strengthening the bonding and increasing its potency and stereoselectivity. Since the conformation of PDE4 does not influence the affinity of RP 73401, its inhibitory potency is not affected by solubilization or V/GSH.

conformational states of PDE4 (see Fig. 12.3). As shown in Table 12.2, the inhibitory potency of RP 73401 is not affected by either solubilization or V/GSH treatment, indicating that, like trequinsin and dipyridamole, its interaction with PDE4 is uninfluenced by the enzyme conformation. Schematic models for the differential effects of these treatments on the inhibitory potencies of rolipram and RP 73401 are presented in Figs. 12.2 and 12.3.

3.3 Possible Role of the High-Affinity Rolipram-Binding Site

That PDE4 can exist in two or more conformational states received support in studies on the stereoselective behaviour of rolipram in eosinophils (Souness and Scott, 1993). In untreated membranes, only slight stereoselectivity is observed with rolipram, the $R(-)$ enantiomer being approximately 3-fold more potent than the $S(+)$ enantiomer. Solubilization of eosinophil PDE4 or treatment of membranes with V/GSH increases the enantiomeric selectivity (~20-fold). This level of stereoselectivity is similar to that observed on the high-affinity rolipram binding site in brain (Schneider *et al.*, 1987). Indeed, a strong correlation exists between displacement of [^{3}H]rolipram from brain membranes by a range of PDE inhibitors and their abilities to inhibit solubilized eosinophil PDE4 (Souness and Scott, 1993). This finding tempts speculation that, if the two-site hypothesis is correct, the putative Sr is similar to the high-affinity rolipram binding site in brain (see Chapter 13). Detection of specific [^{3}H]rolipram binding to eosinophil membranes has been unsuccessful thus far; however, a high-affinity binding site is co-expressed on the human recombinant (hr) PDE4A cloned from a monocyte cDNA library (Torphy *et al.*, 1992). Interestingly, the rank orders of potency of compounds in inhibiting hrPDE4A and displacing [^{3}H]rolipram are distinct (Torphy *et al.*, 1992). Furthermore, the affinity of rolipram for the high-affinity binding site displays marked stereoselectivity with $R(-)$ rolipram exhibiting >20-fold greater potency than $S(+)$ rolipram, whereas only a slight (~3-fold) enantiomeric potency difference is seen for inhibition of PDE4 (Torphy *et al.*, 1992). The nature of this binding site is currently uncertain, although it is known that histidine residues, which are critical in the catalytic activity of PDE4 (Jin *et al.*, 1992) also exert an important influence on [^{3}H]rolipram binding to hrPDE4A (Jacobitz *et al.*, 1994).

$R(-)$ Rolipram is 10-fold more potent than $S(+)$ rolipram in increasing the cAMP content in intact eosinophils (Souness and Scott, 1993). This finding, together with the very tight correlation between inhibition of solubilized PDE4 by a range of structurally

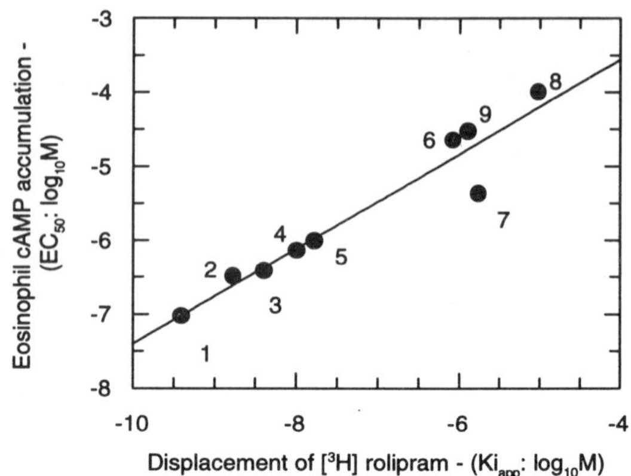

Figure 12.4 Enhancement of isoprenaline-induced cyclic AMP accumulation in eosinophils as a function of displacement of [^{3}H]rolipram binding to brain membranes. Eosinophil cAMP data are expressed as EC_{50} values (log concentration) and rolipram binding-displacement data as apparent Ki values (Ki_{app}, log concentration). Regression analysis demonstrates that stimulation of cAMP accumulates by PDE4 inhibitors as a function of rolipram binding displacement is highly significant ($r = 0.96$). Each point is the mean of at least three determinations. 1, RP 73401; 2, rolipram; 3, denbufylline; 4, ibudilast; 5, Ro 20–1724; 6, IBMX; 7, trequinsin; 8, dipyridamole; 9, AH 21–132.

unrelated compounds and cAMP accumulation in intact eosinophils, suggests that the native PDE4 may exist in a conformation similar to that induced by solubilization and V/GSH (Souness and Scott, 1993). Furthermore, a very close correlation is observed between displacement of [^{3}H]rolipram binding to brain membranes and elevation of eosinophil cAMP levels (Souness and Scott, 1993) (see Fig. 12.4). Whether the high-affinity stereoselective rolipram binding site represents a unique site on PDE4 distinct from the catalytic site or is a manifestation of a particular conformational state of the enzyme is uncertain. Its existence does provide a plausible explanation for the anomalies detailed above between the relative potencies of rolipram and RP 73401 on whole cell responses and against PDE4 in cell-free preparations, since the potency difference between the two compounds is almost identical (~4-fold) for their interactions at the brain high-affinity rolipram binding site and in potentiating isoprenaline-induced cAMP accumulation in intact eosinophils (Souness *et al.*, 1995) (see Fig. 12.4)

4. *Inhibition of PDE4 from Other Cells and Tissues*

4.1 Enzyme Data

RP 73401 displays similar potencies against PDE4s isolated from eosinophils, smooth muscle, macrophages and monocytes (Table 12.2). In contrast, rolipram is several fold more potent against PDE4 from eosinophils (untreated, particulate), macrophages and monocytes than against the partially purified smooth muscle PDE4 preparations (Table 12.2). Previous studies demonstrated that the potencies of rolipram, denbufylline and Ro 20-1724 against eosinophil PDE4 are 3–10-fold higher than those reported against PDE4 preparations from cardiac ventricular myocardium, smooth muscle, neutrophils and endothelial cells, but similar to those observed against PDE4 isolated from thymocytes and mast cells, as well as hrPDE4A and PDE4B (Giembycz and Souness, 1994).

An obvious explanation for the different potencies of rolipram but not RP 73401 against PDE4 preparations from different cells and tissues is that distinct PDE4 subtypes are expressed and rolipram, but not RP 73401, can discriminate between them. Certainly the properties of PDE4 from different cells and tissues are not uniform. For example, unlike the eosinophil enzyme, partially purified PDE4 from smooth muscle and cardiac muscle have been reported to be soluble (apparently) and display linear kinetics (Giembycz and Souness, 1994). The relatively weak inhibition displayed by rolipram against partially purified smooth muscle PDE4 is competitive (in contrast to the eosinophil PDE4) and non-stereoselective (Souness and Scott, 1993; Souness *et al.*, 1995).

The diverse properties of PDE4 in different tissues may be explained, at least in part, by the existence of multiple subtypes. Four different rat and human genes encoding PDE4 subtypes with highly conserved central sequences corresponding to the catalytic domain but with less homologous N- and C-termini have been identified (Swinnen *et al.*, 1989; Conti and Swinnen, 1990; Bolger *et al.*, 1993; Bolger, 1994). Northern blotting and reverse-transcriptase polymerase chain reaction (RT-PCR) studies have demonstrated that transcripts of the four PDE4 variants of different sizes are differentially expressed between tissues (Engels *et al.*, 1994; Monaco *et al.*, 1994). The molecular size of purified PDE4, as revealed by separation on polyacrylamide gel electrophoresis, varies greatly (Conti and Swinnen, 1990; Monaco *et al.*, 1994). Alternative splicing of newly transcribed nuclear RNA is responsible – at least in part – for producing the heterogeneity in the sizes of PDE4 subtypes (Monaco *et al.*, 1994). Different N-termini on PDE4 subtypes determine susceptibility to short-term regulation (Sette *et al.*, 1994) and may influence subcellular localization (Shakur *et al.*, 1993). RT-PCR studies have been used to determine the PDE4 subtype mRNA expression in different cells and tissues. In general, expression of multiple PDE4 subtypes in individual cell types appears to be the rule rather than the exception (Engels *et al.*, 1994), thus making it difficult to relate the widespread

Table 12.2 Inhibitory potencies of RP 73401 and rolipram against PDE4 from different cells and tissues and in displacing [^{3}H]rolipram from its high-affinity binding site in brain membranes

Source of PDE4	Nature of enzyme preparation	IC_{50}/Ki_{app}[a] (nM) RP 73401		Rolipram	
Pig aorta	Partially purified	1.0±0.1	(n=3)	1470±90	(n=3)
Bovine trachea	Partially purified	1.1±0.09		1967±203	(n=3)
Guinea-pig eosinophil	Tightly membrane-bound				
No treatment		2.0±1.0	(n=4)	230±20	(n=10)
Solubilization		1.9±1.1	(n=4)	14±3.0	(n=9)
V/GSH		2.6±0.5	(n=5)	20±10	(n=5)
Frozen[b]		1.0±0.2	(n=4)	685±0.1	(n=4)
Guinea-pig macrophage	Particulate				
No treatment		2.0	(n=2)	205±120	(n=4)
Human monocyte[c]	Cytosolic	1.0	(n=2)	320	(n=2)
Brain membranes	Particulate				
Ki_{app} for displacement of [^{3}H]rolipram binding		0.4±0.1	(n=3)	1.7±0.3	(n=3)

[a] Ki_{app} = apparent antagonist binding affinity.
[b] Particulate enzyme prepared from cells stored at −80°C.
[c] Monocyte PDE4 is located in both the cytosolic and particulate fractions. Neither form is greatly affected by V/GSH. RP 73401 and rolipram exhibit similar potencies against both the soluble and particulate PDE4s (measured in the presence of 10 μM siguazodan).
Data from Souness *et al.* (1995), Ashton *et al.* (1994) and unpublished observations.

potencies of rolipram against PDE4 preparations from different cells and tissues to subtype selectivity.

Limited information is available on the PDE4 subtype selectivity of inhibitors. However, in one study where compounds were tested against human PDE4A (also known as DPDE2), PDE4B (DPDE4) and PDE4D (DPDE3), little subtype selectivity was demonstrated by either rolipram (IC_{50} values: 0.18–0.5 μM) or denbufylline (IC_{50} values: 0.1–0.22 μM) (Bolger *et al.*, 1993). Caution must be exercised in determining the subtype selectivity of compounds based on data with recombinant PDE4 since it would appear that the systems used to express a particular PDE4 subtype can influence the potency of some inhibitors: for example, rolipram is at least 10-fold more potent against hrPDE4 expressed in yeast ($IC_{50} < 0.1$ μM) compared to COS cells ($IC_{50} \approx 1$ μM) (Livi *et al.*, 1990; Torphy *et al.*, 1992). Perhaps post-translational processing (folding) of PDE4A differs between the two expression systems and this might influence the inhibitory potency of rolipram. Based on the preceding arguments, it seems unlikely that the wide-ranging potencies of rolipram against PDE4 from different cells and tissues can be explained satisfactorily by subtype selectivity.

Alternatively, the properties of PDE4 may be influenced by the preparative procedures used to isolate it, which, in turn, may influence the potency of certain inhibitors. Cell/tissue disruption, by releasing proteases which clip an amino acid sequence necessary for high-affinity binding, or chromatography procedures, by altering the conformational state of the enzyme (Souness and Scott, 1993), although having little impact on catalytic activity, may influence the nature of rolipram's interaction with partially purified PDE4 and thus change its potency. If the native PDE4 in different cells and tissues is essentially similar (see below), it is clear that, in contrast to rolipram, preparative procedures do not influence the potency of RP 73401; this supports the view that the two compounds exert their inhibitory effects through distinct interactions with the enzyme. The influence of preparative procedures on the interaction of rolipram with PDE4 is demonstrated by the finding that the conditions used to store the enzyme can affect its potency. For example, when tested against the membrane-bound PDE4 prepared from eosinophils stored frozen at −80°C, the potency of rolipram ($IC_{50} = 660$ nM) is decreased almost 4-fold compared to the particulate enzyme isolated from freshly prepared cells (Souness *et al.*, 1995). In contrast, the potency of RP 73401 is not affected by freezing (Souness *et al.*, 1995).

4.2 WHOLE CELL DATA

As discussed above, a number of studies point to a role for the high-affinity rolipram binding site in regulating eosinophil function. Close correlations have also been reported between the central nervous system (CNS) actions (antagonism of reserpine-induced hypothermia in mice, induction of head twitches in rats, etc.) of PDE4 inhibitors and their potencies in displacing [^{3}H]rolipram from brain membranes (Schneider, 1984; Schultz and Schmidt, 1986; Schultz and Folkers, 1988; Schmiechen *et al.*, 1990). Other peripheral actions of PDE inhibitors also appear to be linked to an interaction with the high-affinity rolipram binding site. In guinea-pig trachea, for example, although a poor correlation exists between PDE4 inhibition and inhibition of histamine-induced contraction, a very strong correlation between inhibitor relaxation and displacement of [^{3}H]rolipram from a high-affinity binding site in brain has been documented (Harris *et al.*, 1989). As in *in vitro* studies, no correlation exists between inhibition of PDE4 activity and suppression of histamine-induced bronchoconstriction in anaesthetized guinea pigs; in contrast, an excellent relationship is obtained when suppression of histamine-induced bronchoconstriction is correlated with the ability of PDE4 inhibitors to displace [^{3}H]rolipram from brain membranes (Harris *et al.*, 1989). The four-fold potency difference between RP 73401 and rolipram in antagonizing the contractile effects of methacholine in guinea-pig trachea lends further weight to the hypothesis that the high-affinity rolipram binding site is involved in regulating airway smooth muscle tone.

Interestingly, recent studies (Mirza *et al.*, 1994; Pollock *et al.*, 1995) demonstrate that, in certain inflammatory cells, RP 73401 is much more potent than rolipram in suppressing functional responses (Table 12.3). For example, RP 73401 ($IC_{50} = 4$ nM) is 50-fold more potent than rolipram ($IC_{50} = 200$ nM) in inhibiting LPS-induced TNF-α release from human monocytes (Pollock *et al.*, 1995). The relative inhibitory potencies of RP 73401 and rolipram on monocyte TNF-α release are reflected in their enhancement of prostaglandin E_2-induced cAMP accumulation (IC_{50} values: RP 73401, 7 nM; rolipram, 300 nM) and, interestingly, the freshly isolated monocyte PDE4 (see Table 12.2). An even greater potency difference is observed in anti-CD3-differentiated mouse splenocytes, where RP 73401 ($IC_{50} = 1$ nM) is 140-fold more potent than rolipram ($IC_{50} = 140$ nM) in suppressing *Staphylococcus aureus* enterotoxin-induced IL-2 production (Pollock *et al.*, 1995) (see Table 12.3).

Thus, it would seem that the interactions of RP 73401 and rolipram with native PDE4 in monocytes, T lymphocytes and eosinophils may differ. Whether enzyme conformational differences between these cell types or interactions of the compounds with distinct PDE4 subtypes provide an explanation for this is uncertain. As discussed above, the subtype hypothesis is unlikely to offer a coherent explanation: eosinophils express PDE4D (Souness *et al.*, 1995), whereas monocytes express PDE4A and PDE4B (PDE4D is weakly

Table 12.3 Comparison of the inhibitory effects of RP 73401 and rolipram on intact cell/tissue functional responses

Functional response	*IC_{50} (nM)*			
	RP 73401		*Rolipram*	
Methacholine-induced contraction of guinea-pig trachealis	34 ± 7	(*n* = 6)	99 ± 9	(*n* = 6)
LTB_4-induced MBP release from guinea-pig eosinophils	115 ± 10	(*n* = 4)	602 ± 29	(*n* = 4)
LPS-induced TNF-*α* release from human monocytes	4.2 ± 0.3	(*n* = 3)	330	(*n* = 2)
Staphylococcus aureus enterotoxin-induced IL-2 release from CD3-differentiated splenocytes	1 ± 0.5	(*n* = 4)	140 ± 50	(*n* = 4)

Data from Souness *et al.* (1995) and Pollock *et al.* (1995).

expressed) (J. Souness *et al.*, unpublished observations). Alternatively, although high-affinity rolipram binding has been measured on both of the predominant PDE4 subtypes expressed in human monocytes (Torphy *et al.*, 1992; McLaughlin *et al.*, 1993), it is possible that an interaction at this "site" is less important in regulating functional responses in these cells (and T lymphocytes).

If this is the case it might be expected that, in contrast to eosinophils, little or no rolipram stereoselectivity would be observed on functional responses in monocytes and T lymphocytes. Limited – and somewhat equivocal – published data are available which, perhaps, could be interpreted as supporting this contention. (−)Rolipram is only 5-fold more potent than (+)rolipram in inhibiting LPS-induced TNF-*α* release from human mononuclear cells (Semmler *et al.*, 1993) and, in human peripheral blood mononuclear cells, only a slight enantiomeric potency difference is observed in the proliferative response to ragweed, a specific T-helper type 2 cell (T_H2) stimulus (Essayan *et al.* 1994). Lack of rolipram stereoselectivity has also been documented in its inhibition of antigen-induced bronchospasm and PGD_2 production in tracheal rings from ovalbumin-sensitized guinea pigs (Underwood *et al.*, 1993).

Although potency differences between RP 73401 and rolipram may provide insight into the nature of native PDE4, it is important to remember that factors other than intrinsic PDE4 inhibitory activity will influence the absolute potencies of inhibitors in eliciting functional responses in intact cells. For example, whereas the IC_{50} values of RP 73401 against PDE4 preparations from a variety of sources are very similar, its potencies in eliciting whole-cell responses vary greatly: RP 73401 is 115-fold more potent in inhibiting *Staphylococcus aureus* enterotoxin-induced IL-2 release from murine spleenocytes than LTB_4-stimulated MBP release from guinea-pig eosinophils, for instance (Table 12.3). This might be due to differences between cell types in the proportion of total PDE4 activity required to be inhibited for elevation of intracellular cAMP to occur, which, in turn, will be dependent on the relative intracellular rates of cAMP synthesis and hydrolysis (turnover). Unequal uptake of PDE4 inhibitors may also be a factor to consider when comparing their potencies on the functional responses in different cells and tissues. Unfortunately, little documented information is available to assess its importance.

5. *Possible Therapeutic Implications*

The possibility that interactions of inhibitors with native PDE4(s) differ between cell types may have important therapeutic implications. PDE4 inhibitors elicit a number of side-effects when administered to experimental animals and in the clinic. The most prominent of these are gastrointestinal disturbances, including nausea and vomiting, which have been observed in human subjects following oral administration of several PDE4 inhibitors, including rolipram (Treese and Rhein, 1990). Emesis may be a consequence of PDE4 actions on the chemoreceptor trigger zone in the area postrema of the brain (Carpenter *et al.*, 1988) and possibly exacerbated by local effects in the gastrointestinal tract. Whether the well-documented stimulatory effects of PDE4 inhibitors on parietal cell acid secretion in the stomach (Puurunen *et al.*, 1978) contribute to the gastric side-effects is uncertain. It was recently proposed (Barnette *et al.*, 1995b) that the conformation of PDE4 in inflammatory cells may differ from that in certain cell types associated with inhibitor-associated side-effects. The molecular basis for the different hypothesized states of PDE4 was not elucidated. The proposal was based on pharmacological data for compounds exhibiting different relative potencies against human recombinant PDE4A and in displacing [^{3}H]rolipram from its high-affinity receptor in brain cytosol (Barnette *et al.*, 1995b). The former activity was suggested to represent a low-affinity binding site on PDE4 (LPDE IV) and the latter a high-affinity binding site (HPDE IV). It was suggested that these different forms of PDE4 are non-interconvertible. In a range of studies with compounds which show great selectivity for HPDE IV (e.g. rolipram, denbufylline) or which exhibit slightly greater potency on LPDE IV or do not discriminate between the two sites (e.g. trequinsin, dipyridamole), it was concluded that anti-inflammatory effects, including (surprisingly) eosinophil O_2^- production, were associated, by and large, with actions on LPDE IV whereas emesis, as well as stimulation of acid secretion, were more closely linked with actions on HPDE IV (Barnette *et al.*, 1995a,b). SB 207499, which is generally equipotent with rolipram on functional responses in several inflammatory cells, is far less effective in stimulating acid secretion in parietal cells (Barnette *et al.*, 1994). This compound, although three fold more potent than rolipram against hrPDE4A is 28-fold less potent in displacing [^{3}H]rolipram from its binding site. The HPDE IV/LPDE IV ratio of 1.1 differed considerably from that for rolipram (0.013) (Barnette *et al.*, 1995b).

RP 73401 displays similar potencies on PDE4, no matter what the source, and in displacing rolipram from its high-affinity binding site in brain membranes (Souness *et al.*, 1995). Thus, if the above hypothesis is correct, it would be anticipated that, at therapeutically effective doses, RP 73401 would be less likely to induce emesis and certain other gastrointestinal side-effects than rolipram-type compounds.

6. Conclusions

RP 73401 is a novel, potent PDE4 inhibitor which exhibits properties suggestive of potential in the treatment of inflammatory disorders such as asthma. Unlike rolipram, it inhibits PDE4 from several cell types with similar potencies and procedures used to prepare the enzyme, which alter the inhibitory activity of rolipram, do not influence the ability of RP 73401 to inhibit PDE4. As discussed above, these findings may indicate that, whereas rolipram acts preferentially against certain conformational states of PDE4, RP 73401 does not discriminate between them. This, together with its great potency and selectivity compared to other PDE isozymes, make it an important tool with which to unravel the emerging roles of PDE4 in regulating cAMP metabolism and biological responses in cellular systems.

7. References

Ashton, M.J., Cook, D.C., Fenton, G., Karlsson, J.-A., Palfreyman, M.N., Raeburn, D., Ratcliffe, A.J., Souness, J.E., Thurairatnam, S. and Vicker, N. (1994). Selective type IV phosphodiesterase inhibitors as anti-asthmatic agents: the synthesis and biological activities of 3-(cyclopentyloxy)-4-methoxybenzamides and analogues. J. Med. Chem. 37, 1696–1703.

Barnette, M.S., Christensen, S.B., Essayan, D.M., Esser, K.M., Grous, M., Huang, S.-K., Manning, C.D., Prabhaker, U., Rush, J. and Torphy, T.J. (1994). SB 207449, a potent and selective phosphodiesterase (PDE) IV inhibitor, suppresses activities of several immune and inflammatory cells. Am. J. Resp. Crit. Care Med. 149, A209. [Abstract]

Barnette, M.S., Torphy, T.J. and Christensen, S.B. (1995a). WO Patent, 00139.

Barnette, M.S., Manning, C.D., Cieslinski, L.B., Burman, M., Christensen, S.B. and Torphy, T.J. (1995b). The ability of phosphodiesterase IV inhibitors to suppress superoxide production in guinea-pig eosinophils is correlated with inhibition of phosphodiesterase IV catalytic activity. J. Pharmacol. Exp. Ther. 273, 674–679.

Bolger, G. (1994). Molecular biology of the cyclic AMP-specific cyclic nucleotide phosphodiesterases: a diverse family of regulatory enzymes. Cell. Signal. 6, 851–859.

Bolger, G., Michaeli, T., Martins, T., St. John, T., Steiner, B., Rodgers, L., Riggs, M., Wigler, M. and Ferguson, K. (1993). A family of human phosphodiesterases homologous to the *dunce* learning and memory gene product of *Drosophila melanogaster* are potential targets for anti-depressant drugs. Mol. Cell. Biol. 13, 6558–6571.

Carpenter, D.O., Briggs, D.B., Knox, A.P. and Strominger, N. (1988). Excitation of area postrema neurons by transmitters, peptides, and cyclic nucleotides. J. Neurophys. 59, 358–369.

Conti, M. and Swinnen, J.V. (1990). Structure and function of the rolipram-sensitive, low-K_m cyclic AMP phosphodiesterases: a family of highly related enzymes. In "Cyclic Nucleotide Phosphodiesterases: Structure, Regulation and Drug Action" (eds. J. Beavo and M.D. Houslay), pp. 243–266. Wiley, Chichester.

Dent, G., Giembycz, M.A., Rabe, K.F. and Barnes, P.J. (1991). Inhibition of eosinophil cyclic nucleotide PDE activity and opsonised zymosan-induced respiratory burst by "type IV"-selective PDE inhibitors. Br. J. Pharmacol. 103, 1339–1346.

Engels, P., Fichtel, K. and Lubbert, H. (1994). Expression and regulation of human and rat phosphodiesterase type IV isogenes. FEBS Lett. 350, 291–295.

Essayan, D.M., Huang, S.-K., Undem, B.J., Kagey-Sobotka, A. and Lichtenstein, L.M. (1994). Antigen- and mitogen-induced proliferative responses of peripheral blood mononuclear cells by non-selective and isozyme selective cyclic nucleotide phosphodiesterase inhibitors. J. Immunol. 153, 3408–3416.

Genain, C.P., Roberts, T., Davis, R.L., Nguyen, M.-H., Uccelli, A., Faulds, D., Li, Y., Hedgpeth, J. and Hauser, S.L. (1995). Prevention of autoimmune demyelination in non-human primates by a cAMP-specific phosphodiesterase inhibitor. Proc. Natl Acad. Sci. USA 92, 3601–3605.

Giembycz, M.A. and Souness, J.E. (1994). Characteristics and properties of the cyclic AMP-specific phosphodiesterase in eosinophil leukocytes: a potential target for asthma therapy? In "Bronchitis V" (eds. D. Postma and J. Gerritsen), pp. 319–332. Van Gorcum, Assen.

Harris, A.L., Connell, M.J., Ferguson, E.W., Wallace, A.M., Gordon, R.J., Pagani, E.D. and Silver, P.J. (1989). Role of low Km cyclic AMP phosphodiesterase inhibition in tracheal relaxation and bronchodilation in the guinea-pig. J. Pharmacol. Exp. Ther. 251, 199–206.

Hatzelmann, A., Tenor, H. and Schudt, C. (1995). Differential effects of non-selective and selective phosphodiesterase inhibitors on human eosinophil functions. Br. J. Pharmacol. 114, 821–831.

Jacobitz, S., McLaughlin, M.M., Livi, G.P., Ryan, M.D. and Torphy, T.J. (1994). The role of conserved histidine residues on cAMP hydrolyzing activity and rolipram binding of human phosphodiesterase IV. FASEB J. 8, A371. [Abstract]

Jin, S.-L., Swinnen, J.V. and Conti, M. (1992). Characterization of the structure of a low Km, rolipram-sensitive cAMP phosphodiesterase: mapping of the catalytic domain. J. Biol. Chem. 267, 18929–18939.

Karlsson, J.-A., Souness, J.E., Webber, S.E., Pollock, K., Raeburn, D., Palfreyman, M.N. and Ashton, M.J. (1994). Suppression of mediator release from granulocytes by RP 73401, a novel, selective PDE IV inhibitor. Am. Rev. Resp. Dis. 149, A947. [Abstract]

Livi, G.P., Kmetz, P., McHale, M., Cieslinski, L.B., Sathe, G.M., Taylor, D.J., Davis, R.L., Torphy, T.J. and Balcarek, J.M. (1990). Cloning and expression of cDNA for a human low-*Km*, rolipram sensitive cyclic AMP phosphodiesterase. Mol. Cell. Biol. 10, 2678–2686.

Marivet, M.C., Bourguignon, J.-J., Lugnier, C., Mann, A., Stoclet, J.-C. and Wermuth, C.-G. (1989). Inhibition of cyclic adenosine-3′,5′-monophosphate phosphodiesterase from vascular smooth muscle by rolipram analogues. J. Med. Chem. 32, 1450–1457.

McLaughlin, M.M., Cieslinski, L.B., Burman, M. Torphy, T.J. and Livi, G.P. (1993). A low-Km, rolipram-sensitive, cyclic AMP-specific phosphodiesterase from human brain: cloning and expression of cDNA, biochemical characterisation of recombinant protein, and tissue distribution of mRNA. J. Biol. Chem. 268, 6470–6476.

Mirza, S., Withnall, M.T. and Karlsson, J.-A. (1994). Elevated cyclic AMP inhibits IL-4 and IL-5 release from mixed mouse splenocytes by blocking IL-2 release. Immunology 83 (Suppl. 1), 79. [Abstract]

Monaco, L., Vicini, E. and Conti, M. (1994). Structure of two rat genes coding for a closely related rolipram-sensitive cyclic AMP phosphodiesterase: multiple mRNA variants originate from alternative splicing and multiple start sites. J. Biol. Chem. 269, 347–357.

Némoz, G., Mouequit, M., Prigent, A.-F. and Pacheco, H. (1989). Isolation of similar rolipram-inhibitable cyclic AMP-specific phosphodiesterases from rat brain and heart. Eur. J. Biochem. 184, 511–520.

Nicholson, C.D. and Shahid, M. (1994). Inhibitors of cyclic nucleotide phosphodiesterase isoenzymes – their potential utility in the therapy of asthma. Pulmonary Pharmacol. 7, 1–17.

Palfreyman, M.N. and Souness, J.E. (1996). Phosphodiesterase type IV inhibitors. Prog. Med. Chem. 33, 1–52.

Pollock, K., Ebsworth, K., Buckley, G., Woodman, V., Raeburn, D. and Karlsson, J.-A. (1995). Effects of the novel type IV PDE inhibitor RP 73401 on cytokine production *in vitro* and *in vivo*. Second World Congress on Inflammation (Brighton).

Puurunen, J., Lucke, C. and Schwabe, U. (1978). Effect of the phosphodiesterase inhibitor 4-(3-cyclopentyloxy-4-methoxyphenyl)-2-pyrrolidone (ZK 62711) on gastric secretion and gastric mucosal cyclic AMP. Naunyn-Schmiedebergs Arch. Pharmacol. 304, 69–75.

Raeburn, D., Underwood, S.L., Lewis, S.A., Woodman, V.R., Battram, C.H., Tomkinson, A., Sharma, S., Jordan, R., Souness, J.E., Webber, S.E. and Karlsson, J.-A. (1994). Anti-inflammatory and bronchodilator properties of RP 73401, a novel and selective phosphodiesterase type IV inhibitor. Br. J. Pharmacol. 113, 1423–1431.

Rupert, D. and Weithmann, K.U. (1982). HL 725, an extremely potent inhibitor of platelet phosphodiesterase and induced platelet aggregation in vitro. Life Sci. 31, 2037–2043.

Schmiechen, R., Schneide, H.H. and Wachtel, H. (1990). Close correlation between behavioral response and binding *in vivo* for inhibitors of the rolipram-sensitive phosphodiesterase. Psychopharmacology 102, 17–20.

Schneider, H.H. (1984). Brain cAMP response to phosphodiesterase inhibitors in rats killed by microwave irradiation or decapitation. Biochem. Pharmacol. 33, 1690–1693.

Schneider, H.H., Schmiechen, R., Brezinski, M. and Seidler, J. (1987). Stereospecific binding of the anti-depressant rolipram to brain-specific structures. Eur. J. Pharmacol. 127, 105–115.

Schultz, J.E. and Folkers, G. (1988). Unusual stereoselectivity of the potential antidepressant rolipram on the cyclic AMP generating system from rat brain cortex. Pharmacopsychiatry 21, 83–86.

Schultz, J.E. and Schmidt, B.H. (1986). Rolipram, a stereospecific inhibitor of calmodulin-independent phosphodiesterase, causes beta-adrenoceptor subsensitivity in rat cerebral cortex. Naunyn-Schmiedebergs Arch. Pharmacol. 333, 23–30.

Sekut, L., Yarnall, D., Stimpson, S.A., Noel, L.S., Batemanfite, R., Clark, R.L., Brackeen, M.F., Menius, J.A. and Connolly, K.M. (1995). Anti-inflammatory activity of phosphodiesterase (PDE) IV inhibitors in acute and chronic models of inflammation. Clin. Exp. Immunol 100, 126–132.

Semmler, J., Wachtel, H. and Endres, S. (1993). The selective type IV phosphodiesterase inhibitor rolipram suppresses tumour necrosis factor-α production by human mononuclear cells. Int. J. Immunopharmacol. 15, 409–413.

Sette, C., Vicini, E. and Conti, M. (1994). The rat PDE3/IVd phosphodiesterase gene codes for multiple proteins differentially activated by cyclic AMP-dependent protein kinase. J. Biol. Chem. 269, 18271–18274.

Shakur, Y., Pryde, J. and Houslay, M.D. (1993). Engineered deletion of the unique N-terminal domain of the cyclic AMP-specific phosphodiesterase RD1 prevents plasma membrane association and the attainment of enhanced thermostability without altering its sensitivity to inhibition by rolipram. Biochem. J. 292, 677–686.

Sommer, N., Loschmann, P.-A., Northoff, G.H., Weller, M., Steinbrecher, A., Steinbach, J.P., Lichtenfels, R., Meyermann, R., Riethmuller, A., Fontana, A., Dichgans, J. and Martin, R. (1995). The antidepressant rolipram suppresses cytokine production and prevents autoimmune encephalomyelitis. Nat. Med. 1, 244–248.

Souness, J.E. and Giembycz, M.A. (1994). Cyclic nucleotide phosphodiesterases in airways smooth muscle. In "Airways Smooth Muscle: Biochemical Control Of Contraction And Relaxation" (eds. D. Raeburn and M.A. Giembycz), pp. 271–308. Birkhäuser, Basel.

Souness, J.E. and Scott, L.C. (1993). Stereospecificity of rolipram actions on eosinophil cyclic AMP-specific phosphodiesterase. Biochem. J. 291, 389–395.

Souness, J.E., Thompson, W.J. and Strada, S.J. (1985). Adipocyte cyclic nucleotide phosphodiesterase activation by vanadate. J. Cyclic Nucleotide Protein Phosphorylation Res. 10, 383–396.

Souness, J.E., Carter, C.M., Diocee, B.K., Hassall, G.A., Wood, L.J. and Turner, N.C. (1991). Characterization of guinea-pig eosinophil phosphodiesterase activity: assessment of its involvement in regulating superoxide generation. Biochem. Pharmacol. 42, 937–945.

Souness, J.E., Maslen, C. and Scott, L.C. (1992). Effects of solubilization and vanadate/glutathione complex on inhibitor potencies against eosinophil cyclic AMP-specific phosphodiesterase. FEBS Lett. 302, 181–184.

Souness, J.E., Maslen, C., Webber, S., Foster, M., Raeburn, D., Palfreyman, M.N., Ashton, M.J. and Karlsson, J.-A. (1995). Suppression of eosinophil function by RP 73401, a potent and selective inhibitor of cyclic AMP-specific phosphodiesterase: comparison with rolipram. Br. J. Pharmacol. 114, 39–46.

Swinnen, J.V., Joseph, D.R. and Conti, M. (1989). Molecular cloning of rat homologs of the *Drosophila melanogaster* dunce cyclic AMP phosphodiesterase: evidence for a family of genes. Proc. Natl Acad. Sci. USA 86, 5325–5329.

Thompson, W.J., Pratt, M.L. and Strada, S.J. (1984). Biochemical properties of high affinity cyclic AMP phosphodiesterase. Adv. Cyclic Nucleotide Protein Phosphorylation Res. 16, 137–148.

Thompson, W.J., Tan, B.H. and Strada, S.J. (1991). Activation of rabbit liver high-affinity cAMP (type IV) phosphodiesterase by vanadyl-glutathione complex. J. Biol. Chem. 266, 17011–17019.

Torphy, T.J. (1994). Inhibitors of phosphodiesterase isozymes: new therapeutic possibilities for asthma. In "Bronchitis V" (eds. D. Postma and J. Gerritson), pp. 303–316. Van Gorcum, Assen.

Torphy, T.J. and Undem, B.J. (1991). Phosphodiesterase inhibitors: new opportunities for the treatment of asthma. Thorax 46, 512–523.

Torphy, T.J., Burman, M., Huang, L.B.F. and Tucker, S.S. (1988). Inhibition of the low Km cyclic AMP phosphodiesterase in intact canine trachealis by SK&F 94836: mechanical and biochemical responses. J. Pharmacol. Exp. Ther. 246, 843–850.

Torphy, T.J., Stadel, J.M., Burman, M., Cieslinski, L.B., McLaughlin, M.M., White, J.R. and Livi, G.P. (1992). Coexpression of human cyclic AMP-specific phosphodiesterase activity and high-affinity rolipram binding in yeast. J. Biol. Chem. 267, 1798–1804.

Torphy, T.J., De Wolf, W.E., Green, D.W. and Livi, G.P. (1993). Biochemical characteristics and cellular regulation of phosphodiesterase IV. Agents Actions Suppl. 43, 51–71.

Treese, N. and Rhein, S. (1990). Phosphodiesterase inhibitors: clinical experience. Drugs News Perspect. 3, 99–105.

Underwood, D.C., Osborn, R.R., Novak, L.B., Matthews, J.K., Newsholme, S.J., Undem, B.J., Hand, J.M. and Torphy, T.J. (1993). Inhibition of antigen-induced bronchoconstriction and eosinophil infiltration in the guinea-pig by the cyclic AMP-specific phosphodiesterase inhibitor, rolipram. J. Pharmacol. Exp. Ther. 266, 306–313.

13. *Molecular Aspects of Inhibitor Interaction with PDE4*

Siegfried B. Christensen, Walter E. DeWolf, Jr, M. Dominic Ryan *and* Theodore J. Torphy

1. Introduction 185
   1.1 Chemistry and Characteristics of the Phosphodiesterases 185
   1.2 Historical Perspective on Isoenzyme-selective PDE Inhibitors 186
2. Molecular Biology of PDE4 190
   2.1 Subtypes and mRNA Splice Variants 190
   2.2 Functional Domains of PDE4 191
   2.3 Role of Conserved Histidines 192
3. Rolipram Binding Site 193
   3.1 Historical Perspective 193
   3.2 Nature and Function of the High-Affinity Rolipram-binding Site 194
   3.3. Proposals on the Nature and Function of the High-Affinity Rolipram-binding Site 194
   3.4 Biological Significance of High-Affinity Rolipram Binding 195
4. Mechanistic Enzymology 195
   4.1 Kinetic Behaviour of PDE4s 195
   4.2 Inhibition by *R*-Rolipram 197
   4.3 Binding of *R*-Rolipram to $Met^{265-886}$ 199
5. Structure–Activity Relationships 199
   5.1 Introduction 199
   5.2 Rolipram and Lead PDE4 Inhibitors 200
   5.3 Rolipram and Derivatives 201
   5.4 Overlay Model of PDE4 Inhibition 201
6. Summary and Conclusions 202
7. References 203

1. *Introduction*

1.1 CHEMISTRY AND CHARACTERISTICS OF THE PHOSPHODIESTERASES

Cyclic nucleotide phosphodiesterases (PDEs) are intracellular enzymes which catalyse the hydrolysis of the 3′-phosphoester bond of the ubiquitous "second messengers" adenosine 3′,5′-cyclic phosphate (cAMP) and guanosine 3′,5′-cyclic phosphate (cGMP) to produce their respective inactive 5′-monophosphates (Fig. 13.1). Beginning with the work of Thompson and Appleman (1971), isolation and classical enzymological characterization of PDEs from diverse tissue and cell samples of numerous species has revealed that there are multiple families of PDE isoenzymes. These isoenzymes are distinguished biochemically on the basis of substrate specificity (or lack thereof), kinetic properties, regulation by allosteric modulators (e.g. Ca^{2+}/calmodulin (CaM), insulin or cGMP) and selective inhibition by pharmacological agents (Weishaar *et al.*, 1985; Beavo, 1988; Beavo and Reifsnyder, 1990; see also Chapters 1 and 2).

To date, seven isoenzyme families, composed of a total of more than thirty individual members (subtypes and splice variants), have been characterized on the basis of immunological reactivity, primary protein sequences and RNA (complementary DNA) sequence data (Beavo *et al.*, 1994; see Chapter 1). Molecular genetic techniques have led to the cloning, expression and purification of quantities of a number of these PDE

Phosphodiesterase Inhibitors
ISBN 0-12-210720-9

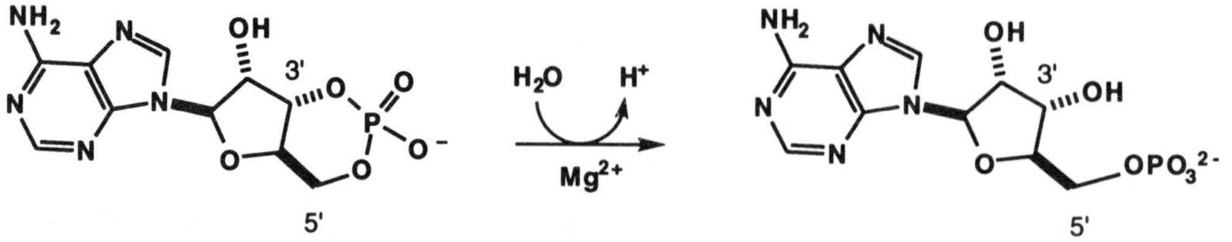

Family	Isozyme	K_m (μM)		Selective Inhibitors
		cAMP	cGMP	
1	Ca^{2+}/CaM-stimulated	1-30	3	vinpocetine, KS-505a
2	cGMP-stimulated	50	50	EHNA
3	cGMP-inhibited	0.2	0.3	amrinone, cilostamide
4	cAMP-specific	4	>3000	rolipram, denbufylline
5	cGMP-specific	150	1	zaprinast, dipyridamole
6	photoreceptor	60	2000	zaprinast, dipyridamole
7	high affinity cAMP-specific	0.2	>1000	none identified

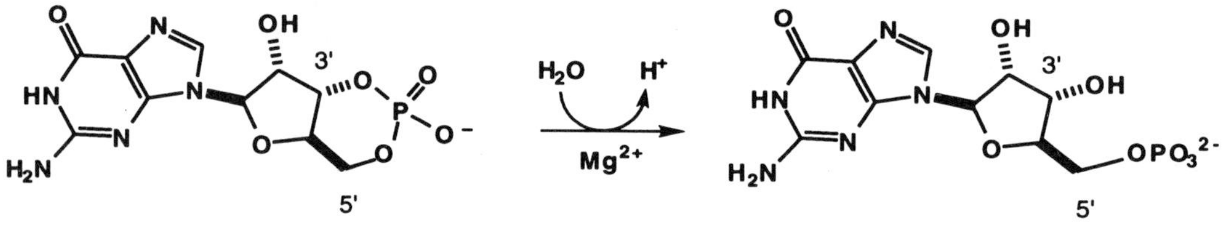

Figure 13.1 Hydrolysis of cyclic nucleotides and characteristics of PDE isozymes. Nomenclature and values adapted from Beavo *et al.* (1994).

subtypes. This has allowed detailed biochemical evaluation of kinetically pure preparations of PDE isozymes and their subtypes which are important in the pathophysiology of a variety of disease states.

As indicated in Fig. 13.1, selective PDE inhibitors have been identified for all families except PDE7. This has stimulated interest in PDEs as drug targets because the differential cellular distribution of the various family members allows specific targeting by compounds of the isoenzyme that predominates in the tissue or cell of interest, potentially reducing the unfavourable side-effect profile of classical, non-selective PDE inhibitors, such as theophylline (see Chapter 3).

After a brief perspective on the historical development of isoenzyme-selective PDE inhibitors, the focus of this chapter will be the molecular characterization of the subtypes of PDE4 with respect to their primary protein sequence, biochemical properties and inhibition profile by a series of known inhibitors and by a homologous series of novel *N*-benzylpyrrolidinone inhibitors.

1.2 HISTORICAL PERSPECTIVE ON ISOENZYME-SELECTIVE PDE INHIBITORS

The non-selective PDE inhibitors include the classical isoquinoline papaverine and the xanthines theophylline and 3-isobutyl-1-methylxanthine (IBMX) (Weishaar *et al.*, 1985), as well as the more recent anti-hypertensive agent trequinsin (Lal *et al.*, 1984; Marivet *et al.*, 1989) (Fig. 13.2). Early chemical efforts to identify selective PDE inhibitors focused on 7- and 8-substituted xanthines, which were found to be reasonably selective for PDE1 (Garst *et al.*, 1976; Wells *et al.*, 1981). More recently, a variety of partially selective analogues of

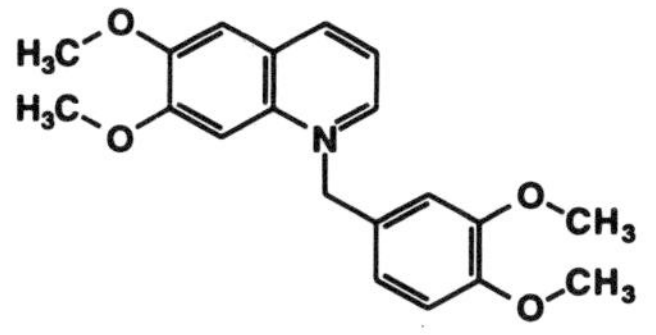

Papaverine 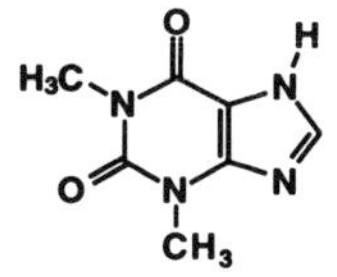 Theophylline 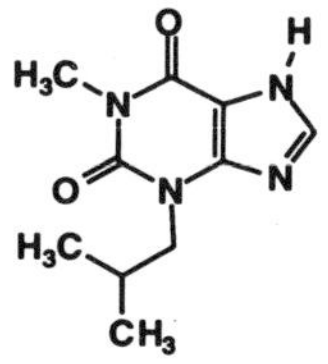 IBMX 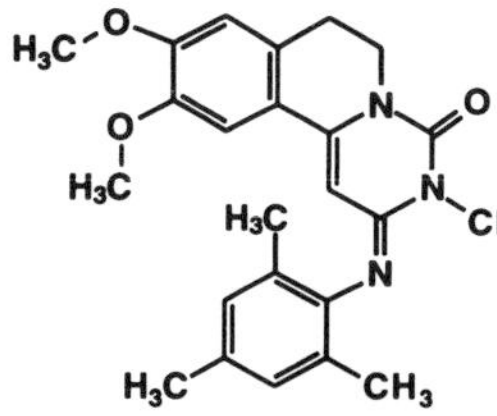 Trequinsin

Figure 13.2 Non-selective PDE inhibitors.

cAMP and cGMP have been developed primarily as probes of the cyclic nucleotide binding sites of PDEs (Genieser *et al.*, 1989; Grant *et al.*, 1990; Beltman *et al.*, 1995; Butt *et al.*, 1995). Structures of representative examples of isoenzyme selective PDE inhibitors, discussed below, are indicated in Fig. 13.3.

No recent directed chemical efforts targeting selective inhibitors of PDE1, the Ca^{2+}/CaM-stimulated isoenzyme, have been conducted. Reported inhibitors are the vascular smooth muscle relaxant vinpocetine (Ahn *et al.*, 1989) and the novel *Streptomyces* metabolite KS-505a (Nakanishi *et al.*, 1992) (Fig. 13.3). The only known selective inhibitor of PDE2 is the recently reported adenosine deaminase inhibitor, erythro-9-(2-

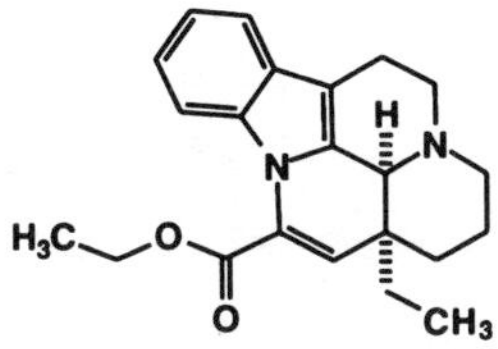

Vinopcetine 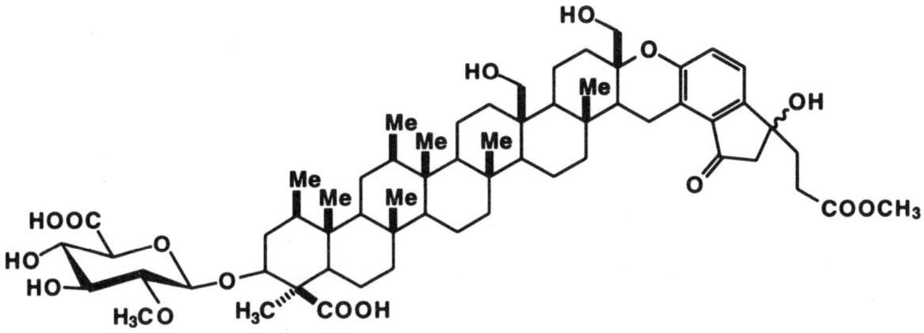 KS-505a

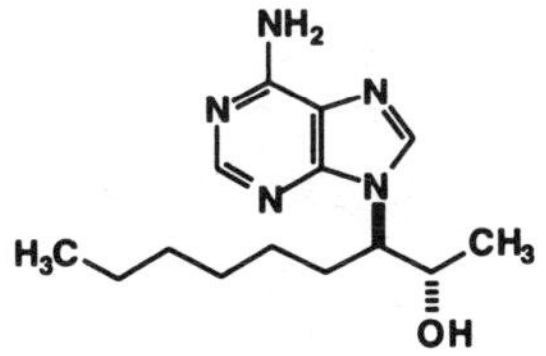

EHNA 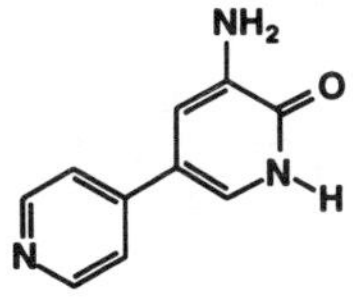 Amrinone 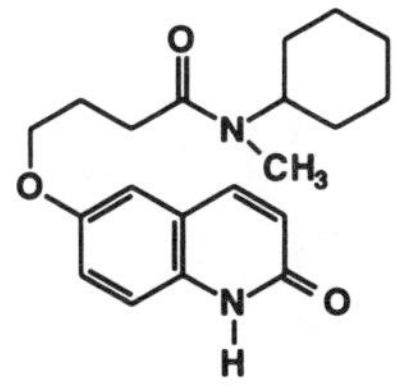 Cilostamide

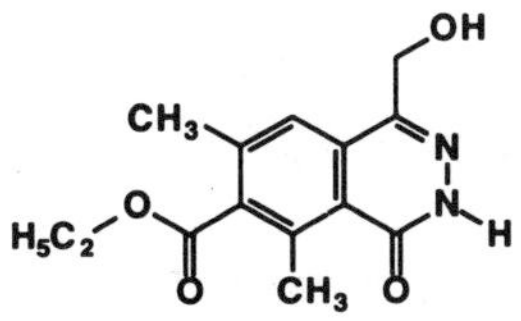

Phthalazinol 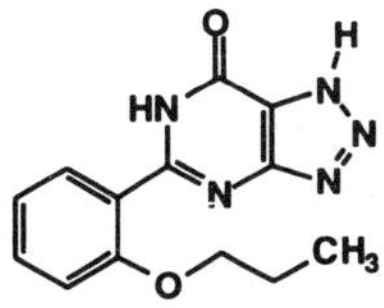 Zaprinast 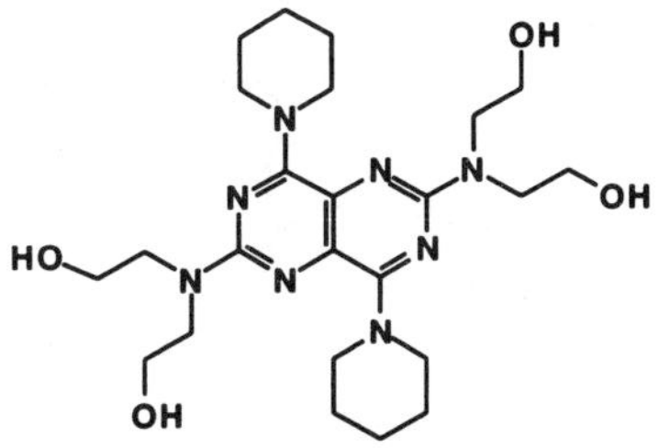 Dipyridamole

Figure 13.3 Selective inhibitors of PDE1, PDE2, PDE3 and PDE5.

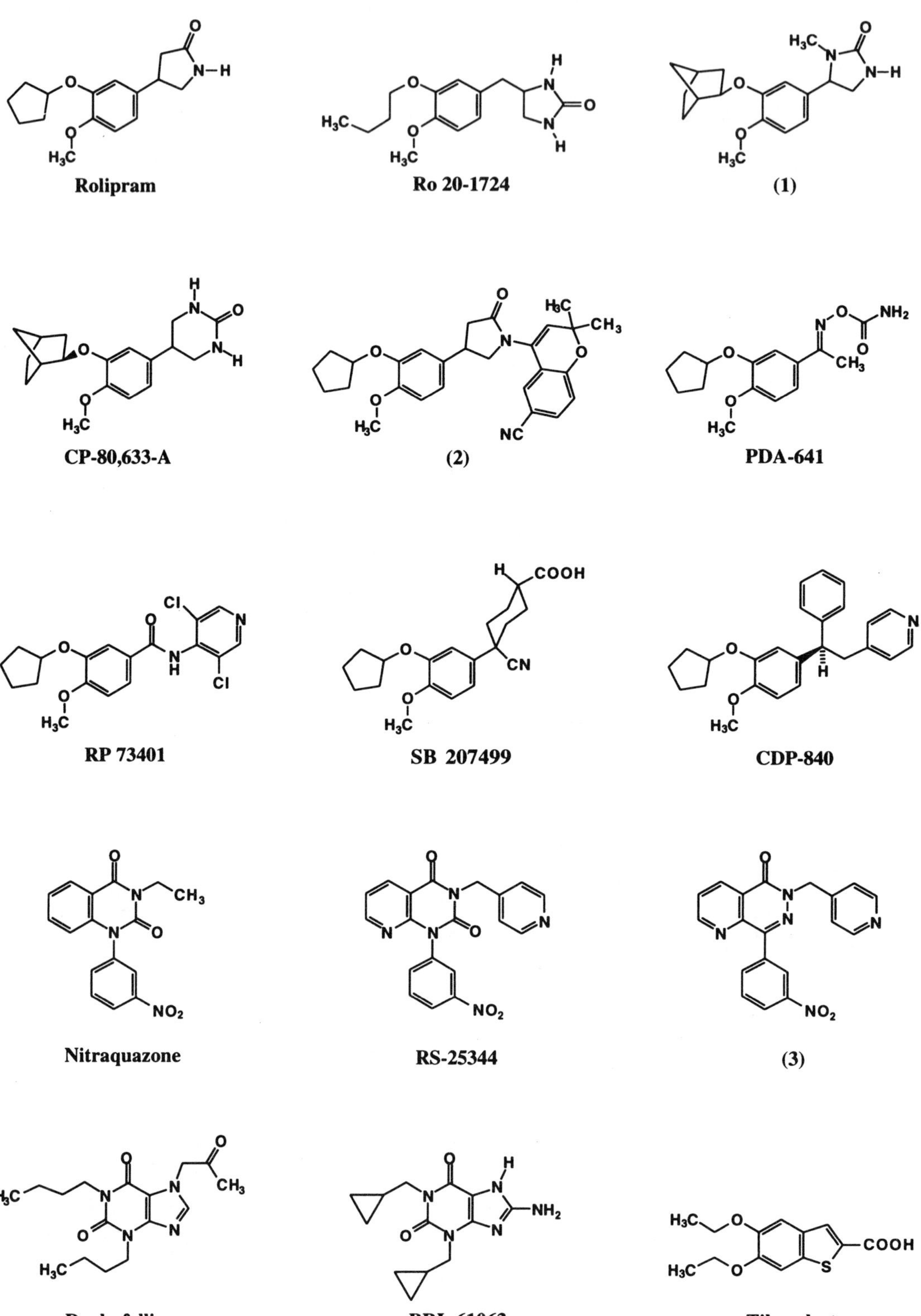

Figure 13.4 Lead selective PDE4 inhibitors.

hydroxy-3-nonyl)adenine (EHNA, Méry *et al.*, 1995; see Chapter 5).

Historically, the discovery of amrinone (Farah and Alousi, 1978; Weishaar *et al.*, 1983) as a novel cardiotonic agent and the discoveries of both phthalazinol (Asano *et al.*, 1977) and cilostamide (Hidaka *et al.*, 1979) (Fig. 13.3) as inhibitors of platelet aggregation stimulated a major focus in the pharmaceutical industry on selective inhibitors of PDE3 as inotropes, peripheral vasodilators and antithrombotics (Weishaar *et al.*, 1985; Erhardt, 1987; Nicholson *et al.*, 1991). More recently, the potential bronchorelaxant activity of these compounds has also been of interest (Torphy and Undem, 1991). Detailed reviews of the structure–activity relationships developed within the wide variety of structural classes encompassed in this work have appeared (Meanwell and Seiler, 1990; Erhardt, 1990).

The pharmacological profile of the selective inhibitors of PDE5/6, zaprinast and the classical antithrombolytic dipyridamole, has suggested the utility of PDE5 inhibitors as vaso- and bronchodilators (Murray, 1993) (Fig. 13.3). Studies resulting from directed chemical efforts to identify novel structural classes of PDE5 inhibitors have appeared recently (Takase *et al.*, 1994; Saeki *et al.*, 1995; Lee *et al.*, 1995; see also Chapter 8 and Chapter 9). No selective inhibitors have been reported for PDE7.

PDE4 is widespread in both brain and peripheral tissues, and reports of its distribution and characteristics have appeared in the literature for some time (Conti and Swinnen, 1990). Initial interest in this family stemmed from the therapeutic potential of rolipram (Fig. 13.4), an anti-depressant, which inhibits PDE4 activity in the central nervous system (Schwabe *et al.*, 1976; Wachtel, 1983a,b; Schultz and Schmidt, 1986). Studies with rolipram and Ro 20-1724 (Gruenman and Hoffer, 1975; Sheppard and Wiggan, 1971) (Fig. 13.4) have allowed demonstration of the widespread distribution and predominant functional role of PDE4 in regulation of cAMP levels in airway smooth muscle and in human inflammatory cells, stimulating interest in the therapeutic potential of PDE4 inhibitors as anti-asthmatic agents (Torphy and Undem, 1991; Nicholson *et al.*, 1991; Torphy *et al.*, 1993b; Christensen and Torphy, 1994) and thus prompting synthetic programmes focused on identification of novel PDE4 inhibitors.

Lead structures of the more recent PDE4 inhibitors, illustrated in Fig. 13.4, include: the *N*-(methyl)-imidazolidinone compound (1) (Saccomano *et al.*, 1991); the tetrahydropyrimidone CP-80,633-A (Cohan *et al.*, 1995); the N-(benzopyranyl)pyrrolidinone compound (2) (Pinto *et al.*, 1993); the oxime carbamate PDA-641 (Lombardo, 1992; Lombardo *et al.*, 1993); the *N*-(pyridinyl)benzamide RP 73401 (Ashton *et al.*, 1994; see Chapter 12); the cyclohexane carboxylic acid SB 207499 (Christensen, 1993; Barnette *et al.*, 1994); the phenethylpyridine CDP 840 (Warrellow *et al.*, 1994; Higgs, 1995); the quinazoline diones nitraquazone (Glaser and Traber, 1984; Lowe *et al.*,

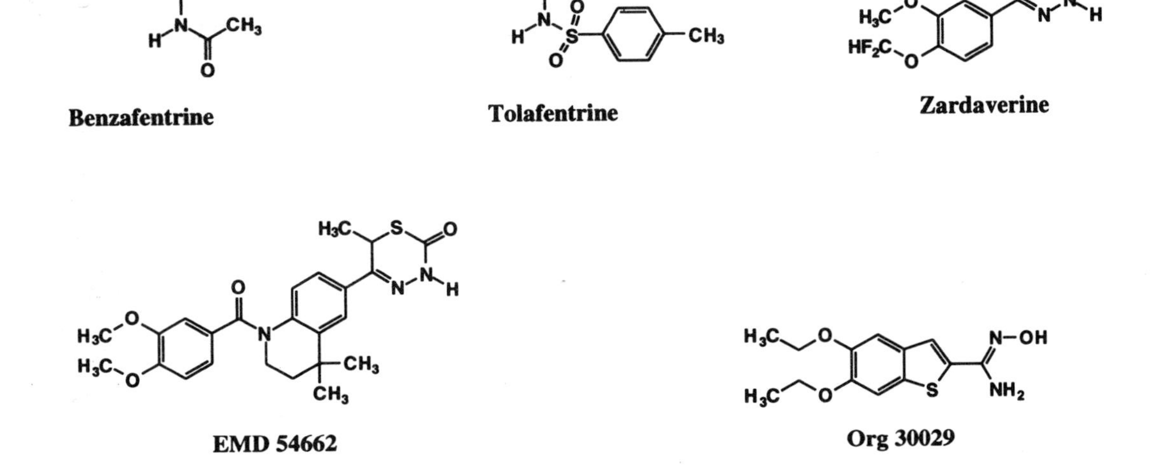

Benzafentrine

Tolafentrine

Zardaverine

EMD 54662

Org 30029

Figure 13.5 Dual PDE3/4 inhibitors.

1991) and RS-25344 (Kaneko *et al.*, 1995; see Chapter 11); the pyridopyridazinone compound (3) (Wilhelm *et al.*, 1993); the xanthines denbufylline (Brenner *et al.*, 1980; Nicholson *et al.*, 1989) and BRL 61063 (Buckle *et al.*, 1994); and the benzothiophene carboxylic acid tibenelast (Ho *et al.*, 1990). Several PDE3/4 dual inhibitors (Fig. 13.5) have also been identified, including: the benzonaphthyridines benafentrine (AH 21–132) (Elliott *et al.*, 1991; Giembycz and Barnes, 1991) and tolafentrine (B9004–070) (Schudt *et al.*, 1993; Hatzelmann *et al.*, 1995), the thiadiazinone EMD 54662 (Klockow and Jonas, 1989) and the benzothiophene hydroxyamidine Org 30029 (Shahid and Nicholson, 1990). This latter group of drugs is discussed in Chapter 10.

Selected examples of the PDE4 inhibitors will be discussed in detail in section 5, together with the pyridazinone PDE3/4 dual inhibitor, zardaverine (Fig. 13.5) (Amschler, 1987; Schudt *et al.*, 1991).

2. *Molecular Biology of PDE4*

2.1 SUBTYPES AND mRNA SPLICE VARIANTS

Several PDE isoenzyme families contain multiple subtypes. Four distinct subtypes, designated PDE4A through PDE4D, make up the PDE4 family (Beavo *et al.*, 1994; Bolger, 1994; see Chapter 1) (Table 13.1). These subtypes are encoded by distinct genes with the following human chromosomal assignments: PDE4A, 19; PDE4B, 1q31; PDE4C, 19; PDE4D, 5q12 (Milatovich *et al.*, 1994). cDNAs encoding these subtypes were initially cloned from rat libraries (Colicelli *et al.*, 1989; Davis *et al.*, 1989; Swinnen *et al.*, 1989, 1991) and, subsequently, their human homologues were isolated (Bolger *et al.*, 1993; McLaughlin *et al.*, 1993; Obernolte *et al.*, 1993; Engels *et al.*, 1995) (Table 13.1). These subtypes have similar kinetic characteristics (cAMP K_m = 1.5–18 μM) (Bolger *et al.*, 1993; Engels *et al.*, 1995; Livi *et al.*, 1990; McLaughlin *et al.*, 1993). Each is highly selective for cAMP (K_m > 1000 μM for cGMP) and sensitive to rolipram and other selective PDE4 inhibitors (Bolger *et al.*, 1993; McLaughlin *et al.*, 1993; Engels *et al.*, 1995). However, the activity of these enzymes can be regulated differentially at the level of transcription or via post-translational modification (see Conti *et al.*, 1991; Torphy *et al.*, 1994).

Additional diversity in the PDE4 family is produced by alternate mRNA splicing of the PDE4B and PDE4D loci. Thus far, at least two splice variants have been identified for human PDE4B and PDE4D (Bolger, 1994; Beavo *et al.*, 1994). Several more splice variants have been identified for the corresponding rat subtypes as well as for rat PDE4A. The alternative splice points generally occur at one of two consensus regions in the 5′ end of the transcripts (Bolger *et al.*, 1994). The functional consequences of these amino-terminal variants with respect to kinetic, structural and regulatory properties have yet to be examined rigorously, although studies with rat immunoprecipitated PDE4B1 and PDE4B2 suggest no difference between these variants with respect to their cAMP K_m or rolipram sensitivity (Lobban *et al.*, 1994). Interestingly, the removal of the

Table 13.1 Human PDE4 subtypes. Listed are the human PDE4 subtypes and splice variants identified to date (taken from Bolger, 1994 and Beavo *et al.*, 1994). Only full-length clones or the longest reported partial clone for a specific splice variant are listed. Nomenclature for the GenBank name (e.g., HSPDE4A5) is assigned as follows: (1) species (HS, *Homo sapiens*); (2) Human Genome Project name for family (PDE4); (3) subtype (A, B, C or D); (4) mRNA transcript (e.g. 5); and (5) chronological order of independently isolated clones produced from the same mRNA (A, B or C).

Human genome project name	*GenBank name*	*Accession number*	*Chromosomal locus*[a]	*Amino acids*	*Reference*
PDE4A	HSPDE4A5	L20965	19	886	Bolger *et al.* (1993)
	HSPDE4A8	U18088		[b]	Horton *et al.* (1995)
PDE4B	HSPDE4B1	L20966	1q31	736	Bolger *et al.* (1993)
	HSPDE4B2A	M97515		564	McLaughlin *et al.* (1993)
	HSPDE4B2B	L20971		564	Bolger *et al.* (1993)
	HSPDE4B2C	L12686		564	Obernolte *et al.* (1993)
PDE4C	HSPDE4C1B	Z46632	19	712	Engels *et al.* (1995)
PDE4D	HSPDE4D3	L20970	5q12	673	Bolger *et al.* (1993)
	HSPDE4D4	L20969		451[c]	Bolger *et al.* (1993)

[a] From Milatovich *et al.* (1994).
[b] 34 base pair insert in catalytic domain causes premature truncation of the protein; expressed protein is inactive.
[c] cDNA does not encode a full-length protein.

first 67 nucleotides from RD1, a cDNA encoding the rat homologue of human PDE4A, yields an expressed protein that is primarily cytosolic, whereas the full-length protein is membrane-bound (Shakur *et al.*, 1993). This suggests that the N-terminus is important for targeting the enzyme to subcellular organelles and that proteins produced through alternative mRNA splicing may have different subcellular localizations. This speculation is supported by recent evidence indicating a differential subcellular distribution of two rat PDE4B splice variants (Lobban *et al.*, 1994).

A critical finding that may have profound implications for drug discovery relates to the tissue and cellular distribution of PDE4 subtypes. Numerous studies employing antibodies (Torphy *et al.*, 1995; Manning *et al.*, 1996), RNase protection or Northern blot analyses (Swinnen *et al.*, 1989; McLaughlin *et al.*, 1993; Bolger *et al.*, 1994; Baecker *et al.*, 1994; Engels *et al.*, 1995), or reverse transcriptase-polymerase chain reaction (RT-PCR) (Engels *et al.*, 1994, 1995) indicate that PDE4 subtypes have unique tissue distributions. For example, RNase protection or RT-PCR experiments using various human neuronal cell lines or mRNA isolated from whole human brain indicate that PDE4C is abundant in neuronal tissue (Bolger *et al.*, 1993). In contrast, RT-PCR experiments indicate that this subtype is absent from immune and inflammatory cells (Engels *et al.*, 1994). Message for PDE4A appears to be distributed ubiquitously (Livi *et al.*, 1990; Bolger *et al.*, 1993; Obernolte *et al.*, 1993; Engels *et al.*, 1994), although in monocytes and perhaps other cells the expression of this subtype is typically very low (Lobban *et al.*, 1994; Manning *et al.*, 1996; Torphy *et al.*, 1995). PDE4B is expressed in heart, brain, skeletal muscle and lung, but not in placenta, liver, kidney or pancreas (McLaughlin *et al.*, 1993).

Messenger RNA splice variants of selected PDE4 subtypes also have unique tissue distributions. In a survey of splice variants of PDE4 subtypes expressed in the rat CNS, Bolger *et al.* (1994) used RNase protection studies to determine that all variants are expressed but at different relative levels. Moreover, the mRNA splice variants display different expression patterns with respect to the specific region of the brain being examined (Bolger *et al.*, 1994). These results were extended using peptide antibodies directed against products of two PDE4B mRNA splice variants to detect protein expression in specific regions of the rat brain (Lobban *et al.*, 1994). PDE4B1 is located in the hypothalamus, striatum, hippocampus and cortex, whereas this variant is absent in the cerebellum, brain stem, mid-brain and pituitary. In contrast, PDE4B2 is present in all of these regions with the exception of the pituitary. Moreover, PDE4B1 is restricted to the cytosol whereas PDE4B2 is found exclusively in membranes, thus supporting the contention that the N-terminal domain is critical for subcellular targeting.

The growing body of evidence indicating that PDE4 subtypes and alternative mRNA splice variants have distinct tissue distributions prompts speculation that the physiological roles of these proteins differ. If this is the case, then the opportunity exists to incorporate an exquisite degree of tissue or cellular selectivity into yet another generation of PDE4 inhibitors by targeting the appropriate subtype or splice variant. As yet, compounds possessing such selectivity have not been reported.

2.2 FUNCTIONAL DOMAINS OF PDE4

Sequence analysis and comparison of PDE4 cDNAs isolated from several species led to the proposal that the subtypes within this enzyme family contain three conserved domains (Bolger *et al.*, 1993). These domains include two upstream conserved regions (UCR1 and UCR2) as well as a catalytic region (Fig. 13.6). Each UCR contains 180–240 base pairs (bp), whereas the catalytic domain is much larger, containing approximately 1170 bp (Bolger *et al.*, 1993; Jacobitz *et al.*, 1996).

A substantial degree of sequence homology exists among the PDE4 subtypes. Compared to PDE4A, the other PDE4 subtypes have amino acid identities of 70–74% across the entire protein sequence (McLaughlin *et al.*, 1993; Jacobitz *et al.*, 1996). This sequence identity increases to 80–84% when the comparison is restricted to the catalytic region. Indeed, considerable sequence homology exists among all PDE isozyme families, with members of one family sharing 20–30% sequence identity with members of another family (Beavo and Reifsnyder, 1990; Bolger, 1994; see also Chapter 1). Predictably, most of the identity exists within the catalytic domain. Interestingly, the catalytic domains of PDE4A, PDE4B and PDE4D contain a perfectly conserved seven amino acid sequence (Glu-Leu-Ala-Leu-Met-Tyr-Asn) that also appears in the cAMP-binding domain of the RIIa regulatory subunit of cAMP-dependent protein kinase (PKA) (Livi *et al.*, 1990; Scott *et al.*, 1987).

Recent studies have mapped the catalytic domains of rat PDE4D1 (Jin *et al.*, 1992) and human PDE4A (Jacobitz *et al.*, 1996) by employing site-directed mutagenesis to prepare truncated variants of these recombinant proteins. For rat PDE4D1, which contains 584 amino acids, the smallest catalytically active fragment is a protein starting at Leu^{146} and terminating at Gln^{485}. The smallest active fragment of human PDE4A begins at Met^{332} and ends at Glu^{722}. Thus, as reflected in Fig. 13.6, the catalytic domains of these subtypes range in size from ~35 to 40 kD and comprise 44% (PDE4A) to 58% (PDE4D1) of the full-length protein.

The roles of the UCRs have yet to be elucidated fully. Evidence suggests, however, that they participate in the targeting of enzymes to subcellular organelles and in the

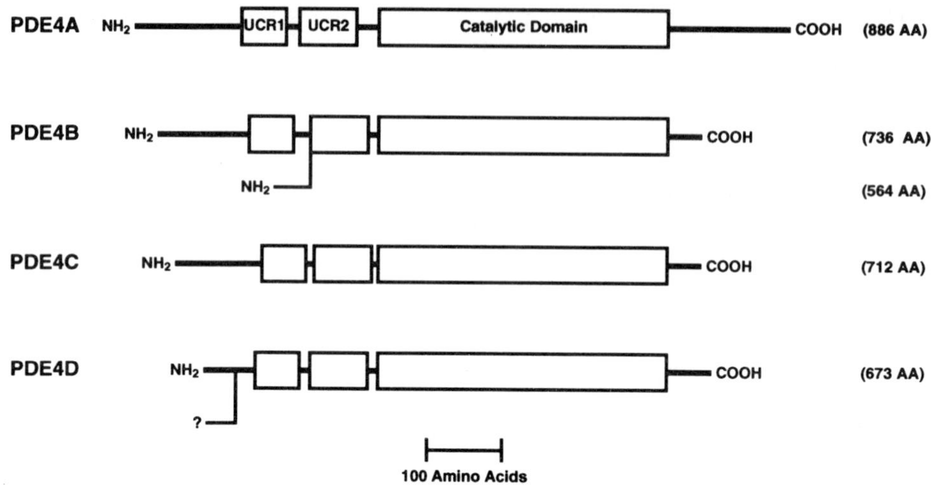

Figure 13.6 Human PDE4 subtypes. This diagram illustrates the 420–450 amino acid catalytic domain and two 60–100 amino acid upstream conserved regions (UCR) common to the four subtypes. Catalytically active mRNA alternative splice variants have been identified for PDE4B and PDE4D. The complete sequence of the N-terminus of one of the PDE4D variants has yet to be confirmed. An alternative splice variant has also been identified for PDE4A (not illustrated), although this does not encode a catalytically active protein. The number of amino acid (AA) residues in each protein appears to the right of the schematic diagram.

allosteric regulation of enzyme activity. For example, Shakur *et al.* (1993) demonstrated that a unique, 25-amino acid N-terminal sequence on a rat PDE4A (RD1) (Davis *et al.*, 1989) targets the protein for membrane attachment. This N-terminal domain is encoded by an alternatively spliced transcript and starts approximately half-way through the second UCR Similarly, alterations in the N-terminus of PDE4B produced by alternatively spliced mRNA also regulate the targeting of protein to membrane or cytosolic compartments (Lobban *et al.*, 1994).

Analysis of the N-termini of PDE4 subtypes indicates the presence of several consensus sequences for phosphorylation by a variety of protein kinases (e.g. PKA, protein kinase C) (Beltman *et al.*, 1993). Thus, by analogy to work done on other PDE isoenzyme families, the activity of PDE4 may be subject to allosteric regulation via phosphorylation pathways (Conti *et al.*, 1991; Manganiello *et al.*, 1992; see also Chapters 1, 4 and 6). Indeed, PKA-mediated phosphorylation of Ser^{54} in rat PDE4D3 – the subtype of PDE4D that contains a full-length N-terminal region – stimulates catalytic activity (Sette and Conti 1995). This activation occurs both with isolated recombinant enzyme and in intact MA-10 cells transfected with PDE4D3.

2.3 ROLE OF CONSERVED HISTIDINES

Studies employing site-directed mutagenesis highlight the importance of conserved histidine residues in the catalytic activity of PDE4 (Fig. 13.7). In rat PDE4D1, changing His^{278} to Ala abolishes catalytic activity, as does changing His^{311} to Asn, Tyr or Asp (Jin *et al.*, 1992). Similar results are obtained when the homologous residues are altered in human PDE4A. Specifically, substituting Gln for His^{473} (analogous to His^{278} in rat PDE4D1) abolishes catalytic activity of human PDE4A, whereas replacement with Arg or Ser reduces activity by 80% (Jacobitz *et al.*, 1994). Replacing either His^{505} or His^{506} (the counterpart to His^{311} in rat PDE4D1) in human PDE4A with Asn yields an inactive enzyme. These results suggest a multi-step mechanism of phosphodiester hydrolysis that requires base-catalysed addition of water to form a phosphorane intermediate followed by acid-catalysed hydrolysis of the phosphorane. His, Ser and Arg are capable of serving both the basic and acidic functions, whereas Asn, Asp or Gln cannot.

Interestingly, the loss of enzyme activity produced by replacing His^{505} or His^{506} is not accompanied by a loss in high-affinity [^{3}H]rolipram binding (Jacobitz *et al.*, 1994). Moreover, cAMP still competes for high-affinity rolipram binding in these mutants. Thus, assuming that this high-affinity rolipram-binding site is the catalytic site, these data suggest that both His^{505} and His^{506} are critical for catalysing the hydrolysis of cAMP, but not for the binding of substrate or enzyme inhibitors.

As summarized in Fig. 13.7, changing other conserved histidines in human PDE4A has a variable effect on activity (S. Jacobitz and T.J. Torphy, unpublished observations). Moreover, the losses in catalytic activity produced by these changes are parallelled by losses in high-affinity rolipram binding activity. Thus, it is impossible to determine whether the functional changes

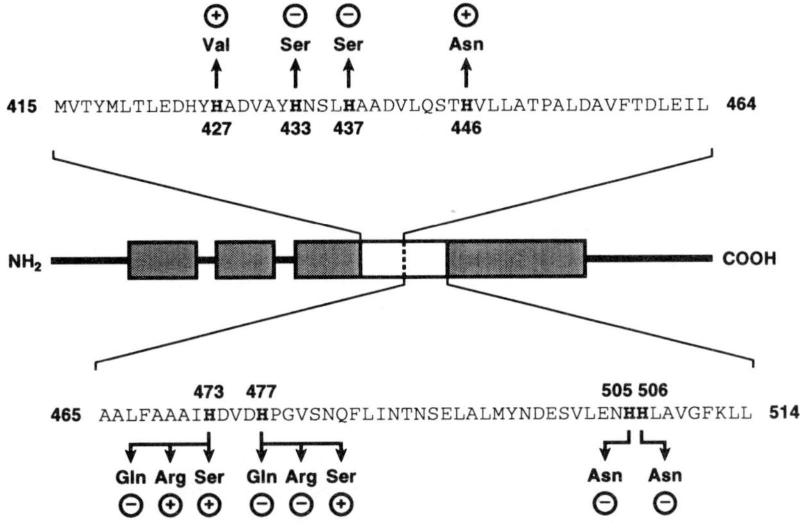

Figure 13.7 Role of conserved histidine residues in human PDE4A. Individual histidines within the catalytic core of PDE4A were replaced with the indicated amino acids using site-directed mutagenesis. The symbols above or below the amino acid substitutions denote whether the mutated enzyme is active (+) or inactive (−). Replacing histidines at position 427 or 446 have no effect, whereas alterations at positions 433, 437, 505 or 506 abolish activity. The effect of mutating position 473 or 477 varies depending on the substitution made. Positions 505 and 506 are particularly interesting: replacing either of these histidines abolishes catalytic activity but has no effect on the binding of inhibitors or cyclic AMP.

observed are related to the loss of key points of interaction between the enzyme and substrate, or whether the loss of activity of the mutant protein is simply a consequence of gross alterations in the tertiary structure of the active site.

3. *Rolipram-Binding Site*

3.1 HISTORICAL PERSPECTIVE

In addition to catalysing the hydrolysis of cAMP, PDE4 also binds rolipram – a PDE4 inhibitor previously under development as an anti-depressant (Wachtel, 1983a; Fleischhacker *et al.*, 1992) – with high affinity. This rolipram-binding activity was first described by Schneider *et al.* (1986), who identified and characterized a saturable, stereoselective and high-affinity (K_d = 2 nM) rolipram-binding site in human and rat brain homogenates. Subsequent autoradiographic studies revealed that this binding activity is differentially distributed among distinct brain substructures (Kaulen *et al.*, 1989).

Perhaps the most straightforward explanation for this high-affinity site is that it simply represents the catalytic site of PDE4. However, results from early studies cast doubt on this proposal. For example, although peripheral tissues contain substantial amounts of PDE4 catalytic activity, most contain few or no high-affinity rolipram-binding sites (Schneider *et al.*, 1986). Thus, high-affinity rolipram-binding activity is not invariably found along with PDE4 catalytic activity. Moreover, there is a poor correlation between the potency of a variety of compounds as PDE4 inhibitors versus their potency as competitors for [^{3}H]rolipram binding (Koe *et al.*, 1990; Torphy *et al.*, 1992). Most striking is the discrepancy between binding affinity and PDE4 inhibitory potency of *R*-rolipram. Although *R*-rolipram binds to its high-affinity site with a K_d of 2 nM (Schneider *et al.*, 1986; Torphy *et al.*, 1992), it inhibits PDE4 catalytic activity at concentrations that are two to three orders of magnitude greater (Fougier *et al.*, 1986; Némoz *et al.*, 1989; Koe *et al.*, 1990; Torphy *et al.*, 1992). Finally, clinical studies indicate that rolipram produces an anti-depressant effect at plasma concentrations ranging from 2 to 20 nM (Krause *et al.*, 1990). These concentrations are far less than those required to inhibit catalytic activity, but are in the range required for interaction with the high-affinity rolipram-binding site.

The information detailed above called into question the identity and function of the high-affinity rolipram-binding site. Interpreting the lack of correlation between the binding activity of compounds and their activity against PDE4, Koe *et al.* (1990) stated that "... the structural requirements of binding affinity for [^{3}H]rolipram-binding sites (possibly localized on PDE4) do not necessarily overlap with those for inhibition of PDE4-catalyzed hydrolysis of cAMP". This statement not only questions the role of the high-affinity binding site in regulating PDE4 catalytic activity, but goes so far as to hint that this site may not be located on

PDE4 ("... possibly localized on PDE4 ..."). Indeed, with the information available in 1990 it was reasonable to suggest that the high-affinity rolipram-binding site may not be PDE4, but may instead represent a distinct, unknown receptor that is enriched in the CNS. If true, such a proposal would have profound implications regarding the mechanism by which rolipram exerts various biological activities (e.g. psychotropic activity versus anti-inflammatory activity).

3.2 NATURE AND FUNCTION OF THE HIGH-AFFINITY ROLIPRAM-BINDING SITE

Doubts about the identity of the high-affinity rolipram-binding site were largely put to rest by the demonstration that this site is expressed along with PDE catalytic activity in yeast genetically engineered to express human recombinant (hr) PDE4A (Torphy *et al.*, 1992) or hrPDE4B (McLaughlin *et al.*, 1993). Thus, the high-affinity rolipram-binding site is indeed a component of PDE4.

Despite the knowledge that the high-affinity rolipram-binding site exists on PDE4, its role in regulating catalytic activity remains uncertain. As mentioned previously, rolipram binds to PDE4 with K_d = 1–6 nM (Schneider *et al.*, 1986; Koe *et al.*, 1990; Torphy *et al.*, 1992; McLaughlin *et al.*, 1993). In contrast, rolipram is considerably less potent as an inhibitor of PDE4 catalytic activity (Ki or IC_{50} = 60–2000 nM) (Némoz *et al.*, 1989; Torphy *et al.*, 1992; Bolger *et al.*, 1993; Pillai *et al.*, 1993; Sullivan *et al.*, 1994). If it is assumed that rolipram is a competitive inhibitor that binds solely to the catalytic site and that PDE4 obeys simple Michaelis–Menten kinetics, then by definition the K_d for rolipram binding should be equal to its Ki. Moreover, although both binding and catalytic activity are inhibited by structurally diverse PDE inhibitors, the relative potencies of these compounds for the two functions differ (Koe *et al.*, 1990; Saccomano *et al.*, 1991; Torphy *et al.*, 1992; Buckle *et al.*, 1994; Barnette *et al.*, 1995a,b, 1996). Thus, the relationship between the interaction of compounds with the high-affinity rolipram-binding site and inhibition of catalytic activity is ambiguous.

3.3 PROPOSALS ON THE NATURE AND FUNCTION OF THE HIGH-AFFINITY ROLIPRAM-BINDING SITE

Three proposals have been advanced regarding the nature of the high-affinity rolipram-binding site. These include the possibility that this site is: (i) an allosteric site (Koe *et al.*, 1990; Torphy *et al.*, 1992; Souness and Scott, 1993); (ii) the catalytic site on a subpopulation of the recombinant enzyme that exists in a distinct conformation (Torphy *et al.*, 1992, 1993a); or (iii) a distinct site within PDE4 that is functionally irrelevant with respect to catalytic activity (Torphy *et al.*, 1992). Before elaborating on these proposals, it is critical to point out that kinetic data from a number of reports suggest that the inhibitory activity of rolipram against hrPDE4 (Torphy *et al.*, 1992; McLaughlin *et al.*, 1993; Michaeli *et al.*, 1993) or guinea-pig eosinophil PDE4 (Souness and Scott, 1993) is not consistent with simple Michaelis–Menten behaviour, a topic which will be discussed in detail in section 4. For the purposes of the present discussion, it is sufficient to point out that in the above studies rolipram inhibited PDE4 catalytic activity over a very broad concentration range, from 3 nM to 10 μM.

Superficially, the proposal that the high-affinity rolipram-binding site represents an allosteric site is attractive. Several pieces of evidence, however, argue against this. First, the Hill coefficient (n_H) for *R*-[^{3}H]rolipram binding is 1.0 (Schneider *et al.*, 1986; Torphy *et al.*, 1992, 1993a), suggesting that there is no allosteric interaction between the high-affinity rolipram-binding site and the catalytic site. Secondly, the potency of PDE4 inhibitors against catalytic activity is not altered in the presence of nanomolar concentrations of rolipram (Torphy *et al.*, 1992). Thirdly, the stoichiometry of high-affinity rolipram binding is much less than 1.0 (Torphy *et al.*, 1993a), in contrast to the theoretically expected stoichiometry of one molecule of rolipram per allosteric site per molecule of enzyme.

An alternative possibility is that PDE4 can exist in two catalytically active conformations, one that is inhibited by low (1–10 nM) concentrations of rolipram (this is the form that binds rolipram with high affinity) and a second that is inhibited by higher concentrations (100–1000 μM). Presumably, these two conformations are either non-interconvertible or interconvert very slowly. Furthermore, rolipram and certain other – but not all – PDE4 inhibitors may have different affinities for the catalytic sites of the two states. As examples, these enzyme states might be differentiated on the basis of their phosphorylation status or the presence of monomeric versus homodimeric subunit organization.

Recently, we have used site-directed mutagenesis to evaluate the structural determinants of rolipram binding and catalytic activity of hrPDE4A (Jacobitz *et al.*, 1995). Employing a modified radioligand binding assay, these studies indicate that the full-length hrPDE4A (Met^{1-886}) contains both high ($K_d \approx 1$ nM) and low ($K_d \approx 100$ nM) affinity *R*-[^{3}H]rolipram binding sites present in relative ratio of ~1 : 10. Similar characteristics were seen in a $Met^{265-886}$ fragment of hrPDE4A. However, deletion of an additional 57 N-terminal amino acids ($Met^{322-886}$) yielded a protein that lost its ability to bind rolipram with high affinity, but retained full catalytic activity along with the low-affinity rolipram-binding activity. Importantly, the catalytic activity of the $Met^{322-886}$ construct was still inhibited by rolipram, albeit at slightly greater concentrations than required for inhibiting the $Met^{265-886}$ construct. These data indicate that high-affinity rolipram-binding activity is not required for catalytic activity or its inhibition. In another study (Torphy *et al.*, 1992), cAMP was shown to compete for high-affinity *R*-[^{3}H]rolipram binding with an IC_{50} of 11 μM. This value is consistent with the K_m for cAMP, which supports the concept that high-affinity rolipram binding takes place at the catalytic site. Overall, these data lend support to the "two conformation" model rather than the "allosteric site" model.

It should be pointed out that, in contrast to the reports described above, several studies on hrPDE4 (Sullivan *et al.*, 1994; Engels *et al.*, 1995) or native PDE4 isolated from tissue sources (Némoz *et al.*, 1989; Torphy *et al.*, 1992) failed to uncover non-Michaelis–Menten behaviour of rolipram. Instead, data obtained in these studies suggest that PDE4 obeys simple Michaelis–Menten kinetics and that rolipram inhibits in a purely competitive fashion with Ki of 0.5–5 μM Using this information, one would have to conclude that high-affinity rolipram binding has no influence on the catalytic activity of PDE4, a scenario that we believe is unlikely. An alternative possibility is that the high-affinity binding state of PDE4 could not be formed in the aforementioned studies, either because of the procedures used to express and isolate the enzyme or because of the conditions under which enzyme activity was assessed.

3.4 BIOLOGICAL SIGNIFICANCE OF HIGH-AFFINITY ROLIPRAM BINDING

Regardless of the molecular basis for the high-affinity rolipram-binding site, a central question is whether this phenomenon has relevance to the functional actions of PDE4 inhibitors. To facilitate this discussion it is useful to consider the "two conformation" proposal as a working model. That is, rolipram has differential inhibitory potencies (Ki $\approx$ 5 nM vs. Ki $\approx$ 1 μM) against two co-existing conformational subpopulations of PDE4. Additional assumptions that must be made are that the relative proportions of these two forms differ among cell types and that the rank-order potency of various inhibitors for the high-affinity rolipram-binding form of PDE4 is distinct from that of the other form. If these assumptions are valid, then the functional role of the form of PDE4 that binds rolipram with high affinity can be evaluated using standard pharmacological approaches. Thus, as judged by the rank-order potency of a variety of PDE4 inhibitors for binding versus functional effects, inhibition of the form of PDE4 that binds rolipram with high affinity appears to be associated with the following functional effects: (i) behavioural activity in rodents (Schmiechen *et al.*, 1990); (ii) elevation of cAMP content in guinea-pig eosinophils (Souness and Scott, 1993); (iii) relaxation of guinea-pig airway (Harris *et al.*, 1989); (iv) stimulation of acid secretion from rabbit parietal cells (Barnette *et al.*, 1995a); (v) inhibition of superoxide production from human neutrophils (Barnette *et al.*, 1996). In contrast, by the same criteria, inhibition of the form of PDE4 that does not bind rolipram with high affinity is linked with inhibiting the following processes: (i) guinea-pig mast cell degranulation (Underwood *et al.*, 1993); (ii) antigen-driven proliferation of human T lymphocytes (Essayan *et al.*, 1994); (iii) tumour necrosis factor (TNF-α) generation in human monocytes (Semmler *et al.*, 1993; Barnette *et al.*, 1996); (iv) superoxide production from guinea-pig eosinophils (Barnette *et al.*, 1996).

The basis for the apparent differential rank-order potency of PDE4 inhibitors in producing these effects remains a matter of conjecture. Notwithstanding this caveat, the demonstration that PDE4 inhibitors produce different pharmacological effects with distinct rank-orders of potency has obvious and important implications for refining the biological profiles of a new generation of compounds (Barnette *et al.*, 1995a, 1996).

4. *Mechanistic Enzymology*

4.1 KINETIC BEHAVIOUR OF PDE4S

Lineweaver–Burk plots of $1/V$ versus 1/[cAMP] for PDE4A display non-classical kinetic behaviour over a wide range of substrate concentrations (0.05–1000 μM) as evidenced by a slight downward curvature as the concentration of substrate is increased (Fig. 13.8A,B). This result is consistent with the presence of at least two non-independent catalytic sites interacting in a negatively co-operative fashion. $Met^{265-886}$ gives a similar result (Fig. 13.8C,D), indicating that this behaviour is not regulated by the N-terminal domain, which has been reported to contain a potential PKA phosphorylation site (DeWolf *et al.*, 1996). A homodimeric quaternary structure is deduced from the subunit molecular weight of 69 300 predicted from the gene sequence of $Met^{265-886}$ and a native molecular mass of 151 kD estimated from the sedimentation coefficient and Stokes' radius of highly enriched $Met^{265-886}$ (De Wolf *et al.*, 1995). $Met^{265-886}$ is similar to other cyclic nucleotide phosphodiesterases with regard to its dimeric structure (Beavo, 1988), providing a structural basis for understanding the co-operative kinetic behaviour of this subtype. A computer fit of the experimental data to Equation 13.1, an equation describing two co-operatively interacting catalytic sites, yielded the "best fit"

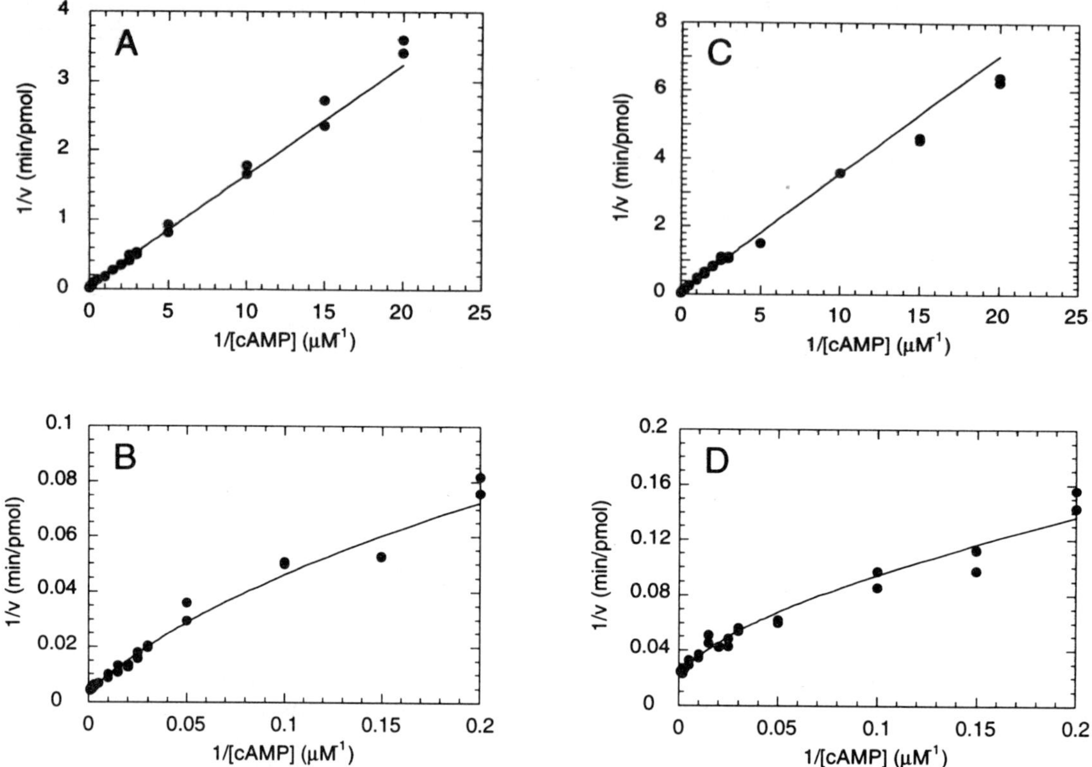

Figure 13.8 Negatively co-operative behaviour of recombinant PDE4A and purified Met$^{265-886}$. cAMP phosphodiesterase activity was assessed using substrate concentrations ranging from 0.05 to 1000 μM (De Wolf *et al.*, 1996) and fit by computer to Equation 13.1 (see text). A,C: double-reciprocal plots for PDE4A and Met$^{265-886}$, respectively, covering the full range of substrate concentrations. B,D: expanded scale double-reciprocal plots to illustrate the fit of the negatively co-operative model at the higher substrate concentrations (5–1000 μM) for PDE4A and Met$^{265-886}$, respectively.

lines shown in Fig. 13.8 and the kinetic parameters reported in Table 13.2.

$$v = \frac{Vm_1[S] + Vm_2\dfrac{[S]^2}{aK_s}}{K_s + 2[S] + \dfrac{[S]^2}{aK_s}} \qquad (13.1)$$

As can be seen, a good fit of the experimental data to the model is obtained over the entire range of substrate concentrations. Although the magnitudes of the parameters differ somewhat for PDE4A and Met$^{265-886}$ (Table 13.2), the results for both enzymes are qualitatively the same. Fig. 13.9 gives a pictorial representation of the binding and catalytic events that could lead to this

Table 13.2. Kinetic parameters for recombinant PDE4 subtypes[a]

Parameter	*Purified Met$^{265-886}$*	*PDE4A*	*PDE4B*	*PDE4D*	*Met$^{332-886}$*
V_{m_1} (pmol/min)	24±2	21±2	3.3±0.3	80±3	67±7
V_{m_2} (pmol/min)	260±10	45±1	5.5±0.2	61±2	168±3
K_S (μM)	3.8±0.3	7.4±0.6	2.7±0.3	6.5±0.3	4.1±0.5
a	23±1	5.9±0.4	3.8±0.4	13±5	5.0±0.3
R-rolipram K_i (nM)	26±1	ND	ND	ND	ND
R-[^{3}H]rolipram K_D (nM)	33±9[b]	ND	ND	ND	ND
R-[^{3}H]rolipram n	0.62±0.06[b]	ND	ND	ND	ND

[a] The numbers listed represent the computer-determined value for each parameter ± the standard error obtained in a typical experiment.

[b] Value reported as mean ± standard deviation for nine determinations.

ND = Not determined.

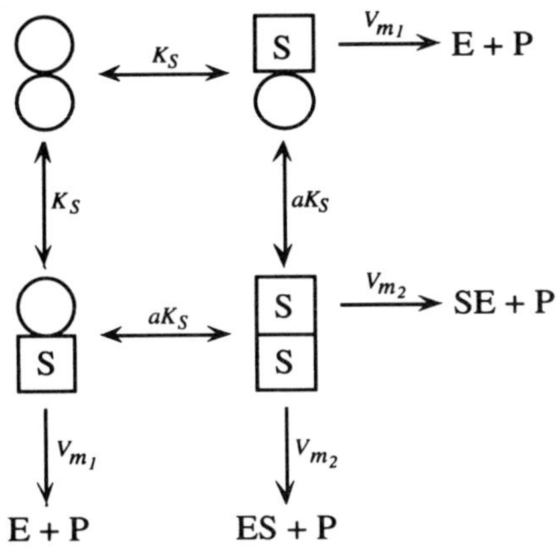

Figure 13.9 Schematic depicting two co-operatively interacting substrate sites of a homodimeric enzyme.

type of behaviour. The first molecule of substrate is proposed to bind with a dissociation constant or K_m characterized by and to be hydrolysed at a rate described by V_{m1}. This binding event induces a conformational change in the dimeric structure causing the affinity of the paired catalytic site for cAMP to be reduced by a factor *a*. This gives rise to the observed negative co-operativity. The induced conformational change also causes an increase in the catalytic efficiency such that, when the second substrate molecule binds, the rate of hydrolysis is characterized by a different, higher maximal velocity V_{m2}. Met$^{332-886}$, which is devoid of high-affinity rolipram binding, also displays co-operative behaviour (Fig. 13.10A,B; Table 13.2). This observation rules out the possibility that this type of kinetic behaviour results from the influence of that binding site.

Like PDE4A, PDE4B shows negative co-operativity with respect to saturation by cAMP (Fig. 13.11A,B); however, PDE4D shows much less co-operative (i.e. more closely hyperbolic) behaviour than subtypes A and B, as can be seen by the linear double reciprocal plots of Fig. 13.11(C,D). Computer fits of the data to Equation 13.1 for PDE4B and PDE4D gave the kinetic constants reported in Table 13.2. The primary determinant for the lack of significant co-operative behaviour by PDE4D appears to be the low value of V_{m2} relative to V_{m1}. Structural studies with PDE4B and PDE4D that could support a dimeric versus monomeric quaternary structure as the basis for these results have not yet been undertaken.

4.2 INHIBITION BY R-ROLIPRAM

Although initial studies suggested that rolipram is a simple competitive inhibitor of PDE4 (Reeves *et al.*, 1987; Livi *et al.*, 1990; Torphy and Cieslinski 1990), more recent work demonstrates that this conclusion requires re-evaluation (Souness *et al.*, 1992; Torphy *et al.*, 1992; McLaughlin *et al.*, 1993). In order to address this issue, we evaluated inhibition data with *R*-rolipram in light of the two-site co-operative model presented above to determine whether the homodimeric quaternary structure of Met$^{265-886}$ might explain the lack of adherence to a simple competitive model. Thus, an experiment was conducted in which the concentration of substrate cAMP was varied over a wide range in the presence of various concentrations of *R*-rolipram from 0 to 45 nM. The raw data (Fig. 13.12) clearly show significant deviation from simple competitive inhibition in a non-co-operative model (straight lines intersecting on the vertical axis). Instead, significant curvature is observed, especially at the higher concentrations of cAMP and *R*-rolipram (Fig. 13.12B). One can expand the two-site co-operative model (Fig. 13.9) to include a competitive inhibitor by incorporating enzyme forms that bind one or two molecules of the inhibitor and one molecule each of substrate and inhibitor. One possible model is shown schematically in Fig. 13.13, which

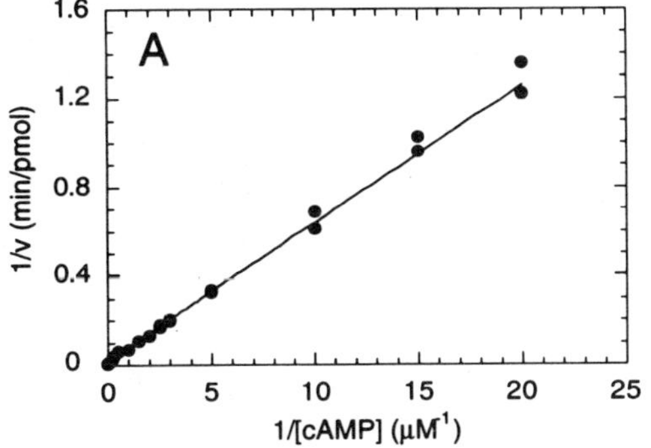

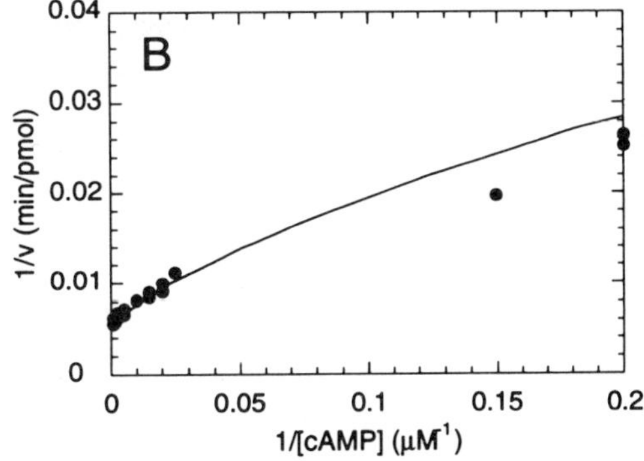

Figure 13.10 Negatively co-operative behaviour of recombinant Met$^{332-886}$. cAMP phosphodiesterase activity was assessed using substrate concentrations ranging from 0.05 to 1000 μM (De Wolf *et al.*, 1996) and fit by computer to Equation 13.1. A, double-reciprocal plot for covering the full range of substrate concentrations. B, expanded scale double-reciprocal plot to illustrate the fit of the negatively co-operative model at the higher substrate concentrations (5–1000 μM).

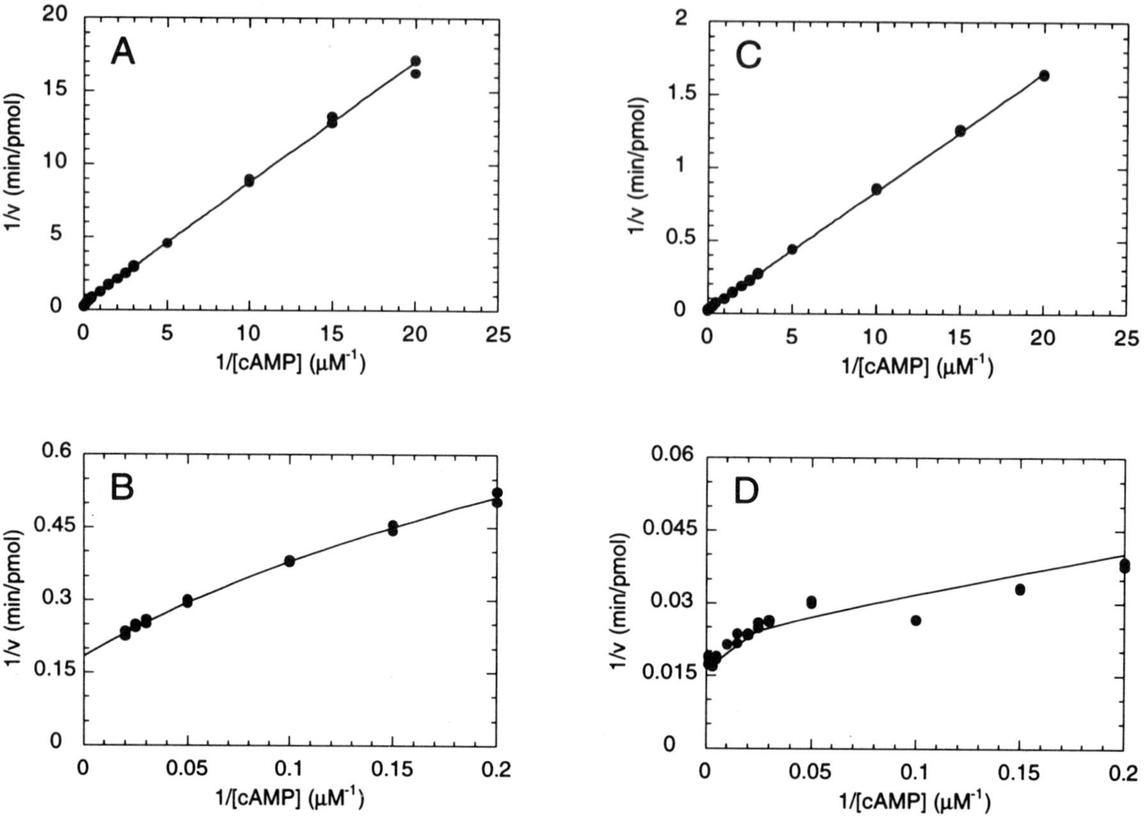

Figure 13.11 Negatively co-operative behaviour of recombinant PDE4B and PDE4D. cAMP phosphodiesterase activity was assessed using substrate concentrations ranging from 0.05 to 50 μM (PDE4B) or 0.05 to 1000 μM (PDE4D) (De Wolf *et al.*, 1996) and fit by computer to Equation 13.1. A,C: double-reciprocal plots for PDE4B and PDE4D, respectively, covering the full range of substrate concentrations. B,D: expanded scale double-reciprocal plots to illustrate the fit of the negatively co-operative model at the higher substrate concentrations for PDE4B (5–50 μM) and PDE4D (5–1000 μM), respectively.

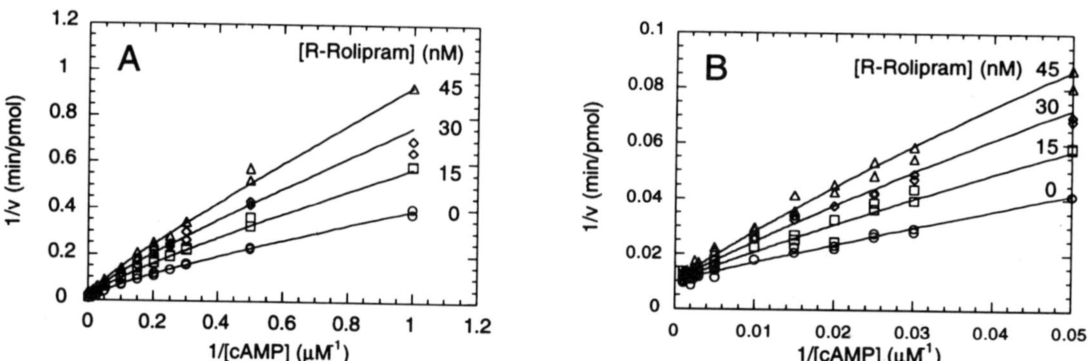

Figure 13.12 Competitive inhibition of $Met^{265-886}$ by *R*-rolipram. $Met^{265-886}$ was assayed in the presence of the various concentrations of *R*-rolipram shown and concentrations of cAMP ranging from 1 to 1000 μM as previously described (DeWolf *et al.*, 1995). Double-reciprocal plots are shown covering the entire range of substrate concentrations used (A) and the range of 20–1000 μM cAMP to better visualize the fit. The lines represent the result of fitting the experimental data to Equation 13.2 (see text), a model describing competitive inhibition in the presence of a co-operatively binding substrate. From the computer fit, a Ki value of 26 ± 1 nM was obtained.

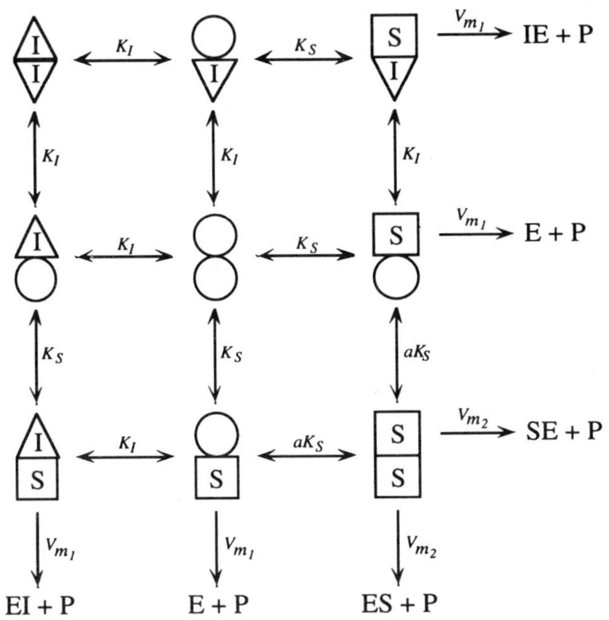

Figure 13.13 Schematic depicting competitive, non-co-operative binding of inhibitor.

assumes a competitive, non-co-operative binding of inhibitor. From this model, Equation 13.2 was derived and fit by non-linear least squares to the data to obtain the "best fit" lines of Fig. 13.12.

$$v = \frac{V_{m_1}\left[[S] + \frac{[S][I]}{\text{Ki}}\right] + V_{m_2}\frac{[S]^2}{a\text{K}_s}}{\text{K}_s + 2[S] + \frac{[S]^2}{a\text{K}_s} + \frac{2[S][I]}{\text{Ki}} + \frac{2\text{K}_s[I]}{\text{Ki}} + \frac{\text{K}_s[I]^2}{\text{Ki}^2}} \quad (13.2)$$

As can be seen, good agreement was obtained between the experimental data and the proposed model in which *R*-rolipram does indeed compete with the negatively co-operative binding of cAMP at the catalytic site. From this, a Ki for *R*-rolipram of 26 ± 1 nM was obtained, which is consistent with the previously reported value analysed with the simple, linear competitive model (Torphy *et al.*, 1992).

4.3 BINDING OF R-ROLIPRAM TO MET$^{265-886}$

The direct binding of *R*-[^{3}H]rolipram to Met$^{265-886}$ was studied by equilibrium ultrafiltration in the N_2-pressurized manifold described by DeWolf *et al.* (1996). A representative double-reciprocal plot (Fig. 13.14) of the concentration of bound ligand (1/r) versus the concentration of the free ligand (1/[L_f]), calculated from the difference between the total and bound *R*-[^{3}H]rolipram, gives a straight line whose vertical and horizontal intercepts correspond to $1/n$ and $-1/\text{K}_d$ respectively. A computer fit of the data to Equation 13.3 gives a binding stoichiometry (n) of 0.62 ± 0.06 and an estimated K_d of 33 ± 9 nM (Table 13.2).

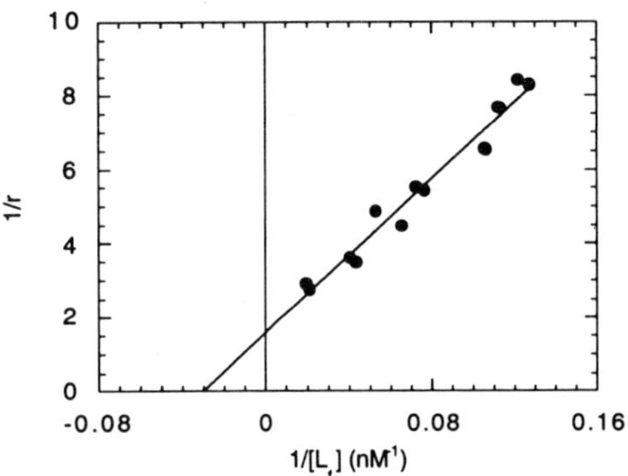

Figure 13.14 Equilibrium ultrafiltration binding of *R*-[^{3}H]rolipram to Met$^{265-886}$. Binding of *R*-[^{3}H]rolipram was measured to Met$^{265-886}$ as previously described (De Wolf *et al.*, 1996) and the resulting data were fit to Equation 13.3 (see text). The plot of the experimental data and the resulting fit is shown in double-reciprocal form where 1/r (moles of enzyme per mole of bound ligand) is plotted versus 1/[L_f] (reciprocal of the calculated concentration of free ligand). In this graphical representation, the vertical intercept equals 1/*n*, where *n* is the moles of ligand bound per subunit at saturating ligand, and the horizontal intercept is $-1/\text{K}_d$, where is the dissociation constant for the ligand. This analysis gave a value of 0.62 ± 0.06 for n and a K_d of 33 ± 9 nM for *R*-rolipram (Table 13.2).

$$r = \frac{n[L_f]}{K_d + [L_f]} \quad (13.3)$$

The value agrees with the estimated Ki from kinetic studies (see above). No evidence for co-operative binding of *R*-rolipram is observed in these studies.

5. *Structure–Activity Relationships*

5.1 INTRODUCTION

The early literature detailing efforts to identify structure–activity relationships among PDE inhibitors relied on extracts or homogenates of tissue samples containing multiple cell types and, therefore, multiple PDE isoenzymes. With improvements in the purification of cell populations from animal tissues and the ability to generate homogenous populations of cells from immortal or induced cell lines, cleaner preparations of PDE isoenzymes have been prepared and used for assay, though such studies have been hampered by the presence in the assay of active, biochemically distinct proteolytic fragments of the major isoenzyme of interest. This approach, coupled with improved protein purification procedures, has provided small amounts of kinetically pure populations of individual isoenzymes for screening purposes. The final step in the evolution of this process has been the use of molecular genetic techniques to clone and express large quantities of structurally homogeneous protein that can be highly purified for both structure–activity relationship assays and structural studies of the individual PDE subtypes and splice variants. Use of such purified proteins and standardization of biochemical assays allows more consistent biochemical results reported for the same compounds both within the same laboratory using different preparations of enzymes over time and between different laboratories.

To this end, a series of PDE4 inhibitors has been assayed using assay conditions reported previously (De Wolf *et al.*, 1996) against four human recombinant PDE4 enzymes (i) the $Met^{265-886}$ fragment of hrPDE4A (Livi *et al.*, 1990); (ii) full-length hrPDE4A (DeWolf *et al.*, 1996); (iii) full-length hrPDE4B (McLaughlin *et al.*, 1993); (iv) hrPDE4D (De Wolf *et al.*, 1996). Selected compounds were also assayed against the mouse brain preparation commonly used in the literature (Schneider *et al.*, 1986) to assess the ability of a compound to compete for high-affinity *R*-[^{3}H]rolipram binding, using assay conditions reported previously (Torphy *et al.*, 1992, 1993a). Tool PDE4 inhibitors were synthesized in the laboratories of SmithKline Beecham according to the individual published procedures referenced in section 1.2. The N-substituted pyrrolidinone derivatives discussed below were synthesized according to procedures analogous to those reported previously for compounds 6 and 7 (Table 13.4) (Baures *et al.*, 1993).

Table 13.3 Inhibition of PDE4 by lead inhibitors

	Inhibition (IC_{50}, nM)			
Compound	*$Met^{265-886}$*	*PDE4A*	*PDE4B*	*PDE4D*
RP 73401	0.8	0.9	0.8	0.6
CDP-840	3.8	5.2	7.5	1.6
SB 207499	46	93	100	10
R-Rolipram	47	52	321	109
PDA-641	96	202	288	71
Trequinsin	132	201	497	226
Denbufylline	266	176	545	50
Nitraquazone	321	222	1850	55
Zardaverine	495	811	1000	96
Ro 20-1724	689	876	1900	1550

5.2 ROLIPRAM AND LEAD PDE4 INHIBITORS

Comparative inhibition data for eight selective PDE4 inhibitors, as well as the dual PDE3/4 inhibitor zardaverine and the non-selective PDE inhibitor trequinsin, are summarized in Table 13.3, ranked in order of decreasing potency against the $Met^{265-886}$ fragment of hrPDE4A. In general, individual compounds inhibit

Table 13.4 Inhibition of PDE4 by rolipram and N-(substituted)pyrrolidin-2-one enantiomers

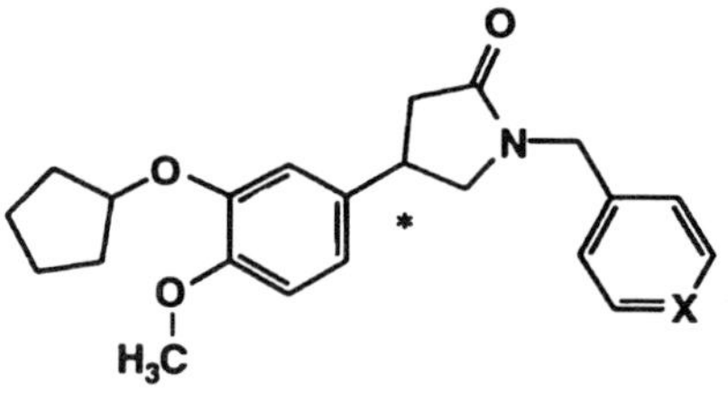

			Inhibition (IC_{50}, nM)				*[^{3}H]rolipram binding (IC_{50}, nM)*
Compound	*Configuration (*)*	*X*	*$Met^{265-886}$*	*PDE4A*	*PDE4B*	*PDE4D*	
rolipram	*R*	–	47	52	321	109	2
rolipram	*S*	–	315	455	1130	83	45
4	*R*	N	1410	1520	3800	2100	650
5	*S*	N	225	268	402	112	1600
6	*R*	C-Br	1480	1880	1530	1580	3500
7	*S*	C-Br	631	907	1690	822	120
8	*R*	C-NH_2	195	220	555	467	40
9	*S*	C-NH_2	52	96	87	48	320
10	*R*	C-$NHCOCH_3$	503	907	1690	744	1300
11	*S*	C-$NHCOCH_3$	18	44	33	18	65

$Met^{265-886}$ and hrPDE4A with IC_{50} values within a factor of 2. RP 73401 is the most potent of the inhibitors examined and is non-selective with respect to the PDE4 subtypes. Although rank-orders of potency for inhibition of PDE4B are similar to those for $Met^{265-886}$, nearly all the compounds are somewhat less potent against PDE4B than against the other two subtypes. SB 207499, zardaverine and nitraquazone display a degree of selectivity for PDE4D over PDE4A and PDE4B.

5.3 ROLIPRAM AND DERIVATIVES

With the exception of PDE4D, where inhibitory potencies are roughly equal, *R*-rolipram is more potent than its *S*-enantiomer, with eudismic ratios ranging from 3 to 9 (Table 13.4). Upon substitution of the nitrogen atom of the rolipram pyrrolidinone ring, the stereochemical preference for PDE4 inhibition is reversed in favour of the *S*-isomers (Table 13.4). The preference is small for the *N*-[4-(bromo)benzyl]-derivatives, compounds 7 and 8 (eudismic ratios 1–2), which are comparatively poor inhibitors; this observation is in sharp contrast to the eudismic ratio of >400 reported for this compound against the partially purified 40 kD human monocyte-derived PDE4 (Baures *et al.*, 1993). In the case of the *N*-[4-(acetamido)benzyl]-derivatives (compounds 10 and 11) the stereochemical preference for PDE4 inhibition by compound 11 is quite strong, illustrating good molecular complementarity with the enzyme. The stereochemical preference displayed by the *S*-isomers of the *N*-[4-(pyridinyl)methyl]- and the *N*-[4-(amino)benzyl]-derivatives (compounds 5 and 9) for PDE4 inhibition is intermediate between those observed for compounds 7 and 11. These compounds are essentially subtype non-selective.

With respect to *R*-[^{3}H]rolipram-binding activity, there is no clear trend. Rolipram and its *N*-[4-(pyridyl)methyl]- and *N*-[4-(amino)benzyl]-derivatives (compounds 4 and 8) display a selectivity for *R*-[^{3}H]rolipram-binding activity favouring the *R*-isomers. For the *N*-[4-(bromo)benzyl]- and *N*-[4-(acetamido)benzyl]-derivatives (compounds 7 and 11) the selectivity for *R*-[^{3}H]rolipram-binding activity is reversed, favouring the S-isomers. Although these results clearly illustrate the existence of divergent structure–activity relationships for *R*-[^{3}H]rolipram-binding activity vis-à-vis PDE4 inhibition, as has been observed in previous studies (Koe *et al.*, 1990; Saccomano *et al.*, 1991; Torphy *et al.*, 1992; Buckle *et al.*, 1994), no obvious explanation is apparent for the changes in stereochemical preference in this case. Nonetheless, these and related compounds have proved to be invaluable probes of the functional effects of these two biochemical properties (Barnette *et al.*, 1995a,b, 1996).

5.4 OVERLAY MODEL OF PDE4 INHIBITION

The compounds summarized in Tables 13.3 and 13.4 may be grouped loosely into three classes: (i) the "benzyl rolipram" series consists of compounds 4–11; (ii) the "rolipram mimics" consist of RP 73401, CDP 840, SB 207499, WAY-641, zardaverine and Ro 20-1724; (iii) the "fused-ring" compounds consist of denbufylline and nitraquazone. Using Macromodel® in conjunction with molecular mechanics calculations, we have developed a pharmacophore model based initially on compound 11, in which the activity is improved by nearly 20-fold relative to *S*-rolipram against $Met^{265-886}$, and subsequently refined this model by incorporation of overlays of other structurally more rigid classes of PDE4 inhibitors (M.D. Ryan *et al.*, manuscript in preparation).

Part of the pharmacophore model is shown in Plate 13.1 (top). Here, the dialkoxyphenyl moiety of compound 11 adopts a pseudo-equatorial arrangement about the pyrrolidinone ring, with a dihedral angle between the rings (measured from the phenyl plane to the carbon alpha to the pyrrolidinone carbonyl group) of approximately 90°. The 4-acetamidophenyl ring adopts an extended conformation, with a distance between the pyrrolidinone amide oxygen and the acetamido oxygen of about 5.5 Å. The acetamido and phenyl groups of this substituent are coplanar.

Overlay of the common dialkoxyphenyl moiety of compound 11 and RP 73401 (Plate 13.1, top) is assumed. The amide carbonyl of RP 73401 adopts a conformation wherein it is twisted with respect to the dialkoxyphenyl ring by 23°. The dichloropyridine ring cannot adopt a conformation coplanar with the amide moiety owing to the steric hindrance of the chlorines and instead is twisted with respect to the amide plane by 72° based on ***ab initio*** molecular orbital calculations (6–31G*//6–31G*). The pyrrolidinone ring oxygen of compound 11 maps to one of the chlorine substituents of the RP 73401 pyridine ring, with an angle between the mean plane of the pyrrolidinone and pyridine rings of approximately 40°. *R*-rolipram itself can be accommodated in this overlay by a 180° rotation about the dialkoxyphenyl-pyrrolidinone ring bond (compound not shown).

CDP 840 (Plate 13.1, top) underscores the importance of the RP 73401 pyridinyl nitrogen as a pharmacophoric point, since a nearly exact overlay between the pyridinyl groups of both compounds is possible. In this orientation, the phenyl group of CDP 840 points in the same general direction as the amide carbonyl group of RP 73401, but diverges by about 25°.

Zardaverine has little conformational mobility. The pyridazinone oxygen overlays with the pyridinyl nitrogen of RP 73401, although the fit is not as good as overlays for other compounds (Plate 13.1, centre). This may account for the comparative decrease in potency observed with this compound.

The importance of the pyridinyl nitrogens in RP 73401 and CDP 840, as seen by their correspondence with the carbonyl oxygen in zardaverine, together with the spatial position of the two chlorine groups around the pyridine ring, suggests an overlay with xanthines. Denbufylline contains a hydrogen bond acceptor nitrogen (N-9) which we propose overlays with the RP 73401 pyridinyl nitrogen (Plate 13.1, centre), with the C-6 oxygen atom corresponding to the RP 73401 chlorine used in the overlay with compound 11. The C-2 oxygen of denbufylline then corresponds to either the RP 73401 amide oxygen or the second chlorine, imposing coplanarity on the overlay between xanthine and the pyridine rings. Of the two choices, mapping of the xanthine C-2 carbonyl group to the RP 73401 amide carbonyl oxygen (Plate 13.1, centre) permits a better overall fit with the other compounds, since the chlorine to oxygen distance of 4.3 Å is a better fit to the O–O distance of 4.5 Å in denbufylline than the chlorine–chlorine distance of 5.4 Å in RP 73401.

This, in turn, allows incorporation of nitraquazone into the overlay (Plate 13.1, bottom), with the quinazoline oxygens of nitraquazone overlaying corresponding xanthine carbonyl oxygens in debufylline. In addition, the nitro group is placed in proximity to the 4-acetamido moiety of compound 11.

Trequinsin has two symmetry-related conformers that differ only by a ring pucker about the benzylic position. As such, the spatial orientation of the pharmacophoric points differ little between the two conformers. The primary overlay is between the dialkoxy phenyl group and the dialkoxy phenyl group of the rolipram series

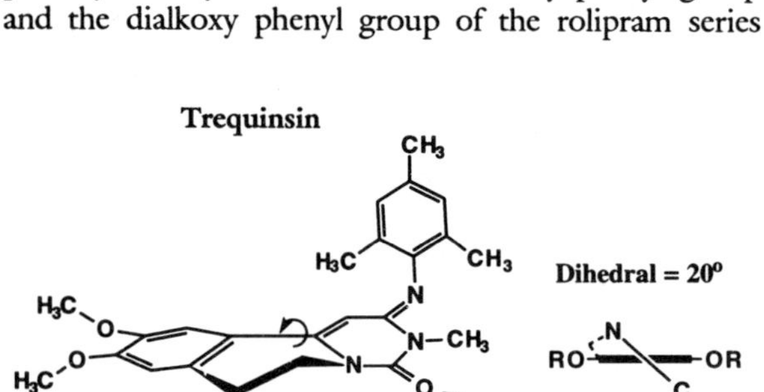

Trequinsin

CH_3 H_3C CH_3 N H_3C O O N–CH_3 N O H_3C

Dihedral = 20°

N RO OR C

map to RP chloro

Figure 13.15 Two-dimensional illustration of the conformation of trequinsin used in overlay. Two ring puckers are possible in the central ring of trequinsin. The conformer that fits the pharmacophore model has the methylene-attached nitrogen "up". This is illustrated on the right, showing the twist about the bond indicated on the left by an arrow in which the aryl plane is seen in the rear. In the model, the aryl groups of trequensin and other inhibitors are superimposed. The urea oxygen of trequinsin maps to the rolipram amide oxygen and the RP 73401 chloro group, as depicted for the latter two compounds in Plate 13.1 (top).

compounds (illustrated two-dimensionally in Fig. 13.15). Overlay of the trequinsin carbonyl oxygen with the RP 73401 chlorine used above results in placement of the trimethylphenyl group in proximity to a butyl group of denbufylline. Incorporation of this molecule into the overlay imposes additional constraints which allow subtle refinement of the conformations of several of the overlaid compounds consistent with their low energy conformers, completing the model.

Rational drug design is preferably based on an X-ray crystal structure of the target enzyme or receptor. The PDE4 subtypes are large proteins that have been expressed in pure form only recently and no crystal structures are available. Despite this, traditional pharmacophore analysis is capable of providing valuable insight into the design of ligands. The pharmacophore overlay presented herein integrates multiple classes of inhibitors, allowing identification of common structural features important for PDE4 inhibition as well as identification of those unique structural features that modulate the inhibitory potency, either positively or negatively, of individual molecules. This model forms a broad basis for the design of novel inhibitors of PDE4 that will be reported elsewhere in the future.

6. *Summary and Conclusions*

Research conducted during the past 25 years has provided significant insights into the biochemistry and physiology of PDEs. The differential cellular and tissue distribution of these isoenzymes, coupled with the identification of isoenzyme-selective PDE inhibitors, has stimulated considerable excitement and effort within several laboratories to identify novel, highly-selective inhibitors as therapeutic agents. This has led to evaluation in the clinic of selective PDE3 inhibitors for the treatment of congestive heart failure and thrombosis (see Chapter 6), selective PDE4 inhibitors for the treatment of depression, dementia and, most recently, asthma (see Chapter 7), and dual PDE3/4 inhibitors for the treatment of asthma (see Chapter 10).

Molecular biology has proved to be an invaluable tool that has allowed a major advance in our recent understanding of the biochemistry of PDEs. Thus, the cloning and expression of kinetically and/or physically pure PDE4 subtypes has provided sufficient quantities of the enzymes for more extensive mechanistic enzymology, stoichiometry and structural studies. Evaluation of genetically engineered mutant PDE4 enzymes has allowed determination of the importance of single or multiple amino acids to the catalytic process. Similarly, studies with defined, truncated mutant PDE4 enzymes have allowed determination of the minimum sequences required for both catalytic and *R*-[^{3}H]rolipram-binding activity.

The existence of differential structure–activity relationships displayed by certain compounds for PDE4 inhibition vis-à-vis R-[^{3}H]rolipram-binding activity has provided tools to probe the functional effects of these two biochemical properties, which may be related to the therapeutic index of PDE4 inhibitors. In turn, differential tissue distribution of PDE subtypes leads to the possibility of developing even more selective pharmacological agents if differential inhibitor structure–activity relationships can be identified. Pharmacophore models provide powerful tools to assist in the design of such differential inhibitors in the absence of subtype structural data.

7. *References*

Ahn, H.S., Crim, W., Romano, M., Sybertz, E. and Pitts, B. (1989). Effects of selective inhibitors on cyclic nucleotide phosphodiesterases of rabbit aorta. Biochem. Pharmacol. 38, 3331–3339.

Amschler, H. (1987). 6-(Polyfluoroalkylphenyl)pyridazinones, their compositions, synthesis and use. US Patent 4,665,074.

Asano, T., Ochiai, Y. and Hidaka, H. (1977). Selective inhibition of separated forms of human platelet cyclic nucleotide phosphodiesterase by platelet aggregation inhibitors. Mol. Pharmacol. 13, 400–406.

Ashton, M.J., Cook, D.C., Fenton, G., Karlsson, J.-A., Palfreyman, M.N., Raeburn, D., Ratcliffe, A.J., Souness, J.E., Thurairatnam, S. and Vicker, N. (1994). Selective type IV phosphodiesterase inhibitors as antiasthmatic agents: the synthesis and biological activities of 3-(cyclopentyloxy)-4-methoxybenzamides and analogues. J. Med. Chem. 37, 1696–1703.

Baecker, P.A., Obernolte, R., Bach, C., Yee, C. and Shelton, E.R. (1994). Isolation of a cDNA encoding a human rolipram-sensitive cyclic AMP phosphodiesterase (PDE IVD). Gene 138, 253–256.

Barnette, M.S., Christensen, S.B., Essayan, D.M., Esser, K.M., Grous, M., Huang, S.-K., Manning, C.D., Prabhakar, U., Rush, J. and Torphy, T.J. (1994). SB 207499, a potent and selective phosphodiesterase inhibitor, suppresses activities of several immune and inflammatory cells. Am. Rev. Respir. Dis. 149, A762. [Abstract]

Barnette, M.S., Grous, M., Cieslinski, L.B., Burman, M., Christensen, S.B. and Torphy, T.J. (1995). Inhibitors of phosphodiesterase IV (PDE IV) increase acid secretion in rabbit isolated gastric glands: correlation between function and interaction with a high-affinity rolipram binding site. J. Pharmacol. Exp. Ther. 273, 1369–1402.

Barnette, M.S., O'Leary Bartus, J., Burman, M., Christenaen, S.B., Cieslinski, L.B., Esser, K.M., Prabhakar, U.S., Rush, J.A. and Torphy, T.J. (1996). Association of the anti-inflammatory activity of phosphodiesterase 4 (PDE4) inhibitors with either inhibition of PDE4 catalytic activity or competition for [^{3}H]-rolipram binding. Biochem. Pharmacol. 51, 949–956.

Barnette, M.S., Manning, C.D., Cieslinski, L.B., Burman, M., Christensen, S.B. and Torphy, T.J. (1996b). The ability of phosphodiesterase IV inhibitors to suppress superoxide production in guinea pig eosinophils is correlated with inhibition of phosphodiesterase IV catalytic activity. J. Pharmacol. Exp. Ther. 273, 674–679.

Baures, P.W., Eggleston, D.S., Erhard, K.F., Cieslinski, L.B., Torphy, T.J. and Christensen, S.B. (1993). The crystal structure, absolute configuration and phosphodiesterase inhibitory activity of (+)-1-(4-bromobenzyl)-4-(3-cyclopentyloxy-4-methoxyphenyl)pyrrolidin-2-one. J. Med. Chem. 36, 3274–3277.

Beavo, J.A. (1988). Multiple isozymes of cyclic nucleotide phosphodiesterase. Adv. Second Messenger Phosphoprotein Res. 22, 1–38.

Beavo, J.A. and Reifsnyder, D.H. (1990). Primary sequence of cyclic nucleotide phosphodiesterase isozymes and the design of selective inhibitors. Trends Pharmacol. Sci. 11, 150–155.

Beavo, J.A., Conti, M. and Heaslip, R.J. (1994). Multiple cyclic nucleotide phosphodiesterases. Mol. Pharmacol. 46, 399–405.

Beltman, J., Sonnenburg, W.K. and Beavo, J.A. (1993). The role of protein phosphorylation in the regulation of cyclic nucleotide phosphodiesterases. Mol. Cell. Biochem. 127/128, 239–253.

Beltman, J., Becker, D.E., Butt, E., Jensen, G.S., Rybalkin, S.D., Jastorff, B. and Beavo, J.A. (1995). Characterization of cyclic nucleotide phosphodiesterases with cyclic GMP analogs: topology of the catalytic domains. Mol. Pharmacol. 47, 330–339.

Bolger, G.B. (1994). Molecular biology of the cyclic AMP-specific cyclic nucleotide phosphodiesterases: a diverse family of regulatory enzymes. Cell. Signal. 6, 851–859.

Bolger, G., Michaeli, T., Martins, T., St John, T., Steiner, B., Rodgers, L., Riggs, M., Wigler, M. and Ferguson, K. (1993). A family of human phosphodiesterases homologous to the dunce learning and memory gene product of *Drosophila melanogaster* are potential targets for antidepressant drugs. Mol. Cell. Biol. 13, 6558–6571.

Bolger, G.B., Rodgers, L. and Riggs, M. (1994). Differential CNS expression of alternative mRNA isoforms of the mammalian genes encoding cAMP-specific phosphodiesterases. Gene 149, 237–244.

Brenner, G., Unterbach, E., Goring, J., Khan, E.A., Rohte, O. and Tauscher, M. (1980) 7-(Oxoalkyl)-1,3,-dialkyl xanthines, and medicaments containing them. US Patent 4,242,345.

Buckle, D.R., Arch, J.R.S., Connolly, B.J., Fenwick, A.E., Foster, K.A., Murray, K.J., Readshaw, S.A., Smallridge, M. and Smith, D.G. (1994). Inhibition of cyclic nucleotide phosphodiesterase by derivatives of 1,3-bis(cyclopropylmethyl)xanthine. J. Med. Chem. 37, 476–485.

Butt, E., Beltman, J., Becker, D.E., Jensen, G.S., Rybalkin, S.D., Jastorff, B. and Beavo, J.A. (1995). Characterization of cyclic nucleotide phosphodiesterases with cyclic AMP analogs: topology of the catalytic sites and comparison with other cyclic AMP-binding proteins. Mol. Pharmacol. 47, 340–347.

Christensen, S.B. (1993). Compounds useful for treating allergic and inflammatory diseases. Patent Cooperation Treaty International Patent Application WO 93/19749.

Christensen, S.B. and Torphy, T.J. (1994). Isozyme-selective phosphodiesterase inhibitors as antiasthmatic agents. Ann. Rep. Med. Chem. 29, 185–194.

Cohan, V.L., Showell, H.J., Pettipher, E.R., Fischer, D.A., Pazoles, C.J., Watson, J.W., Turner, C.R. and Cheng, J.B.

(1995). *In vitro* pharmacology of the novel type IV phosphodiesterase (PDE_{IV}) inhibitor, CP-80,633. J. Allergy Clin. Immunol. 95, 350. [Abstract]

Colicelli, J., Birchmeier, C., Michaeli, T., O'Neill, K., Riggs, M. and Wigler, M. (1989). Isolation and characterization of a mammalian gene encoding a high-affinity cAMP phosphodiesterase. Proc. Natl Acad. Sci. USA 86, 3599–3603.

Conti, M. and Swinnen, J.V. (1990). Structure and function of the rolipram-sensitive, low-K_m cyclic AMP phosphodiesterases: a family of highly related enzymes. In "Cyclic Nucleotide Phosphodiesterases: Structure, Regulation and Drug Action" (eds. J. Beavo and M.D. Houslay), pp. 317–332. Wiley, Chichester.

Conti, M., Jin, S.-L.C., Monaco, L., Repaske, D.R. and Swinnen, J.V. (1991). Hormonal regulation of cyclic nucleotide phosphodiesterases. Endocr. Rev. 12, 218–234.

Davis, R.L., Takayasu, H., Eberwine, M. and Myres, J. (1989). Cloning and characterization of mammalian homologs of the *Drosophila* dunce$^+$ gene. Proc. Natl Acad. Sci. USA 86, 3604–3608.

DeWolf, W.E., Jr., McLaughlin, M.M., Green, D.W., Fisher, S.M., Hensley, P., Doughty, E.O., Livi, G.P. and Torphy, T.J. (1995). Kinetic analysis of purified human cAMP phosphodiesterase 4A: equivalence of full length and N-terminally truncated forms. J. Biol. Chem. 270. [In press]

Elliott, K.R.F., Berry, J.L., Bate, A.J., Foster, R.W. and Small, R.C. (1991). The isozyme selectivity of AH 21–132 as an inhibitor of cyclic nucleotide phosphodiesterase activity. J. Enzyme Inhibition 4, 245–251.

Engels, P., Fichtel, K. and Lübbert, H. (1994). Expression and regulation of human and rat phosphodiesterase type IV isogenes. FEBS Lett. 350, 291–295.

Engels, P., Sullivan, M., Müller, T. and Lübbert, H. (1995). Molecular cloning and functional expression in yeast of a human cAMP-specific phosphodiesterase subtype (PDE IV-C). FEBS Lett. 358, 305–310.

Erhardt, P.W. (1987). In search of the digitalis replacement. J. Med. Chem. 30, 231–237.

Erhardt, P.W. (1990). Second-generation phosphodiesterase inhibitors: structure–activity relationships and receptor models. In "Cyclic Nucleotide Phosphodiesterases: Structure, Regulation and Drug Action" (eds. J. Beavo and M.D. Houslay), pp. 317–332. Wiley, Chichester.

Essayan, D.M., Huang, S.-K., Undem, B.J., Kagey-Sobotka, A. and Lichtenstein, L.M. (1994). Modulation of antigen- and mitogen-induced proliferative responses of peripheral blood mononuclear cells by nonselective and isozyme selective cyclic nucleotide phosphodiesterase inhibitors. J. Immunol. 153, 3408–3416.

Farah, A.E. and Alousi, A.A. (1978). New cardiotonic agents: a search for digitalis substitute. Life Sci. 22, 1139–1148.

Fleischhacker, W.W., Hinterhuber, H., Bauer, H., Pflug, B., Berner, P., Simhandl, C., Wolf, R., Gerlach, W., Jaklitsch, H., Sastre-y-Hernández, M., Schmeding-Wiegel, H., Sperner-Unterweger, B., Voet, B. and Schubert, H. (1992). A multicenter double-blind study of three different doses of the new cAMP-phosphodiesterase inhibitor rolipram in patients with major depressive disorder. Neuropsychobiology 26, 59–64.

Fougier, S., Némoz, G., Prigent, A.F., Marivet, M., Bourguignon, J.J., Wermuth, C. and Pacheco, H. (1986). Purification of cAMP-specific phosphodiesterase from rat heart by affinity chromatography on immobilized rolipram. Biochem. Biophys. Res. Commun. 138, 205–214.

Garst, J.E., Kramer, G.L., Wu, Y.J. and Wells, J.N. (1976). Inhibition of separated forms of phosphodiesterases from pig coronary arteries by uracils and by 7-substituted derivatives of 1-methyl-3-isobutylxanthine. J. Med. Chem. 19, 499–503.

Genieser, H.-G., Butt, E., Bottin, U., Dostmann, W. and Jastorff, B. (1989). Synthesis of the 3′,5′-cyclic phosphates from unprotected nucleosides. Synthesis 1, 53–54.

Giembycz, M.A. and Barnes, P.J. (1991). Selective inhibition of high affinity type IV cyclic AMP phosphodiesterase in bovine trachealis by AH 21–132: relevance to the spasmolytic and anti-spasmogenic actions of AH 21–132 in the intact tissue. Biochem. Pharmacol. 42, 663–667.

Glaser, T. and Traber, J. (1984). TVX 2706 – a new phosphodiesterase inhibitor with antiinflammatory action. Agents Actions 15, 341–348.

Grant, P.G., De Camp, D.L., Bailey, J.M., Colman, R.W. and Colman, R.F. (1990). Three new cAMP affinity labels: inactivation of human platelet low K_m cAMP phosphodiesterase by 8-[(4-bromo-2,3-dioxobutyl)thio]adenosine 3′,5′-cyclic monophosphate. Biochemistry 29, 887–894.

Gruenman, V. and Hoffer, M. (1975). *N*-[(1-Cyano-2-phenyl)ethyl]carbamates. US Patent 3,923,833.

Harris, A.L., Connell, M.J., Ferguson, E.W., Wallace, A.M., Gordon, R.J., Pagani, E.D. and Silver, P.J. (1989). Role of low K_m cyclic AMP phosphodiesterase inhibition in tracheal relaxation and bronchodilation in the guinea pig. J. Pharmacol. Exp. Ther. 251, 199–207.

Hatzelmann, A., Tenor, H. and Schudt, C. (1995). Differential effects of non-selective and selective phosphodiesterase inhibitors on human eosinophils. Br. J. Pharmacol. 114, 821–831.

Hidaka, H., Hayashi, H., Kohri, H., Kimura, Y., Hosokawa, T., Igawa, T. and Saitoh, Y. (1979). Selective inhibitor of platelet cyclic adenosine monophosphate phosphodiesterase, cilostamide, inhibits platelet aggregation. J. Pharmacol. Exp. Ther. 211, 26–30.

Higgs, G. (1995). The development of a PDE IV inhibitor for the treatment of bronchial asthma. Abstracts, William Harvey Research Conferences, London, June 21–22.

Ho, P.P.K., Wang, L.Y., Towner, R.D., Hayes, S.J., Pollock, D., Bowling, N., Wyss, V. and Panetta, J.A. (1990). Cardiovascular effect and stimulus-dependent inhibition of superoxide generation from human neutrophils by tibenelast, 5,6-diethoxy-benzo(b)thiophene-2-carboxylic acid, sodium salt (LY 186655). Biochem. Pharmacol. 40, 2085–2097.

Horton, Y.M., Sullivan, M. and Houslay, M.D. (1995). Molecular cloning of a novel splice variant of human type-IVA (PDE-IVA) cyclic-AMP phosphodiesterase and localization of the gene to the p13.1–q12 region of human chromosome-19. Biochem. J. 308, 683–691.

Jacobitz, S., McLaughlin, M.M., Livi, G.P., Ryan, M.D. and Torphy, T.J. (1994). The role of conserved histidine residues on cAMP hydrolyzing activity and rolipram binding of human phosphodiesterase IV. FASEB J. 8, A371. [Abstract]

Jacobitz, S., McLaughlin, M.M., Livi, G.P., Burman, M. and Torphy, T.J. (1996). Mapping the functional domains of human recombinant phosphodiesterase 4A: structural requirements for catalytic activity and rolipram binding. Mol. Pharmacol. [In press]

Jin, S.-L.C., Swinnen, J.V. and Conti, M. (1992). Characterization of the structure of a low K_m rolipram-sensitive cAMP phosphodiesterase. J. Biol. Chem. 267, 18929–18939.

Kaneko, T., Alvarez, R., Ueki, I.F. and Nadel, J.A. (1995). Elevated intracellular cyclic AMP inhibits chemotaxis in human eosinophils. Cell. Signal. 7, 527–534.

Kaulen, P., Brüning, G., Schneider, H.H., Sarter, M. and Baumgarten, H.G. (1989). Autoradiographic mapping of a selective cyclic adenosine monophosphate phosphodiesterase in rat brain with the antidepressant [^{3}H]rolipram. Brain Res. 503, 229–245.

Klockow, M. and Jonas, R. (1989). Particulate cAMP-specific phosphodiesterase (P-PDE) in cardiac ventricle of guinea pig. Naunyn-Schmiedebergs Arch. Pharmacol. 339, R53. [Abstract]

Koe, B.K., Lebel, L.A., Nielsen, J.A., Russo, L.L., Saccomano, N.A., Vinick, F.J. and Williams, I.H. (1990). Effects of novel catechol ether imidazolidinones on calcium-independent phosphodiesterase activity, [^{3}H]rolipram binding, and reserpine-induced hypothermia in mice. Drug Dev. Res. 21, 135–142.

Krause, W., Kühne, G. and Sauerbrey, N. (1990). Pharmacokinetics of (+)-rolipram and (−)-rolipram in healthy volunteers. Eur. J. Clin. Pharmacol. 38, 71–75.

Lal, B., Dohadwalla, A N., Dadkar, N.K., D'Sa, A. and de Sousa, N.J. (1984). Trequinsin, a potent new antihypertensive vasodilator in the series of 2-(arylimino)-3-alkyl-9, 10-dimethoxy-3, 4, 6, 7-tetrahydro-2*H*-pyrimido[6,1-*a*]isoquinolin-4-ones. J. Med. Chem. 27, 1470–1480.

Lee, S.J., Konishi, Y., Yu, D.T., Miskowski, T.A., Riviello, C.M., Macina, O.T., Frierson, M.R., Kondo, K., Sugitani, M., Sircar, J.C. and Blazejewski, K.M. (1995). Discovery of potent cyclic GMP phosphodiesterase inhibitors: 2-pyridyl- and 2-imidazolylquinazolones possesses cyclic GMP phosphodiesterase and thromboxane synthesis inhibitory activities. J. Med. Chem. 38, 3547–3557.

Livi, G.P., Kmetz, P., McHale, M.M., Cieslinski, L.B., Sathe, G.M., Taylor, D.P., Davis, R.L., Torphy, T.J. and Balcarek, J.M. (1990). Cloning and expression of cDNA for a human low-K_m, rolipram-sensitive cyclic AMP phosphodiesterase. Mol. Cell. Biol. 10, 2678–2686.

Lobban, M., Shakur, Y., Beattie, J. and Houslay, M.D. (1994). Identification of two splice variant forms of type IV_B cyclic AMP phosphodiesterase, DPD (rPDE-IV_{B1}) and PDE-4 (rPDE-IV_{B2}) in brain: selective localization in membrane and cytosolic compartments and differential expression in various brain regions. Biochem. J. 304, 399–406.

Lombardo, L.J. (1992). Oxime-carbamates and oxime-carbonates as bronchodilators and anti-inflammatory agents. European Patent Application 0470805A1.

Lombardo, L.J., Bagli, J.F., Golankiewicz, J.M., Heaslip, R.J. and Mudrick, J.K. (1993). Novel oxime-carbamates and oxime-carbonates as potent, isozyme-selective inhibitors of PDE IV. Abstracts, A.C.S. 205th Natl. Meeting, Denver, March 28–April 2, MEDI No. 128.

Lowe, J.A., III, Archer, R.L., Chapin, D.S., Cheng, J.B., Helwig, D., Johnson, J.L., Koe, B.K., Lebel, L.A., Moore, P.F., Nielsen, J.A., Russo, L.L. and Shirley, J.T. (1991). Structure–activity relationships of quinazolinedione inhibitors of calcium-independent phosphodiesterase. J. Med. Chem. 34, 624–628.

Manganiello, V.C., Degerman, E., Smith, C.J., Vasta, V., Tornqvist, H. and Belfrage, P. (1992). Mechanisms for activation of the rat adipocyte particulate cyclic-GMP-inhibited cyclic AMP phosphodiesterase and its importance in the antilipolytic action of insulin. Adv. Second Messenger Phosphoprotein Res. 25, 147–164.

Manning, C.D., McLaughlin, M.M., Livi, G.P., Cieslinski, L.B., Torphy, T.J. and Barnette, M.S. (1996). Prolonged β-adrenoceptor stimulation upregulates cAMP phosphodiesterase activity in human monocytes by increasing mRNA and protein for phosphodiesterases 4A and 4B. J. Pharmacol. Exp. Ther. 273, 810–818.

Marivet, M.C., Bourguignon, J.-J., Lugnier, C., Mann, A., Stoclet, J.-C. and Wermuth, C.-G (1989). Inhibition of cyclic adenosine-3′,5′-phosphate phosphodiesterase from vascular smooth muscle by rolipram analogues. J. Med. Chem. 32, 1450–1457.

McLaughlin, M.M, Cieslinski, L.B., Burman, M., Torphy, T.J. and Livi, G.P. (1993). A low-Km, rolipram-sensitive, cAMP-specific phosphodiesterase from human brain. J. Biol. Chem. 268, 6470–6476.

Meanwell, N.A. and Seiler, S.M. (1990). Inhibitors of platelet cAMP phosphodiesterase. Drugs Fut. 15, 369–390.

Méry, P.-F., Pavoine, C., Pecker, F. and Fischmeister, R. (1995). Erythro-9-(2-hydroxy-3-nonyl)adenine inhibits cyclic GMP-stimulated phosphodiesterase in isolated cardiac myocytes. Mol. Pharmacol. 48, 121–130.

Michaeli, T., Bloom, T.J., Martins, T., Loughney, K., Ferguson, K., Riggs, M., Rodgers, L., Beavo, J.A. and Wigler, M. (1993). Isolation and characterization of a previously undetected human cAMP phosphodiesterase by complementation of cAMP phosphodiesterase-deficient *Saccharomyces cerevisiae*. J. Biol. Chem. 268, 12925–12932.

Milatovich, A., Bolger, G., Michaeli, T. and Francke, U. (1994). Chromosome localizations of genes for five cAMP-specific phosphodiesterases in man and mouse. Somatic Cell Mol. Gen. 20, 75–86.

Murray, K.J. (1993). Phosphodiesterase V_A inhibitors. Drugs News Perspectives 6, 150–156.

Nakanishi, S., Osowa, K., Saito, Y., Kawamoto, I., Kuroda, K. and Kase, H. (1992). KS-505a, a novel inhibitor of bovine brain Ca^{2+} and calmodulin-dependent cyclic-nucleotide phosphodiesterase from *Streptomyces argenteolus*. J. Antibiotics 45, 341–347.

Némoz, G., Moueqqit, M., Prigent, A.F. and Pacheco, H. (1989). Isolation of similar rolipram-inhibitable cyclic-AMP-specific phosphodiesterases from rat brain and heart. Eur. J. Biochem. 184, 511–520.

Nicholson, C.D., Jackman, S.A. and Wilke, R. (1989). The ability of denbufylline to inhibit cyclic nucleotide phosphodiesterase and its affinity for adenosine receptors and the adenosine re-uptake site. Br. J. Pharmacol. 97, 889–897.

Nicholson, C.D., Challiss, R.A.J. and Shahid, M. (1991). Differential modulation of tissue function and therapeutic potential of selective inhibitors of cyclic nucleotide phosphodiesterase isozymes. Trends Pharmacol. Sci. 12, 19–27.

Obernolte, R., Bhakta, S., Alvarez, R., Bach, C., Zuppan, P., Mulkins, M., Jarnagin, K. and Shelton, E.R. (1993). The cDNA of a human lymphocyte cyclic-AMP phosphodiesterase (PDEIV) reveals a multigene family. Gene 129, 239–247.

Pillai, R., Kytle, K., Reyes, A. and Colicelli, J. (1993). Use of a yeast expression system for the isolation and analysis of drug-resistant mutants of a mammalian phosphodiesterase. Proc. Natl Acad. Sci. USA 90, 11970–11974.

Pinto, I.L., Buckle, D.R., Readshaw, S.A. and Smith, D.G. (1993). The selective inhibition of phosphodiesterase IV by benzopyran derivatives of rolipram. Bio. Med. Lett. 3, 1743–1746.

Reeves, M.L., Leigh, B.K. and England, P.J. (1987). The identification of a new cyclic nucleotide phosphodiesterase activity in human and guinea-pig cardiac ventricle: implications for the mechanism of action of selective phosphodiesterase inhibitors. Biochem. J. 241, 535–541.

Saccomano, N.A., Vinick, F.J., Koe, K., Nielsen, J.A., Whalen, W.M., Meltz, M., Phillips, D., Thadieo, P.F., Jung, S., Chapin, D.S., Lebel, L.A., Russo, L.L., Helwig, D.A., Johnson, J.L., Jr, Ives, J.L. and Williams, I.H. (1991). Calcium-independent phosphodiesterase inhibitors as putative antidepressants: [3-(bicycloalkoxy)-4-methoxyphenyl]-2-imidazolidinones. J. Med. Chem. 34, 291–298.

Saeki, T., Adachi, H., Takase, Y., Yoshitake, S., Souda, S. and Saito, I. (1995). A selective type V phosphodiesterase inhibitor, E4021, dilates porcine large coronary artery. J. Pharmacol. Exp. Ther. 272, 825–831.

Schmiechen, R., Schneider, H.H. and Watchel, H. (1990). Close correlation between behavioral response and binding *in vivo* for inhibitors of the rolipram-sensitive phosphodiesterase. Psychopharmacology 102, 17–20.

Schneider, H.H., Schmiechen, R., Brezinski, M. and Seidler, J. (1986). Stereospecific binding of the antidepressant rolipram to brain protein structures. Eur. J. Pharmacol. 127, 105–115.

Schudt, C., Winder, S., Muller, B. and Ukena, D. (1991). Zardaverine as a selective inhibitor of phosphodiesterase isoenzymes. Biochem. Pharmacol. 42, 153–162.

Schudt, C., Tenor, H., Wendel, A., Eltze, M., Magnussen, H. and Rabe, K.F. (1993). Influence of the PDEIII/IV inhibitor B9004–070 on contraction and PDE activities in airway and vascular smooth muscle. Am. Rev. Respir. Dis. 147, A184. [Abstract]

Schultz, J.E. and Schmidt, B.H. (1986). Rolipram, a stereospecific inhibitor of calmodulin-independent phosphodiesterase, causes β-adrenoceptor subsensitivity in rat cerebral cortex. Naunyn-Schmiedebergs Arch. Pharmacol. 333, 23–30.

Schwabe, U., Miyake, M., Ohga, Y. and Daly, J. (1976). 4-(3-Cyclopentyloxy-4-methoxyphenyl)-2-pyrrolidone (ZK 62711): a potent inhibitor of adenosine cyclic 3′,5′-monophosphate phosphodiesterase in homogenates and tissue slices from rat brain. Mol. Pharmacol. 12, 900–910.

Scott, J.D., Glaccum, M.B., Zoller, M.J., Uhler, M.D., Helfman, D.M, McKnight, G.S. and Krebs, E.G. (1987). The molecular cloning of a type II regulatory subunit of the cAMP-dependent protein kinase from rat skeletal muscle and mouse brain. Proc. Natl Acad. Sci. USA 84, 5192–5196.

Semmler, J., Wachtel, H. and Endres, S. (1993). The specific type IV phosphodiesterase inhibitor rolipram suppresses tumor necrosis factor-α production by human mononuclear cells. Int. J. Immunopharmacol. 15, 409–413.

Sette, C. and Conti, M. (1995). Phosphorylation and activation of a cAMP-specific phosphodiesterase by the cAMP-dependent protein kinase: mapping of the phosphorylation sites. FASEB J. 9, A1262. [Abstract]

Shahid, M. and Nicholson, C.D. (1990). Comparison of cyclic nucleotide phosphodiesterase isozymes in rat and rabbit ventricular myocardium: positive inotropic and phosphodiesterase inhibitory effects of Org 30029, milrinone and rolipram. Naunyn-Schmiedebergs Arch. Pharmacol. 342, 698–705.

Shakur, Y., Pryde, J.G. and Houslay, M.D. (1993). Engineered deletion of the unique N-terminal domain of the cyclic AMP-specific phosphodiesterase RD1 prevents plasma membrane association and the attainment of enhanced thermostability without altering its sensitivity to inhibition by rolipram. Biochem. J. 292, 677–686.

Sheppard, H. and Wiggan, G. (1971). Analogues of 4-(3,4-dimethoxybenzyl)-2-imidazolidinone as potent inhibitors of rat erythrocyte adenosine cyclic 3′,5′-phosphate phosphodiesterase. Mol. Pharmacol. 7, 111–115.

Souness, J.E. and Scott, L.C. (1993). Stereospecificity of rolipram actions on eosinophil cyclic AMP-specific phosphodiesterase. Biochem. J. 291, 389–395.

Souness, J.E., Maslen, C. and Scott, L.C. (1992). Effects of solubilization and vanadate/glutathione complex on inhibitor potencies against eosinophil cyclic AMP-specific phosphodiesterase. FEBS Lett. 302, 181–184.

Sullivan, M., Egerton, M., Shakur, Y., Marquardsen, A. and Houslay, M.D. (1994). Molecular cloning and expression, in both COS-1 cells and *S. cerevisiae*, of a human cytosolic type-IV_A, cyclic AMP specific phosphodiesterase (hPDE-$IV_{A\text{-}b6.1}$). Cell. Signal. 6, 793–812.

Swinnen, J.V., Joseph, D.R. and Conti, M. (1989). Molecular cloning of rat homologues of the *Drosophila melanogaster* dunce cAMP phosphodiesterase: evidence for a family of genes. Proc. Natl Acad. Sci. USA 86, 5325–5329.

Swinnen, J.V., Tsikalas, K.E. and Conti, M. (1991). Properties and hormonal regulation of two structurally related cAMP phosphodiesterases from the rat sertoli cell. J. Biol. Chem. 266, 18370–18377.

Takase, Y., Saeki, T., Watanabe, N., Adachi, H., Souda, S. and Saito, I. (1994). Cyclic GMP phosphodiesterase inhibitors. 2. Requirements of 6-substitution of quinazoline derivatives for potent and selective inhibitory activity. J. Med. Chem. 37, 2106–2111.

Thompson, W.J. and Appleman, M.M. (1971). Multiple cyclic nucleotide phosphodiesterase activities from rat brain. Biochemistry 10, 311–316.

Torphy, T.J. and Cieslinski, L.B. (1990). Characterization and selective inhibition of cyclic nucleotide phosphodiesterase isozymes in canine tracheal smooth muscle. Mol. Pharmacol. 37, 206–214.

Torphy, T.J. and Undem, B.J. (1991) Phosphodiesterase inhibitors: new opportunities for the treatment of asthma. Thorax 46, 512–523.

Torphy, T.J., Stadel, J.M., Burman, M., Cieslinski, L.B., McLaughlin, M.M., White, J. and Livi, G.P. (1992). Coexpression of human cAMP-specific phosphodiesterase activity and high affinity rolipram binding in yeast. J. Biol. Chem. 267, 1798–1804.

Torphy, T.J., DeWolf, W.E., Jr, Green, D.W. and Livi, G.P. (1993a). Biochemical characteristics and cellular regulation of phosphodiesterase IV. Agents Actions Suppl. 43, 51–71.

Torphy, T.J., Livi, G.P. and Christensen, S.B. (1993b). Novel phosphodiesterase inhibitors for the therapy of asthma. Drugs News Perspect. 6, 203–214.

Torphy, T.J., Murray, K.J. and Arch, J.R.S. (1994). Selective phosphodiesterase isozyme inhibitors. In "Drugs and the Lung" (eds. C.P. Page and W.J. Metzger), pp. 397–447. Raven, New York.

Torphy, T.J., Zhou, H.-L., Foley, J.J., Sarau, H.M., Manning, C.D. and Barnette, M.S. (1995). β-Adrenoceptor agonists up-regulate PDE4 activity and induce a heterologous desensitization of U937 cells to prostaglandin E_2. J. Biol. Chem. 270, 23598–23604.

Underwood, D.C., Osborn, R.R., Novak, L.B., Matthews, J.K., Newsholme, S.J., Undem, B.J., Hand, J.B. and Torphy, T.J. (1993). Inhibition of antigen-induced bronchoconstriction and eosinophil infiltration in the guinea pig by the cyclic AMP-specific phosphodiesterase inhibitor, rolipram. J. Pharmacol. Exp. Ther. 266, 306–313.

Wachtel, H. (1983a). Potential antidepressant activity of rolipram and other selective cyclic adenosine 3′,5′-monophosphate phosphodiesterase inhibitors. Neuropharmacology 22, 267–272.

Wachtel, H. (1983b). Neurotropic effects of the optical isomers of the selective adenosine cyclic 3′,5′-monophosphate phosphodiesterase inhibitor rolipram in rats *in vivo*. J. Pharm. Pharmacol. 35, 440–444.

Warrellow, G.J., Boyd, E.C. and Alexander, R.P. (1994). Trisubstituted phenyl derivatives as phosphodiesterase inhibitors. Patent Cooperation Treaty International Patent Application WO 94/14742.

Weishaar, R.E., Quade, M., Boyd, D., Schenden, J., Marks, S. and Kaplan, H.R. (1983). The effect of several "new and novel" cardiotonic agents on key subcellular processes involved in regulation of myocardial contractility: implications for mechanism of action. Drug Dev. Res. 3, 517–534.

Weishaar, R.E., Cain, M.H. and Bristol, J.A. (1985). A new generation of phosphodiesterase inhibitors: multiple molecular forms of phosphodiesterase and the potential for drug selectivity. J. Med. Chem. 28, 537–545.

Wells, J.N., Garst, J.E. and Kramer, G.L. (1981). Inhibition of separated forms of phosphodiesterases from pig coronary arteries by 1,3-disubstituted and 1,3,8-trisubstituted xanthines. J. Med. Chem. 24, 954–958.

Wilhelm, R.S., Loe, B.E., Devens, B.H., Alverez, R. and Martin, M.G. (1993). Patent Cooperation Treaty International Patent Application WO 93/07146.

Rolipram

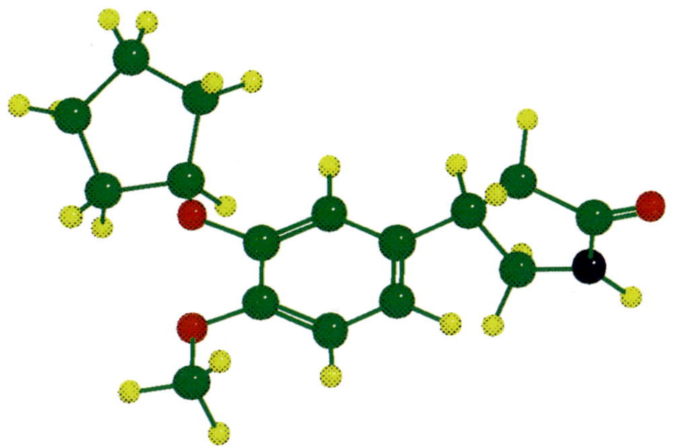

RP 73401

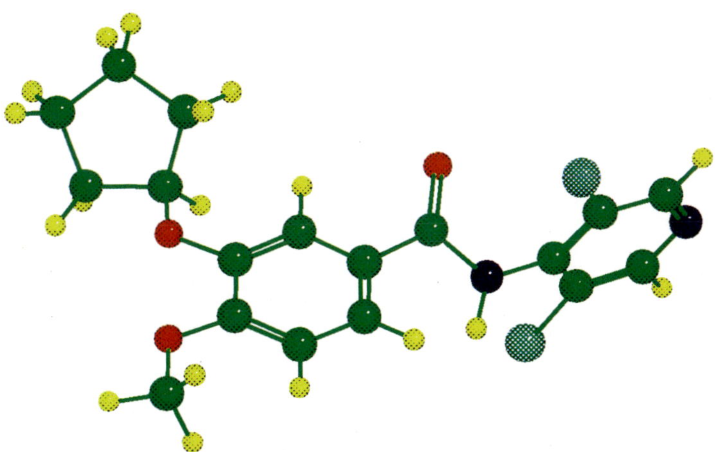

Plate 12.1 Stick and ball models of the structures of RP 73401 and rolipram. Green = C; yellow = H; red = O; purple = N; stippled green = Cl.

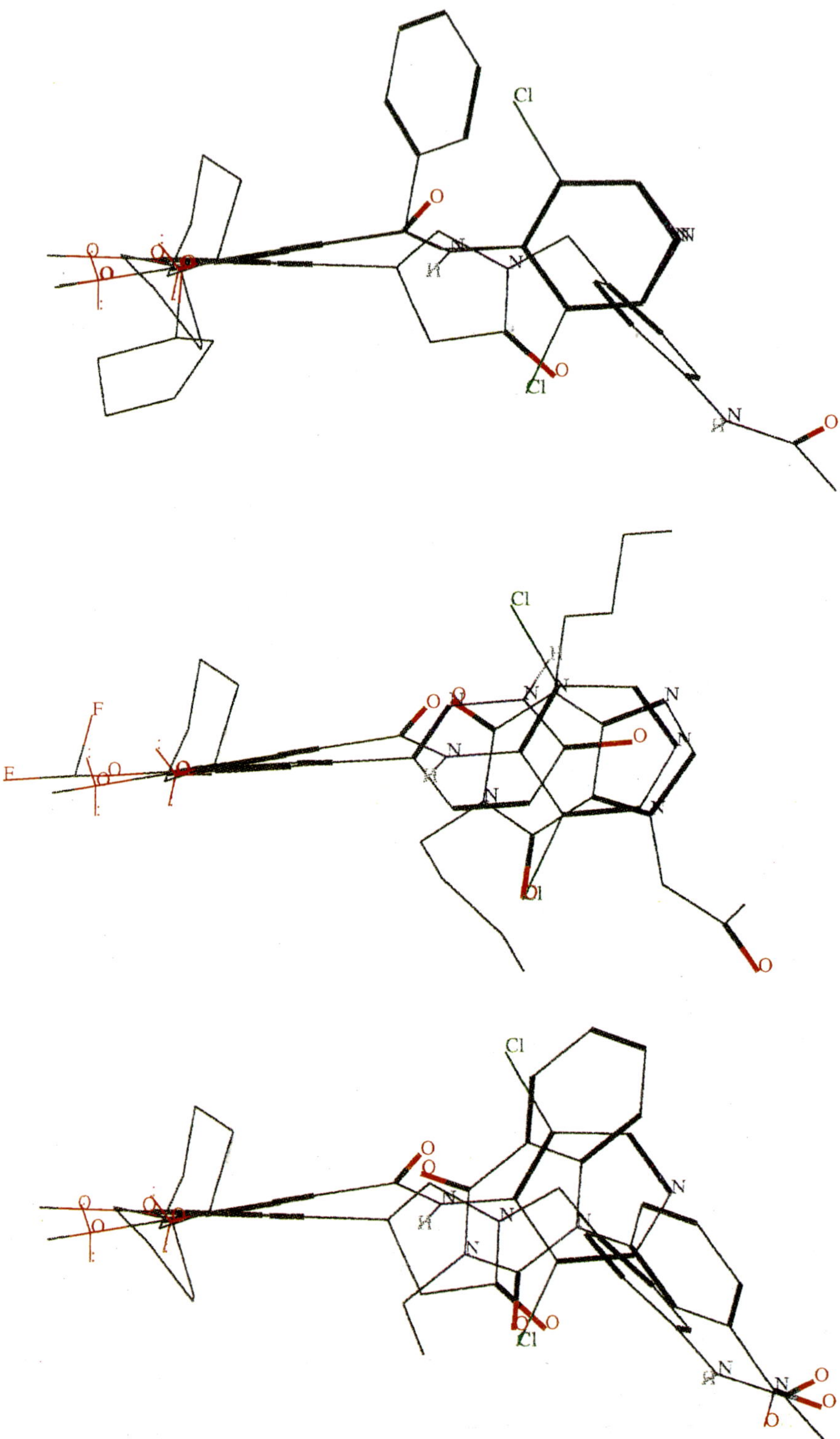

Plate 13.1 Overlay of inhibitors. Top, Compound 11, RP 73401 and CDP-840. The dialkoxyphenyl ring of each compound is superimposed and oriented perpendicular to the plane of the page, appearing on the left of the figure as a thickened black line. Centre, RP 73401, zardaverine and denbufylline. Bottom, Compound 11, RP 73401 and nitraquazone.

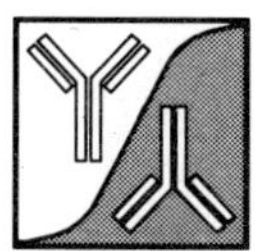

Glossary

Notes: This glossary is up to date for the current volume only and will be supplemented with each subsequent volume.

α_1, α_2 receptors Adrenoceptor subtypes
α_1-ACT α_1-Antichymotrypsin
α_1-AP α_1-antiproteinase *also known as* α_1-antitrypsin and α_1-proteinase inhibitor
α_1-AT α_1-Antitrypsin inhibitor *also known as* α_1-antiproteinase and α_1-proteinase inhibitor
α_1-PI α_1-Proteinase inhibitor *also known as* α_1-antitrypsin and α_1-antiproteinase
α_2-M α_2-macroglobulin
A Absorbance
AI, AII Angiotensin I, II
Å Angstrom
AA Arachidonic acid
aa Amino acids
AAb Autoantibody
ABAP 2′,2′-azobis-2-amidino propane
Ab Antibody
Ab1 Idiotype antibody
Ab2 Anti-idiotype antibody
Ab2α Anti-idiotype antibody which binds outside the antigen binding region
Ab2β Anti-idiotype antibody which binds to the antigen binding region
Ab3 Anti-anti-idiotype antibody
Abcc Antibody dependent cellular cytotoxicity
ABA-L-GAT Arsanilic acid conjugated with the synthetic polypeptide L-GAT
AC Adenylate cyclase
ACAT Acyl-co-enzyme-A acyltransferase
ACAID Anterior chamber-associated immune deviation
ACE Angiotensin-converting enzyme
ACh Acetylcholine
ACTH Adrenocorticotrophin hormone
ADCC Antibody-dependent cellular cytotoxicity
ADH Alcohol dehydrogenase
Ado Adenosine
ADP Adenosine diphosphate
ADPRT Adenosine diphosphate ribosyl transferase
AES Anti-eosinophil serum
Ag Antigen
AGE Advanced glycosylation end-product
AGEPC 1-*O*-alkyl-2-acetyl-*sn*-glyceryl-3-phosphocholine; *also known as* PAF and APRL
AH Acetylhydrolase
AID Autoimmune disease
AIDS Acquired immune deficiency syndrome
A/J A Jackson inbred mouse strain
ALP Anti-leukoprotease
ALS Amyotrophic lateral sclerosis
cAMP Cyclic adenosine monophosphate *also known as* adenosine 3′,5′-phosphate
AM Alveolar macrophage
AML Acute myelogenous leukaemia
AMP Adenosine monophosphate
AMVN 2,2′-azobis (2,4-dimethylvaleronitrile)
ANAb Anti-nuclear antibodies
ANCA Anti-neutrophil cytoplasmic auto antibodies
cANCA Cytoplasmic ANCA
pANCA Perinuclear ANCA
AND Anaphylactic degranulation
ANF Atrial natriuretic factor
ANP Atrial natriuretic peptide
Anti-I-A, Anti-I-E Antibody against class II MHC molecule encoded by I-A locus, I-E locus
anti-Ig Antibody against an immunoglobulin
anti-RTE Anti-tubular epithelium
AP-1 Activator protein-1
APA B-azaprostanoic acid
APAS Antiplatelet antiserum
APC Antigen-presenting cell
APD Action potential duration
apo-B Apolipoprotein B
APRL Anti-hypertensive polar renal lipid *also known as* PAF
APUD Amine precursor uptake and decarboxylation
AR Aldose reductase
AR-CGD Autosomal recessive form of chronic granulomatous disease
ARDS Adult respiratory distress syndrome
AS Ankylosing spondylitis
ASA Acetylsalicylic acid *also known as* aspirin
4-ASA, 5-ASA 4-, 5-aminosalicylic acid
ATHERO-ELAM A monocyte adhesion molecule
ATL Adult T cell leukaemia
ATP Adenosine triphosphate
ATPase Adenosine triphosphatase
ATPγs Adenosine 3′ thiotriphosphate
AITP Autoimmune thrombocytopenic purpura
AUC Area under curve
AVP Arginine vasopressin

β_1, β_2 receptors Adrenoceptor subtypes
β_2 (CD18) A leucocyte integrin
β_2M β_2-Microglobulin
β-TG β-Thromboglobulin
B_7/BB_1 *Known to be* expressed on B cell blasts and immunostimulatory dendritic cells
BAF Basophil-activating factor
BAL Bronchoalveolar lavage
BALF Bronchoalveolar lavage fluid
BALT Bronchus-associated lymphoid tissue
B cell Bone marrow-derived lymphocyte
BCF Basophil chemotactic factor
B-CFC Basophil colony-forming cell
BCG Bacillus Calmette-Guérin
BCNU 1,3-bis(2-chloroethyl)-1-nitrosourea
bFGF Basic fibroblast growth factor
Bg Birbeck granules
BHR Bronchial hyperresponsiveness
BHT Butylated hydroxytoluene
b.i.d. *Bis in die* (twice a day)
Bk Bradykinin
Bk_1, Bk_2 receptors Bradykinin receptor subtypes *also known as* B_1 and B_2 receptors
Bl-CFC Blast colony-forming cells

B-lymphocyte Bursa-derived lymphocyte
BM Bone marrow
BMCMC Bone marrow cultured mast cell
BMMC Bone marrow mast cell
BOC-FMLP Butoxycarbonyl-FMLP
bp Base pair
BPB Para-bromophenacyl bromide
BPI Bacterial permeability-increasing protein
BSA Bovine serum albumin
BSS Bernard-Soulier Syndrome

^{51}Cr Chromium^{51}
C1, C2 ... C9 The 9 main components of complement
C1 inhibitor A serine protease inhibitor which inactivates C1r/C1s
C1q Complement fragment 1q
C1qR Receptor for C1w; facilitates attachment of immune complexes to mononuclear leucocytes and endothelium
C3a Complement fragment 3a (anaphylatoxin)
$C3a_{72-77}$ A synthetic carboxyterminal peptide C3a analogue
C3aR Receptor for anaphylatoxins, C3a, C4a, C5a
C3b Complement fragment 3b (anaphylatoxin)
C3bi Inactivated form of C3b fragment of complement
C4b Complement fragment 4b (anaphylatoxin)
C4BP C4 binding protein; plasma protein which acts as co-factor to factor I inactivate C3 convertase
C5a Complement fragment 5a (anaphylatoxin)
C5aR Receptor for anaphylatoxins C3a, C4a and C5a
C5b Complement fragment 5b (anaphylatoxin)
$C_\epsilon 2$, $C_\epsilon 3$, $C_\epsilon 4$ Heavy chain of immunoglobulin E: domains 2, 3 and 4
Ca *The chemical symbol for* calcium
$[CA^{2+}]_i$ Intracellular free calcium concentration
CAH Chronic active hepatitis
CALLA Common lymphoblastic leukaemia antigen
CALT Conjunctival associated lymphoid tissue
CaM Calmodulin
CAM Cell adhesion molecule
cAMP Cyclic adenosine monophosphate *also known as* adenosine 3′,5′-phosphate
CaM-PDE Ca^{2+}/CaM-dependent PDE
CAP57 Cationic protein from neutrophils
CAT Catalase
CatG Cathepsin G
CB Cytochalasin B
CBH Cutaneous basophil hypersensitivity
CBP Cromolyn-binding protein
CCK Cholecystokinin
CCR Creatinine clearance rate
CD Cluster of differentiation (a system of nomenclature for surface molecules on cells of the immune system); cluster determinant
CD1 Cluster of differentiation 1 *also known as* MHC class I-like surface glycoprotein
CD1a Isoform a *also known as* non-classical MHC class I-like surface antigen; present on thymocytes and dendritic cells
CD1b *Known to be* present on thymocytes and dendritic cells
CD1c Isoform c *also known as* non-classical MHC class I-like surface antigen; present on thymocytes
CD2 Defines T cells involved in antigen non-specific cell activation
CD3 *Also known as* T cell receptor-associated surface glycoprotein on T cells
CD4 Defines MHC class II-restricted T cell subsets
CD5 *Known to be* present on T cells and a subset of B cells; *also known as* Lyt 1 in mouse
CD7 Cluster of differentiation 7; present on most T cells and NK cells
CD8 Defines MHC class I-restricted T cell subset; present on NK cells
CD10 *Known to be* common acute leukaemia antigen
CD11a *Known to be* an α chain of LFA-1 (leucocyte function antigen-1) present on several types of leucocyte and which mediates adhesion
CD11c *Known to be* a complement receptor 4 α chain
CD13 Aminopeptidase N; present on myeloid cells
CD14 *Known to be* a lipid-anchored glycoprotein; present on monocytes
CD15 *Known to be* Lewis X, fucosyl-*N*-acetyllactosamine
CD16 *Known to be* Fcγ receptor III
CD16-1, CD16-2 Isoforms of CD16
CD19 Recognizes B cells and follicular dendritic cells
CD20 *Known to be* a pan B cell
CD21 C3d receptor
CD23 Low affinity FcεR
CD25 Low affinity receptor for interleukin-2
CD27 Present on T cells and plasma cells
CD28 Present on resting and activated T cells and plasma cells
CD30 Present on activated B and T cells
CD31 *Known to be* on platelets, monocytes, macrophages, granulocytes, B-cells and endothelial cells; *also known as* PECAM
CD32 Fcγ receptor II
CD33^{+} *Known to be* a monocyte and stem cell marker
CD34 *Known to be* a stem cell marker
CD35 C3b receptor
CD36 *Known to be* a macrophage thrombospondin receptor
CD40 Present on B cells and follicular dendritic cells
CD41 *Known to be* a platelet glycoprotein
CD44 *Known to be* a leucocyte adhesion molecule; *also known as* hyaluronic acid cell adhesion molecule (H-CAM), Hermes antigen, extracellular matrix receptor III (ECMIII); present on polymorphonuclear leucocytes
CD45 *Known to be* a pan leucocyte marker
CD45RO *Known to be* the isoform of leukosialin present on memory T cells
CD46 *Known to be* a membrane cofactor protein
CD49 Cluster of differentiation 49
CD51 *Known to be* vitronectin receptor alpha chain
CD54 *Known to be* Intercellular adhesion molecule-1 *also known as* ICAM-1
CD57 Present on T cells and NK subsets
CD58 A leucocyte function-associated antigen-3, *also known to be* a member of the β-2 integrin family of cell adhesion molecules
CD59 *Known to be* a low molecular weight HRf present to many haematopoietic and non-haematopoietic cells
CD62 *Known to be present on* activated platelets and endothelial cells; *also known as* P-selectin
CD64 *Known to be* Fcγ receptor I
CD65 *Known to to* fucoganglioside
CD68 Present on macrophages
CD69 *Known to be* an activation inducer molecule; present on activated lymphocytes
CD72 Present on B-lineage cells
CD74 An invariant chain of class II B cells
CDC Complement-dependent cytotoxicity
cDNA Complementary DNA
CDP Choline diphosphate
CDR Complementary-determining region

CD_{xx} Common determinant *xx*
CEA Carcinoembryonic antigen
CETAF Corneal epithelial T cell activating factor
CF Cystic fibrosis
Cf Cationized ferritin
CFA Complete Freund's adjuvant
CFC Colony-forming cell
CFU Colony-forming unit
CFU-Mk Megakaryocyte progenitors
CFU-S Colony-forming unit, spleen
CGD Chronic granulomatous disease
cGMP Cyclic guanosine monophosphate *also known as* guanosine 3′,5′-phosphate
CGRP Calcitonin gene-related peptide
CH2 Hinge region of human immunoglobulin
CHO Chinese hamster ovary
CI Chemical ionization
CIBD Chronic inflammatory bowel disease
CK Creatine phosphokinase
CKMB The myocardial-specific isoenzyme of creatine phosphokinase
Cl *The chemical symbol for* chlorine
CL Chemiluminescent
CLA Cutaneous lymphocyte antigen
CL18/6 Anti-ICAM-1 monoclonal antibody
CLC Charcot–Leyden crystal
CMC Critical micellar concentration
CMI Cell mediated immunity
CML Chronic myeloid leukaemia
CMV Cytomegalovirus
CNS Central nervous system
CO Cyclooxygenase
CoA Coenzyme A
CoA-IT Coenzyme A – independent transacylase
Con A Concanavalin A
COPD Chronic obstructive pulmonary disease
COS Fibroblast-like kidney cell line established from simian cells
CoVF Cobra venom
CP Creatine phosphate
Cp Caeruloplasmin
c.p.m. Counts per minute
CPJ Cartilage/pannus junction
Cr *The chemical symbol for* chromium
CR Complement receptor
CR1, CR2 & CR4 Complement receptor types 1, 2 and 4
CR3-α Complement receptor type 3-α
CRF Corticotrophin-releasing factor
CRH Corticotrophin-releasing hormone
CRI Cross-reactive idiotype
CRP C-reactive protein
CSA Cyclosporin A
CSF Colony-stimulating factor
CSS Churg–Strauss syndrome
CT Computed tomography
CTAP-III Connective tissue-activating peptide
CTD Connective tissue diseases
C terminus Carboxy terminus of peptide
CThp Cytotoxic T lymphocyte precursors
CTL Cytotoxic T lymphocyte
CTLA-4 *Known to be* co-expressed with CD20 on activated T cells
CTMC Connective tissue mast cell
CVF Cobra venom factor

2D Second derivative
Da Dalton (the unit of relative molecular mass)
DAF Decay-accelerating factor
DAG Diacylglycerol
DAO Diamine oxidase
D-Arg D-Arginine
DArg-[Hyp^3,DPhe^7]-BK A bradykinin B_2 receptor antagonist. Peptide derivative of bradykinin
DArg-[Hyp^3,Thi^5,DTic^7,Tic^8]-BK A bradykinin B_2 receptor antagonist. Peptide derivative of bradykinin
DBNBS 3,5-dibromo-4-nitroso-benzenesulphonate
DC Dendritic cell
DCF Oxidized DCFH
DCFH 2′,7′-dichlorofluorescin
DEC Diethylcarbamazine
DEM Diethylmaleate
desArg^9-BK Carboxypeptidase N product of bradykinin
desArg^{10}KD Carboxypeptidase N product of kallidin
DETAPAC Diethylenetriaminepentaacetic acid
DFMO α1-Difluoromethyl ornithine
DFP Diisopropyl fluorophosphate
DFX Desferrioxamine
DGLA Dihomo-γ-linolenic acid
DH Delayed hypersensitivity
DHA Docosahexaenoic acid
DHBA Dihydroxybenzoic acid
DHR Delayed hypersensitivity reaction
DIC Disseminated intravascular coagulation
DL-CFU Dendritic cell/Langerhans cell colony forming
DLE Discoid lupus erythematosus
DMARD Disease-modifying anti-rheumatic drug
DMF *N*,*N*-dimethylformamide
DMPO 5,5-dimethyl-l-pyrroline *N*-oxide
DMSO Dimethyl sulfoxide
DNA Deoxyribonucleic acid
D-NAME D-Nitroarginine methyl ester
DNase Deoxyribonuclease
DNCB Dinitrochlorobenzene
DNP Dinitrophenol
Dpt4 *Dermatophagoides pteronyssinus* allergen 4
DGW2, DR3, DR7 HLA phenotypes
DREG-56 (Antigen) L-selectin
DREG-200 A monoclonal antibody against L-selectin
ds Double-stranded
DSCG Disodium cromoglycate
DST Donor-specific transfusion
DTH Delayed-type hypersensitivity
DTPA Diethylenetriamine pentaacetate
DTT Dithiothreitol
dv/dt Rate of change of voltage within time

ε Molar absorption coefficient
EA Egg albumin
EACA Epsilon-amino-caproic acid
EAE Experimental autoimmune encephalomyelitis
EAF Eosinophil-activating factor
EAR Early phase asthmatic reaction
EAT Experimental autoimmune thyroiditis
EBV Epstein-Barr virus
EC Endothelial cell
ECD Electron capture detector
ECE Endothelin-converting enzyme
E-CEF Eosinophil cytotoxicity enhancing factor
ECF-A Eosinophil chemotactic factor of anaphylaxis
ECG Electrocardiogram
ECGF Endothelial cell growth factor
ECGS Endothelial cell growth supplement
E. coli *Escherichia coli*
ECP Eosinophil cationic protein
EC-SOD Extracellular superoxide dismutase
EC-SOD C Extracellular superoxide dismutase C
ED_{35} Effective dose producing 35% maximum response
ED_{50} Effective dose producing 50% maximum response
EDF Eosinophil differentiation factor
EDL Extensor digitorum longus
EDN Eosinophil-derived neurotoxin
EDRF Endothelium-derived relaxing factor
EDTA Ethylenediamine tetraacetic acid *also known as* etidronic acid
EE Eosinophilic eosinophils
EEG Electroencephalogram
EET Epoxyeicosatrienoic acid
EFA Essential fatty acid
EFS Electical field stimulation
EG1 Monoclonal antibody specific for the cleaved form of eosinophil cationic peptide

EGF Epidermal growth factor
EGTA Ethylene glycol-bis(β-aminoethyl ether) N,N,N',N'-tetraacetic acid
EHNA Erythro-9-(2-hydroxy-3-nonyl)-adenine
EI Electron impact
EIB Exercise-induced bronchoconstriction
eIF-2 Subunit of protein synthesis initiation factor
ELAM-1 Endothelial leucocyte adhesion molecule-1
ELF Respiratory epithelium lung fluid
ELISA Enzyme-linked immunosorbent assay
EMS Eosinophilia-myalgia syndrome
ENS Enteric nervous system
EO Eosinophil
EO-CFC Eosinophil colony-forming cell
EOR Early onset reaction *also known as* EAR
EPA Eicosapentaenoic acid
EpDIF Epithelial-derived inhibitory factor *also known as* epithelium-derived relaxant factor
EPO Eosinophil peroxidase
EPOR Erythropoietin receptor
EPR Effector cell protease
EPX Eosinophil protein X
ER Endoplasmic reticulum
ERCP Endoscopic retrograde cholangiopancreatography
E-selectin Endothelial selectin *formerly known as* endothelial leucocyte adhesion molecule-1 (ELAM-1)
ESP Eosinophil stimulation promoter
ESR Erythrocyte sedimentation rate
e.s.r. Electron spin resonance
ET, ET-1 Endothelin, -1
ETYA Eicosatetraynoic acid

FA Fatty acid
FAB Fast-electron bombardment
Fab Antigen binding fragment
F(ab')2 Fragment of an immunoglobulin produced pepsin treatment
FACS Flow activated cell sorter
factor B Serine protease in the C3 converting enzyme of the alternative pathway
factor D Serine protease which cleaves factor B
factor H Plasma protein which acts as a co-factor to factor I
factor I Hydrolyses C3 converting enzymes with the help of factor H
FAD Flavine adenine dinucleotide
FapyAde 5-formamido-4,6-diamino-pyrimidine
FapyGua 2,6-diamino-4-hydroxy-5-formamidopyrimidine
FBR Fluorescence photobleaching recovery
Fc Crystallizable fraction of immunoglobulin molecule
Fcγ Receptor for Fc portion of IgG
FcγRI Ig Fc receptor I *also known as* CD64
FcγRII Ig Fc receptor II *also known as* CD32
FcγRIII Ig Fc receptor III *also known as* CD16
Fc$_\varepsilon$RI High affinity receptor for IgE
Fc$_\varepsilon$RII Low affinity receptor for IgE
FcR Receptor for Fc region of antibody
FCS Foetal calf (bovine) serum
FEV$_1$ Forced expiratory volume in 1 second
Fe-TPAA Fe(III)-tris[N-(2-pyridylmethyl)-2-aminoethyl]amine
Fe-TPEN Fe(II)-tetrakis-N,N,N',N'-(2-pyridyl methyl-2-aminoethyl)amine
FFA Free fatty acids
FGF Fibroblast growth factor
FID Flame ionization detector
FITC Fluorescein isothiocyanate
FKBP FK506-binding protein
FLAP 5-lipoxygenase-activating protein
FMLP N-Formyl-methionyl-leucyl-phenylalanine
FNLP Formyl-norleucyl-leucyl-phenylalanine
FOC Follicular dendritic cell
FPLC Fast protein liquid chromatography
FPR Formyl peptide receptor
FS cell Folliculo-stellate cell
FSG Focal sequential glomerulosclerosis
FSH Follicle stimulating hormone
FX Ferrioxamine
5-FU 5-fluorouracil

Ga G-protein
G6PD Glucose 6-phosphate dehydrogenase
GABA γ-Aminobutyric acid
GAG Glycosaminoglycan
GALT Gut-associated lymphoid tissue
GAP GTPase-activating protein
GBM Glomerular basement membrane
GC Guanylate cyclase
GC-MS Gas chromatography mass spectroscopy
G-CSF Granulocyte colony-stimulating factor
GDP Guanosine 5'-diphosphate
GEC Glomerular epithelial cell
GF-1 An insulin-like growth factor
GFR Glomerular filtration rate
GH Growth hormone
GH-RF Growth hormone-releasing factor
Gi Family of pertussis toxin sensitive G-proteins
GI Gastrointestinal
GIP Granulocyte inhibitory protein
GlyCam-1 Glycosylation-dependent cell adhesion molecule-1
GMC Gastric mast cell
GM-CFC Granulocyte-macrophage colony-forming cell
GM-CSF Granulocyte-macrophage colony-stimulating factor
GMP Guanosine monophosphate (guanosine 5'-phosphate)
Go Family of pertussis toxin sensitive G-proteins
GP Glycoprotein
gp45–70 Membrane co-factor protein
gp90^{MEL} 90 kD glycoprotein recognized by monoclonal antibody MEL-14; *also known as* L-selectin
GPIIb-IIIa Glycoprotein IIb-IIIa *known to be* a platelet membrane antigen
GppCH$_2$P Guanyl-methylene diphosphanate *also known as* a stable GTP analogue
GppNHp Guanylyl-imidiodiphosphate *also known as* a stable GTP analogue
GRGDSP Glycine–arginine–glycine–aspartic acid–serine–proline
Gro Growth-related oncogene
GRP Gastrin-related peptide
Gs Stimulatory G protein
GSH Glutathione (reduced)
GSHPx Glutathione peroxidase
GSSG Glutathione (oxidized)
GT Glanzmann Thrombasthenia
GTP Guanosine triphosphate
GTP-γ-S Guanosine 5'O-(3-thiotriphosphate)
GTPase Guanosine triphosphatase
GVHD Graft-versus-host-disease
GVHR Graft-versus-host-reaction

H Histamine
H$_1$, H$_2$, H$_3$ Histamine receptor types 1, 2 and 3
H_2O_2 *The chemical symbol for* hydrogen peroxide
Hag Haemagglutinin
Hag-1, Hag-2 Cleaved haemagglutinin subunits-1, -2
H & E Haematoxylin and eosin
hIL Human interleukin
Hb Haemoglobin
HBBS Hank's balanced salt solution
HCA Hypertonic citrate
H-CAM Hyaluronic acid cell adhesion molecule
HDC Histidine decarboxylase
HDL High-density lipoprotein

HEL Hen egg white lysozyme
HEPE Hydroxyeicosapentanoic acid
HEPES *N*-2-Hydroxylethylpiperazine-*N'*-2-ethane sulphonic acid
HES Hypereosinophilic syndrome
HETE 5,8,9,11,12 and 15 Hydroxyeicosatetraenoic acid
5(S)HETE A stereo isomer of 5-HETE
HETrE Hydroxyeicosatrienoic acid
HEV High endothelial venule
HFN Human fibronectin
HGF Hepatocyte growth factor
HHTrE 12(*S*)-Hydroxy-5,8,10-heptadecatrienoic acid
HIV Human immunodeficiency virus
HL60 Human promyelocytic leukaemia cell line
HLA Human leucocyte antigen
HLA-DR2 Human histocompatability antigen class II
HMG CoA Hydroxylmethylglutaryl coenzyme A
HMW High molecular weight
HMT Histidine methyltransferase
HMVEC Human microvascular endothelial cell
HNC Human neutrophil collagenase (MMP-8)
HNE Human neutrophil elastase
HNG Human neutrophil gelatinase (MMP-9)
HODE Hydroxyoctadecanoic acid
HO· Hydroxyl radical
HO_2· Perhydroxyl radical
HPETE, 5-HPETE & 15-HPETE 5 and 15 Hydroperoxyeicosatetraenoic acid
HPETrE Hydroperoxytrienoic acid
HPODE Hydroperoxyoctadecanoic acid
HPLC High-performance liquid chromatography
HRA Histamine-releasing activity
HRAN Neutrophil-derived histamine-releasing activity
HRf Homologous-restriction factor
HRF Histamine-releasing factor
HRP Horseradish peroxidase
HSA Human serum albumin
HSP Heat-shock protein
HS-PG Heparan sulphate proteoglycan
HSV, HSV-1 Herpes simplex virus, -1
^{3}HTdR Tritiated thymidine
5-HT 5-Hydroxytryptamine *also known as* Serotonin
HTLV-1 Human T-cell leukaemia virus-1
HUVEC Human umbilical vein endothelial cell
[Hyp^3]-BK Hydroxyproline derivative of bradykinin
[Hyp^4]-KD Hydroxyproline derivative of kallidin

^{111}In Indium^{111}
Ia Immune reaction-associated antigen
Ia+ Murine class II major histocompatibility complex antigen
IB4 Anti-CD18 monoclonal antibody
IBD Inflammatory bowel disease
IBMX 3-isobutyl-1-methylxanthine
IBS Inflammatory bowel syndrome
iC3 Inactivated C3
iC4 Inactivated C4
IC_{50} Concentration producing 50% inhibition
I_{Ca} Calcium current
ICAM Intercellular adhesion molecules
ICAM-1, ICAM-2, ICAM-3 Intercellular adhesion molecules-1, -2, -3
cICAM-1 Circulating form of ICAM-1
ICE IL-1β-converting enzyme
i.d. Intradermal
ID_{50} Dose of drug required to inhibit response by 50%
IDC Interdigitating cell
IDD Insulin-dependent (type 1) diabetes
IEL Intraepithelial leucocyte
IELym Intraepithelial lymphocytes
IFA Incomplete Freund's adjuvant
IFN Interferon
IFNα, IFNβ, IFNγ Interferons α, β, γ
Ig Immunoglobulin
IgA, IgE, IgG, IgM Immunoglobulins A, E, G, M
IgG1 Immunoglobulin G class 1
IgG_{2a} Immunoglobulin G class 2a
IGF-1 Insulin-like growth factor
Ig-SF Immunoglobulin supergene family
IGSS Immuno-gold silver stain
IHC Immunohistochemistry
IHES Idiopathic hypereosinophilic syndrome
IκB NFκB inhibitor protein
IL Interleukin
IL-1, Il-2 ... IL-8 Interleukins-1, 2 ... -8
IL-1α, IL-1β Interleukin-1α, -1β
ILR Interleukin receptor
IL-1R, IL-2R; IL-3R–IL-6R Interleukin 1–6 receptors
IL-1Ra Interleukin-1 receptor antagonist
IL-2Rβ Interleukin-2 receptor β
IMF Integrin modulating factor
IMMC Intestinal mucosal mast cell
i.p. Intraperitoneally
IP_1 Inositol monophosphate
IP_2 Inositol biphosphate
IP_3 Inositol 1,4,5-trisphosphate
IP_4 Inositol tetrakisphosphate
IPF Idiopathic pulmonary fibrosis
IPO Intestinal peroxidase
IpOCOCq Isopropylidene OCOCq
I/R Ischaemia-reperfusion
IRAP IL-1 receptor antagonist protein
IRF-1 Interferon regulatory factor 1
I_{sc} Short-circuit current
ISCOM Immune-stimulating complexes
ISGF3 Interferon-stimulated gene Factor 3
ISGF3α, ISGFγ α, γ subunits of ISGF3
IT Immunotherapy
ITP Idiopathic thrombocytopenic purpura
i.v. Intravenous

K *The chemical symbol for* potassium
K_a Association constant
kb Kilobase
20KDHRF A homologous restriction factor; binds to C8
65KDHRF A homologous restriction factor, also known as C8 binding protein; interferes with cell membrane pore-formation by C5b-C8 complex
Kcat Catalytic constant; a measure of the catalytic potential of an enzyme
K_d dissociation constant
kD Kilodalton
KD Kallidin
K_i Antagonist binding affinity
K*i*67 Nuclear membrane antigen
KLH Keyhole limpet haemocyanin
K_m Michaelis constant
KOS KOS strain of herpes simplex virus

λ_{max} Wavelength of maximum absorbance
LAD Leucocyte adhesion deficiency
LAK Lymphocyte-activated killer (cell)
LAM, LAM-1 Leucocyte adhesion molecule, -1
LAR Late-phase asthmatic reaction
L-Arg L-Arginine
LBP LPS binding protein
LC Langerhans cell
LCF Lymphocyte chemoattractant factor
LCR Locus control region
LDH Lactate dehydrogenase
LDL Low-density lipoprotein
LDV Laser Doppler velocimetry
Le^x(Lewis X) Leucocyte ligand for selectin
LFA Leucocyte function-associated antigen
LFA-1 Leucocyte function-associated antigen-1; *also known to be* a member of the β-2 integrin family of cell adhesion molecules

LG β-Lactoglobulin
LGL Large granular lymphocyte
LH Luteinizing hormone
LHRH Luteinizing hormone-releasing hormone
LI Labelling index
LIS Lateral intercellular spaces
LMP Low molecular mass polypeptide
LMW Low molecular weight
L-NOARG L-Nitroarginine
LO Lipoxygenase
5-LO, 12-LO, 15-LO 5-, 12-, 15-Lipoxygenases
LP(a) Lipoprotein(a)
LPS Lipopolysaccharide
L-selectin Leucocte selectin, *formerly known as* monoclonal antibody that recognizes murine L-selectin (MEL-14 antigen), leucocyte cell adhesion molecule-1 (LeuCAM-1), lectin cell adhesion molecule-1 (LeCAM-1 or LecCAM-1), leucocyte adhesion molecule-1 (LAM-1)
LT Leukotriene
LTA_4, LTB_4, LTC_4, LTD_4, LTE_4 Leukotrienes A_4, B_4, C_4, D_4 and E_4
L_y-1 $^+$ (Cell line)
LX Lipoxin
LXA_4, LXB_4, LXC_4, LXD_4, LXE_4 Lipoxins A_4, B_4, C_4, D_4 and E_4

M Monocyte
M3 Receptor Muscarinic receptor subtype 3
M-540 Merocyanine-540
mAb Monoclonal antibody
mAb IB4, mAb PB1.3, mAb R 3.1, mAb R 3.3, mAb 6.5, mAb 60.3 Monoclonal antibodies IB4, PB1.3, R 3.1, R 3.3, 6.5, 60.3
MABP Mean arterial blood pressure
MAC Membrane attack molecule
Mac Macrophage (also abbreviated to MΦ)
Mac- Macrophage-1 antigen; a member of the β-2 integrin family of cell adhesion molecules (also abbreviated to MΦ1), *also known as* monocyte antigen-1 (M-1), complement receptor-3 (CR3), CD11b/CD18
MAF Macrophage-activating factor
MAO Monoamine oxidase
MAP Monophasic action potential
MAPTAM An intracellular Ca^{2+} chelator
MARCKS Myristolated, alanine-rich C kinase substrate; specific protein kinase C substrate
MBP Major basic protein
MBSA Methylated bovine serum albumin
MC Mesangial cells
MCAO Middle cerebral artery occlusion
M cell Microfold or membranous cell of Peyer's patch epithelium
MCP Membrane co-factor protein
MCP-1 Monocyte chemotactic protein-1
M-CSF Monocyte/macrophage colony-stimulating factor
MC_T Tryptase-containing mast cell
MC_{TC} Tryptase- and chymase-containing mast cell
MDA Malondialdehyde
MDGF Macrophage-derived growth factor
MDP Muramyl dipeptide
MEA Mast cell growth-enhancing activity
MEL Metabolic equivalent level
MEM Minimal essential medium
MG Myasthenia gravis
MGSA Melanoma-growth-stimulatory activity
MHC Major histocompatibility complex
MI Myocardial ischaemia
MIF Migration inhibition factor
mIL Mouse interleukin
MIP-1α Macrophage inflammatory protein 1α
MI/R Myocardial ischaemia/reperfusion
MIRL Membrane inhibitor of reactive lysis
mix-CFC Colony-forming cell mix
Mk Megakaryocyte
MLC Mixed lymphocyte culture
MLymR Mixed lymphocyte reaction
MLR Mixed leucocyte reaction
mmLDL Minimally modified low-density lipoprotein
MMC Mucosal mast cell
MMCP Mouse mast cell protease
MMP, MMP1 Matrix metalloproteinase, -1
MNA 6-Methoxy-2-napthylacetic acid
MNC Mononuclear cells
MΦ Macrophage (also abbreviated to Mac)
MPG *N*-(2-mercaptopropionyl)-glycine
MPO Myeloperoxidase
MPSS Methyl prednisolone
MPTP *N*-methyl-4-phenyl-1,2,3,6-tetrahydropyridine
MRI Magnetic resonance imaging
mRNA Messenger ribonucleic acid
MS Mass spectrometry
MSS Methylprednisolone sodium succinate
MT Malignant tumour
MW Molecular weight

Na *The chemical symbol for* sodium
NA Noradrenaline *also known as* norepinephrine
NAAb Natural autoantibody
NAb Natural antibody
NAC *N*-acetylcysteine
NADH Reduced nicotinamide adenine dinucleotide
NADP Nicotinamide adenine diphosphate
NADPH Reduced nicotinamide adenine dinucleotide phosphate
NAF Neutrophil activating factor
L-NAME L-Nitroarginine methyl ester
NANC Non-adrenergic, non-cholinergic
NAP Neutrophil-activating peptide
NAPQI *N*-acetyl-*p*-benzoquinone imine
NAP-1, NAP-2 Neutrophil-activating peptides -1 and -2
NBT Nitro-blue tetrazolium
NC1 Non-collagen 1
N-CAM Neural cell adhesion molecule
NCEH Neutral cholesteryl ester hydrolase
NCF Neutrophil chemotactic factor
NDGA Nordihydroguaretic acid
NDP Nucleoside diphosphate
Neca 5′-(*N*-ethyl carboxamido)-adenosine
NED Nedocromil sodium
NEP Neutral endopeptidase (EC 3.4.24.11)
NF-AT Nuclear factor of activated T lymphocytes
NF-κB Nuclear factor-κB
NgCAM Neural-glial cell adhesion molecule
NGF Nerve growth factor
NGPS Normal guinea-pig serum
NIH 3T3 (fibroblasts) National Institute of Health 3T3-Swiss albino mouse fibroblast
NIMA Non-inherited maternal antigens
NIRS Near infrared spectroscopy
Nk Neurokinin
NK Natural killer
Nk-1, Nk2, NK-3 Neurokinin receptor subtypes 1, 2 and 3
NkA Neurokinin A
NkB Neurokinin B
NLS Nuclear location sequence
NMDA *N*-methyl-D-aspartic acid
L-NMMA L-Nitromonomethyl arginine
NMR Nuclear magnetic resonance
NNA Nω-nitro-L-arginine
1,N^2-NET β-(2-Naphthyl)-1,N^2-etheno
1,N^2-PET β-Phenyl-1,N^2-etheno
NO *The chemical symbol for* nitric oxide
NOD Non-obese diabetic
NOS Nitric oxide synthase
c-NOS Ca^{2+}-dependent constitutive form of NOS

i-NOS Inducible form of NOS
NPK Neuropeptide K
NPY Neuropeptide Y
NRS Normal rabbit serum
NSAID Non-steroidal anti-inflammatory drug
NSE Nerve-specific enolase
NT Neurotensin
N terminus Amino terminus of peptide

$^1\Delta O_2$ Singlet Oxygen (Delta form)
$^1\Sigma O_2$ Singlet Oxygen (Sigma form)
$O_2^{\cdot -}$ *The chemical symbol for* the superoxide anion radical
OA Osteoarthritis
OAG Oleoyl acetyl glycerol
OD Optical density
ODC Ornithine decarboxylase
ODFR Oxygen-derived free radical
ODS Octadecylsilyl
OH^- *The chemical symbol for* hydroxyl ion
·OH *The chemical symbol for* hydroxyl radical
8-OH-Ade 8-hydroxyadenine
6-OHDA 6-hydroxyguanine
8-OH-dG 8-hydroxydeoxyguanosine *also known as* 7,8-dihydro-8-oxo-2′-deoxyguanosine
8-OH-Gua 8-hydroxyguanine
OHNE Hydroxynonenal
4-OHNE 4-hydroxynonenal
OT Oxytocin
OVA Ovalbumin
ox-LDL Oxidized low-density lipoprotein
OZ Opsonized zymosan

Ψa Apical membrane potential
P Probability
P Phosphate
P_aO_2 Arterial oxygen pressure
P_i Inorganic phosphate
p150,95 A member of the β-2-integrin family of cell adhesion molecules; *also known as* CD11c
PA Phosphatidic acid
pA_2 Negative logarithm of the antagonist dissociation constant
PAEC Pulmonary artery endothelial cells
PAF Platelet-activating factor *also known as* APRL and AGEPC
PAGE Polyacrylamide gel electrophoresis
PAI Plasminogen activator inhibitor
PA-IgG Platelet associated immunoglobulin G
PAM Pulmonary alveolar macrophages
PAS Periodic acid–Schiff reagent
PBA Polyclonal B cell activators
PBC Primary biliary cirrhosis
PBL Peripheral blood lymphocytes
PBMC Peripheral blood mononuclear cells
PBN *N-tert*-butyl-α-phenylnitrone
PBS Phosphate-buffered saline
PC Phosphatidylcholine
PCA Passive cutaneous anaphylaxis
pCDM8 Eukaryotic expression vector
PCNA Proliferating cell nuclear antigen
PCR Polymerase chain reaction
PCT Porphyria cutanea tarda
p.d. Potential difference
PDBu 4α-phorbol 12,13-dibutyrate
PDE Phosphodiesterase
PDGF Platelet-derived growth factor
PDGFR Platelet-derived growth factor receptor
PE Phosphatidylethanolamine
PECAM-1 Platelet endothelial cell adhesion molecule-1; *also known as* CD31
PEG Polyethylene glycol
PET Positron emission tomography
PEt Phosphatidylethanolamine
PF_4 Platelet factor 4
PG Prostaglandin
PGAS Polyglandular autoimmune syndrome
PGD_2 Prostaglandin D_2
PGE1, PGE_2, PGF_2, $PGF_{2\alpha}$, PGG_2, PGH_2 Prostaglandins E_1, E_2, F_2, $F_{2\alpha}$, G_2, H_2
PGF, PGH Prostaglandins F and H
PGI_2 Prostaglandin I_2 *also known as* prostacyclin
P_aO_2 Arterial oxygen pressure
PGP Protein gene-related peptide
Ph^1 Philadelphia (chromosome)
PHA Phytohaemagglutinin
PHD PHD[8(1-hydroxy-3-oxo-propyl)-9,12-dihydroxy-5,10 heptadecadienic acid]
PHI Peptide histidine isoleucine
PHM Peptide histidine methionine
P_i Inorganic phosphate
pI Isoelectric point
PI Phosphatidylinositol
PI-3,4-P2 Phosphatidylinositol 3, 4-biphosphate
PI-3,4,5-P3 Phosphatidylinositol 3, 4, 5-trisphosphate
PI-3-kinase Phosphatidylinositol-3-kinase
PI-4-kinase Phosphatidylinositol-4-kinase
PI-3-P Phosphatidylinositol-3-phosphate
PI-4-P Phosphatidylinositol-4-phosphate
PI-4,5-P2 Phosphatidylinositol 4,5-biphosphate
PIP Phosphatidylinositol monophosphate
PIP_2 Phosphatidylinositol biphosphate
PK Protein kinase
PKA, PKC, PKG Protein kinases A, C and G
PL Phospholipase
PLA, PLA_2, PLC, PLD Phospholipases A, A_2, C and D
PLN Peripheral lymph node
PLNHEV Peripheral lymph node HEV
PLP Proteolipid protein
PLT Primed lymphocyte typing
PMA Phorbol myristate acetate
PMC Peritoneal mast cell
PMN Polymorphonuclear neutrophil
PMSF Phenylmethylsulphonyl fluoride
PNAd Peripheral lymph node vascular addressin
PNH Paroxysmal nocturnal hemoglobinuria
PNU Protein nitrogen unit
p.o. *Per os* (by mouth)
POBN α-4-pyridyl-oxide-*N-t*-butyl nitrone
PPD Purified protein derivative
PPME Polymeric polysaccharide rich in mannose-6-phosphate moieties
PQ Phenylquinone
PRA Percentage reactive activity
PRD, PRDII Positive regulatory domain, -II
PR3 Proteinase-3
PRBC Parasitized red blood cell
proET-1 Proendothelin-1
PRL Prolactin
PRP Platelet-rich plasma
PS Phosphatidylserine
P-selectin Platelet selectin *formerly known as* platelet activation-dependent granule external membrane protein (PADGEM), granule membrane protein of MW 140 kD (GMP-140)
PT Pertussis toxin
PTCA Percutaneous transluminal coronary angioplasty
PTCR Percutaneous transluminal coronary recanalization
Pte-H_4 Tetrahydropteridine
PUFA Polyunsaturated fatty acid
PUMP-1 Punctuated metalloproteinase *also known as* matrilysin
PWM Pokeweed mitogen
Pyran Divinylether maleic acid

q.i.d. *Quater in die* (four times a day)
QRS Segment of electrocardiogram

·R Free radical
R15.7 Anti-CD18 monoclonal antibody
RA Rheumatoid arthritis

RANTES A member of the IL8 supergene family (*R*egulated on *a*ctivation, *n*ormal *T* *e*xpressed and *s*ecreted)
RAST Radioallergosorbent test
R_{aw} Airways resistance
RBC Red blood cell
RBF Renal blood flow
RBL Rat basophilic leukaemia
RC Respiratory chain
RE RE strain of herpes simplex virus type 1
REA Reactive arthritis
REM Relative electrophoretic mobility
RER Rough endoplasmic reticulum
RF Rheumatoid factor
RFL-6 Rat foetal lung-6
RFLP Restriction fragment length polymorphism
RGD Arginine–glycine–asparagine
rh- Recombinant human – (prefix usually referring to peptides)
RIA Radioimmunoassay
RMCP, RMCPII Rat mast cell protease, -II
RNA Ribonucleic acid
RNase Ribonuclease
RNHCl *N*-Chloramine
RNL Regional lymph nodes
ROM Reactive oxygen metabolite
RO· *The chemical symbol for* alkoxyl radical
ROO· *The chemical symbol for* peroxy radical
ROP Retinopathy of prematurity
ROS Reactive oxygen species
R-PIA *R*-(1-methyl-1-phenyltheyl)-adenosine
RPMI 1640 Roswell Park Memorial Institute 1640 medium
RS Reiter's syndrome
RSV Rous sarcoma virus
RTE Rabbit tubular epithelium
RTE-a-5 Rat tubular epithelium antigen a-5
r-tPA Recombinant tissue-type plasminogen activator
RT-PCR Reverse transcriptase/polymerase chain reaction
RW Ragweed

S Svedberg (unit of sedimentation density)
SALT Skin-associated lymphoid tissue
SaR Sacroplasmic reticulum
SAZ Sulphasalazine
SC Secretory component
SCF Stem cell factor
SCFA Short-chain fatty acid
SCG Sodium cromoglycate *also known as* DSCG
SCID Severe combined immunodeficiency syndrome
sCR1 Soluble type–1 complement receptors
SCW Streptococcal cell wall
SD Standard deviation
SDS Sodium dodecyl sulphate
SDS-PAGE Sodium dodecyl sulphate-polyacrylamide gel electrophoresis
SEM Standard error of the mean
SGAW Specific airway conductance
SHR Spontaneously hypertensive rat
SIM Selected ion monitoring
SIN-1 3-Morpholinosydnonimine
SIRS Soluble immune response suppressor
SIV Simian immunodeficiency virus
SK Streptokinase
SLE Systemic lupus erythematosus
SLe^x Sialyl Lewis X antigen
SLO Streptolysin-O
SLPI Secretory leucocyte protease inhibitor
SM Sphingomyelin
SNAP *S*-Nitroso-*N*-acetylpenicillamine
SNP Sodium nitroprusside
SOD Superoxide dismutase
SOM Somatostatin *also known as* somatotrophin release-inhibiting factor
SOZ Serum-opsonized zymosan
SP Sulphapyridine
SR Systemic reaction
sr Sarcoplasmic reticulum
sR_{aw} Specific airways resistance
SRBC Sheep red blood cells
SRS Slow-reacting substance
SRS-A Slow-reacting substance of anaphylaxis
STZ Streptozotocin
Sub P Substance P

T Thymus-derived
α-TOC α-Tocopherol
$t_{1/2}$ Half-life
T84 Human intestinal epithelial cell line
TauNHCl Taurine monochloramine
TBA Thiobarbituric acid
TBAR Thiobarbituric acid-reactive product
TBM Tubular basement membrane
TBN di-*tert*-Butyl nitroxide
tBOOH *tert*-Butylhydroperoxide
TCA Trichloroacetic acid
T cell Thymus-derived lymphocyte
TCR T cell receptor α/β or γ/δ heterodimeric forms
TDI Toluene diisocyanate
TEC Tubular epithelial cell
TF Tissue factor
Tg Thyroglobulin
TGF Transforming growth factor
TGFα, TGFβ, TGFβ_1 Transforming growth factors α, β, and β_1
T_H T helper cell
T_Ho T helper o
T_Hp T helper precursor
T_H0, T_H1, T_H2 Subsets of helper T cells
THP-1 Human monocytic leukaemia
Thy 1+ Murine T cell antigen
t.i.d. Ter in die (three times a day)
TIL Tumour-infiltrating lymphocytes
TIMP Tissue inhibitors of metalloproteinase
TIMP-1, TIMP-2 Tissue inhibitors of metalloproteinases 1 and 2
Tla Thymus leukaemia antigen
TLC Thin-layer chromatography
TLCK Tosyl-lysyl-CH_2Cl
TLP Tumour-like proliferation
Tm T memory
TNF, TNF-α Tumour necrosis factor, -α
tPA Tissue-type plasminogen activator
TPA 12-*O*-tetradeconylphorbol-13-acetate
TPCK Tosyl-phenyl-CH_2Cl
TPK Tyrosine protein kinases
TPP Transpulmonary pressure
TRAP Thrombospondin related anomalous protein
Tris Tris(hydroxymethyl)-aminomethane
TSH Thyroid-stimulating hormone
TSP Thrombospondin
TTX Tetrodotoxin
TX Thromboxane
TXA_2, TXB_2 Thromboxane A_2, B_2
Tyk2 Tyrosine kinase

U937 (cells) Histiocytic lymphoma, human
UC Ulcerative colitis
UCR Upstream conserved region
UDP Uridine diphosphate
UPA Urokinase-type plasminogen activator
UTP Uridine triphosphate
UV Ultraviolet
UVA Ultraviolet A
UVB Ultraviolet B
UVR Ultraviolet irradiation
UW University of Wisconsin (preserving solution)

VAP Viral attachment protein
VC Veiled cells
VCAM, VCAM-1 Vascular cell adhesion molecule, -1, *also known as* inducible cell adhesion molecule MW 110 kD (INCAM–110)
VF Ventricular fibrillation
V/GSH Vanadate/glutathione complex
VIP Vasoactive intestinal peptide
VLA Very late activation antigen beta chain; *also known as* CD29
VLA α2 Very late activation antigen alpha 2 chain; *also known as* CD49b

VLA α4 Very late activation antigen alpha 4 chain; *also known as* CD49d
VLA α6 Very late activation antigen alpha 6 chain; *also known as* CD49f
VLDL Very low-density lipoprotein
***V* max** Maximal velocity
***V* min** Minimal velocity
VN Vitronectin
VO_4^- *The chemical symbol for* vanadate
vp Viral protein
VP Vasopressin
VPB Ventricular premature beat
VT Ventricular tachycardia
vWF von Willebrand factor

W Murine dominant white spotting mutation
WBC White blood cell
WGA Wheat germ agglutinin
WI Warm ischaemia

XD Xanthine dehydrogenase
XO Xanthine oxidase

Y1/82A A monoclonal antibody detecting a cytoplasmic antigen in human macrophages

ZA Zonulae adherens
ZAP Zymosan-activated plasma
ZAS Zymosan-activated serum
zLYCK Carboxybenzyl-Leu-Tyr-CH_2Cl
ZO Zonulae occludentes

Key to Illustrations

Helper lymphocyte

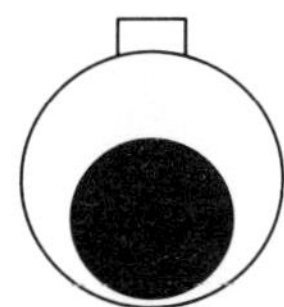

Suppressor lymphocyte

Killer lymphocyte

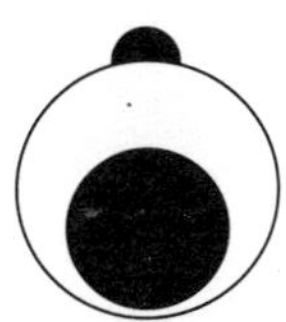

Plasma cell

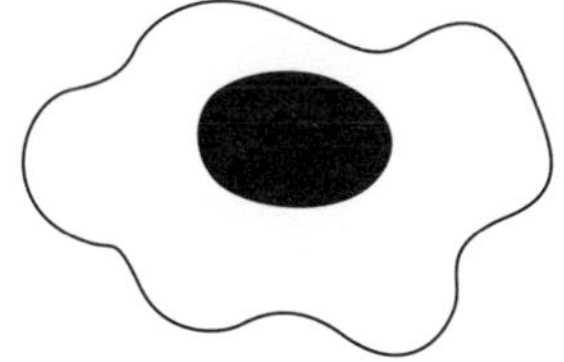

Bacterial or Tumour cell

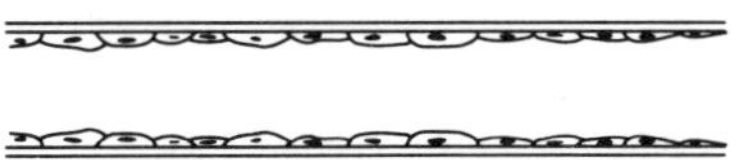

Blood vessel lumen

Eosinophil passing through vessel wall

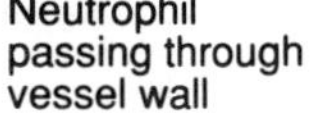

Neutrophil passing through vessel wall

Resting neutrophil

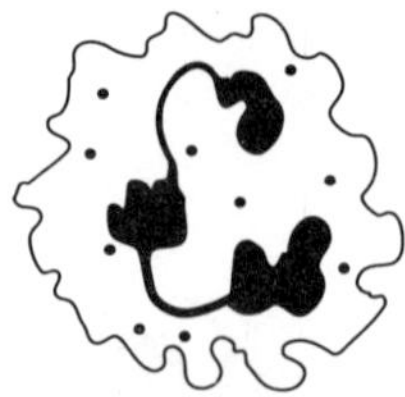
Activated neutrophil

Resting eosinophil

Activated eosinophil

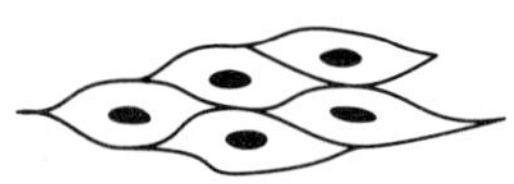
Smooth muscle

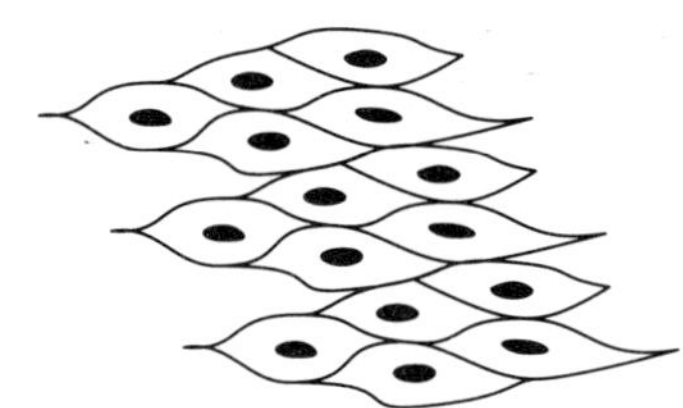
Smooth muscle thickening

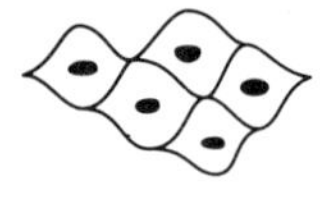
Smooth muscle contraction

Normal blood vessel

Endothelial cell permeability

Resting macrophage

Activated macrophage

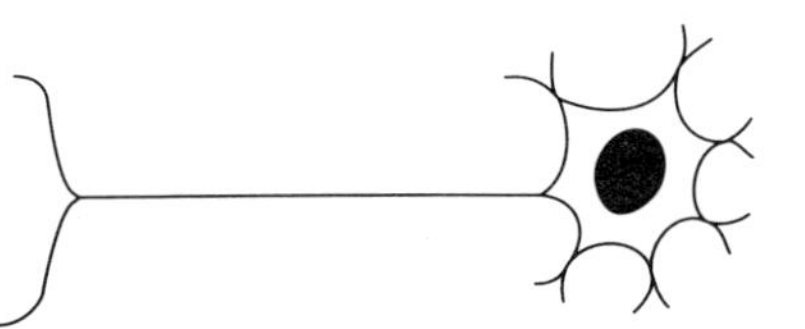
Nerve

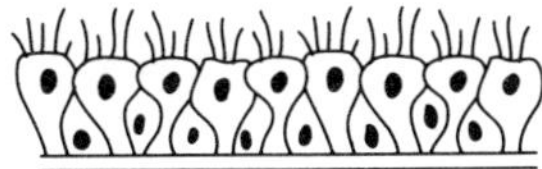

Intact epithelium

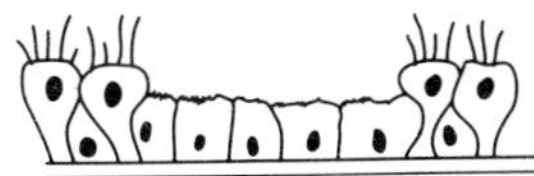

Damaged epithelium

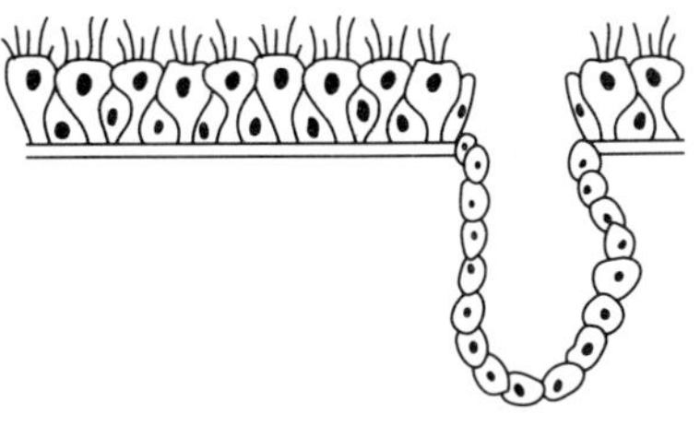

Intact epithelium with submucosal gland

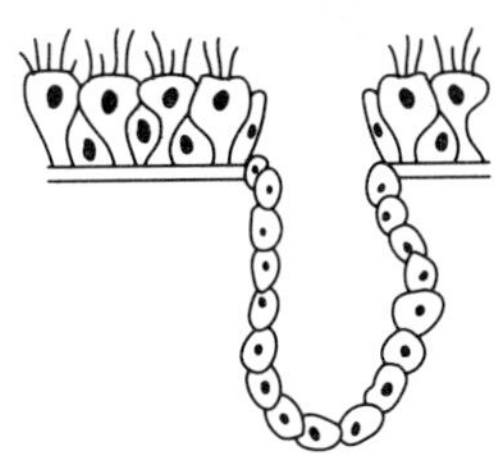

Normal submucosal gland

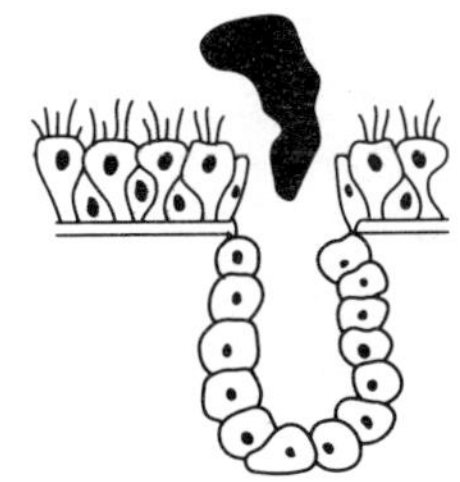

Hypersecreting submucosal gland

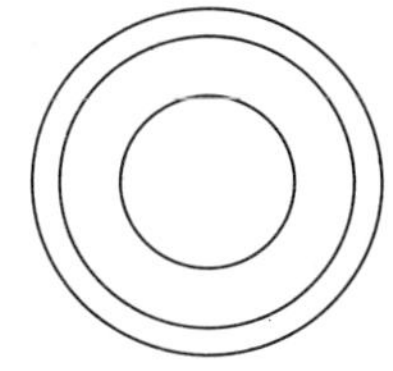

Normal airway

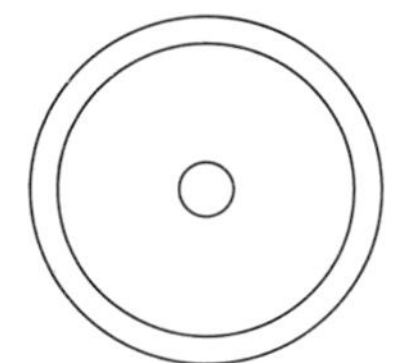

Oedema

Bronchospasm

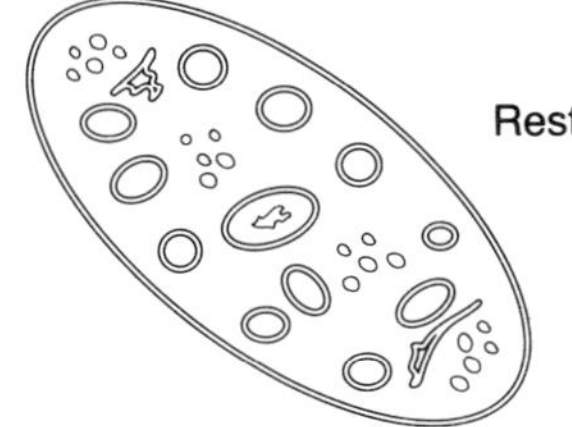

Resting platelet

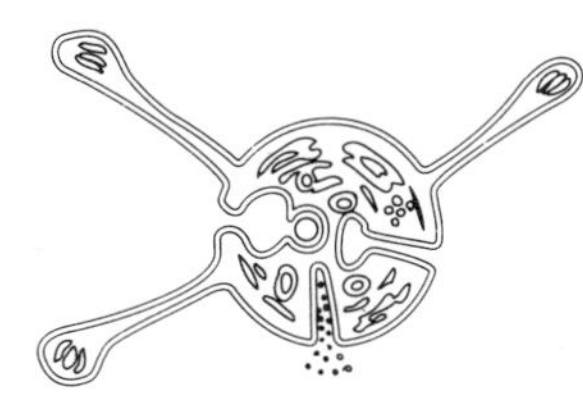

Activated platelet

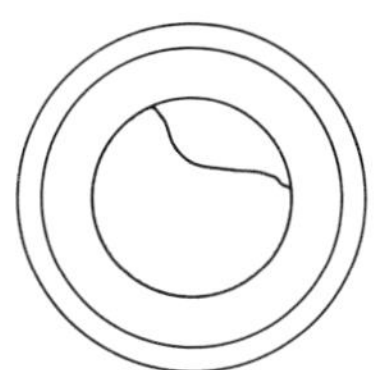

Airway hypersecreting mucus

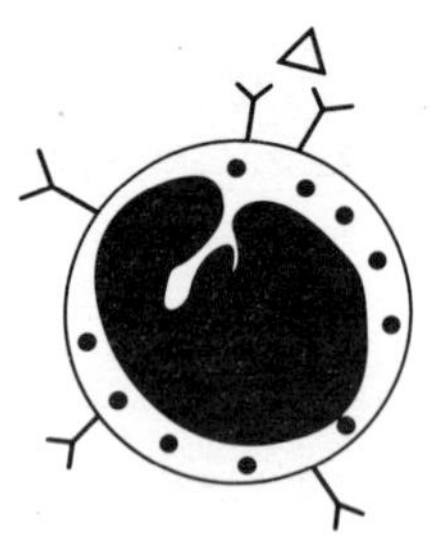

Resting basophil

Activated basophil

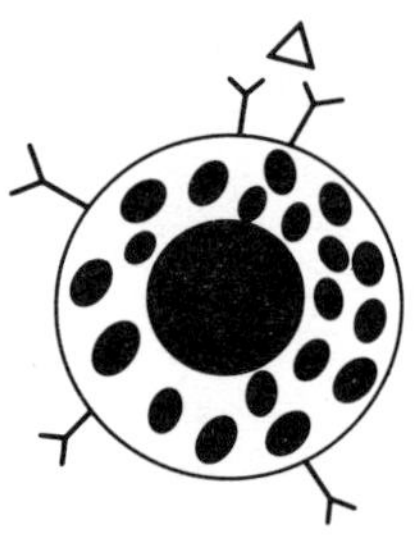

Resting mast cell

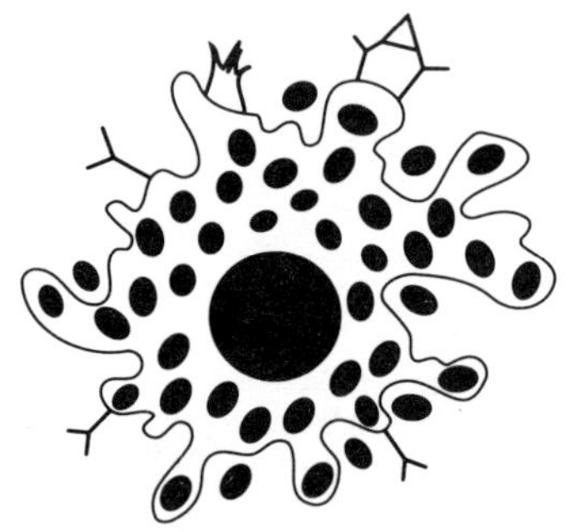

Activated mast cell

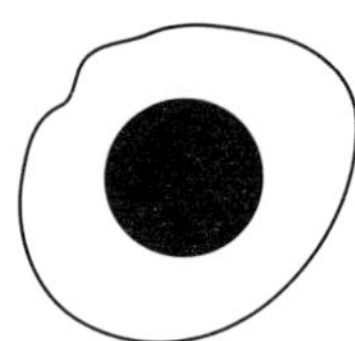

Resting chondrocyte

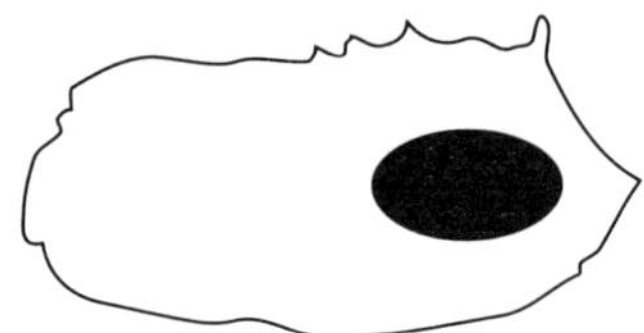

Activated chondrocyte

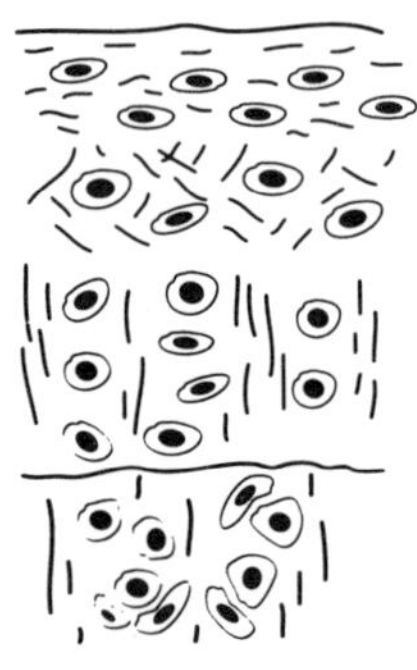

Cartilage

Fibroblast

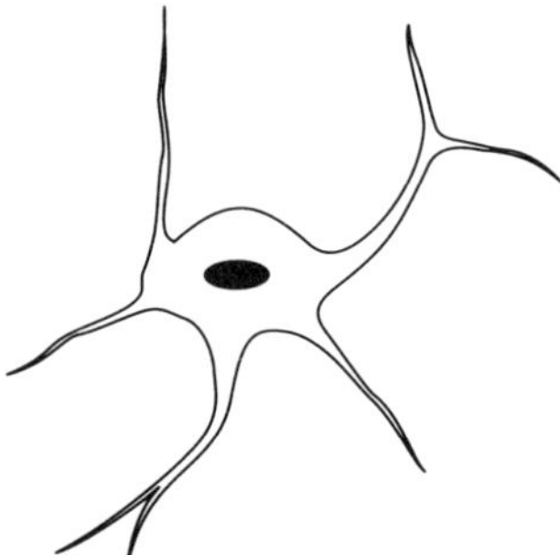

Dendritic cell/ Langerhans cell

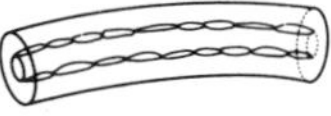

Arteriole

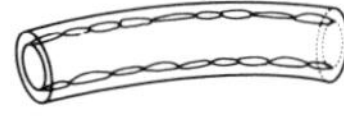

Venule

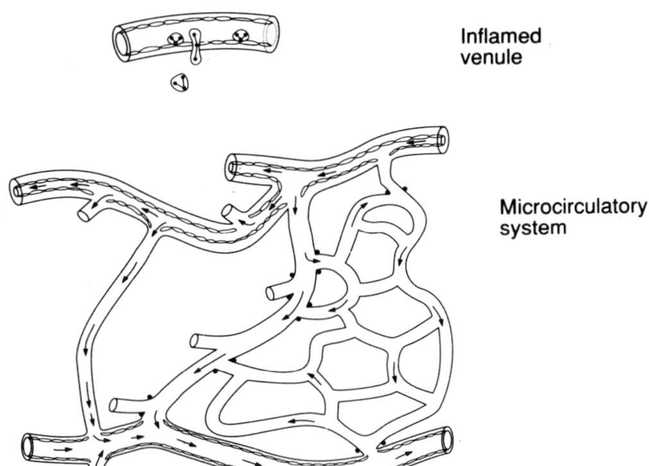

Inflamed venule

Microcirculatory system

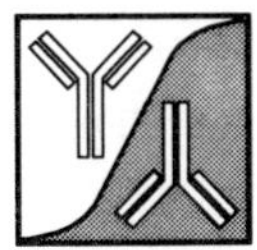

Index

43D cells *see* lymphocytes B-cells
A549 cells *see* epithelium
adenosine 41–42, 47, 50, 52–55, 86, 113, 116
adipocytes 32, 91–93, 95–96
  3T3-L1 7, 93
adipose tissue 7, 90, 93
adrenal gland 6, 31
α-adrenoceptor agonists 98
β-adrenoceptor agonists 31–33, 42, 47, 49–51, 83, 85, 95, 98–99, 116, 120, 152, 174–175
AH 21-132 (benafentrine) 113, 118, 149–150, 152–153, 178, 189–190
  inhibition of PDE isoenzymes 149
aminophylline *see* theophylline
amrinone 91–92, 97, 101, 186–187, 189
anagrelide 91–92, 94, 99–100, 161
anaphylaxis 165, 168–169
ANF/ANP *see* atrial natriuretic peptide
angiotensin 73
asthma 21, 41–42, 44–45, 47–48, 50, 54, 56, 119–120, 147–148, 152–158, 168, 173, 202
atopic dermatitis 28, 45, 119–120
atopy 33–34, 119–120
atrial natriuretic peptide 30, 127–130, 132
autoimmune encephalomyelitis 46

B-cells *see* lymphocytes
B9004-070 *see* tolafentrine
basophils 27, 52–53, 161
  effects of PDE inhibitors 27, 53–54, 117
  PDE isoenzyme expression 27, 112, 117
bemoradan 117
benafentrine *see* AH 21-132
blood vessels 7, 29, 96, 98, 127, 129–132, 140, 157; *see also* endothelium, microvasculature *and* muscle smooth
brain 4–7, 9–12, 14, 44, 55, 65, 96, 119, 149, 181, 191, 193
BRL 30892 *see* denbufylline
BRL 61063 113, 188, 190
bronchus *see* muscle smooth

calcium/calmodulin-dependent PDE *see* PDE1
calmodulin 23, 31–32, 65, 68–72, 74, 120
cancer 74
cDNA 2–3, 6, 12–13, 185–186
  cloning 3–4, 6–9, 12–13, 190–191
CDP 840 113, 173, 188–189, 200–202
chronic obstructive pulmonary disease (COPD) 155
CI-914 *see* imazodan
CI-930 90, 92, 98–99, 118, 148, 151
cilostamide (OPC 3689) 89–92, 94, 97, 100, 151, 186–187, 189
cilostazol 91–92, 100
CP 80,633 113, 188–189
cyclic AMP 1, 9–11, 27, 29–33, 41–42, 45–51, 53–55, 69–71, 74, 81–84, 86, 91–101, 112, 115–117, 120, 132, 135–136, 150, 161–163, 167, 170, 175, 178, 180–181, 186, 190, 192, 195, 197
cyclic AMP-responsive element 11
cyclic AMP-specific PDE *see* PDE4 *and* PDE7
cyclic GMP 1, 23, 25–26, 31, 42, 54–55, 81–82, 84–86, 89–90, 92–94, 97–100, 127–132, 135–136, 139–140, 144, 186
  analogues 136–137, 142–145
cyclic GMP-inhibited PDE *see* PDE3
cyclic GMP-specific PDE *see* PDE5 *and* PDE6
cyclic GMP-stimulated PDE *see* PDE2

denbufylline (BRL 30892) 113–114, 116, 119, 151, 165–166, 175, 77–181, 186, 188, 190, 200–202
development 4, 7, 10, 96
diabetes mellitus 47, 52, 96
dipyridamole 41, 54, 138, 140, 177–178, 181, 186–187, 189
dopamine 5

Drosophila melanogaster 1, 7, 111

EHNA (erythro-9-[2-hydroxyl-3-nonyl]-adenine) 82–87, 186–187, 189
EMD 54622 149, 152, 189–190
  inhibition of PDE isoenzymes 152
endothelium 7, 27, 30–31, 98, 116, 129, 131, 179
  effects of PDE inhibitors 27, 46, 151
  PDE isoenzyme expression 27–28, 148
enoximone (MDL 17,043) 91–92, 97–98, 101, 157
enprofylline 43, 47, 53–54
eosinophils 10, 23–24, 44, 49, 118–119, 161, 165, 175–182, 195
  effects of PDE inhibitors 24, 49–50, 116, 150–152, 174–175, 181, 195
  PDE isoenzyme expression 24, 112, 148, 176
epithelium 5, 7, 28
  effects of PDE inhibitors 28
  PDE isoenzyme expression 28–29, 96, 148
erythro-9-(2-hydroxyl-3-nonyl)-adenine *see* EHNA
eye 7, 13, 28, 52, 96, 118; *see also* photoreceptors

fibroblasts
  3T3-L1 7
follicle-stimulating hormone 11, 30, 32–33
forskolin 11, 32–33, 42, 49, 96, 120

gastro-intestinal tract 49, 119, 149, 181, 195
ginsenosides 73–74
glial cells 11, 29
glucagon 95
glutathione *see* PDE4 enzymology

HaCaT cells *see* keratinocytes
heart 4–7, 10, 14, 44, 65, 90, 152, 162, 191; *see also* muscle cardiac

heart (*continued*)
effects of PDE inhibitors 30, 55, 90–91, 97–98, 101, 127, 132, 157
HIV 51
HL-60 cells 10
HL-725 *see* trequinsin
HUVEC *see* endothelium

ibudilast 178
IBMX 11, 32–33, 42–45, 47, 49–56, 82–84, 136, 144–145, 178, 186–187
analogues 136–141, 143–145
inhibition of PDE isoenzymes 140, 166
inhibition of PDE isoenzymes 42–44, 73, 140
ICI 1233188 91–92
imazodan (CI-914) 91–92, 97–99
indolidan (LY 195115) 91, 97–99
insulin 7, 32, 55, 90–93, 95–96
Internet site 2
isbufylline 43, 50
isomazole 152, 157

Jurkat cells *see* lymphocytes T-cells

keratinocytes 14, 28, 33, 120
effects of PDE inhibitors 28, 46
PDE isoenzyme expression 28–29
kidney 4–7, 10, 14, 32, 55, 96, 111, 119, 128, 130, 149
KS 505a 73, 186–187

LAS 31025 113, 173
leucocytes *see* basophils, eosinophils, lymphocytes, macrophages, monocytes *and* neutrophils
tissue infiltration 118, 150–152, 165, 168–169, 175
liver 4–7, 10, 49, 55, 89, 91–92, 95–96, 151
lixazinone (RS 82856) 91–92, 94, 97, 100, 161, 165–166
lung 4–6, 10, 29, 42, 49–51, 53, 115–116, 118–119, 131, 148, 150–158, 165, 168–169, 175, 181
PDE isoenzyme expression 42–43, 68, 93, 127, 191
LY 186655 *see* tibenelast
LY 195115 see indolidan
lymphocytes 42, 45, 118, 120, 161
B-cells 45, 112, 120
43D 163, 165–167
effects of PDE inhibitors 45, 115
Namwala 10
PDE isoenzyme expression 165
effects of PDE inhibitors 151
natural killer cells 45, 47
PDE isoenzyme expression 25–26, 65, 100
T-cells 6, 25, 33, 45, 181; *see also* thymocytes
alteration of PDE activity 33–34, 119
effects of PDE inhibitors 33, 45–47, 92, 100, 115–116, 180–181, 195
Hut78 14, 26
Jurkat 10–11, 26, 33, 115–116
K30a-3.3 5
PDE isoenzyme expression 26, 90, 112, 148
S49.1 5
T_H1-cells 25–26, 115–116, 120
T_H2-cells 25–26, 30, 115–116, 120, 181

M&B 22,948 *see* zaprinast
macrophages 24, 50–51, 118, 178–179
effects of PDE inhibitors 51–52, 116–117, 151–152
PDE isoenzyme expression 25, 90, 116, 148
RAW 264.7 116
mast cells 26, 52–53, 161, 179
effects of PDE inhibitors 27, 53, 117, 195
PDE isoenzyme expression 26–27, 112
MDL 17,043 *see* enoximone
megakaryocytes 7, 54, 96
MEP-1 *see* EHNA
microvasculature 118–119, 175
milrinone 30, 83–84, 90–92, 97–101, 117
monocytes 24, 50–51, 179–181
alteration of PDE activity 11, 30, 32–34, 119–120
effects of PDE inhibitors 51–52, 116, 174, 180–181, 195
Mono Mac6 11, 25, 33
PDE isoenzyme expression 24–25, 65, 112, 116, 148, 191
U937 10–11, 25, 32–33, 120, 169
Mono Mac6 cells *see* monocytes
mortality
platelet activating factor-induced 164–165, 168–169
motapizone 22–23, 92, 99
mRNA 33, 96, 111, 115–116, 179
splice variants 2–14, 24, 95, 111, 179, 190–191
muscle *see also* myoblasts
cardiac 7, 30–31, 81–82, 97, 179
effects of PDE inhibitors 31, 55, 82–86, 97–98, 101, 148, 189
PDE isoenzyme expression 30, 82, 93, 96
skeletal 10, 14, 191
smooth 7, 29, 32, 33, 44, 96, 98, 127, 131, 135–136, 149, 151, 178–179
effects of PDE inhibitors 29, 41, 55, 90, 92, 98–101, 118, 129, 131–132, 143–144, 148–152, 173, 175, 180–181, 189, 195
PDE isoenzyme expression 29, 90, 99, 147
myoblasts 11, 33
myocardium *see* muscle cardiac

natural killer cells *see* lymphocytes
neurones 65; *see also* brain
effects of PDE inhibitors 29
N18TG2 29
olfactory 6, 10, 12, 136
PDE isoenzyme expression 29
SH-SY5Y 11, 29
neutrophils 10, 24, 44, 47, 118–119, 149, 151, 161, 179
effects of PDE inhibitors 24, 47–49, 116–117, 151, 195
PDE isoenzyme expression 24, 112, 148
nitraquazone (TVX 2706) 113–114, 165, 188–189, 200–202
nitric oxide 31–32, 84–85, 92, 97–98, 100, 116–117, 127–131
nitroglycerine 131–132

oedema 119, 164, 168–169
oocytes 7, 91, 96
OPC 3689 *see* cilostamide
OPC 3911 90–92, 99
OPC 8212 *see* vesnarinone
Org 20241 113, 116, 149, 152
inhibition of PDE isoenzymes 152
Org 30029 98, 100, 149, 152, 189–190
inhibition of PDE isoenzymes 152
Org 9935 100

pancreas 4–5, 10, 14, 55, 65, 119
papaverine 11, 41, 45, 51, 55, 186–187
parietal cells *see* gastro-intestinal tract
PDE1 3–5, 22–23, 65–75, 162
alteration in disease 74
calcium/calmodulin binding 3–4, 68–72
enzymology 30–31, 66, 68–69
genes 3–4
inhibitors 73–74, 99, 132, 141, 186–187; *see also* KS 505a, vinpocetine
isoforms 2–5, 66–74
phosphorylation 32, 69–73
structure 3–4, 66–67
subcellular localization 4, 25, 28, 30, 66
tissue expression 4–5, 25, 27–30, 43, 65, 68, 99, 116, 148
PDE2 5–6, 22–23, 81–87, 136, 162

cGMP binding 5, 31, 82, 86, 136
enzymology 31
gene 5–6
inhibitors 82, 141, 186–187; *see also* EHNA
structure 5–6
subcellular localization 28, 30, 85–86
tissue expression 6, 25, 27–30, 43, 99, 148
PDE3 6–7, 22–23, 89–101, 162
enzymology 30–32, 89–90, 94
genes 6–7, 92–93, 96
mapping 7, 92–93
inhibitors 90–92, 94, 96–101, 115, 117, 132, 141, 147–158, 186–187, 189; *see also* AH 21–132, amrinone, anagrelide, bemoradan, CI-930, cilostamide, cilostazol, EMD 54622, enoximone, ICI 1233188, imazodan, indolidan, isomazole, lixazinone, milrinone, motapizone, OPC 3911, Org 20241, Org 30029, Org 9935, phthalazinol, pimobendan, piroximone, R 80122, RS 82856, RX RA 69, SDZ MKS 492, siguazodan, SK&F 94120, SK&F 95654, tolafentrine, trequinsin, vesnarinone, Y–590 *and* zardaverine
isoforms 6, 90, 92, 94–96, 152
phosphorylation 32, 95–96
structure 6–7, 92–96
subcellular localization 25–28, 30, 95–96
tissue expression 7, 24–30, 43, 90, 93, 96, 99–100, 116–117, 148, 157
up-regulation 92
PDE4 7–12, 22–23, 98–99, 111–112, 162–163, 190–199
alteration in disease 33–34, 119–120
conformational states 177–178, 194–195
enzymology 24, 30, 112, 176, 179, 194–198
effects of solubilization and vanadate/glutathione 176–177
genes 7–9, 33, 111, 179, 190
cloning 163–164, 167–168, 194
mapping 7, 9, 190
inhibitors 99–100, 112–121, 132, 147–158, 162, 173–174, 186, 188–189, 193, 197–203; *see also* AH 21–132, BRL 61063, CDP 840, CP 80,633, denbufylline, EMD 54622, LAS 31025, nitraquazone, Org 20241, Org 30029, Ro 20–1724, rolipram, RP 73401, RS 14203, RS 14491, RS 25344, SB 207499, tibenelast, tolafentrine, WAY-PDA-641 *and* zardaverine
isoforms 7–10, 24–26, 29, 32–33, 111–112, 116, 119–120, 162–164, 166–170, 179–181, 190–192, 194, 196–198, 200–201
phosphorylation 24, 32–33, 112, 162, 164, 168–169, 192
structure 7–9, 177, 191–195
subcellular localization 8, 24–28, 30, 112, 176, 179, 190–192
tissue expression 9–12, 24–30, 43, 99, 111–112, 116–117, 147–148, 157, 176, 189, 191
up-regulation 11, 32–34, 112, 120, 169
PDE5 12, 22–23
enzymology 30, 128, 144
gene 12
inhibitors 99, 127–132, 136–145, 186–187, 189; *see also* cyclic GMP analogues, dipyridamole, IBMX analogues, WIN 58237 *and* zaprinast
phosphorylation 32
structure 12
subcellular localization 27–28
tissue expression 27–29, 99, 116–117
up-regulation 32
PDE6 12–13
alterations in disease *see* photoreceptors
genes 12–13
mapping 13
inhibitors 186, 189
isoforms 12
structure 12–13
tissue expression 13
PDE7 13–14, 23, 111
gene 13–14
cloning 164
mapping 14
structure 13
tissue expression 14, 26
pentoxifylline 43, 45–46, 49, 54–56, 115
phosphodiesterase 1, 21, 42, 89, 135, 185
genes 1–2, 3, 7
inhibitors 22, 186; *see also* enprofylline, ibudilast, IBMX, isbufylline, papaverine, pentoxifylline *and* theophylline
isoenzymes 1, 21, 89, 185
nomenclature 2, 135, 161, 186
structure 2
tissue distribution 23–30
measurement 22–23, 162–163
regulation 30–34
photoreceptor PDE *see* PDE6
photoreceptors 12, 136
and disease 13
phthalazinol 187, 189
piclamilast *see* RP 73401
pimobendan (UK-CG115) 92, 101, 152
piroximone 91–92, 97
placenta 10, 90, 93
platelets 7, 27, 32, 42, 44, 54, 96, 128, 149–150, 162
effects of PDE inhibitors 31, 54, 90, 92, 96–97, 100
PDE isoenzyme expression 27, 90, 112, 128, 165
prostacyclin 32, 92
prostaglandin E1/E2 31–32, 34, 42, 49–50, 52, 98, 120, 169, 180
protein kinase 2, 70, 112
calcium/phospholipid-dependent (PKC) 32, 192
calmodulin-dependent 69–72
cAMP-dependent (PKA) 9, 11, 32–33, 45, 52–53, 69, 92, 95–96, 98, 116–117, 135–136, 150, 162, 164, 175, 191–192
cGMP-dependent (PKG) 32, 98, 127, 136, 144
insulin-dependent PDE3 kinase 32, 95–96
phosphatidylinositol-3-kinase 95
protein phosphatase 32, 70–72, 96
protein phosphatase inhibitor 72

R 80122 92, 100
RAW 264.7 cells *see* macrophages
Ro 20–1724 25, 33, 83–84, 90, 98–99, 111, 116–118, 120, 162, 165–166, 177–179, 188–189, 200–201
rolipram 7, 22–24, 33, 98–100, 113–120, 148, 151, 162, 165–166, 168–169, 173, 175–182, 186, 188–189, 190, 193–194, 196–198, 200–201
high affinity binding site 24, 178–182, 192–196, 199–201
inhibition of PDE isoenzymes 174, 200
RP 73401 (piclamilast) 24, 113–114, 116, 119, 173–182, 188–189, 201–202
inhibition of PDE isoenzymes 174, 200–201
RS 14203 166–167, 169
RS 14491 166–167, 169
RS 25344 113, 118–119, 164–170, 188, 190
inhibition of PDE isoenzymes 165–168
RS 82856 *see* lixazinone
RX RA 69 165–166

Saccharomyces cerevisiae 1, 13

sarcoplasmic reticulum 30, 95, 97
SB 207499 113–114, 173, 182, 188–189, 200–201
SDZ MKS 492 113
Sertoli cells *see* testis
SH-SY5Y cells *see* neurones
siguazodan 92, 99–100
SIN-1 (3-morpholinosydnonimine) 31, 54, 85, 98; *see also* nitric oxide
skin 50, 53, 118; *see also* atopic dermatitis *and* keratinocytes
SK&F 94120 54, 91–92, 98–100, 151
SK&F 95654 115
smooth muscle *see* muscle
sodium nitroprusside 31, 54, 85, 100, 129; *see also* nitric oxide
spermatocytes 7, 10, 65–66, 96
spleen 6, 115–116, 162, 174–175, 180
splice variants *see* mRNA

T-cells *see* lymphocytes
testis 10–12, 32–33, 68
theobromine 52
theophylline 21, 24, 41–56, 116, 119–120, 147, 157, 162, 186–187
  inhibition of PDE isoenzymes 42–44
thymocytes 179
  effects of PDE inhibitors 31–32, 45–46
thyroid-stimulating hormone 32
tibenelast (LY 186655) 113, 116, 157, 188, 190
tolafentrine (B9004-070) 113, 116, 149, 151–152, 189–190
  inhibition of PDE isoenzymes 151
trachea *see* muscle smooth
transducin *see* photoreceptors
trequinsin (HL-725) 100, 161, 165–166, 175, 177–178, 181, 186–187, 200, 202
trifluoperazine 74
TVX 2706 *see* nitraquazone

U937 cells *see* monocytes
UK-CG115 *see* pimobendan

vanadate *see* PDE4 enzymology
vesnarinone (OPC 8212) 92, 101
vinpocetine 128, 186–187

WAY-PDA-641 113, 115, 116, 173, 188–189, 200–201
WIN 58237 129–132
writhing 165, 168–169

xanthine derivatives 41–42, 113; *see also* IBMX analogues

Y-590 91–92

zaprinast (M&B 22,948) 22–23, 32, 99, 128–132, 138, 140, 157, 186–187, 189
zardaverine 23–24, 99, 113, 116, 118, 149, 151, 153–156, 189–190
  inhibition of PDE isoenzymes 151, 200–202